Multivariable Calculus

MONOGRAPHS AND TEXTBOOKS IN PURE AND APPLIED MATHEMATICS

1. *K. Yano,* Integral Formulas in Riemannian Geometry (1970) *(out of print)*
2. *S. Kobayashi,* Hyperbolic Manifolds and Holomorphic Mappings (1970) *(out of print)*
3. *V. S. Vladimirov,* Equations of Mathematical Physics (A. Jeffrey, editor; A. Littlewood, translator) (1970)
4. *B. N. Pshenichnyi,* Necessary Conditions for an Extremum (L. Neustadt, translation editor; K. Makowski, translator) (1971)
5. *L. Narici, E. Beckenstein, and G. Bachman,* Functional Analysis and Valuation Theory (1971)
6. *D. S. Passman,* Infinite Group Rings (1971)
7. *L. Dornhoff,* Group Representation Theory (in two parts). Part A: Ordinary Representation Theory. Part B: Modular Representation Theory (1971, 1972)
8. *W. Boothby and G. L. Weiss (eds.),* Symmetric Spaces: Short Courses Presented at Washington University (1972)
9. *Y. Matsushima,* Differentiable Manifolds (E. T. Kobayashi, translator) (1972)
10. *L. E. Ward, Jr.,* Topology: An Outline for a First Course (1972) *(out of print)*
11. *A. Babakhanian,* Cohomological Methods in Group Theory (1972)
12. *R. Gilmer,* Multiplicative Ideal Theory (1972)
13. *J. Yeh,* Stochastic Processes and the Wiener Integral (1973) *(out of print)*
14. *J. Barros-Neto,* Introduction to the Theory of Distributions (1973) *(out of print)*
15. *R. Larsen,* Functional Analysis: An Introduction (1973)
16. *K. Yano and S. Ishihara,* Tangent and Cotangent Bundles: Differential Geometry (1973)
17. *C. Procesi,* Rings with Polynomial Identities (1973)
18. *R. Hermann,* Geometry, Physics, and Systems (1973)
19. *N. R. Wallach,* Harmonic Analysis on Homogeneous Spaces (1973)
20. *J. Dieudonné,* Introduction to the Theory of Formal Groups (1973)
21. *I. Vaisman,* Cohomology and Differential Forms (1973)
22. *B.-Y. Chen,* Geometry of Submanifolds (1973)
23. *M. Marcus,* Finite Dimensional Multilinear Algebra (in two parts) (1973, 1975)
24. *R. Larsen,* Banach Algebras: An Introduction (1973)
25. *R. O. Kujala and A. L. Vitter (eds),* Value Distribution Theory: Part A; Part B. Deficit and Bezout Estimates by Wilhelm Stoll (1973)
26. *K. B. Stolarsky,* Algebraic Numbers and Diophantine Approximation (1974)
27. *A. R. Magid,* The Separable Galois Theory of Commutative Rings (1974)
28. *B. R. McDonald,* Finite Rings with Identity (1974)
29. *J. Satake,* Linear Algebra (S. Koh, T. Akiba, and S. Ihara, translators) (1975)

30. *J. S. Golan,* Localization of Noncommutative Rings (1975)
31. *G. Klambauer,* Mathematical Analysis (1975)
32. *M. K. Agoston,* Algebraic Topology: A First Course (1976)
33. *K. R. Goodearl,* Ring Theory: Nonsingular Rings and Modules (1976)
34. *L. E. Mansfield,* Linear Algebra with Geometric Applications: Selected Topics (1976)
35. *N. J. Pullman,* Matrix Theory and Its Applications (1976)
36. *B. R. McDonald,* Geometric Algebra Over Local Rings (1976)
37. *C. W. Groetsch,* Generalized Inverses of Linear Operators: Representation and Approximation (1977)
38. *J. E. Kuczkowski and J. L. Gersting,* Abstract Algebra: A First Look (1977)
39. *C. O. Christenson and W. L. Voxman,* Aspects of Topology (1977)
40. *M. Nagata,* Field Theory (1977)
41. *R. L. Long,* Algebraic Number Theory (1977)
42. *W. F. Pfeffer,* Integrals and Measures (1977)
43. *R. L. Wheeden and A. Zygmund,* Measure and Integral: An Introduction to Real Analysis (1977)
44. *J. H. Curtiss,* Introduction to Functions of a Complex Variable (1978)
45. *K. Hrbacek and T. Jech,* Introduction to Set Theory (1978)
46. *W. S. Massey,* Homology and Cohomology Theory (1978)
47. *M. Marcus,* Introduction to Modern Algebra (1978)
48. *E. C. Young,* Vector and Tensor Analysis (1978)
49. *S. B. Nadler, Jr.,* Hyperspaces of Sets (1978)
50. *S. K. Sehgal,* Topics in Group Rings (1978)
51. *A. C. M. van Rooij,* Non-Archimedean Functional Analysis (1978)
52. *L. Corwin and R. Szczarba,* Calculus in Vector Spaces (1979)
53. *C. Sadosky,* Interpolation of Operators and Singular Integrals: An Introduction to Harmonic Analysis (1979)
54. *J. Cronin,* Differential Equations: Introduction and Quantitative Theory (1980)
55. *C. W. Groetsch,* Elements of Applicable Functional Analysis (1980)
56. *I. Vaisman,* Foundations of Three-Dimensional Euclidean Geometry (1980)
57. *H. I. Freedman,* Deterministic Mathematical Models in Population Ecology (1980)
58. *S. B. Chae,* Lebesgue Integration (1980)
59. *C. S. Rees, S. M. Shah, and Č. V. Stanojević,* Theory and Applications of Fourier Analysis (1981)
60. *L. Nachbin,* Introduction to Functional Analysis: Banach Spaces and Differential Calculus (R. M. Aron, translator) (1981)
61. *G. Orzech and M. Orzech,* Plane Algebraic Curves: An Introduction Via Valuations (1981)
62. *R. Johnsonbaugh and W. E. Pfaffenberger,* Foundations of Mathematical Analysis (1981)

63. *W. L. Voxman and R. H. Goetschel,* Advanced Calculus: An Introduction to Modern Analysis (1981)
64. *L. J. Corwin and R. H. Szczarba,* Multivariable Calculus (1982)
65. *V. I. Istrăţescu,* Introduction to Linear Operator Theory (1981)
66. *R. D. Järvinen,* Finite and Infinite Dimensional Linear Spaces: A Comparative Study in Algebraic and Analytic Settings (1981)
67. *J. K. Beem and P. E. Ehrlich,* Global Lorentzian Geometry (1981)
68. *D. L. Armacost,* The Structure of Locally Compact Abelian Groups (1981)
69. *J. W. Brewer and M. K. Smith, editors,* Emmy Noether: A Tribute to Her Life and Work (1981)

Other Volumes in Preparation

Multivariable Calculus

LAWRENCE J. CORWIN
Rutgers University
New Brunswick, New Jersey

ROBERT H. SZCZARBA
Yale University
New Haven, Connecticut

CRC Press
Taylor & Francis Group
Boca Raton London New York

CRC Press is an imprint of the
Taylor & Francis Group, an **informa** business

First published 1982 by Marcel Dekker, Inc.

Published 2019 by CRC Press
Taylor & Francis Group
6000 Broken Sound Parkway NW, Suite 300
Boca Raton, FL 33487-2742

First issued in paperback 2019

ISBN 13: 978-0-367-45197-4 (pbk)
ISBN 13: 978-0-8247-6962-8 (hbk)

Visit the Taylor & Francis Web site at
http://www.taylorandfrancis.com

and the CRC Press Web site at
http://www.crcpress.com

Library of Congress Cataloging in Publication Data

Corwin, Lawrence J., [date]
Multivariable calculus.

(Monographs and textbooks in pure and applied mathematics; v. 64)
Includes index.
1. Calculus. I. Szczarba, Robert Henry, [date]
II. Title. III. Series.
QA303.C824 515 81–5399
ISBN 0–8247–6962–7 AACR2

All figures were previously published in *Calculus in Vector Spaces* by L. J. Corwin and R. H. Szczarba.

To Amy and Arlene

Preface

This text is basically an introduction to calculus of functions of several variables. It is intended for use in either a traditional advanced calculus course (emphasizing rigor or not) or in a course for good students who have had one year of calculus. The notions covered include continuity, differentiation, multiple integrals, line and surface integrals, differential forms, and infinite series. The level of rigor is high; with a few exceptions, every result in the book is proved. On the other hand, the material is organized so that the proofs can be de-emphasized or even omitted. In addition, numerous examples and exercises are given which illustrate the use of the concepts.

After a chapter of preliminary material, we begin in Chapter 2 with the linear algebra of Euclidean spaces. This chapter can be covered systematically or, preferably, can be referred back to as needed. Chapter 3 deals with the basic properties of continuous functions and Chapter 4 with differentiability. We introduce the derivative of a function between Euclidean spaces as an approximating linear transformation, develop some of its properties, and describe it in terms of partial derivatives.

The next two chapters continue to explore the notion of the derivative. Chapter 5 gives some geometric applications, and in Chapter 6, higher order derivatives are defined and Taylor's theorem is proved.

In Chapter 7, the notions of compactness and connectedness are introduced. These are combined in Chapter 8 with earlier results to study the maxima and minima of real-valued functions of several variables. Chapter 9 treats the inverse and implicit function theorems.

The next six chapters deal, in one way or another, with

integration. Chapters 10 and 11 study multiple integration; the first develops the integral from upper and lower sums and the second shows how to compute multiple integrals in terms of iterated integrals. Line and surface integrals are dealt with in Chapters 12 and 13. In particular, Green's theorem is proved in Chapter 12 and Stokes' theorem stated in Chapter 13. Differential forms are introduced in Chapter 14 and the integration of differential forms is carried out in Chapter 15. The general Stokes' theorem is proved and Green's theorem, Stokes' theorem, and the Gauss theorem are derived from it.

The final two chapters deal with series, the first with numerical series and the second with series of functions. In this latter chapter we discuss Taylor series and give an introduction to Fourier series.

A one-semester course at Yale for students with two years of calculus covered the material on differentiation in Chapters 4, 5, and 8, the integration theory in Chapters 10 and 11, and line and surface integrals (roughly Chapters 12 and 13, plus Gauss' theorem). There are, of course, many other ways to use the book.

In each chapter, propositions, theorems, and lemmas are numbered by section and by order in the section; thus in Chapter 7, Proposition 1.1 is the first result of Section 1 and Theorem 1.2 is the second. We refer to results in this same way within a chapter. To refer to results from other chapters, we merely cite the result by double number and state the chapter number.

We have indicated the less routine exercises by asterisks. These exercises often develop material not discussed in the text proper; instructors with sufficient time may wish to regard them as outlines for lecture topics. Occasionally an exercise anticipates material to be presented later.

Many people have helped us with this book, and we are happy to acknowledge their aid. A number of classes at Yale University have used preliminary versions of the text and made helpful comments; in particular, Ed Odell and Paul Blanchard made a number of valuable suggestions, and William Massey and George Seligman have been sources of much valuable advice. Donna Belli typed both the first and the last versions of the text efficiently, accurately, and with great good humor. We do not believe that the book would ever have been finished without her help.

Lawrence J. Corwin
Robert H. Szczarba

Contents

Multivariable Calculus

1

Some Preliminaries

We collect here some basic information about topics which, while not really part of multivariable calculus, are necessary to it. In addition to providing a convenient reference for this material, this chapter serves to establish some of our notation. Since most of the ideas discussed here will be familiar to the reader, we suggest that this chapter not be worked through in detail to begin with but rather be referred back to when necessary.

1. The Rudiments of Set Theory

In this section, we review some of the ideas and notation from set theory. We also briefly discuss the notion of a function.

A *set* is a collection of objects; the objects are the *elements* of the set. One specifies a set by one of two methods:

(a) Naming the elements (usually listing them inside curly brackets)
(b) Giving a criterion for membership in the set (again in curly brackets)

Thus the set whose elements are 2, 4, 6, 8, and 10 can be given as

$$\{2, 4, 6, 8, 10\}$$

or as
$$\{x : x \text{ is an even integer and } 2 \leq x \leq 10\}$$

or as

$$\{x : x \text{ satisfies } x^5 - 30x^4 + 340x^3 - 1800x^2 + 4384x - 3840 = 0\}$$

All of these describe the *same* set; what matters is not the description, but the actual elements of the set.

Sets are usually denoted by capital letters. The symbol $\in$ is used to denote membership in a set: $x \in S$ means "x is an element of the set S." Similarly, $\notin$ denies membership: $x \notin S$ means "x is not an element of the set S."

The simplest set of all is the *empty set*, usually denoted by $\varnothing$. This set has no elements; it is characterized by the property that $x \notin \varnothing$, whatever x is.

A set S is called a *subset* of T if every element of S is also an element of T; that is, if $x \in T$ whenever $x \in S$. Thus the subsets of $\{2, 4, 6\}$ are $\varnothing$, $\{2\}$, $\{4\}$, $\{6\}$, $\{2, 4\}$, $\{2, 6\}$, $\{4, 6\}$, $\{2, 4, 6\}$.Notice that $\varnothing$ is a subset of every set: regardless of the set T, $x \in T$ whenever $x \in \varnothing$, since x is never in $\varnothing$. (If this bit of logic seems too fancy, read the next section for an explanation.)

We write "A is a subset of B" as $A \subset B$, and "A is not a subset of B" as "$A \not\subset B$."

Two sets are *equal* if they contain the same elements. It should be clear that two sets A and B are equal ($A = B$) exactly when $A \subset B$ and $B \subset A$.

There are three basic ways of making new sets from old; by union, by intersection, and by complementation. The *union* of two sets A and B, written $A \cup B$, is defined by

$$A \cup B = \{x : x \in A \text{ or } x \in B\}^*$$

For instance, if $A = \{1, 2, 3, 4, 5\}$ and $B = \{2, 4, 6, 8, 10\}$, then

$$A \cup B = \{1, 2, 3, 4, 5, 6, 8, 10\}$$

The *intersection* of A and B, $A \cap B$, is defined by

$$A \cap B = \{x : x \in A \text{ and } x \in B\}$$

If, for instance, $A = \{1, 2, 3, 4, 5\}$ and $B = \{2, 4, 6, 8, 10\}$, as before, then

$$A \cap B = \{2, 4\}$$

The *complement* of B in A, $A - B$, is defined by

$$A - B = \{x : x \in A \text{ and } x \notin B\}$$

Thus if A and B are as above, $A - B = \{1, 3, 5\}$ and $B - A = \{6, 8, 10\}$.

The notions of union and intersection can both be extended. Suppose, for instance, that $A_1, A_2, \ldots, A_n$ are sets. We define

$$\bigcup_{j=1}^{n} A_j = \{x : x \in A_j \text{ for some } j,\ 1 \leq j \leq n\}$$
$$= \{x : x \in A_1, \text{ or } x \in A_2, \ldots, \text{ or } x \in A_n\}$$

and

$$\bigcap_{j=1}^{n} A_j = \{x : x \in A_j \text{ for every } j,\ 1 \leq j \leq n\}$$
$$= \{x : x \in A_1, \text{ and } x \in A_2, \ldots, \text{ and } x \in A_n\}$$

*Or both. In mathematics, *or* usually means "one or the other or both."

That is, $\cup_{j=1}^{n} A_j$ contains every element in at least one of the A's, while $\cap_{j=1}^{n} A_j$ consists of the elements common to all the A's.

We can carry this idea one step farther. Suppose that we have sets $A_1, A_2, A_3, \ldots$. We define

$$\bigcup_{n=1}^{\infty} A_n = \{x : x \in A_n \text{ for some } n = 1, 2, \ldots\}$$
$$= \{x : x \in A_1, \text{ or } x \in A_2, \ldots\}$$

and

$$\bigcap_{n=1}^{\infty} A_n = \{x : x \in A_n \text{ for all } n = 1, 2, \ldots\}$$
$$= \{x : x \in A_1, \text{ and } x \in A_2, \ldots\}$$

In fact, we can extend these notions still further. For instance, suppose J is any set and we are given, for each $j \in J$, a set A_j. We can then define

$$\bigcup_{j \in J} A_j = \{x : x \in A_j \text{ for some } j \in J\}$$
$$\bigcap_{j \in J} A_j = \{x : x \in A_j \text{ for all } j \in J\}$$

We sometimes write

$$\bigcup_j A_j \qquad \bigcap_j A_j$$

if the "index" set J is either clear from the context or not important.

There is one special case of complementation that often arises. In most situations, we are concerned with various subsets of some fixed set S. (For instance, we might be working with various sets of integers; all these sets are subsets of the set $\mathbf{Z}$ of all integers.) The complement of a set A in S is then written A': $A' = S - A$. Since we do not mention S in this notation, it should be clear in advance what the set S is. If, for instance, $S = \{2, 4, 6, 8, 10\}$ and $A = \{2, 4, 6\}$, then $A' = \{8, 10\}$. But if $S = Z$, the set of all integers, and $A = \{2, 4, 6\}$, then $A' = \{x : x \text{ is an integer, and } x \neq 2, 4, \text{ or } 6\}$.

The notions of union, intersection, and complementation are connected by two formulas, known as *De Morgan's laws*:

$$\left(\bigcup_j A_j\right)' = \bigcap_j A_j' \tag{1.1}$$

$$\left(\bigcap_j A_j\right)' = \bigcup_j A_j' \tag{1.2}$$

The proofs are similar. For instance, if $x \in (\cup_j A_j)'$, then $x \notin \cup_j A_j$; hence $x \notin A_j$ for every j, or $x \in A_j'$ for every j. Therefore $x \in \cap_j A_j'$, and this proves that $(\cup_j A_j)' \subset \cap_j A_j'$. The reverse inclusion and the proof of (1.2) are left as Exercise 9.

We mention one other useful term in connection with sets. The sets A and B are said to be *disjoint* if $A \cap B = \varnothing$. Thus $\{1, 2, 3\}$ and $\{4, 5, 6\}$ are disjoint; $\{1, 2, 3\}$ and $\{3, 4, 5\}$ are not. Note that A and A' are always disjoint.

If A and B are sets, their *Cartesian product*, $A \times B$, is the set consisting of all ordered pairs (a, b), with $a \in A$ and $b \in B$. If, for instance, $A = \{1, 2\}$ and $B = \{1, 2, 3\}$, then

$$A \times B = \{(1, 1), (1, 2), (1, 3), (2, 1), (2, 2), (2, 3)\}$$

and

$$B \times A = \{(1, 1), (2, 1), (3, 1), (1, 2), (2, 2), (3, 2)\}$$

Notice that two ordered pairs (a, b) and (c, d) are equal exactly when $a = c$ and $b = d$. Thus the order of listing elements in an ordered pair matters (though the order of listing elements of a set does not).

Similarly, the set $A \times B \times C$ (where C is a third set) consists of ordered triples (a, b, c), with $a \in A$, $b \in B$, and $c \in C$. It should be clear how to generalize this notion.

The reason for the name "Cartesian product" is that Cartesian coordinates in the plane let one regard the plane as $\mathbf{R} \times \mathbf{R}$. We generally write $\mathbf{R} \times \mathbf{R}$ as $\mathbf{R}^2$. Similarly, $\mathbf{R}^n = \mathbf{R} \times \mathbf{R} \times \cdots \times \mathbf{R}$ (n factors) is the set of ordered n-tuples of real numbers.

Let A and B be sets. A *function* $f: A \to B$ is a rule which associates to each element $x \in A$ a unique element $f(x)$ in B. The key points here are that the function must be defined for every x in A and that $f(x)$ must be uniquely specified. (For instance, if $B = \mathbf{R}$, the set of real numbers, and $A = \{x \in \mathbf{R} : 0 \leq x \leq 1\}$, then $f(x) = \pm\sqrt{x}$ is not a function because one has a choice of values for $f(x)$. But $f(x) = \sqrt{x}$, the positive square root of x, is a function.)

There are two ways to specify the function $f: A \to B$: by listing the value of $f(x)$ for each $x \in A$ or by describing a rule used to compute $f(x)$. The first method is usually unwieldy, but it makes one important fact clear: a function is defined not by the exact phrasing of the rule, but by the actual correspondence between x and $f(x)$. Thus $f(x) = (x + 1)^2$ and $g(x) = x^2 + 2x + 1$ define the same function, since $f(x) = g(x)$ for every real number x.

This may be a good time to bring up a point which usually causes confusion: the way one describes some common functions. Lots of common functions have names (e.g., sin, tan, and ln), and the value of sin at x, for instance, is $\sin x$. Lots of others do not. For instance, one common function assigns to x the value x^2. This function is often called *the function* x^2, which is fine much of the time but which can cause confusion when one wants to distinguish between a function (which, as we have seen, is a law of correspondence) and its value at a point. We shall usually use the phrase "the function f, where $f(x) = x^2$" to describe this function, which is accurate but rather pedantic-sounding. Another way out is to write "the function $x \mapsto x^2$," which is shorter but which does not give the function a name. We could give this function the name "the squaring function," but that would not solve the problem for the function $x \mapsto x^2 + 3x - 7$. All in all, pedantry seems the best way out.

Finally, a function $f: X \to Y$ is said to be *injective* (or 1–1 or *one-to-one*) if $f(x) = f(x')$ implies $x = x'$ and *surjective* (or *onto*) if, given any $y \in Y$, there is an $x \in X$ with $f(x) = y$. The function f is *bijective* if it is both injective and surjective. The *image* of f is $\{y \in Y : f(x) = y \text{ for some } x \in X\}$.

Exercises

1. Let $A = \{1, 2, 3, 5, 7, 10\}$, $B = \{1, 3, 4, 6, 20, 21\}$, and $C = \{2n : n \text{ a positive integer}\}$. Determine the following:
   (a) $A \cup B$ (b) $A \cap C$
   (c) $(A \cup B) \cap C$ (d) $A \cup B \cup C$
   (e) $A - C$ (f) $C - B$
2. Determine all subsets of $X = \{1, 2, a, b\}$.
3. Suppose A and B are sets with $A \subset B$. Prove that
   (a) $A \cup B = B$
   (b) $A \cap B = A$
   (c) $A - B = \varnothing$
4. Let A and B be sets. Prove that
   (a) If $A \cup B = B$, then $A \subset B$.
   (b) If $A \cap B = A$, then $A \subset B$.
   (c) If $A - B = \varnothing$, then $A \subset B$.
5. Show that if A and B are sets, then
   (a) $A \cap B \subset B$, $A \cap B \subset A$
   (b) $A \subset A \cup B$, $B \subset A \cup B$
6. Show that if A, B, and C are sets, then
   (a) $A \cap (B \cup C) = (A \cap B) \cup (A \cap C)$
   (b) $A \cup (B \cap C) = (A \cup B) \cap (A \cup C)$
7. Show that if A and B are sets, then
   (a) $A = (A \cap B) \cup (A - B)$
   (b) $\varnothing = (A \cap B) \cap (A - B)$
8. Which of the following are functions?
   (a) $F: \mathbf{R} \to \mathbf{R}$ by $F(x) = \sqrt[3]{x}$.
   (b) $G: \mathbf{R} \to \mathbf{R}$ by $G(x) = y$, where $y^3 - y + x = 0$.
   (c) $H: \mathbf{R} \to \mathbf{R}$, where $H(x)$ is the biggest integer $\leq x$.
9. Prove the De Morgan laws [equations (1.1) and (1.2)].

2. Some Logic

Mathematics is distinguished from all other sciences by its insistence on rigor: every statement must be derived by logical rules from clearly stated assumptions. We cannot go into all the subtleties of logic here, but we shall give a brief outline of some basic material.

A rigorous treatment of our subject would start with a list of the axioms which we assume. We shall treat this as unnecessary, taking it for granted that the reader knows some algebra, geometry, trigonometry, and calculus. We shall often use facts from those fields without special comment.

For the most part, however, we shall be developing new results, generally by proving lemmas, propositions, thorems, and corollaries. There is no strong distinction among the terms. We generally reserve the term *theorem* for the most important results; *propositions* are somewhat less important or interesting. A *lemma* often has no particular interest in itself, but is useful in proving some other result. A *corollary* is an easy consequence of a previous result.

Every mathematical statement is assumed to be either true or false. The statements we deal with will be either simple statements (whose truth or falsity is presumably known) or statements built of simple statements in certain easily describable ways. The principal means of building are:

(1) *Conjunction.* If P and Q are statements, then "P and Q" is a statement. Thus the conjunction of "$\pi > 3$" and "$e > 4$" is "$\pi > 3$ and $e > 4$." The conjunction of two statements is true only when both statements are true. (Thus "$\pi > 3$ and $e > 4$" is false, since $e \leq 4$.)

(2) *Disjunction.* If P and Q are statements, then "P or Q" is a statement, and it is true if either or both of the two statements is true. (The disjunction of the two statements in the previous example is "$\pi > 3$ or $e > 4$," and it is true because $\pi > 3$.) In common speech, "P or Q" often implies that exactly one of P and Q is true; in mathematics, there is no such implication, and both may be true.

(3) *Negation.* If P is a statement, so is "not P"; "not P" is false if P is true and true if P is false. "Not P," of course, is the statement which says the opposite of P. (The negation of "$\pi > 3$" is "$\pi \not> 3$," or "$\pi \leq 3$.") One sometimes writes "$\sim P$" for "not P."

(4) *Implication.* If P and Q are statements, then "P implies Q," or "if P, then Q," is a statement. (It is the sort of statement one expects to see in theorems.) The statement "if P, then Q" is true unless P is true and Q is false. (Thus "if P, then Q" means only that either P is false or Q is true.) In ordinary speech, implications are expected to involve a rational connection between the two statements; thus "if it rains hard tomorrow, the reservoir will overflow" is a reasonable sort of implication (though its truth depends on the state of the reservoir), while "if it rains hard tomorrow, then Napoleon lost at Waterloo" is not. In mathematics, no such connection is required: "if $e > 4$, then $\pi > 3$" is a true implication, though there does not seem to be any causal connection between the sizes of e and π.

To prove the implication "P implies Q," we need to rule out the possibility that P is true and Q is false. We can do this directly, by showing that Q is true whenever P is, or indirectly, by showing that if Q is false, then P is false, or (also indirectly) by showing that assuming P is true and Q is false leads to a contradiction.

One sometimes writes "$P \Rightarrow Q$" for "P implies Q."

(5) *Equivalence.* "P is equivalent to Q" means "P implies Q and Q implies P." This holds if P and Q are both true or both false; it is false if one is true and the other is false. Another common way of expressing equivalence is "P if and only if Q" and one sometimes writes $P \Leftrightarrow Q$ as well.

To prove that $P \Leftrightarrow Q$, one ordinarily needs to prove two statements: $P \Rightarrow Q$ and $Q \Rightarrow P$. The statements "$P \Rightarrow Q$" and "$Q \Rightarrow P$" are called *converses* of one another. The converse of a true statement need *not* be true. The *contrapositive* of "$P \Rightarrow Q$" is "not $Q \Rightarrow$ not P," and the contrapositive of a true statement is true. (The reason is that "not $Q \Rightarrow$ not P" is false only if "not Q" is true and "not P" is false, or if Q is false and P is true; that is exactly the condition which makes "$P \Rightarrow Q$" false.)

It is easy to check (see Exercise 3) that if $P \Rightarrow Q$ and $Q \Rightarrow R$, then $P \Rightarrow R$. This fact leads to another way of proving the equivalence of a number of statements: we can reason in a cycle. For example, one possible way of proving statements P, Q, and R equivalent is to show $P \Rightarrow Q \Rightarrow R \Rightarrow P$. (It might well turn out not to be the simplest way.)

We now consider another method of complicating simple statements: *quantification*. It is met with in calculus, in the definition of continuity at a point: the function f is continuous at a real number x if for all real numbers $\varepsilon > 0$, there is a number $\delta > 0$ such that if y is any number with $|x - y| < \delta$, then $|f(x) - f(y)| < \varepsilon$.

There are two quantifiers: the *universal* ("for all," written $\forall$) and the *existential* ("there exists," or $\exists$). By using these symbols, the definition of continuity given above can be abbreviated as

$$(\forall \varepsilon > 0)(\exists \delta > 0)(\forall y)\, |x - y| < \delta \Rightarrow |f(x) - f(y)| < \varepsilon$$

If we begin with a statement $P(x)$ involving some unspecified term x, then "for all x, $P(x)$" is true only if the statement $P(x)$ is true when we specify x in any way. (Ordinarily x will be restricted, explicitly or otherwise, to some fixed set; in the definition of continuity, for instance, y, ε, and δ are tacitly assumed to be real numbers.) One counterexample makes the statement false. Thus "$(\forall x)x^2 > 0$" is false, since $0^2 \leq 0$. By contrast, "There exists an x such that $P(x)$" is true if we can find one number x for which $P(x)$ is true. Thus "$(\exists x)x^2 > 0$" is true; let $x = 1$, for instance.

The order of quantifiers is very important. Consider, for instance, the statements

$$(\forall x)(\exists y \neq x)\, y > x$$

and

$$(\exists x)(\forall y \neq x)\, y > x$$

The first statement says that given any number x, there is a bigger number y; that is, there is no biggest number. The second says that there is one number x smaller than every other number y; that is, there is a smallest number. The first statement is true; the second false.

We have seen that one example can make a statement involving $\forall x$ false and one involving $\exists x$ true. This state of affairs is reflected in the way we work out the negation of a statement involving quantifiers. Suppose, for instance, that the statement $(\forall x)\, P(x)$ is false. Then we can find some x such that $P(x)$ is false; that is, the statement $(\exists x) \sim P(x)$ is true. Similarly, if the statement $(\exists x)\, P(x)$ is false, then for every x we choose, $P(x)$ must be false. That means that the statement $(\forall x) \sim P(x)$ is true. In short, to form the negation of a statement involving a quantifier, change the quantifier ($\forall$ to $\exists$, or vice versa) and put the negation inside. For instance, to find the negation of

$$(\forall \varepsilon > 0)(\exists \delta > 0)(\forall y)(|x - y| < \delta \Rightarrow |f(x) - f(y)| < \varepsilon)$$

proceed from

$$\sim(\forall \varepsilon > 0)(\exists \delta > 0)(\forall y)(|x - y| < \delta \Rightarrow |f(x) - f(y)| < \varepsilon)$$

through

$$(\exists \varepsilon > 0) \sim (\exists \delta > 0)(\forall y)(|x - y| < \delta \Rightarrow |f(x) - f(y)| < \varepsilon)$$

and

$$(\exists \varepsilon > 0)(\forall \delta > 0) \sim (\forall y)(|x - y| < \delta \Rightarrow |f(x) - f(y)| < \varepsilon)$$

to

$$(\exists \varepsilon > 0)(\forall \delta > 0)(\exists y) \sim (|x - y| < \delta \Rightarrow |f(x) - f(y)| < \varepsilon)$$

or

$$(\exists \varepsilon > 0)(\forall \delta > 0)(\exists y)(|x - y| < \delta \text{ and } |f(x) - f(y)| \geq \varepsilon)$$

We mentioned that one cannot interchange an existential quantifier and a universal one without possibly changing the meaning of the statement. However, one can interchange two existential quantifiers or two universal quantifiers freely.

In principle, every mathematical statement which is not an axiom or a definition should be proved. In practice, any sufficiently obvious statement is merely asserted, on the grounds that any proof would be tedious, possibly confusing, and unnecessary (since anybody could supply one if required). The problem which arises, of course, is that not everyone agrees about what is obvious.

Exercises

1. Show that if P is any statement, then "P or $\sim P$" is true and "P and $\sim P$" is false.
2. Show that the statements "$\sim(P$ and $Q)$" and "$(\sim P)$ or $(\sim Q)$" are equivalent.
3. Show that if $P \Rightarrow Q$ and $Q \Rightarrow R$, then $P \Rightarrow R$.
4. Let $a_1, a_2, \ldots, a_n, \ldots$ be a sequence of real numbers. This sequence *converges* to

the number a if for any $\varepsilon > 0$, there is an integer N such that $|a_n - a| < \varepsilon$ whenever $n > N$. Write out what it means for a sequence not to converge to the number a.

5. A function $f: \mathbf{R} \to \mathbf{R}$ is *uniformly continuous* if for any $\varepsilon > 0$, there is a $\delta > 0$ such that for all x, y, if $|x - y| < \delta$, then $|f(x) - f(y)| < \varepsilon$. Write out what it means for a function not to be uniformly continuous.

3. Mathematical Induction

Mathematical induction is a somewhat specialized, but very useful, method of proof. Suppose that we want to prove a statement true for every positive integer n. It suffices to show that

(a) The statement is true for $n = 1$.
(b) Whenever the statement is true for $n = k$, it is true for $n = k + 1$.

An example will probably help. We shall prove the following result:

Proposition 3.1 *For all positive integers n,*

$$1 + 2 + \cdots + n = \frac{n(n+1)}{2} \tag{3.1}$$

Proof. For $n = 1$, the formula reads

$$1 = \frac{1(1+1)}{2}$$

This result is certainly true, since both sides are 1.

Now suppose that (3.1) holds for $n = k$:

$$1 + 2 + \cdots + k = \frac{k(k+1)}{2} \tag{3.2}$$

We need to prove the corresponding result for $n = k + 1$:

$$1 + 2 + \cdots + (k+1) = \frac{(k+1)[(k+1)+1]}{2} \tag{3.3}$$

This is not hard, since we can use (3.2):

$$\begin{aligned} 1 + 2 + \cdots + (k+1) &= (1 + 2 + \cdots + k) + (k+1) \\ &= \frac{k(k+1)}{2} + (k+1) \\ &= (k+1)\left(\frac{k}{2} + 1\right) = (k+1)\frac{(k+2)}{2} \end{aligned}$$

which is the same as (3.3). Thus the proposition is proved.

To see intuitively why the proposition holds, notice that we proved it for $n = 1$. Letting $k = 1$, we see that the proposition holds for $n = 2$. Now let $k = 2$; the proposition must hold for $n = 3$, and so on. By proceeding in this way, it seems clear that we can prove formula (3.1) for any given positive integer. Mathematical induction expresses a basic property of the set of natural numbers $\mathbf{N}$.

One can use mathematical induction to prove many other propositions besides formulas (though most early examples of the method involve formulas). Here is another example.

Proposition 3.2 *If n is a positive integer, then $x^n - y^n$ is divisible by $x - y$.*

Proof. When $n = 1$, $x^1 - y^1 = x - y$ is certainly divisible by $x - y$. Next, suppose that the proposition holds for $n = k$; that is, $x^k - y^k$ is divisible by $x - y$. Then $x^{k+1} - y^{k+1} = x(x^k - y^k) + y^k(x - y)$, and both terms are divisible by $x - y$. Therefore $x^{k+1} - y^{k+1}$ is divisible by $x - y$. The proposition follows, by mathematical induction.

It will often be convenient in what follows to use the so-called *sigma notation* for sums. If $a_1, \ldots, a_k$ are numbers (or elements in any set in which addition makes sense), we define

$$\sum_{i=1}^{k} a_i = a_1 + a_2 + \cdots + a_k \tag{3.4}$$

The left side of (3.4) is read "the sum of the a_i as i goes from 1 to k." For example, equation (3.1) can be written

$$\sum_{i=1}^{n} i = \frac{n(n+1)}{2}$$

Exercises

1. Prove that $\sum_{j=1}^{n} j^2 = n(n+1)(2n+1)/6$.
2. Prove that $\sum_{j=0}^{n} r^j = (1 - r^{n+1})/(1 - r)$ (if $r \neq 1$).
3. Prove that $\sum_{j=1}^{n} j^3 = n^2(n+1)^2/4$.
4. Prove that $n^5 - n$ is divisible by 5 for any positive integer n.
5. What is wrong with the following inductive proof?

   Theorem. In any set S with n elements, all the elements are equal.

   Proof. If $n = 1$, there is only one element, and the result is obvious. Now assume the result for $n = k$; we prove it for $n = k + 1$. Let the elements of S be $x_1, \ldots, x_{k+1}$.

   By the inductive hypothesis, we know that

   $$x_1 = x_2 = \cdots = x_k$$

We also know (again by the inductive hypothesis) that

$$x_2 = x_3 = \cdots = x_{k+1}$$

Therefore $x_1 = x_2 = \cdots = x_{k+1}$, and the inductive step is proved.

6. Let n and k be integers and define quantities $\binom{n}{k}$ by

$$\binom{n}{k} = \begin{cases} \dfrac{n!}{k!\,(n-k)!} & \text{if } 0 \le k \le n \\ 0 & \text{if } k < 0 \text{ or } k > n \end{cases}$$

[Here $n! = 1 \cdot 2 \cdot 3 \cdots (n-1) \cdot n$ if $n > 0$ and $0! = 1$.]

(a) Show that for each nonnegative integer n, and for all integers k,

$$\binom{n}{k-1} + \binom{n}{k} = \binom{n+1}{k}$$

(b) Prove by induction that $\binom{n}{k}$ is an integer whenever n is a nonnegative integer and k is any integer.

(c) If n is an *arbitrary* integer (i.e., not necessarily nonnegative), define

$$\binom{n}{k} = \frac{n(n-1)(n-2)\cdots(n-k+1)}{k!}$$

if $k \ge 0$ and $\binom{n}{k} = 0$ if $k < 0$. Show that this definition agrees with the one above for nonnegative integers n. Prove that this $\binom{n}{k}$ is an integer for all integers n and k.

(d) Prove the *binomial theorem*: *If a and b are any numbers, and if n is any nonnegative integer, then*

$$\begin{aligned}(a+b)^n &= \sum_{k=0}^{n} \binom{n}{k} a^k b^{n-k} \\ &= b^n + \binom{n}{1} ab^{n-1} + \binom{n}{2} a^2 b^{n-2} + \cdots + \binom{n}{n} a^n\end{aligned}$$

(e) Prove *Leibniz' formula*: *If n is a nonnegative integer, and if $f(x)$ and $g(x)$ are two functions, each differentiable at least n times, set $D^0(f) = f$, $D^k(f) = f^{(k)}(x) = d^k f/dx^k$, and likewise for g. Then*

$$D^n(f \cdot g) = \sum_{k=0}^{n} \binom{n}{k} D^k(f)\, D^{n-k}(g)$$

*7. Discover and prove results like (d) and (e) in Exercise 6 for $(a_1 + a_2 + \cdots + a_r)^n$ and for $D^n(f_1 \cdot f_2 \cdot f_3 \cdots f_r)$.

*8. (a) Show that if $|S|$ = number of elements in $S = n$, the number of subsets T of S with $|T| = k$ is $\binom{n}{k}$.

(b) Show that if $|S| = n$ and if T is another set with $|T| = k$, then the number of mappings of S into T is k^n, and the number of one-to-one ("injective") mappings is $\binom{k}{n} n!$. [It might be easier to do (b) before (a).]

9. Let i be the imaginary number with $i^2 = -1$ and prove *Demoivre's theorem* $(\cos\theta + i\sin\theta)^n = \cos n\theta + i\sin n\theta$.

4. Inequalities and Absolute Value

In what follows, it will often be necessary to manipulate inequalities involving absolute values. We include a brief review here.

To begin with, we use the standard notation for expressing inequality between two numbers:

$a < b$ for "a is less than b"
$a \le b$ for "a is less than or equal to b"
$a > b$ for "a is greater than b"
$a \ge b$ for "a is greater than or equal to b"

These relations have the following properties. (We state these for $<$, but they also hold for $\le$, $>$, $\ge$.)

(4.1) If $a < b$ and $b < c$, then $a < c$.

(4.2) If $a < b$ and c is any number, then $a + c < b + c$.

(4.3) If $a < b$ and $c > 0$, then $ac < bc$.

(4.4) If $a < b$ and $c < 0$, then $ac > bc$.

The *absolute value of a real number* x, $|x|$, is defined by

$$|x| = \begin{cases} x & \text{if } x \ge 0 \\ -x & \text{if } x < 0 \end{cases}$$

Thus $|3| = 3$, $|-\pi| = \pi$, $|\sqrt{2} - 4| = 4 - \sqrt{2}$, $|0| = 0$. Notice that $-|x| \le x \le |x|$.

Since by convention $\sqrt{x}$ is the positive square root of x, we could also define $|x|$ to be $\sqrt{x^2}$. Notice, in particular, that $|x|^2 = x^2$.

For any real numbers x and y, we have

(4.5) $$|xy| = |x|\,|y|$$

and

$$(4.6) \qquad |x + y| \leq |x| + |y| \qquad (\textit{triangle inequality})$$

Inequality (4.5) is easy to see. The triangle inequality is easily proved by checking cases ($x > 0$, $x < 0$, etc.). A faster method is the following: (4.6) is equivalent to

$$|x + y|^2 \leq (|x| + |y|)^2$$

or
$$x^2 + 2xy + y^2 \leq |x|^2 + 2|x|\,|y| + |y^2| = x^2 + 2|x|\,|y| + y^2$$

But since $2xy \leq 2|x|\,|y|$, this last inequality is certainly true, and (4.2) follows.

We shall often meet inequalities like $|2x - 7| < 4$ or $|2x - 7| > 4$. As we see now, these inequalities can be reduced to inequalities not involving the absolute value.

Proposition 4.1 *The inequality $|a| < b$ is equivalent to the two inequalities $a < b$ and $-b < a$.*

For example, the inequality $|2x - 7| < 4$ is equivalent to the two inequalities

$$-4 < 2x - 7 < 4$$

or
$$3 < 2x < 11 \qquad \text{using (4.2)}$$

or
$$\frac{3}{2} < x < \frac{11}{2} \qquad \text{using (4.3)}$$

Proof of Proposition 4.1. We may as well assume $b > 0$, for otherwise the proposition is trivial.

We first show that $|a| < b$ implies $-b < a < b$. If $a \geq 0$, then $|a| = a$; so $a < b$. In addition, $-b < a$ since $-b < 0$. If $a < 0$, then $|a| = -a$ and the inequality $|a| < b$ becomes $-a < b$. Using equation (4.4) with $c = -1$, we have $a > -b$ or $-b < a$. Again $a < 0$; so clearly $a < b$.

Suppose now that $-b < a < b$. We need to prove that $|a| < b$. If $a \geq 0$, then $|a| = a$; so $a < b$ implies $|a| < b$. If $a < 0$, then $|a| = -a$, and $-b < a$ implies $b > -a$ [using equation (4.4)], or $-a = |a| < b$.

In an analogous manner, we can prove the following:

Proposition 4.2 *The inequality $|a| > b$ holds if and only if either $a > b$ or $a < -b$.*

We leave the proof to the reader.

Remark. Proposition 4.1 remains true if $<$ is replaced by $\leq$. Proposition 4.2 remains true if $>$ is replaced by $\geq$.

We note here for future reference some sets in **R** which will arise in the future, the *intervals*. We have

$$(a, b) = \{x : a < x < b\}$$

$$(a, b] = \{x : a < x \leq b\}$$

$$[a, b) = \{x : a \leq x < b\}$$

$$[a, b] = \{x : a \leq x \leq b\}$$

$$(a, \infty) = \{x : x > a\}$$

$$[a, \infty) = \{x : x \geq a\}$$

$$(-\infty, a) = \{x : x < a\}$$

$$(-\infty, a] = \{x : x \leq a\}$$

The principle is that square brackets include the endpoints, while parentheses do not. Of course, **R** and $\varnothing$ are also intervals.

Sets of the form (a, b), (a, ∞), or $(-\infty, a)$ are called *open* intervals; those of the form $[a, b]$, $[a, \infty)$, or $(-\infty, a]$ are called *closed* intervals. (The reason for the names will be explained in Section 3.5.) The sets $[a, b)$ and $(a, b]$ are sometimes called *half-open*.

Exercises

1. Prove equation (4.5) directly from the definition of absolute value.
2. Prove equation (4.6) directly from the definition of absolute value.
3. Prove Proposition 4.2.
4. What intervals do the following inequalities describe?
   (a) $|3x - 5| < 8$
   (b) $|6x + 1| \leq 11$
   (c) $|2 - 3x| < 4$
5. What pairs of intervals do the following inequalities describe?
   (a) $|2x - 7| > 1$
   (b) $|7x + 3| > 14$
   (c) $|4 - 3x| \geq 7$
6. (a) Show that if $x_1, \ldots, x_n$ are real numbers, then there exists a largest number; i.e., that $\exists j$ $(1 \leq j \leq n)$ such that for all i with $1 \leq i \leq n$, $x_i \leq x_j$. (We write $x_j = \max\{x_1, \ldots, x_n\}$.)
   (b) Show also that $\exists k (1 \leq k \leq n)$: for all i with $1 \leq i \leq n$, $x_i \geq x_k$. (We write $x_k = \min\{x_1, \ldots, x_n\}$.) (Hint: use induction.)

5. Equivalence Relations

There is often a need in mathematics to say that two things are equivalent to each other in some sense. For instance, congruent triangles can be regarded as the same for most purposes in geometry; for some purposes, when size is unimportant, similar triangles can be regarded as equivalent. Mathematicians have therefore found it useful to systematize the notion of equivalence.

An *equivalence relation* on a set S is a relation $\sim$ on S which satisfies the following rules:

(1) For any $a \in S$, $a \sim a$ (*reflexive law*).
(2) If $a \sim b$, then $b \sim a$ (*symmetric law*).
(3) If $a \sim b$ and $b \sim c$, then $a \sim c$ (*transitive law*).

Here are some examples:

1. Ordinary equality is an equivalence relation; it is easy to see that properties (1) to (3) hold.
2. Suppose that we define a relation on the integers by decreeing that $a \equiv b$ if $a - b$ is even. Then $\equiv$ is an equivalence relation.
3. More generally, let $n > 1$ be an integer, and write $a \equiv b \bmod n$ ("a is *congruent* to $b \bmod n$") if $a - b$ is divisible by n. Then congruence mod n is an equivalence relation.
4. The relation $a \leq b$ is *not* an equivalence relation; property (2) fails.
5. The relation $a \parallel b$, (a is parallel to b), among lines in the plane, becomes an equivalence relation if one agrees that any line is parallel to itself. Otherwise, property (1) fails.
6. The relation "the angle between a and b is $< 30°$," again between lines in the plane, is not an equivalence relation; property (3) fails.

Given an equivalence relation $\sim$, and an element x, we define

$$R_x = \{y : x \sim y\}$$

Thus R_x is the set of elements equivalent to x; it is often called the *equivalence class* of x.

Proposition 5.1 *If R_x and R_y are two equivalence classes for the relation $\sim$, then either $R_x \cap R_y = \varnothing$ or $R_x = R_y$. (Equivalence classes are either disjoint or identical.)*

Proof. Suppose $R_x \cap R_y \neq \varnothing$; we need to show that $R_x = R_y$. Pick z in $R_x \cap R_y$, and let w be any element of R_x. Then $x \sim z$ and $x \sim w$; hence [properties (2) and (3)] $w \sim z$. Since $z \in R_y$, $z \sim y$; now property (3) says that

$w \sim y$. Therefore $w \in R_y$. It follows that $R_x \subset R_y$. The reverse implication proceeds in the same way.

Corollary *If $\sim$ is an equivalence relation on a set S, then S is the disjoint union of the subsets of R_x, $x \in S$.*

For many purposes, all elements of an equivalence class behave identically. Indeed, it is a standard mathematical trick to regard the equivalence classes as basic objects and to work with them. We shall see a number of examples of this device in the rest of this book.

Exercises

1. Which of the following are equivalence relations?
   (a) $a \sim b$ if a is divisible by b (a, b nonzero integers).
   (b) $(a, b) \sim (c, d)$ if $ad = bc$ [(a, b) and (c, d) are pairs of integers with b and d nonzero].
   (c) $a \sim b$ if $a + b$ is divisible by 3 (a, b nonzero integers).
   (d) $a \sim b$ if $ab > 0$ (a, b real numbers).
   (e) $a \sim b$ if a and b have the same parents.
2. Let p be a fixed positive integer. We say that two integers a and b are *congruent modulo p* (written $a \equiv b \mod p$) if $a - b$ is divisible by p.
   (a) Prove that the relation of being congruent modulo p is an equivalence relation on the set $\mathbf{Z}$ of integers.
   (b) Determine the number of equivalence classes of this equivalence relation.
3. Can property (1) of equivalence relations be derived from the other two? [If it can, give a proof. If a derivation is impossible, give a relation satisfying (2) and (3), but not (1).]

2

Euclidean Spaces and Linear Transformations

Our basic goal in this text is the study of "vector-valued" functions of several variables, that is, functions between Euclidean spaces. In order to accomplish this, we need first to develop the structure of Euclidean spaces themselves. In addition, we need an understanding of some of the simpler functions between Euclidean spaces, the linear transformations. This chapter is devoted to these topics.

We note that the material on linear transformations in Sections 5 through 9 is not needed in Chapter 3 (except for Proposition 3.5 where we prove linear transformations continuous). Some readers may wish to defer these sections until they are ready to begin Chapter 4.

1. The Derivative—Another View

In this section, we reinterpret the derivative of a real-valued function of one variable. This interpretation will form the basis for the definition of the derivative of a vector-valued function of several variables (given in Chapter 4) and will serve to motivate the material in the remainder of the chapter.

Let $f: \mathbf{R} \to \mathbf{R}$ be a real-valued function of one variable. We say that f is *differentiable* at x_0 if the limit

$$\lim_{x \to x_0} \frac{f(x) - f(x_0)}{x - x_0}$$

exists. The value of this limit is called the *derivative* of f at x_0 and denoted by

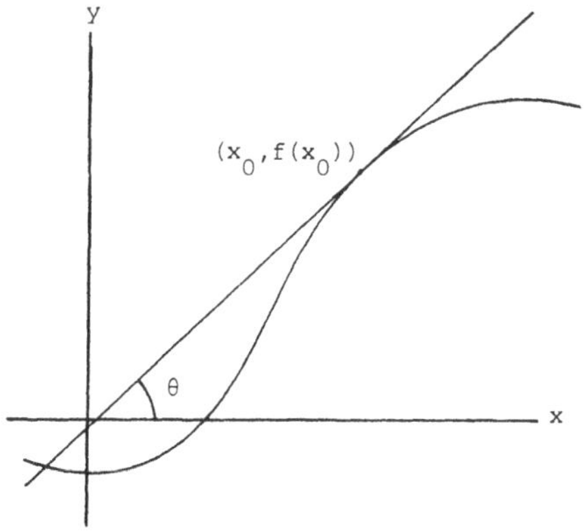

Figure 1.1 The tangent line.

$f'(x_0)$. Geometrically, the number $f'(x_0)$ gives the slope of the tangent line to the graph of $f(x)$ at the point $(x_0, f(x_0))$. (See Figure 1.1.)

According to the definition of the limit,

$$\lim_{x \to x_0} \frac{f(x) - f(x_0)}{x - x_0} = f'(x_0)$$

means roughly that the quantity

$$\frac{f(x) - f(x_0)}{x - x_0}$$

is close to $f'(x_0)$ when x is near x_0. Equivalently, we can say that

$$\left| \frac{f(x) - f(x_0)}{x - x_0} - f'(x_0) \right|$$

is small when x is near x_0. Rewriting this last expression, we see that

$$\text{(1.1)} \qquad \frac{|f(x) - f(x_0) - f'(x_0)(x - x_0)|}{|x - x_0|}$$

is small when x is near x_0. Now let $h: \mathbf{R} \to \mathbf{R}$ be the function defined by

$$h(x) = f'(x_0)(x - x_0) + f(x_0)$$

Then (1.1) becomes

$$\frac{|f(x) - h(x)|}{|x - x_0|}$$

which is small when x is close to x_0. It follows that the function $h(x)$ approximates the function $f(x)$ near x_0. In fact, $h(x)$ approximates $f(x)$ so well that, not only is $|f(x) - h(x)|$ small near x_0, but $|f(x) - h(x)|$ multiplied by the large number

$$\frac{1}{|x - x_0|}$$

(larger as x gets closer to x_0) is small when x is near x_0.

Let us take a closer look at the function $h(x)$. By definition, the graph of $h(x)$ is the set of all pairs (x, y) for which $y = h(x)$ or

$$y = f'(x_0)(x - x_0) + f(x_0)$$

This equation is exactly the equation of the line in the plane through the point $(x_0, f(x_0))$ with slope $f'(x_0)$; that is, the tangent line to the graph of $f(x)$ at the point $(x_0, f(x_0))$. Since the tangent line to the graph of $f(x)$ at the point $(x_0, f(x_0))$ is clearly close to the graph at this point, it is not surprising that $h(x)$ is a good approximation to $f(x)$ near x_0.

The discussion shows that if $f: \mathbf{R} \to \mathbf{R}$ is differentiable at x_0, then there is a function $h: \mathbf{R} \to \mathbf{R}$ whose graph is a straight line and which approximates f near x_0 in the sense that

$$\lim_{x \to x_0} \frac{|f(x) - h(x)|}{|x - x_0|} = 0$$

In fact, the existence of such a function is essentially equivalent to the differentiability of f at x_0, and it is in these terms that we define differentiability of functions of several variables in Chapter 4. In order to carry this out, we need to study the structure of Euclidean spaces. Explicitly, we need to be able to add and subtract the elements of Euclidean space, multiply them by real numbers, take their "absolute value" (or, more accurately, their norm), and introduce the analogue of the functions whose graphs are straight lines. This will be accomplished in the remainder of this chapter.

2. The Cartesian Plane

Anyone reading this text is undoubtedly acquainted with the Cartesian plane; for one thing, graphs of functions of one variable are drawn on it. Nonetheless, we shall review this material, both to ease the path into this book and to emphasize different aspects of the subject.

We shall usually denote the Cartesian plane by $\mathbf{R}^2$; it consists of ordered pairs of real numbers. Geometrically, the point (a, b) represents the point on

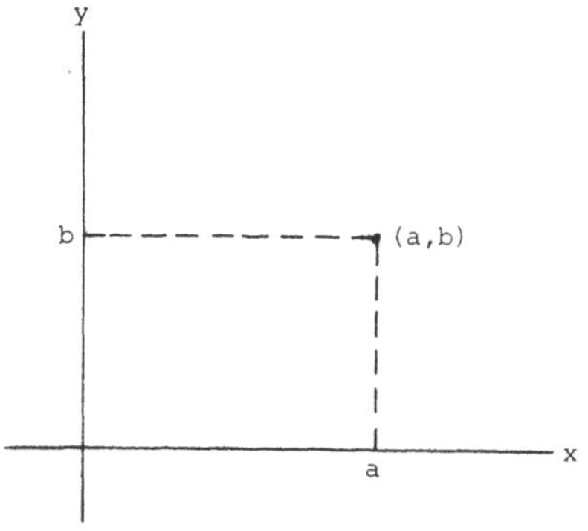

Figure 2.1 Coordinates in the Cartesian plane.

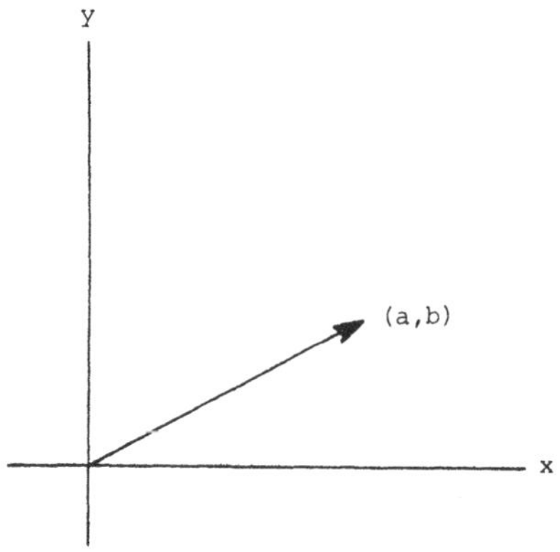

Figure 2.2 A vector in $\mathbf{R}^2$.

the plane whose x coordinate is a and whose y coordinate is b. (See Figure 2.1.)

It is sometimes useful to think of the point (a, b) in other ways. In many applications, (a, b) is regarded as a line segment stretching from the origin to the point (a, b). (See Figure 2.2.) Different points in $\mathbf{R}^2$ give rise to different line segments; more specifically, these segments differ either in length or in direction. For this reason, one sometimes represents the point (a, b) by any line segment with the same length and direction as the segment from (0, 0) to

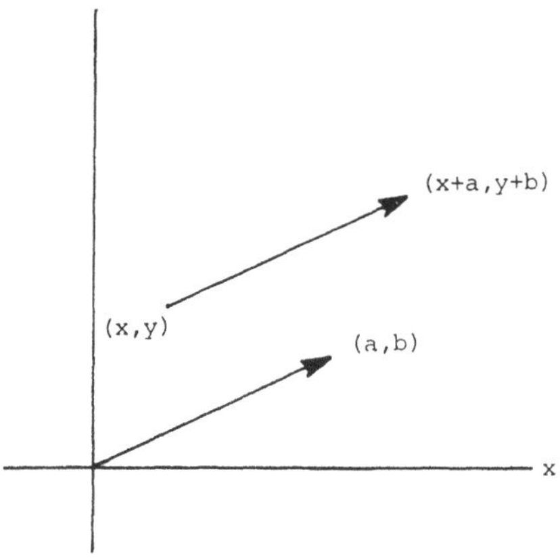

Figure 2.3 Vectors in $\mathbf{R}^2$.

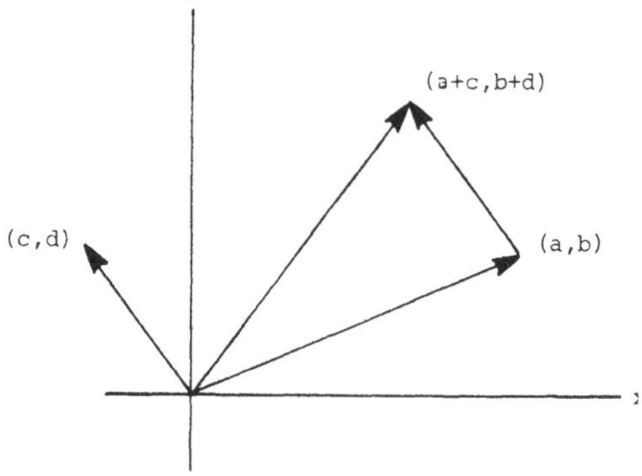

Figure 2.4 The sum of two vectors.

(a, b). For example, if (x, y) is any point in the plane, then the segment from (x, y) to $(x + a, y + b)$ has the same length and direction as that from $(0, 0)$ to (a, b). (See Figure 2.3.) Such directed line segments are usually called *vectors*; in physics, they represent certain kinds of quantities such as force or velocity.

We shall, in what follows, call the elements of $\mathbf{R}^2$ vectors. Instead of always writing a vector as an ordered pair, we shall often represent it by a boldface letter: $\mathbf{v} = (a, b)$. We call the numbers a and b the *components* of $\mathbf{v}$.

There are a number of algebraic operations which can be performed on vectors in $\mathbf{R}^2$; we shall be concerned in this section with two of them. We define the *sum* of two vectors, (a, b) and (c, d), by

(2.1) $$(a, b) + (c, d) = (a + c, b + d)$$

That is, one adds vectors in $\mathbf{R}^2$ by adding their components. For example, $(2, 1) + (4, -2) = (6, -1)$.

We can interpret the sum geometrically if we think of the vectors as directed line segments: to add (a, b) and (c, d), put the tail of the arrow representing (c, d) at (a, b) and see where the head is. (See Figure 2.4.)

The other operation is called *scalar multiplication*; in this operation, one multiplies a vector by a real number. The product of (a, b) by the number r is defined by

(2.2) $$r(a, b) = (ra, rb)$$

Thus one simply multiplies each component by r. For instance, $7(1, -3) = (7, -21)$.

Interpreting the product geometrically is not hard. The vector $2(a, b)$ for instance, points in the same direction as (a, b), but is twice as long; $\frac{1}{3}(a, b)$ points in the same direction as (a, b), but is a third as long. There is one slightly tricky matter to be noted: multiplication by a negative number reverses the direction. Thus $-2(a, b)$ points in the opposite direction from (a, b) and is twice as long. (See Figure 2.5.)

These operations satisfy a number of relations, some of which are the following:

(2.3) If $\mathbf{v}, \mathbf{w} \in \mathbf{R}^2$, then $\mathbf{v} + \mathbf{w} = \mathbf{w} + \mathbf{v}$.

(2.4) If $\mathbf{v}_1, \mathbf{v}_2, \mathbf{v}_3 \in \mathbf{R}^2$, then $(\mathbf{v}_1 + \mathbf{v}_2) + \mathbf{v}_3 = \mathbf{v}_1 + (\mathbf{v}_2 + \mathbf{v}_3)$.

(2.5) If $\mathbf{0} = (0, 0)$, then $\mathbf{v} + \mathbf{0} = \mathbf{0} + \mathbf{v}$, for all $\mathbf{v} \in \mathbf{R}^2$.

(2.6) For each $\mathbf{v} = (x, y)$, the vector $-\mathbf{v} = (-x, -y)$ satisfies $\mathbf{v} + (-\mathbf{v}) = \mathbf{0}$.

(2.7) $1 \cdot \mathbf{v} = \mathbf{v}$ for all vectors $\mathbf{v}$.

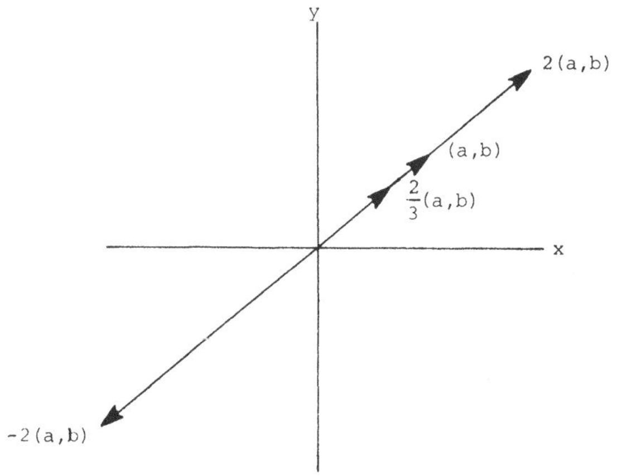

Figure 2.5 Scalar multiples of a vector.

(2.8) If $\mathbf{v}, \mathbf{w} \in \mathbf{R}^2$ and $a \in \mathbf{R}$, then $a(\mathbf{v} + \mathbf{w}) = a\mathbf{v} + a\mathbf{w}$.

(2.9) If $\mathbf{v} \in \mathbf{R}^2$ and $a, b \in \mathbf{R}$, then $(a + b)\mathbf{v} = a\mathbf{v} + b\mathbf{v}$.

(2.10) If $\mathbf{v} \in \mathbf{R}^2$ and $a, b \in \mathbf{R}$, then $(ab)\mathbf{v} = a(b\mathbf{v})$.

Relation (2.3) is called the *commutative law*, (2.4) the *associative law*, and (2.8) and (2.9) two forms of the *distributive law.*

The proofs of these formulas are all easy; one just evaluates both sides. For instance, if $\mathbf{v} = (x, y)$, then the left-hand side of (2.10) is

$$ab(x, y) = (abx, aby)$$

while the right side is

$$a[b(x, y)] = a(bx, by) = (abx, aby)$$

Thus the two sides are equal. We leave the verification of the other formulas as exercises.

Finally, if $\mathbf{v}$ and $\mathbf{w}$ are vectors, we will write $\mathbf{v} - \mathbf{w}$ for $\mathbf{v} + (-\mathbf{w})$. If $\mathbf{v} = (a, b)$ and $\mathbf{w} = (c, d)$, then $-\mathbf{w} = (-c, -d)$, so that $\mathbf{v} - \mathbf{w} = (a - c, b - d)$.

Exercises

1. Perform the following computations:
   (a) $(1, 4) + (2, 8)$ (b) $(2, 3) - (-1, 7)$
   (c) $(4, -2) + (-4, 8)$ (d) $(3, -2) - (-2, 3)$

(e) $(5,-1)+(-5,1)$
(f) $6(1,-3)$
(g) $3(-2,7)$
(h) $-2(4,-5)$
(i) $0(3,3)$
(j) $3(1,2)+(2,7)$
(k) $-1(-2,3)+2(4,7)$
(l) $2(3,8)+(-2)(3,8)$
(m) $3(1,0)+2(0,1)$
(n) $4(1,3)-2(2,6)$
(o) $5(1,2)+3(-1,4)$

2. A *linear combination* of vectors $\mathbf{v}_1, \ldots, \mathbf{v}_k$ is any expression of the form $a_1\mathbf{v}_1 + \cdots + a_k\mathbf{v}_k$, where $a_1, \ldots, a_k$ are real numbers. Express
   (a) $(2, 3)$ as a linear combination of $(1, 1)$ and $(1, 2)$
   (b) $(2, 3)$ as a linear combination of $(1, 1)$ and $(2, 1)$
   (c) $(1, 4)$ as a linear combination of $(1, 1)$, $(2, 1)$, and $(3, 1)$
   (d) (x_1, x_2) as a linear combination of $(1, 0)$ and $(0, 1)$
3. Vectors $\mathbf{v}_1, \ldots, \mathbf{v}_k$ are said to be *linearly dependent* if one of them can be written as a linear combination of the rest, and *linearly independent* otherwise. Show that
   (a) $(1, 1)$, $(2, 1)$, and $(3, 1)$ are linearly dependent.
   (b) $(1, 1)$ and $(1, 2)$ are linearly independent.
4. Which of the following collections of vectors are linearly dependent?
   (a) $(1,4), (4,1)$
   (b) $(0,2), (1,3), (3,1)$
   (c) $(1,0), (0,1)$
   (d) $(1,0), (1,1)$
   (e) $(1,3), (-2,-6)$
   (f) $(1,-2), (2,-5)$
   (g) $(1,4), (2,3), (e, \pi)$
   (h) $(-3,7), (7,-3)$
   (i) $(2,3), (6,9)$
   (j) $(4,2), (2,1)$
5. Let $\mathbf{v} = (x, y)$, $\mathbf{w} = (t, u)$. Show that $\mathbf{v}$ and $\mathbf{w}$ are linearly dependent if and only if $xu - yt = 0$.
6. Show that every vector in $\mathbf{R}^2$ can be written as a linear combination of $(1, 0)$ and $(1, 1)$ in exactly one way.
7. Prove formulas (2. 3) through (2. 6).
8. Prove formulas (2. 7), (2. 9), and (2. 10).
9. Prove by induction that

$$a \sum_{i=1}^{k} \mathbf{v}_i = \sum_{i=1}^{k} a\mathbf{v}_i$$

   where $\mathbf{v}_1, \ldots, \mathbf{v}_k \in \mathbf{R}^2$ and $a \in \mathbf{R}$. (See Section 1.3.)
10. Prove by induction that

$$\left(\sum_{i=1}^{k} a_i\right)\mathbf{v} = \sum_{i=1}^{k} a_i\mathbf{v}$$

   where $a_1, \ldots, a_k \in \mathbf{R}$ and $\mathbf{v} \in \mathbf{R}^2$. (See Section 1.3.)
11. Let $\mathbf{v}$ and $\mathbf{w}$ be two linearly independent vectors in $\mathbf{R}^2$. Show that every vector in $\mathbf{R}^2$ can be written as a linear combination of $\mathbf{v}$ and $\mathbf{w}$ in exactly one way.
12. Suppose $\mathbf{v}$ and $\mathbf{w}$ are vectors in $\mathbf{R}^2$ such that every vector in $\mathbf{R}^2$ is a linear combination of $\mathbf{v}$ and $\mathbf{w}$. (The vectors $\mathbf{v}$ and $\mathbf{w}$ are said to *span* $\mathbf{R}^2$.) Prove that $\mathbf{v}$ and $\mathbf{w}$ are linearly independent.
13. Show that any three vectors in $\mathbf{R}^2$ are linearly dependent.

3. Euclidean Spaces

It may have occurred to the reader that most of the ideas of the previous section can be generalized. For example, a point in ordinary three-dimensional space can be specified by three numbers (representing the distances from three perpendicular planes) and we may therefore regard three-dimensional space as made up of triples (x, y, z) of real numbers. (See Figure 3.1.)

We shall generalize further to *Euclidean n space*, which we denote by $\mathbf{R}^n$. The space $\mathbf{R}^n$ consists of all ordered n-tuples of real numbers $(x_1, \ldots, x_n)$. We call the elements of $\mathbf{R}^n$ *vectors* and often denote them by a single boldface letter; $\mathbf{v} = (x_1, \ldots, x_n)$. The numbers $x_1, \ldots, x_n$ are called the *components* of $\mathbf{v}$. We shall see presently how to perform the same operations on these vectors that we described for vectors in $\mathbf{R}^2$. First, however, a brief digression may be in order: we should explain why working with $\mathbf{R}^n$ may be useful.

We shall deal in this course with functions of several variables, and we shall use $\mathbf{R}^n$ as a way of describing n variables simultaneously. It may help to consider a specific example. Suppose that a company wishes to determine the cost of producing automobiles. This cost depends on a large number of factors: the cost of sheet steel, glass, aluminum, fabric, bolts, and other materials and components; the salaries paid to various classes of workers; the cost of transporting parts to the factory; and taxes, among others. The cost, then, is a function of some large number of variables, and we can regard a point in $\mathbf{R}^n$ (where n is the number of variables) as specifying the value assigned to all the variables simultaneously.

Here is another example, from physics. Consider a space probe traveling

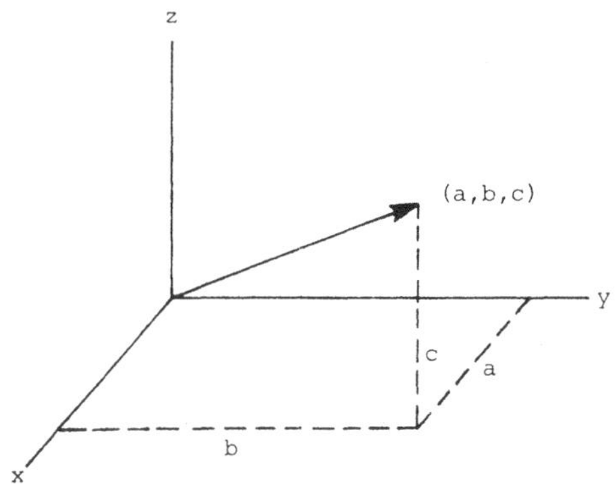

Figure 3.1 A vector in $\mathbf{R}^3$.

from Earth past Jupiter. It is important, when figuring out its path, to know the gravitational forces acting on it at any point in space. These forces depend not only on the position of the probe, but also on the positions of Earth, the sun, Mars, and Jupiter. Since it takes three numbers to specify the position of any of these bodies (we may, after all, regard the universe as three dimensional), we need 15 variables to give the positions of the probe and these four heavenly bodies. We can most easily represent these 15 quantities as vectors in $\mathbf{R}^{15}$. We need, therefore, to be able to deal with spaces of large dimension.

The problem with discussing $\mathbf{R}^n$ is largely psychological. Most of us have little trouble visualizing $\mathbf{R}^2$, and with a bit of work we can understand $\mathbf{R}^3$. Virtually no one, however, has an intuitive picture of, for instance, $\mathbf{R}^{47}$. As a result, we tend to fear working with vectors in $\mathbf{R}^n$ when n is greater than 3. Mathematicians compensate for their lack of insight by being very precise: they lay down exact definitions, state theorems carefully, and prove them rigorously. While we shall certainly be careful in this text, it is not our purpose to give this sort of logical treatment. Instead, we shall often prove results for $\mathbf{R}^2$ or $\mathbf{R}^3$ (where it is possible to "see" why the result is true), give a proof which also works for $\mathbf{R}^n$, and then explain the result for $\mathbf{R}^n$ (where it is true for the "same" reason). In other words, we shall attempt to develop our intuitions for $\mathbf{R}^n$ by stressing its affinities with $\mathbf{R}^2$ and $\mathbf{R}^3$.

We now return to vectors in $\mathbf{R}^n$. Much as in $\mathbf{R}^2$, we can define the sum of two vectors: if $\mathbf{v} = (x_1, \ldots, x_n)$ and $\mathbf{w} = (y_1, \ldots, y_n)$, then

$$\mathbf{v} + \mathbf{w} = (x_1 + y_1, \ldots, x_n + y_n) \tag{3.1}$$

Observe, however, that $\mathbf{v}$ and $\mathbf{w}$ must both be in $\mathbf{R}^n$; we cannot, for instance, add a vector in $\mathbf{R}^4$ to one in $\mathbf{R}^6$.

We can also multiply a vector by a real number: if r is real,

$$r\mathbf{v} = (rx_1, \ldots, rx_n) \tag{3.2}$$

These operations, addition and scalar multiplication, obey the laws (2.3) through (2.10). (One needs to define $\mathbf{0}$ and $-\mathbf{v}$ for $\mathbf{R}^n$; see below.) Once again, the proofs are easy and we leave them as exercises. We mention them because these laws form the basis for the study of *vector spaces*, a topic customarily dealt with in linear algebra courses. We shall not study vector spaces in any detail here.

We conclude this section with some observations about notation:

1. It will often be convenient to denote the vector $(x_1, \ldots, x_n)$ simply by (x_i). With this notation, addition and scalar multiplication are given by

$$(x_i) + (y_i) = (x_i + y_i)$$
$$r(x_i) = (rx_i)$$

and
$$-(x_i) = (-x_i)$$

We also define the difference of two vectors by

$$(x_i) - (y_i) = (x_i - y_i)$$

2. The vector (0, 0, . . . , 0) in $\mathbf{R}^n$ is customarily written $\mathbf{0}$ and called the *zero vector*. The symbol $\mathbf{0}$ can thus represent, for instance, (0, 0), (0, 0, 0), or (0, 0, 0, 0) depending on whether we are concerned with $\mathbf{R}^2$, $\mathbf{R}^3$, or $\mathbf{R}^4$. Ordinarily, this use of the same letter for different vectors causes no confusion. If we need to distinguish the different zeroes, we shall add a subscript giving the dimension; thus $\mathbf{0}_3$ is (0, 0, 0).
3. The space $\mathbf{R}^1$, according to our definition, consists of 1-tuples (x_1). There is (for our purposes, at least) no reason to distinguish (x_1) from the real number x_1; for this reason, we shall generally write $\mathbf{R}$ instead of $\mathbf{R}^1$ and x_1 (or x) for (x_1). Occasionally we shall use the parentheses for elements of $\mathbf{R}$; the reason will usually be to make an analogy between $\mathbf{R}$ and $\mathbf{R}^n$ clearer.
4. In the last section, we wrote a typical vector in $\mathbf{R}^2$ as (x, y); similarly, a typical point in $\mathbf{R}^3$ is often written as (x, y, z). The practice of using a different letter for each coordinate becomes wasteful as n increases—in fact, it is impossible for $\mathbf{R}^{27}$ unless we extend the alphabet—and it is inconvenient when we work with a general space $\mathbf{R}^n$. For this reason, we wrote a typical vector in $\mathbf{R}^n$ as $(x_1, \ldots, x_n)$. For consistency, we shall generally use subscripts in all cases; we usually write a typical vector in $\mathbf{R}^2$ as (x_1, x_2).

Exercises

1. Perform the following operations:
   (a) (3, 9, 4) + (2, 8, 5)
   (b) (1, 2, 3) − (4, 5, 6)
   (c) (8, 2, −5) + (−3, 1, 6)
   (d) (4, 3, 2) − (9, 7, 1) + (3, 0, 4)
   (e) (1, 2, 3) − 2(4, 5, 6) + (7, 8, 9)
   (f) 3(1, 0, 0) − 2(0, 1, 0) + 6(0, 0, 1)
   (g) 4(1, −3, 2) + 2(4, 0, −4) − 3(4, −4, 0)
   (h) (1, 0, 1) − 2(1, 1, 0) + 3(0, 1, 1)
2. Which of the following collections of vectors are linearly dependent? (See Exercise 3, Section 2.)
   (a) (1, 0, 3), (2, 1, 0)
   (b) (2, −1, 5), (−8, 4, −20)
   (c) (1, 2, 3), (0, 0, 2), (0, −1, 5)

(d) (1, 2, 4), (3, −1, 2), (8, 4, −6), (4, 3, −12)
(e) (1, −3, 2), (4, 0, −4), (4, −4, 0)
(f) (1, 0, 0), (0, 1, 0), (0, 0, 1)
(g) (1, 2, 3), (4, 5, 6), (7, 8, 9)
(h) (1, 0, 0), (0, 1, 0), (0, 0, 1), (4, −5, −3)

3. Show that each of the following collections of vectors span $\mathbf{R}^3$. (See Exercise 12, Section 2.)
(a) (1, 0, 0), (0, 1, 0), (0, 0, 1)
(b) (1, 2, 3), (0, 0, 2), (0, −1, 5)
(c) (6, 3, 2), (3, −2, −6), (2, −6, 3)
(d) (1, 0, 1), (1, 1, 1), (1, 1, 2), (2, 1, 2)
4. Show that (1, 0, 1, 1), (0, 1, 2, 3), (0, 0, 1, 4), and (0, 0, 0, 1) span $\mathbf{R}^4$, but that (1, 3, 5, 7), (0, 1, 2, 4), and (0, 0, 1, −2) do not. (See Exercise 3, Section 2.)
5. (a) Show that every vector in $\mathbf{R}^3$ can be written as a linear combination of (1, 0, 0), (0, 1, 0), and (0, 0, 1) in exactly one way.
(b) Do the same for (6, 3, 2), (3, −2, −6), (−2, 6, −3).
6. Verify that (2.3) through (2.6) hold in $\mathbf{R}^n$.
7. Verify that (2.7) through (2.10) hold in $\mathbf{R}^n$.
8. Prove by induction that

$$a \sum_{i=1}^{k} \mathbf{v}_i = \sum_{i=1}^{k} a\mathbf{v}_i$$

where $\mathbf{v}_1, \ldots, \mathbf{v}_k \in \mathbf{R}^n$ and $a \in \mathbf{R}$. (See Section 1.3.)
9. Prove by induction that

$$\left(\sum_{i=1}^{k} a_i\right)\mathbf{v} = \sum_{i=1}^{k} a_i\mathbf{v}$$

where $a_1, \ldots, a_k \in \mathbf{R}$ and $\mathbf{v} \in \mathbf{R}^n$. (See Section 1.3.)
*10. Show that no two vectors can span $\mathbf{R}^3$. [Hint: Suppose that $\mathbf{v}_1$ and $\mathbf{v}_2$ span $\mathbf{R}^3$. Then $\mathbf{e}_1 = (1, 0, 0)$ is a linear combination of $\mathbf{v}_1$ and $\mathbf{v}_2$. Show that one of the vectors $\mathbf{v}_1, \mathbf{v}_2$ ($\mathbf{v}_1$, say) is a linear combination of $\mathbf{e}_1$ and the other, and that $\mathbf{e}_1$ and $\mathbf{v}_2$ span $\mathbf{R}^3$. Repeat this process with $\mathbf{e}_2 = (0, 1, 0)$ to show that $\mathbf{e}_1$ and $\mathbf{e}_2$ span $\mathbf{R}^3$. Show that this is false.]
*11. Show that no $n - 1$ vectors can span $\mathbf{R}^n$. (*Hint*: Use the method suggested in Exercise 10.)
*12. Show that any four vectors in $\mathbf{R}^3$ are linearly dependent. [*Hint*: Let $\mathbf{e}_1 = (1, 0, 0)$, $\mathbf{e}_2 = (0, 1, 0)$, $\mathbf{e}_3 = (0, 0, 1)$, and let $\mathbf{v}_1, \mathbf{v}_2, \mathbf{v}_3, \mathbf{v}_4$ be the four vectors. Suppose that $\mathbf{v}_1, \ldots, \mathbf{v}_4$ are linearly independent. Write $\mathbf{v}_1$ as a linear combination of $\mathbf{e}_1, \mathbf{e}_2, \mathbf{e}_3$; show that one of the $\mathbf{e}$'s ($\mathbf{e}_1$, say) can be written as a linear combination of $\mathbf{v}_1$, $\mathbf{e}_2$, and $\mathbf{e}_3$, so that $\mathbf{v}_1, \mathbf{e}_2$, and $\mathbf{e}_3$ span $\mathbf{R}^3$. Repeat this process with $\mathbf{v}_2$ and $\mathbf{v}_3$ to show that $\mathbf{v}_1, \mathbf{v}_2$, and $\mathbf{v}_3$ span $\mathbf{R}^3$. Conclude that the $\mathbf{v}$'s are linearly dependent after all.]
*13. Show that any $n + 1$ vectors in $\mathbf{R}^n$ are linearly dependent. (*Hint*: This can be done by an extension of the method in Exercise 12.)

4. The Norm in $\mathbf{R}^n$

As we saw earlier, one of the notions needed to define the derivative of a function $f: \mathbf{R} \to \mathbf{R}$ is the absolute value. We now define an analogous notion for Euclidean spaces, the *norm* of a vector in $\mathbf{R}^n$.

Let $\mathbf{v} = (x_1, \ldots, x_n)$ be a vector in $\mathbf{R}^n$. The *norm* of $\mathbf{v}$, $\|\mathbf{v}\|$, is defined by

$$\|\mathbf{v}\| = (x_1^2 + \cdots + x_n^2)^{1/2}$$

Examples

1. If $n = 1$, $\mathbf{v}$ is a real number x_1 and

$$\|\mathbf{v}\| = (x_1^2)^{1/2} = |x_1|$$

Thus the norm and absolute value coincide in this case.

2. If $n = 2$ and $\mathbf{v} = (x_1, x_2)$, then

$$\|\mathbf{v}\| = (x_1^2 + x_2^2)^{1/2}$$

It follows from the Pythagorean theorem that $\|\mathbf{v}\|$ is the length of the vector $\mathbf{v}$ (Figure 4.1).

We now give the important properties of the norm.

Proposition 4.1 (a) *For any vector* $\mathbf{v} \in \mathbf{R}^n$, $\|\mathbf{v}\| \geq 0$. *Furthermore,* $\|\mathbf{v}\| = 0$ *if and only if* $\mathbf{v} = \mathbf{0}$.

(b) *For any vector* $\mathbf{v} \in \mathbf{R}^n$ *and* $r \in \mathbf{R}$, $\|r\mathbf{v}\| = |r| \cdot \|\mathbf{v}\|$.

(c) *For any vectors* $\mathbf{v}, \mathbf{w} \in \mathbf{R}^n$, $\|\mathbf{v} + \mathbf{w}\| \leq \|\mathbf{v}\| + \|\mathbf{w}\|$.

We note that property (c) of Proposition 4.1 is called the *triangle*

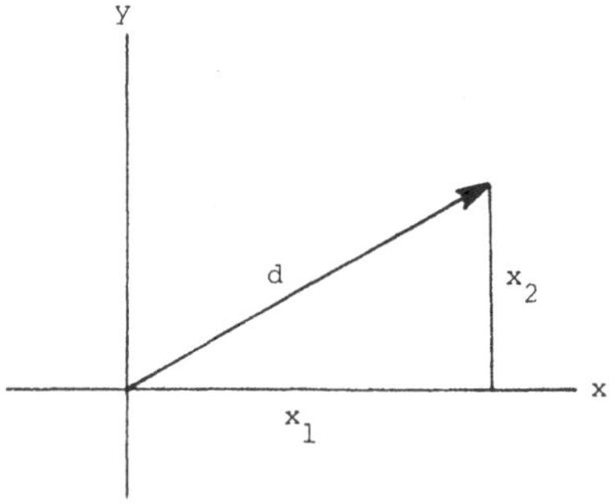

Figure 4.1 $d^2 = x_1^2 + x_2^2$.

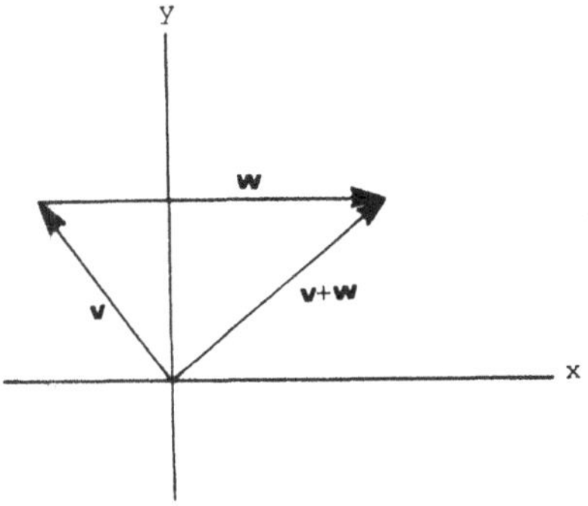

Figure 4.2 A triangle of vectors.

inequality. If $n = 2$, it expresses the fact that the sum of the lengths of any two sides of a triangle is at least as large as the length of the third side. (See Figure 4.2.)

Proof of Proposition 4.1. The proof of (a) is trivial. The norm of $\mathbf{v}$ is nonnegative by definition and $\|\mathbf{0}\|$ is certainly 0. On the other hand, if

$$\|\mathbf{v}\| = (x_1^2 + \cdots + x_n^2)^{1/2} = 0$$

then $x_1^2 + \cdots + x_n^2 = 0$ which clearly implies that $x_1 = x_2 = \cdots = x_n = 0$. (The sum of nonnegative numbers can only be zero if each of them is zero.) Thus $\mathbf{v} = \mathbf{0}$.

To prove (b), we simply compute:

$$\begin{aligned}\|r\mathbf{v}\| &= [(rx_1)^2 + \cdots + (rx_n)^2]^{1/2} \\ &= (r^2x_1^2 + \cdots + r^2x_n^2)^{1/2} \\ &= (r^2)^{1/2}(x_1^2 + \cdots + x_n^2)^{1/2} \\ &= |r| \cdot \|\mathbf{v}\|\end{aligned}$$

The proof of (c) is more difficult; we give it only for $n = 2$ here. The general case follows from the Cauchy inequality, Theorem 2.1 of Chapter 5. We begin with the following.

Lemma 4.2 *For any vectors* $\mathbf{v} = (x_1, x_2)$, $\mathbf{w} = (y_1, y_2)$ *in* $\mathbf{R}^2$, *we have*

$$|x_1y_1 + x_2y_2| \leq \|\mathbf{v}\| \cdot \|\mathbf{w}\|$$

Proof. This result is a consequence of the law of cosines. Consider the triangle in the plane with sides $\mathbf{v}$, $\mathbf{w}$, and $\mathbf{v} - \mathbf{w}$. (See Figure 4.3.) The law of cosines states that

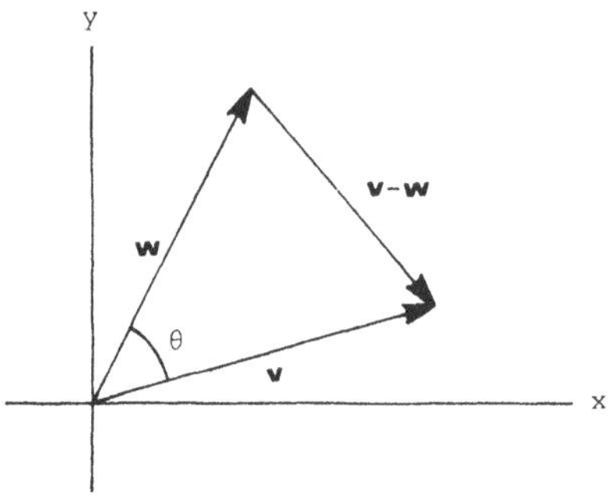

Figure 4.2 The difference between two vectors.

$$\|\mathbf{v} - \mathbf{w}\|^2 = \|\mathbf{v}\|^2 + \|\mathbf{w}\|^2 - 2\|\mathbf{v}\| \cdot \|\mathbf{w}\| \cos\theta$$

where θ is the angle between $\mathbf{v}$ and $\mathbf{w}$. Writing this out in terms of components, we have

$$(4.1)\quad (x_1 - y_1)^2 + (x_2 - y_2)^2 = x_1^2 + x_2^2 + y_1^2 + y_2^2 - 2\|\mathbf{v}\| \cdot \|\mathbf{w}\| \cos\theta$$

Now $\quad (x_1 - y_1)^2 + (x_2 - y_2)^2 = x_1^2 - 2x_1y_1 + y_1^2 + x_2^2 - 2x_2y_2 + y_2^2$

so equation (4.1) becomes

$$-2(x_1y_1 + x_2y_2) = -2\|\mathbf{v}\| \cdot \|\mathbf{w}\| \cos\theta$$

or, taking absolute values,

$$|x_1y_1 + x_2y_2| = \|\mathbf{v}\| \cdot \|\mathbf{w}\|\, |\cos\theta|$$

Since $|\cos\theta| \leq 1$, it follows that

$$\|\mathbf{v}\| \cdot \|\mathbf{w}\|\, |\cos\theta| \leq \|\mathbf{v}\| \cdot \|\mathbf{w}\|$$

and the lemma is proved.

We can now prove statement (c) of Proposition 4.1 when $n = 2$. We begin by computing $\|\mathbf{v} + \mathbf{w}\|^2$:

$$\begin{aligned}\|\mathbf{v} + \mathbf{w}\|^2 &= (x_1 + y_1)^2 + (x_2 + y_2)^2\\ &= x_1^2 + 2x_1y_1 + y_1^2 + x_2^2 + 2x_2y_2 + y_2^2\\ &= \|\mathbf{v}\|^2 + \|\mathbf{w}\|^2 + 2(x_1y_1 + x_2y_2)\\ &\leq \|\mathbf{v}\|^2 + \|\mathbf{w}\|^2 + 2\|\mathbf{v}\| \cdot \|\mathbf{w}\|\end{aligned}$$

by Lemma 4.2. However,

$$\|\mathbf{v}\|^2 + \|\mathbf{w}\|^2 + 2\|\mathbf{v}\| \cdot \|\mathbf{w}\| = (\|\mathbf{v}\| + \|\mathbf{w}\|)^2$$

so we have

$$(4.2) \qquad \|\mathbf{v} + \mathbf{w}\|^2 \leq (\|\mathbf{v}\| + \|\mathbf{w}\|)^2$$

Statement (c) now follows by taking square roots of both sides of the inequality (4.2).

Exercises

1. Compute the norm of the following vectors.
   (a) $(2, -3)$ (b) $(4, 1)$
   (c) $(1, 0, 2)$ (d) $(2, -1, 7)$
   (e) $(0, 1, 1, 7)$ (f) $(4, -1, 1, 2)$
   (g) $(1, 2, 1, 3, 0)$ (h) $(-7, 2, 0, 4, 2)$.
2. Let $\mathbf{v}$ and $\mathbf{w}$ be vectors in $\mathbf{R}^n$ and define the *distance* from $\mathbf{v}$ to $\mathbf{w}$, $d(\mathbf{v}, \mathbf{w})$, by
   $$d(\mathbf{v}, \mathbf{w}) = \|\mathbf{v} - \mathbf{w}\|$$
   Show that this gives the usual distance when $n = 1, 2$.
3. Find the distance between the following vectors.
   (a) $(2, 1), (1, 2)$ (b) $(2, -1), (-1, 3)$
   (c) $(2, 0, 1), (0, -2, 2)$ (d) $(1, 2, 3), (3, 2, 1)$
   (e) $(1, 0, 1, 2), (4, -1, 2, 0)$ (f) $(2, 1, 3, 1), (0, 4, 2, -4)$
4. Use Proposition 4.1 to prove the following properties of the distance function.
   (a) For all $\mathbf{v}, \mathbf{w} \in \mathbf{R}^n$, $d(\mathbf{v}, \mathbf{w}) \geq 0$. Furthermore, $d(\mathbf{v}, \mathbf{w}) = 0$ if and only if $\mathbf{v} = \mathbf{w}$.
   (b) For all $\mathbf{v}, \mathbf{w} \in \mathbf{R}^n$, $d(\mathbf{v}, \mathbf{w}) = d(\mathbf{w}, \mathbf{v})$.
   (c) For all $\mathbf{u}, \mathbf{v}, \mathbf{w} \in \mathbf{R}^n$, $d(\mathbf{u}, \mathbf{v}) \leq d(\mathbf{u}, \mathbf{w}) + d(\mathbf{w}, \mathbf{v})$.
5. Use the triangle inequality to prove that
   $$\big|\, \|\mathbf{v}\| - \|\mathbf{w}\| \,\big| \leq \|\mathbf{v} - \mathbf{w}\|$$
   for all vectors $\mathbf{v}, \mathbf{w} \in \mathbf{R}^n$. *Hint*: This inequality is equivalent to the two inequalities
   $$\|\mathbf{v}\| - \|\mathbf{w}\| \leq \|\mathbf{v} - \mathbf{w}\| \qquad \|\mathbf{w}\| - \|\mathbf{v}\| \leq \|\mathbf{v} - \mathbf{w}\|$$
   (See Proposition 1–4.1.)
6. Prove by induction that
   $$\left\| \sum_{i=1}^{k} \mathbf{v}_i \right\| \leq \sum_{i=1}^{k} \|\mathbf{v}_i\|$$
   (See Section 1.3.)

5. Linear Transformations between Euclidean Spaces

As we saw in the first section of this chapter, a function $f: \mathbf{R} \to \mathbf{R}$ differentiable at a point x_0 can be approximated near x_0 by a function whose graph is a straight line. These functions whose graphs are straight lines have analogues in higher dimensions which we now introduce. More precisely, we describe the analogues of the functions whose graphs are straight lines through the origin, the linear transformations. We will have something to say about the others in Section 5.4.

A function $T: \mathbf{R}^n \to \mathbf{R}^p$ is called a *linear transformation* (or *linear map*) if the following conditions hold:

(5.1) For all vectors $\mathbf{v}, \mathbf{w} \in \mathbf{R}^n$, $T(\mathbf{v} + \mathbf{w}) = T(\mathbf{v}) + T(\mathbf{w})$.

(5.2) For all vectors $\mathbf{v} \in \mathbf{R}^n$ and all real numbers r, $T(r\mathbf{v}) = rT(\mathbf{v})$.

We can paraphrase this definition by saying that the function $T: \mathbf{R}^n \to \mathbf{R}^p$ is linear if it "preserves" addition and scalar multiplication; that is, T takes the sum of two vectors into the sum of their images and the scalar multiple of a vector into the same scalar multiple of the image of the vector.

Here are some examples.

1. Let T: $\mathbf{R} \to \mathbf{R}$ be defined by $T(x) = cx$, where c is a fixed real number. Then

$$T(x_1 + x_2) = c(x_1 + x_2) = cx_1 + cx_2 = T(x_1) + T(x_2)$$

and $$T(rx) = c(rx) = r(cx) = rT(x)$$

so T is a linear transformation. Note that the graph of T is a line through the origin with slope c.

2. Define $S: \mathbf{R}^2 \to \mathbf{R}$ by $S(x_1, x_2) = 2x_1 - 3x_2$. Then

$$\begin{aligned} S((x_1, x_2) + (y_1, y_2)) &= S(x_1 + y_1, x_2 + y_2) \\ &= 2(x_1 + y_1) - 3(x_2 + y_2) \\ &= (2x_1 - 3x_2) + (2y_1 - 3y_2) \\ &= S(x_1, x_2) + S(y_1, y_2) \end{aligned}$$

and $$\begin{aligned} S(r(x_1, x_2)) &= S(rx_1, rx_2) \\ &= 2rx_1 - 3rx_2 \\ &= r(2x_1 - 3x_2) \\ &= rS(x_1, x_2) \end{aligned}$$

Thus S is a linear transformation.

3. Let $L: \mathbf{R}^2 \to \mathbf{R}^2$ be given by

$$L(x_1, x_2) = (x_1 - x_2, 7x_1 + 3x_2)$$

Then $$\begin{aligned} L((x_1, x_2) + (y_1, y_2)) &= L(x_1 + y_1, x_2 + y_2) \\ &= ((x_1 + y_1) - (x_2 + y_2), 7(x_1 + y_1) \\ &\quad + 3(x_2 + y_2)) \end{aligned}$$

$$\begin{aligned}&= ((x_1 - x_2) + (y_1 - y_2), (7x_1 + 3x_2)\\&\quad + (7y_1 + 3y_2))\\&= (x_1 - x_2, 7x_1 + 3x_2)\\&\quad + (y_1 - y_2, 7y_1 + 3y_2)\\&= L(x_1, x_2) + L(y_1, y_2)\end{aligned}$$

so that L satisfies (5.1). We leave the verification of (5.2) to the reader.

4. Define $T: \mathbf{R}^n \to \mathbf{R}^p$ by $T(\mathbf{v}) = \mathbf{0}$ for all $\mathbf{v} \in \mathbf{R}^n$. Then

$$T(\mathbf{v} + \mathbf{w}) = \mathbf{0} = \mathbf{0} + \mathbf{0} = T(\mathbf{v}) + T(\mathbf{w})$$

and
$$T(r\mathbf{v}) = \mathbf{0} = r\mathbf{0} = rT(\mathbf{v})$$

so T is linear. T is called the *zero transformation* for obvious reasons.

5. The *identity transformation* $I: \mathbf{R}^n \to \mathbf{R}^n$ (sometimes denoted I_n) is defined by $I(\mathbf{v}) = \mathbf{v}$ for all $\mathbf{v} \in \mathbf{R}^n$. The verification that I is linear is left as an exercise.

We note that any linear transformation $T: \mathbf{R}^n \to \mathbf{R}^p$ takes the zero vector in $\mathbf{R}^n$ into the zero vector in $\mathbf{R}^p$:

$$\begin{aligned}T(\mathbf{0}) &= T(0 \cdot \mathbf{0})\\&= 0T(\mathbf{0}) \qquad \text{[by (5.2)]}\\&= \mathbf{0}\end{aligned}$$

since multiplying any vector in $\mathbf{R}^p$ by the number 0 gives the zero vector.

We next give a useful characterization of linear transformations which combines (5.1) and (5.2) into a single equation.

Proposition 5.1 *A function $T: \mathbf{R}^n \to \mathbf{R}^p$ is a linear transformation if and only if*

(5.3) $$T(a\mathbf{v} + b\mathbf{w}) = aT(\mathbf{v}) + bT(\mathbf{w})$$

for all vectors $\mathbf{v}$, $\mathbf{w} \in \mathbf{R}^n$ and real numbers a, b.

Proof. Suppose $T: \mathbf{R}^n \to \mathbf{R}^p$ satisfies (5.1) and (5.2). Then

$$\begin{aligned}T(a\mathbf{v} + b\mathbf{w}) &= T(a\mathbf{v}) + T(b\mathbf{w}) \qquad \text{by (5.1)}\\&= aT(\mathbf{v}) + bT(\mathbf{w}) \qquad \text{by (5.2)}\end{aligned}$$

Thus T satisfies (5.3).

Conversely, if $T: \mathbf{R}^n \to \mathbf{R}^p$ satisfies (5.3) for all $\mathbf{v}, \mathbf{w} \in \mathbf{R}^n$ and $a, b \in \mathbf{R}$, we can set $a = b = 1$ to get

$$\begin{aligned} T(\mathbf{v} + \mathbf{w}) &= T(1\mathbf{v} + 1\mathbf{w}) \\ &= 1T(\mathbf{v}) + 1T(\mathbf{w}) \qquad \text{by (5.3)} \\ &= T(\mathbf{v}) + T(\mathbf{w}) \end{aligned}$$

so that (5.1) holds for all $\mathbf{v}, \mathbf{w} \in \mathbf{R}^n$.

Similarly, if we set $b = 0$ in (5.3), we obtain (5.1):

$$\begin{aligned} T(a\mathbf{v}) &= T(a\mathbf{v} + 0\mathbf{w}) \\ &= aT(\mathbf{v}) + 0T(\mathbf{w}) \qquad \text{by (5.3)} \\ &= aT(\mathbf{v}) + \mathbf{0} \\ &= aT(\mathbf{v}) \end{aligned}$$

Therefore T satisfies (5.1) and (5.2) whenever T satisfies (5.3).

The next result follows by induction from Proposition 5.1 (see Section 1.3); we leave the proof as an exercise.

Proposition 5.2 *Let $T : \mathbf{R}^n \to \mathbf{R}^p$ be a linear transformation. Then*

$$T\left(\sum_{i=1}^{k} a_i \mathbf{v}_i\right) = \sum_{i=1}^{k} a_i T(\mathbf{v}_i)$$

for all $\mathbf{v}_1, \ldots, \mathbf{v}_k \in \mathbf{R}^n$ and $a_1, \ldots, a_k \in \mathbf{R}$.

Exercises

1. Which of the following functions from $\mathbf{R}^2$ to $\mathbf{R}$ are linear transformations?
   (a) $A(x_1, x_2) = 2x_1 + 3x_2$
   (b) $B(x_1, x_2) = x_1^2 - x_2^2$
   (c) $C(x_1, x_2) = 7x_1 - 5x_2$
   (d) $D(x_1, x_2) = x_1 x_2$
   (e) $E(x_1, x_2) = 2x_1 - 3x_2 + 4$
   (f) $F(x_1, x_2) = \sqrt{x_1^2 + x_2^2}$
   (g) $G(x_1, x_2) = x_1$
   (h) $H(x_1, x_2) = x_1 + |x_2|$
2. Which of the following functions from $\mathbf{R}^2$ to $\mathbf{R}^2$ are linear transformations?
   (a) $A(x_1, x_2) = (x_1 - 3x_2, x_2 - 3x_1)$
   (b) $B(x_1, x_2) = (2x_2, 3x_1)$
   (c) $C(x_1, x_2) = (\sqrt{x_1^2 + x_2^2}, e^{x_1 + x_2})$
   (d) $D(x_1, x_2) = (x_1 + x_2, x_1 x_2)$
   (e) $E(x_1, x_2) = (\sin x_1, \cos x_2)$
   (f) $F(x_1, x_2) = (x_1 + 1, x_2 + 2)$
   (g) $G(x_1, x_2) = (4x_1 - 5x_2, 0)$
   (h) $H(x_1, x_2) = (3x_1 + 4x_2, 3x_1 + 4x_2)$

3. Which of the following are linear transformations?
   (a) $A : \mathbf{R}^3 \to \mathbf{R}^2$ by $A(x_1, x_2, x_3) = (x_1x_2, x_3)$
   (b) $B : \mathbf{R}^1 \to \mathbf{R}^4$ by $B(x) = (x, 2x, 3x, 4x)$
   (c) $C : \mathbf{R}^2 \to \mathbf{R}^3$ by $C(x_1, x_2) = (x_1, x_2, x_1 - x_2)$
   (d) $D : \mathbf{R}^3 \to \mathbf{R}$ by $D(x_1, x_2, x_3) = x_1 + x_2$
   (e) $E : \mathbf{R} \to \mathbf{R}^2$ by $E(x) = (x, 4)$
   (f) $F : \mathbf{R}^3 \to \mathbf{R}^3$ by $F(x_1, x_2, x_3) = (x_1 + 2x_3, x_2 - x_3, x_3 + x_1)$
   (g) $G : \mathbf{R}^2 \to \mathbf{R}^4$ by $G(x_1, x_2) = (x_1^2, x_1 + x_2, x_2^2, x_2 + x_1)$
   (h) $H : \mathbf{R}^3 \to \mathbf{R}$ by $H(x_1, x_2, x_3) = x_1x_2x_3 + x_2$
4. Prove that the identity function $I: \mathbf{R}^n \to \mathbf{R}^n$ defined by $I(\mathbf{v}) = \mathbf{v}$ is a linear transformation.
5. Show that any linear transformation $T : \mathbf{R} \to \mathbf{R}^n$ has the form $T(x) = x\mathbf{v}$ for some $\mathbf{v} \in \mathbf{R}^n$.
6. Prove Proposition 5.2.
7. Let $T : \mathbf{R}^n \to \mathbf{R}^p$ be a linear transformation. A vector $\mathbf{v}$ is said to be in the *kernel* of T if $T(\mathbf{v}) = \mathbf{0}$. Prove:
   (a) If $\mathbf{v}$ and $\mathbf{w}$ are in the kernel of T, so is $\mathbf{v} + \mathbf{w}$.
   (b) If $\mathbf{v}$ is in the kernel of T, so is $r\mathbf{v}$ (where r is any real number).
8. A linear transformation is said to be *injective* if $T(\mathbf{v}) = T(\mathbf{w})$ implies $\mathbf{v} = \mathbf{w}$ for all $\mathbf{v}$, $\mathbf{w} \in \mathbf{R}^n$. Prove that T is injective if and only if the only vector in the kernel of T is the zero vector.

6. Matrices and Linear Transformations

We continue here with our discussion of linear transformations. We will show that any linear transformation $T : \mathbf{R}^n \to \mathbf{R}^p$ determines and is determined by a "matrix" of np numbers. This provides an important tool in the study of linear transformations.

Let $\mathbf{e}_1, \mathbf{e}_2, \ldots, \mathbf{e}_n$ be the vectors in $\mathbf{R}^n$ defined by

$$\begin{aligned} \mathbf{e}_1 &= (1, 0, \ldots. 0) \\ \mathbf{e}_2 &= (0, 1, 0, \ldots, 0) \\ &\cdots\cdots\cdots\cdots \\ \mathbf{e}_n &= (0, \ldots, 0, 1) \end{aligned}$$

Thus $\mathbf{e}_j$ has a single nonzero component, the jth component, which is equal to 1. For example, the vector $\mathbf{e}_4$ in $\mathbf{R}^7$ is $(0, 0, 0, 1, 0, 0, 0)$. The vectors $\mathbf{e}_1, \ldots, \mathbf{e}_n$ are called the *standard basis* for $\mathbf{R}^n$.

Note that our notation is somewhat ambiguous; we denote by $\mathbf{e}_j$ different vectors in different Euclidean spaces. For instance, $\mathbf{e}_2$ represents the vector $(0, 1, 0)$ in $\mathbf{R}^3$ and the vector $(0, 1, 0, 0, 0)$ in $\mathbf{R}^5$. This ambiguity should, however, cause no difficulty; the context will make clear in which Euclidean space a particular $\mathbf{e}_j$ lies.

The important property of the standard basis for $\mathbf{R}^n$ is expressed in the following observation: if $\mathbf{v} = (x_1, \ldots, x_n)$ is any vector in $\mathbf{R}^n$, then

$$\mathbf{v} = x_1\mathbf{e}_1 + x_2\mathbf{e}_2 + \cdots + x_n\mathbf{e}_n = \sum_{i=1}^{n} x_i\mathbf{e}_i$$

That is, $\mathbf{v}$ can be written as a linear combination of $\mathbf{e}_1, \ldots, \mathbf{e}_n$ in exactly one way.

Our first result shows how the standard basis for $\mathbf{R}^n$ can be used to study linear transformations.

Proposition 6.1 *Let $T : \mathbf{R}^n \to \mathbf{R}^p$ be a linear transformation and let $\mathbf{e}_1, \ldots, \mathbf{e}_n$ be the standard basis for $\mathbf{R}^n$. Then T is completely determined by the vectors $\mathbf{w}_1 = T(\mathbf{e}_1), \ldots, \mathbf{w}_n = T(\mathbf{e}_n)$. In fact,*

$$(6.1) \qquad T(x_1, \ldots, x_n) = x_1\mathbf{w}_1 + x_2\mathbf{w}_2 + \cdots + x_n\mathbf{w}_n = \sum_{i=1}^{n} x_i\mathbf{w}_i$$

for any vector $(x_1, \ldots, x_n) \in \mathbf{R}^n$.

Conversely, if $\mathbf{w}_1, \ldots, \mathbf{w}_n$ are arbitrary vectors in $\mathbf{R}^p$, then equation (6.1) defines the unique linear transformation with $T(\mathbf{e}_1) = \mathbf{w}_1, \ldots, T(\mathbf{e}_n) = \mathbf{w}_n$.

This proposition tells us how linear transformations $T : \mathbf{R}^n \to \mathbf{R}^p$ are constructed; simply select n vectors $\mathbf{w}_1, \ldots, \mathbf{w}_n$ in $\mathbf{R}^p$ and use (6.1) to define T.

Proof. We prove the result for the case $n = 2$; the proof for $n > 2$ is entirely similar and left as an exercise.

Suppose first of all that $T : \mathbf{R}^2 \to \mathbf{R}^p$ and $T(\mathbf{e}_1) = \mathbf{w}_1$, $T(\mathbf{e}_2) = \mathbf{w}_2$. Then, for any vector $(x_1, x_2) \in \mathbf{R}^2$, we have

$$\begin{aligned} T(x_1, x_2) &= T(x_1\mathbf{e}_1 + x_2\mathbf{e}_2) \\ &= x_1 T(\mathbf{e}_1) + x_2 T(\mathbf{e}_2) \\ &= x_1\mathbf{w}_1 + x_2\mathbf{w}_2 \end{aligned}$$

(since T is a linear transformation) by the definition of $\mathbf{w}_1$ and $\mathbf{w}_2$. Thus, the value of T on any vector is determined by the components of the vector and $\mathbf{w}_1$ and $\mathbf{w}_2$.

Now, let $\mathbf{u}_1$ and $\mathbf{u}_2$ be vectors in $\mathbf{R}^p$. If $S : \mathbf{R}^2 \to \mathbf{R}^p$ is a linear transformation with $S(\mathbf{e}_1) = \mathbf{u}_1$ and $S(\mathbf{e}_2) = \mathbf{u}_2$, then S is uniquely determined by the first part of this proposition. We need only check that the formula (6.1) defines a linear transformation [since clearly $S(\mathbf{e}_1) = \mathbf{u}_1$ and $S(\mathbf{e}_2) = \mathbf{u}_2$ for this S]:

$$\begin{aligned} S((x_1,x_2) + (y_1,y_2)) &= S(x_1 + y_1, x_2 + y_2) \\ &= (x_1 + y_1)\mathbf{u}_1 + (x_2 + y_2)\mathbf{u}_2 \\ &= (x_1\mathbf{u}_1 + x_2\mathbf{u}_2) + (y_1\mathbf{u}_1 + y_2\mathbf{u}_2) \\ &= S(x_1, x_2) + S(y_1, y_2) \\ S(r(x_1, x_2)) &= S(rx_1, rx_2) \\ &= rx_1\mathbf{u}_1 + rx_2\mathbf{u}_2 \\ &= r(x_1\mathbf{u}_1 + x_2\mathbf{u}_2) \\ &= rS(x_1, x_2) \end{aligned}$$

Thus S is linear and Proposition 6.1 is proved (when $n = 2$).

Remark. The key property of the vectors $\mathbf{e}_1, \ldots, \mathbf{e}_n$ which we use in this proof is that every vector is a linear combination of the **e**'s in exactly one way. As a result, the definition

$$S(x_1, x_2) = x_1\mathbf{u}_1 + x_2\mathbf{u}_2$$

is complete (S is defined for every vector) and unambiguous.

Let us take a closer look at Proposition 6.1 in the case $n = p = 2$. Suppose $T : \mathbf{R}^2 \to \mathbf{R}^2$ is a linear transformation. Then we know that T is determined by the two vectors $\mathbf{w}_1 = T(\mathbf{e}_1), \mathbf{w}_2 = T(\mathbf{e}_2) \in \mathbf{R}^2$. If $\mathbf{w}_1 = (a, c)$ and $\mathbf{w}_2 = (b, d)$, it follows that T is determined by the four numbers a, b, c, d. It is common practice to write these four numbers in a 2 by 2 (often written 2×2) *matrix* (two rows and two columns):

(6.2) $$A = \begin{bmatrix} a & b \\ c & d \end{bmatrix}$$

This matrix is called the *matrix of the linear transformation T*. Note that the components of $T(\mathbf{e}_1)$ make up the first column of A and the components of $T(\mathbf{e}_2)$ the second column.

Conversely, suppose we have any 2×2 matrix (6.2). Then we can set $\mathbf{w}_1 = (a, c), \mathbf{w}_2 = (b, d)$ and use (6.1) to define T. This linear transformation is called the *linear transformation defined by the matrix A*. It is immediate that the matrix of this linear transformation is A.

We see therefore that there is a correspondence between linear transformations $T : \mathbf{R}^2 \to \mathbf{R}^2$ and 2×2 matrices.

Here are some examples.

1. Let $T : \mathbf{R}^2 \to \mathbf{R}^2$ be the linear transformation defined in Example 3 of Section 5: $T(x_1, x_2) = (x_1 - x_2, 7x_1 + 3x_2)$. Then

$$T(\mathbf{e}_1) = T(1, 0) = (1, 7)$$
$$T(\mathbf{e}_2) = T(0, 1) = (-1, 3)$$

so that

$$\begin{bmatrix} 1 & -1 \\ 7 & 3 \end{bmatrix}$$

is the matrix of T.

2. Suppose A is the matrix

$$A = \begin{bmatrix} 2 & -1 \\ 1 & 4 \end{bmatrix}$$

Then, using (6.1), A defines a linear transformation $T: \mathbf{R}^2 \to \mathbf{R}^2$ with $T(\mathbf{e}_1) = (2, 1)$ and $T(\mathbf{e}_2) = (-1, 4)$. Explicitly,

$$T(x_1, x_2) = x_1(2, 1) + x_2(-1, 4) = (2x_1 - x_2, x_1 + 4x_2)$$

More generally, a $p \times n$ *matrix* is a rectangular array

$$(6.3) \qquad A = \begin{bmatrix} a_{11} & a_{12} & \cdots & a_{1n} \\ a_{21} & a_{22} & \cdots & a_{2n} \\ \vdots & & & \vdots \\ a_{p1} & a_{p2} & \cdots & a_{pn} \end{bmatrix}$$

A $p \times n$ matrix has p rows and n columns. The element a_{ij} is the *component* of A in the ith row and jth column.

For instance,

$$\begin{bmatrix} 1 \\ 6 \end{bmatrix}$$

is a 2×1 matrix,

$$\begin{bmatrix} 1 & 2 & 3 \\ 4 & 5 & 6 \end{bmatrix}$$

is a 2×3 matrix, and

$$\begin{bmatrix} 1 & 0 & 6 & 4 & 2 \\ 4 & -3 & 0 & 2 & 2 \\ 3 & 5 & -1 & 0 & 1 \end{bmatrix}$$

is a 3×5 matrix.

In view of our earlier discussion of linear transformations from $\mathbf{R}^2$ to $\mathbf{R}^2$, one might guess that there is a correspondence between linear transformation $T: \mathbf{R}^n \to \mathbf{R}^p$ and $p \times n$ matrices. This is indeed the case. In fact, if $T(\mathbf{e}_1) = \mathbf{w}_1, \ldots, T(\mathbf{e}_n) = \mathbf{w}_n$, we write

$$(6.4)\qquad \begin{aligned} \mathbf{w}_1 &= (a_{11}, a_{21}, \ldots, a_{p1}) \\ \mathbf{w}_2 &= (a_{12}, a_{22}, \ldots, a_{p2}) \\ &\cdots\cdots\cdots\cdots\cdots \\ \mathbf{w}_n &= (a_{1n}. a_{2n}, \ldots, a_{pn}) \end{aligned}$$

We now use these vectors to form the columns of a $p \times n$ matrix,

$$A = \begin{bmatrix} a_{11} & \cdots & a_{1n} \\ \vdots & & \vdots \\ a_{p1} & \cdots & a_{pn} \end{bmatrix}$$

the matrix of the linear transformation $T : \mathbf{R}^n \to \mathbf{R}^p$.

Conversely, given a matrix A as in (6.3), we can define vectors $\mathbf{w}_1, \ldots, \mathbf{w}_n$ as in (6.4) and define a linear transformation using (6.1). Thus we have established a correspondence between linear transformations $T : \mathbf{R}^n \to \mathbf{R}^p$ and $p \times n$ matrices.

Here are some more examples.

3. Let $T : \mathbf{R}^3 \to \mathbf{R}^2$ be defined by

$$T(x_1, x_2, x_3) = (x_1 - x_2 + x_3, x_2 - 2x_1)$$

Then T is a linear transformation (verify this), and

$$\begin{aligned} T(\mathbf{e}_1) &= T(1, 0, 0) = (1, -2) \\ T(\mathbf{e}_2) &= T(0, 1, 0) = (-1, 1) \\ T(\mathbf{e}_3) &= T(0, 0, 1) = (1, 0) \end{aligned}$$

Thus, the matrix of T is

$$\begin{bmatrix} 1 & -1 & 1 \\ -2 & 1 & 0 \end{bmatrix}$$

4. Suppose A is the 3×2 matrix

$$\begin{bmatrix} 1 & 2 \\ -1 & 3 \\ 0 & 5 \end{bmatrix}$$

Then the columns of A define elements

$$\mathbf{w}_1 = (1, -1, 0) \qquad \mathbf{w}_2 = (2, 3, 5)$$

in $\mathbf{R}^3$. These two elements define a linear transformation $T : \mathbf{R}^2 \to \mathbf{R}^3$ by the formula (6.1):

$$\begin{aligned} T(x_1, x_2) &= x_1\mathbf{w}_1 + x_2\mathbf{w}_2 \\ &= (x_1, -x_1, 0) + (2x_2, 3x_2, 5x_2) \\ &= (x_1 + 2x_2, -x_1 + 3x_2, 5x_2) \end{aligned}$$

5. Let $T : \mathbf{R}^3 \to \mathbf{R}^3$ be the function defined by

$$T(x_1, x_2, x_3) = (x_2 - x_1, 2x_3 - x_1, -x_2)$$

Then T is linear (verify this), and

$$T(\mathbf{e}_1) = (-1, -1, 0) \qquad T(\mathbf{e}_2) = (1, 0, -1) \qquad T(\mathbf{e}_3) = (0, 2, 0)$$

Thus the matrix of T is the 3×3 matrix

$$\begin{bmatrix} -1 & 1 & 0 \\ -1 & 0 & 2 \\ 0 & -1 & 0 \end{bmatrix}$$

6. Let A be the 4×2 matrix

$$\begin{bmatrix} 2 & -1 \\ -2 & 4 \\ 0 & 3 \\ 1 & 0 \end{bmatrix}$$

Then A defines $T : \mathbf{R}^2 \to \mathbf{R}^4$ by

$$\begin{aligned} T(x_1, x_2) &= x_1(2, -2, 0, 1) + x_2(-1, 4, 3, 0) \\ &= (2x_1 - x_2, -2x_1 + 4x_2, 3x_2, x_1) \end{aligned}$$

7. Let $T : \mathbf{R}^n \to \mathbf{R}^p$ be the zero transformation, $T(\mathbf{v}) = \mathbf{0}$ for all $\mathbf{v} \in \mathbf{R}^n$. Then $T(\mathbf{e}_i) = \mathbf{0}$, $1 \leq i \leq n$, and it follows that the matrix of T is the $p \times n$ matrix all of whose entries are 0. Not surprisingly, this matrix is called the $p \times n$ *zero matrix*.
8. Let $I: \mathbf{R}^n \to \mathbf{R}^n$ be the identity linear transformation, $I(\mathbf{v}) = \mathbf{v}$ for all $\mathbf{v} \in \mathbf{R}^n$. Then $I(\mathbf{e}_j) = \mathbf{e}_j$, $1 \leq j \leq n$, so that the matrix of I is given by

$$\begin{bmatrix} 1 & 0 & 0 & \cdots & 0 & 0 \\ 0 & 1 & 0 & \cdots & & 0 \\ \vdots & & & & & \vdots \\ 0 & 0 & 0 & \cdots & 0 & 1 \end{bmatrix}$$

the matrix with ones on the diagonal from upper left to lower right and zeroes elsewhere. This matrix is called the *identity matrix* and denoted by either I or I_n.

Exercises

1. Find the matrices for the following linear transformations from $\mathbf{R}^2$ to $\mathbf{R}$:
   (a) $S(x_1, x_2) = x_1 - 7x_2$
   (b) $T(x_1, x_2) = 2x_1 + x_2$
   (c) $Q(x_1, x_2) = 4x_1 - 6x_2$
   (d) $D(x_1, x_2) = -3(2x_1 - 9x_2)$
   (e) $E(x_1, x_2) = x_2 - 4x_1$
2. Write the matrices for the following linear transformations from $\mathbf{R}^2$ to $\mathbf{R}^2$:
   (a) $A(x_1, x_2) = (4x_1 - x_2, 2x_1 + 3x_2)$
   (b) $B(x_1, x_2) = (x_1 + 2x_2, 3x_2)$
   (c) $C(x_1, x_2) = (0, x_1)$
   (d) $D(x_1, x_2) = (x_1, x_2)$
   (e) $E(x_1, x_2) = (x_2, x_1)$
   (f) $F(x_1, x_2) = (3x_1 - x_2, x_2 - x_1)$
   (g) $G(x_1, x_2) = (x_1 + 2x_2, x_1 + 2x_2)$
   (h) $H(x_1, x_2) = (x_1 - x_2, x_1 + x_2)$
3. Write the matrices for each of these linear transformations.
   (a) $S : \mathbf{R}^3 \to \mathbf{R}^2$ by $S(x_1, x_2, x_3) = (x_1 + 2x_2, 2x_1 - x_2 + x_3)$
   (b) $T : \mathbf{R}^2 \to \mathbf{R}^3$ by $T(x_1, x_2) = (x_2 - x_1, 2x_2 + 3x_1, 0)$
   (c) $Q : \mathbf{R} \to \mathbf{R}^4$ by $Q(x) = (x, 4x, -x, -4x)$
   (d) $D : \mathbf{R}^3 \to \mathbf{R}^3$ by $D(x_1, x_2, x_3) = (x_1 - x_2 + 4x_3, 2x_1 - 3x_2 - x_3, -x_1 + 5x_2 + 6x_3)$
   (e) $E : \mathbf{R}^3 \to \mathbf{R}^2$ by $E(x_1, x_2, x_3) = (x_1 - 4x_3, 6x_3 - x_1)$
4. Determine the linear transformations (as was done in Examples 2, 4, and 6) defined by the following matrices.

(a) $\begin{bmatrix} 2 \\ 3 \end{bmatrix}$ (b) $\begin{bmatrix} 1 & 7 & 4 \end{bmatrix}$

(c) $\begin{bmatrix} 2 & 0 \\ -1 & -2 \end{bmatrix}$ (d) $\begin{bmatrix} 1 & -3 & 3 \\ 7 & 0 & 4 \end{bmatrix}$

(e) $\begin{bmatrix} 2 & 0 \\ 6 & 4 \\ 4 & 6 \end{bmatrix}$ (f) $\begin{bmatrix} 1 & 3 & 5 & 7 \\ 2 & 4 & 6 & 8 \end{bmatrix}$

(g) $\begin{bmatrix} 0 & 1 & 1 \\ 2 & 1 & 7 \\ -4 & 3 & 3 \end{bmatrix}$ (h) $\begin{bmatrix} 0 & 1 & 0 \\ 0 & 0 & 1 \\ 1 & 0 & 0 \end{bmatrix}$

(i) $\begin{bmatrix} 1 & 0 & 0 & 1 \\ 1 & 1 & 0 & 0 \\ 0 & 1 & 1 & 0 \\ 0 & 0 & 1 & 1 \end{bmatrix}$

5. Let $T : \mathbf{R}^2 \to \mathbf{R}^2$ be the linear transformation defined by the matrix (c) of Exercise 4. Find

(a) $T(1, 3)$
(b) $T(0, -6)$
(c) $T(4, -2)$

6. Let $T : \mathbf{R}^4 \to \mathbf{R}^2$ be the linear transformation defined by the matrix (f) of Exercise 4. Find
(a) $T(0, 4, -3, 2)$
(b) $T(1, 0, 1, 1)$
(c) $T(7, -2, 2, 6)$
7. Let $T : \mathbf{R}^4 \to \mathbf{R}^4$ be the linear transformation defined by the matrix (i) of Exercise 4. Find
(a) $T(1, 1, 3, 0)$
(b) $T(2, 0, -1, 7)$
(c) $T(5, 1, -1, 2)$

*8. Prove Proposition 6.1 for arbitrary n and p.

7. Some Operations on Linear Transformations and Matrices

In this section, we define addition and scalar multiplication of linear transformations and matrices. We also relate these operations by showing that the matrix of the sum of two linear transformations is the sum of the matrices of the individual linear transformations and that the matrix of the scalar multiple of a linear transformation is the same scalar multiple of the matrix of the linear transformation.

We begin by defining the sum of two linear transformations. Let $T, S : \mathbf{R}^n \to \mathbf{R}^p$ be linear transformations. The *sum* $T + S : \mathbf{R}^n \to \mathbf{R}^p$ is the function defined by

$$(T + S)(\mathbf{v}) = T(\mathbf{v}) + S(\mathbf{v})$$

For example, if $T : \mathbf{R} \to \mathbf{R}$ is multiplication by a, $T(x) = ax$, and $S : \mathbf{R} \to \mathbf{R}$ is multiplication by b, $S(x) = bx$, then $T + S$ is multiplication by $a + b$:

$$(T + S)(x) = T(x) + S(x) = ax + bx = (a + b)x$$

Thus we see that the sum of two linear transformations from $\mathbf{R}$ to $\mathbf{R}$ is again a linear transformation. The next result generalizes this observation.

Proposition 7.1 *The sum of two linear transformations is a linear transformation.*

Remark. The sum of two linear transformation is defined only when the transformations are from the same space $\mathbf{R}^n$ to the same space $\mathbf{R}^p$.

Proof. According to Proposition 5.1, we need only verify that

$$(T+S)(a\mathbf{v}+b\mathbf{w}) = a(T+S)(\mathbf{v}) + b(T+S)(\mathbf{w})$$

for all $\mathbf{v}$, $\mathbf{w} \in \mathbf{R}^n$ and a, $b \in \mathbf{R}$. This is a straightforward computation:

$$\begin{aligned}(T+S)(a\mathbf{v}+b\mathbf{w}) &= T(a\mathbf{v}+b\mathbf{w}) + S(a\mathbf{v}+b\mathbf{w}) \\ &\quad \text{(by the definition of } T+S\text{)} \\ &= aT(\mathbf{v}) + bT(\mathbf{w}) + aS(\mathbf{v}) + bS(\mathbf{w}) \\ &\quad \text{(by Proposition 5.1 and the linearity of } T \text{ and } S\text{)} \\ &= a[T(\mathbf{v}) + S(\mathbf{v})] + b[T(\mathbf{w}) + S(\mathbf{w})] \\ &= a(T+S)(\mathbf{v}) + b(T+S)(\mathbf{w})\end{aligned}$$

again by the definition of $T+S$.

Suppose now that $T: \mathbf{R}^n \to \mathbf{R}^p$ is a linear transformation and r a real number. We define a function $rT: \mathbf{R}^n \to \mathbf{R}^p$ by

$$(rT)(\mathbf{v}) = rT(\mathbf{v})$$

For example, if $T: \mathbf{R} \to \mathbf{R}$ is multiplication by a, then

$$(rT)(x) = rT(x) = rax$$

so that rT is multiplication by ra. Therefore rT is linear in this case. More generally, we have the following.

Proposition 7.2 *Let $T: \mathbf{R}^n \to \mathbf{R}^p$ be a linear transformation and r a real number. Then $rT: \mathbf{R}^n \to \mathbf{R}^p$ is a linear transformation.*

Proof. We verify (5.3) for rT:

$$\begin{aligned}(rT)(a\mathbf{v}+b\mathbf{w}) &= rT(a\mathbf{v}+b\mathbf{w}) \qquad \text{(by the definition of } rT\text{)} \\ &= r[aT(\mathbf{v}) + bT(\mathbf{w})] \qquad \text{(since } T \text{ is linear)} \\ &= arT(\mathbf{v}) + brT(\mathbf{w}) \\ &= a(rT)(\mathbf{v}) + b(rT)(\mathbf{w})\end{aligned}$$

It follows that rT is a linear transformation.

Now that we know that the sum and scalar multiple of linear transformations are again linear transformations, it is reasonable to attempt to describe the matrices of these transformations in terms of the matrices of the original transformations. We begin with the sum in the case $n = p = 2$.

Suppose that $T, S: \mathbf{R}^2 \to \mathbf{R}^2$ are linear transformations and that A is the matrix of T and B the matrix of S:

$$A = \begin{bmatrix} a_{11} & a_{12} \\ a_{21} & a_{22} \end{bmatrix} \qquad B = \begin{bmatrix} b_{11} & b_{12} \\ b_{21} & b_{22} \end{bmatrix}$$

Thus

$$T(\mathbf{e}_1) = (a_{11}, a_{21})$$
$$T(\mathbf{e}_2) = (a_{12}, a_{22})$$
$$S(\mathbf{e}_1) = (b_{11}, b_{21})$$

and

$$S(\mathbf{e}_2) = (b_{12}, b_{22})$$

It follows that

$$\begin{aligned}(T+S)(\mathbf{e}_1) &= T(\mathbf{e}_1) + S(\mathbf{e}_1)\\ &= (a_{11}, a_{21}) + (b_{11}, b_{21})\\ &= (a_{11} + b_{11}, a_{21} + b_{21})\end{aligned}$$

and

$$\begin{aligned}(T+S)(\mathbf{e}_2) &= T(\mathbf{e}_2) + S(\mathbf{e}_2)\\ &= (a_{12}, a_{22}) + (b_{12}, b_{22})\\ &= (a_{12} + b_{12}, a_{22} + b_{22})\end{aligned}$$

Therefore the matrix of $T + S$ is the 2×2 matrix

(7.1)
$$\begin{bmatrix} a_{11} + b_{11} & a_{12} + b_{12} \\ a_{21} + b_{21} & a_{22} + b_{22} \end{bmatrix}$$

the matrix obtained by adding corresponding components of A and B. This motivates the following definition.

Let A and B be $p \times n$ matrices:

$$A = \begin{bmatrix} a_{11} & \cdots & a_{1n} \\ \vdots & & \vdots \\ a_{p1} & \cdots & a_{pn} \end{bmatrix} \qquad B = \begin{bmatrix} b_{11} & \cdots & b_{1n} \\ \vdots & & \vdots \\ b_{p1} & \cdots & b_{pn} \end{bmatrix}$$

Then $A + B$ is defined to be the $p \times n$ matrix obtained by adding the corresponding components of A and B:

$$A + B = \begin{bmatrix} a_{11} + b_{11} & \cdots & a_{1n} + b_{1n} \\ \vdots & & \vdots \\ a_{p1} + b_{p1} & \cdots & a_{pn} + b_{pn} \end{bmatrix}$$

For example, if A and B are the 2×3 matrices

$$A = \begin{bmatrix} 1 & 0 & 3 \\ 7 & -1 & 4 \end{bmatrix} \qquad B = \begin{bmatrix} 2 & 2 & -3 \\ 0 & 4 & 6 \end{bmatrix}$$

then

$$A + B = \begin{bmatrix} 1+2 & 0+2 & 3-3 \\ 7+0 & -1+4 & 4+6 \end{bmatrix} = \begin{bmatrix} 3 & 2 & 0 \\ 7 & 3 & 10 \end{bmatrix}$$

We see now that the matrix (7.1) is the sum $A + B$. It follows that, if

$T, S : \mathbf{R}^2 \to \mathbf{R}^2$ have matrices A and B, respectively, then $T + S$ has matrix $A + B$. In fact, this holds true in general.

Proposition 7.3 *Let $T, S : \mathbf{R}^n \to \mathbf{R}^p$ be linear transformations, and suppose T has matrix A and S has matrix B. Then the linear transformation $T + S : \mathbf{R}^n \to \mathbf{R}^p$ has matrix $A + B$.*

The proof of this result is essentially the same as the proof given above when $n = p = 2$. We leave it as Exercise 7.

We can make the same analysis for the product of a linear transformation by a real number. Suppose $T : \mathbf{R}^2 \to \mathbf{R}^2$ has matrix A:

$$A = \begin{bmatrix} a_{11} & a_{12} \\ a_{21} & a_{22} \end{bmatrix}$$

Then $$T(\mathbf{e}_1) = (a_{11}, a_{21}) \qquad T(\mathbf{e}_2) = (a_{12}, a_{22})$$

It follows that

$$(rT)(\mathbf{e}_1) = rT(\mathbf{e}_1) = r(a_{11}, a_{21}) = (ra_{11}, ra_{21})$$
$$(rT)(\mathbf{e}_2) = rT(\mathbf{e}_2) = r(a_{12}, a_{22}) = (ra_{12}, ra_{22})$$

Thus the matrix of rT is the 2×2 matrix

$$\begin{bmatrix} ra_{11} & ra_{12} \\ ra_{21} & ra_{22} \end{bmatrix} \tag{7.2}$$

the matrix obtained from A by multiplying each of its components by r. This suggests the following. Let A be a $p \times n$ matrix,

$$A = \begin{bmatrix} a_{11} & \cdots & a_{1n} \\ \vdots & & \vdots \\ a_{p1} & \cdots & a_{pn} \end{bmatrix}$$

and r a real number. Then rA is defined to be the matrix obtained from A by multiplying each of its components by r:

$$rA = \begin{bmatrix} ra_{11} & \cdots & ra_{1n} \\ \vdots & & \vdots \\ ra_{p1} & \cdots & ra_{pn} \end{bmatrix}$$

For example, if A is the 2×4 matrix

$$A = \begin{bmatrix} 2 & 0 & 1 & 3 \\ -1 & 7 & 4 & 2 \end{bmatrix}$$

then $3A$ is the 2×4 matrix

$$3A = \begin{bmatrix} 6 & 0 & 3 & 9 \\ -3 & 21 & 12 & 6 \end{bmatrix}$$

We see now that the matrix (7.2) is just rA so we have proved that the matrix of rT is rA when $n = p = 2$. This holds more generally.

Proposition 7.4 *Let $T : \mathbf{R}^n \to \mathbf{R}^p$ be a linear transformation and r a real number. If A is the matrix of T, then rA is the matrix of rT.*

The proof of this result is essentially the same as the proof given above when $n = p = 2$. We leave it as Exercise 8.

Remarks

1. We note that one can only add two matrices of the same size. It is not possible, for example, to add a 2×3 matrix to a 3×5 matrix.
2. It will sometimes be convenient to denote a $p \times n$ matrix simply as (a_{ij}). With this notation, the sum and scalar product operations are defined by

$$(a_{ij}) + (b_{ij}) = (a_{ij} + b_{ij})$$
$$r(a_{ij}) = (ra_{ij})$$

Exercises

1. Let $Q, S, T : \mathbf{R}^2 \to \mathbf{R}^2$ be the linear transformations defined by

$$Q(x_1, x_2) = (x_1 - x_2, x_1)$$
$$S(x_1, x_2) = (3x_2, 2x_2 - x_1)$$
$$T(x_1, x_2) = (x_1 + x_2, 3x_1 + 5x_2)$$

Determine the following linear transformations:

(a) $Q + S$ (b) $3T$
(c) $2S + T$ (d) $S - 2Q$
(e) $5T + 4Q$ (f) $T + Q - S$
(g) $2S - 4Q + 3T$ (h) $T - 6S + 2Q$

2. Let $S, T : \mathbf{R}^4 \to \mathbf{R}^4$ be defined by

$$S(x_1, x_2, x_3, x_4) = (x_2, x_1, x_4, x_3)$$
$$T(x_1, x_2, x_3, x_4) = (2x_1 - x_3, x_4 + x_2 - x_3, x_1, 2x_2 + x_3)$$

Determine the following linear transformations:

(a) $S + T$ (b) $7S$
(c) $2S + 3T$ (d) $5S - 4T$

3. Let A, B, and C be the 2×2 matrices

$$A = \begin{bmatrix} 1 & 0 \\ 1 & 1 \end{bmatrix} \quad B = \begin{bmatrix} 2 & 1 \\ 1 & 3 \end{bmatrix} \quad C = \begin{bmatrix} 1 & -3 \\ -1 & 2 \end{bmatrix}$$

   Determine the following matrices:
   (a) $A + C$ (b) $7B$
   (c) $3A + 4B$ (d) $2C - 5B$
   (e) $A + B - C$ (f) $2A - 3C + 5B$
4. Let A and B be the 4×3 matrices

$$A = \begin{bmatrix} 1 & 0 & 2 \\ 2 & 1 & -1 \\ 1 & 3 & 1 \\ 0 & 6 & 2 \end{bmatrix} \quad B = \begin{bmatrix} 0 & 1 & 1 \\ 2 & -2 & 2 \\ 6 & 0 & 1 \\ 3 & -1 & -4 \end{bmatrix}$$

   Determine the following matrices:
   (a) $A + B$ (b) $-4B$
   (c) $2A + 3B$ (d) $7A - 2B$
5. Let $S, T, Q : \mathbf{R}^n \to \mathbf{R}^p$ be linear transformations and b, c any real numbers. Prove the following:
   (a) $b(S + T) = bS + bT$
   (b) $(b + c)S = bS + cS$
   (c) $(S + T) + Q = S + (T + Q)$
6. Let A, B, C be $p \times n$ matrices and b, c any real numbers. Prove the following:
   (a) $b(A + B) = bA + bB$
   (b) $(b + c)A = bA + cA$
   (c) $(A + B) + C = A + (B + C)$
7. Prove Proposition 7.3 for arbitrary n and p.
8. Prove Proposition 7.4 for arbitrary n and p.

8. Composites of Linear Transformations and Matrix Multiplication

Suppose that $T : \mathbf{R}^n \to \mathbf{R}^p$ and $S : \mathbf{R}^p \to \mathbf{R}^q$ are linear transformations. We can then define the *composite* of S with T, $S \circ T : \mathbf{R}^n \to \mathbf{R}^q$, to be the function

$$(S \circ T)(\mathbf{v}) = S(T(\mathbf{v}))$$

That is, we first apply T to the vector $\mathbf{v}$ and then apply S to the vector $T(\mathbf{v})$. For example, if T: $\mathbf{R}^2 \to \mathbf{R}^2$ is defined by $T(x_1, x_2) = (x_1 - x_2, x_1 + 3x_2)$ and $S : \mathbf{R}^2 \to \mathbf{R}^1$ is defined by $S(y_1, y_2) = y_1 + 2y_2$,* then

*To avoid confusion, we write (x_1, x_2) for an arbitrary point in the $\mathbf{R}^2$ on which T is defined and (y_1, y_2) for an arbitrary point in the $\mathbf{R}^2$ on which S is defined (and into which T maps).

$$\begin{aligned}(S \circ T)(x_1, x_2) &= S(T(x_1, x_2)) \\ &= S(x_1 - x_2, x_1 + 3x_2) \\ &= x_1 - x_2 + 2(x_1 + 3x_2) \\ &= 3x_1 + 5x_2\end{aligned}$$

It is natural to ask if $S \circ T$ is a linear transformation. (It is in the example above.) Our first result gives an affirmative answer to this question.

Proposition 8.1 Let $T: \mathbf{R}^n \to \mathbf{R}^p$ *and* $S: \mathbf{R}^p \to \mathbf{R}^q$ *be linear transformations. Then* $S \circ T: \mathbf{R}^n \to \mathbf{R}^q$ *is a linear transformation.*

Proof. According to Proposition 5.1, we need only verify that

$$(S \circ T)(a\mathbf{v} + b\mathbf{w}) = a(S \circ T)(\mathbf{v}) + b(S \circ T)(\mathbf{w})$$

for all $\mathbf{v}, \mathbf{w} \in \mathbf{R}^n$ and all $a, b \in \mathbf{R}$. We simply compute:

$$\begin{aligned}(S \circ T)(a\mathbf{v} + b\mathbf{w}) &= S(T(a\mathbf{v} + b\mathbf{w})) && \text{(by definition of } S \circ T) \\ &= S(aT(\mathbf{v}) + bT(\mathbf{w})) && \text{(since } T \text{ is linear)} \\ &= aS(T(\mathbf{v})) + bS(T(\mathbf{w})) && \text{(since } S \text{ is linear)} \\ &= a(S \circ T)(\mathbf{v}) + b(S \circ T)(\mathbf{w})\end{aligned}$$

again by the definition of $S \circ T$. It follows that $S \circ T$ is linear.

Now that we know that $S \circ T$ is a linear transformation, it is reasonable to attempt to express the matrix of $S \circ T$ in terms of the matrices of S and T. This is more complicated than the corresponding question for sums and scalar products. We first carry it out when $n = p = q = 2$.

Suppose, in the notation above, that A is the matrix of S and B the matrix of T:

$$A = \begin{bmatrix} a_{11} & a_{12} \\ a_{21} & a_{22} \end{bmatrix} \qquad B = \begin{bmatrix} b_{11} & b_{12} \\ b_{21} & b_{22} \end{bmatrix}$$

Thus

$$\begin{aligned}S(\mathbf{e}_1) &= (a_{11}, a_{21}) \\ S(\mathbf{e}_2) &= (a_{12}, a_{22}) \\ T(\mathbf{e}_1) &= (b_{11}, b_{21}) = b_{11}\mathbf{e}_1 + b_{21}\mathbf{e}_2\end{aligned}$$

and

$$T(\mathbf{e}_2) = (b_{12}, b_{22}) = b_{12}\mathbf{e}_1 + b_{22}\mathbf{e}_2$$

It follows that

$$\begin{aligned}(S \circ T)(\mathbf{e}_1) &= S(T(\mathbf{e}_1)) \\ &= S(b_{11}\mathbf{e}_1 + b_{21}\mathbf{e}_2)\end{aligned}$$

$$
\begin{aligned}
&= b_{11}S(\mathbf{e}_1) + b_{21}S(\mathbf{e}_2)\\
&= b_{11}(a_{11}, a_{21}) + b_{21}(a_{12}, a_{22})\\
&= (b_{11}a_{11} + b_{21}a_{12}, b_{11}a_{21} + b_{21}a_{22})\\
&= (a_{11}b_{11} + a_{12}b_{21}, a_{21}b_{11} + a_{22}b_{21})
\end{aligned}
$$

and, similarly,

$$(S \circ T)(\mathbf{e}_2) = (a_{11}b_{12} + a_{12}b_{22}, a_{21}b_{12} + a_{22}b_{22})$$

Therefore, the matrix of $S \circ T$ is the 2×2 matrix

$$\begin{bmatrix} a_{11}b_{11} + a_{12}b_{21} & a_{11}b_{12} + a_{12}b_{22} \\ a_{21}b_{11} + a_{22}b_{21} & a_{21}b_{12} + a_{22}b_{22} \end{bmatrix} \tag{8.1}$$

We use the expression (8.1) to define the *product* AB of the 2×2 matrices A and B. Thus, *the matrix of the composite of two linear transformations is the product of the matrices of the linear transformations* (*when* $n = p = q = 2$). In fact, we defined the product of two 2×2 matrices so that this would be true.

To illustrate the notion of matrix multiplication, we compute the product AB, where

$$A = \begin{bmatrix} 2 & 1 \\ 3 & 0 \end{bmatrix} \qquad B = \begin{bmatrix} 2 & -2 \\ 4 & 7 \end{bmatrix}$$

According to (8.1),

$$
\begin{aligned}
\begin{bmatrix} 2 & 1 \\ 3 & 0 \end{bmatrix}\begin{bmatrix} 2 & -2 \\ 4 & 7 \end{bmatrix} &= \begin{bmatrix} 2\cdot 2 + 1\cdot 4 & 2\cdot(-2) + 1\cdot 7 \\ 3\cdot 2 + 0\cdot 4 & 3\cdot(-2) + 0\cdot 7 \end{bmatrix}\\
&= \begin{bmatrix} 8 & 3 \\ 6 & -6 \end{bmatrix}
\end{aligned}
$$

Before giving more examples, we analyze the product of 2×2 matrices in more detail. To begin with, we note the element in the first row and first column of AB, is obtained by multiplying each element in the first row of A,

$$[a_{11} \quad a_{12}]$$

by the corresponding element of the first column of B,

$$\begin{bmatrix} b_{11} \\ b_{21} \end{bmatrix}$$

and then adding:

$$a_{11}b_{11} + a_{12}b_{21}$$

The element in the first row and second column of AB is obtained by multiplying each element of the first row of A with the corresponding element of the *second* column of B and then adding:

$$a_{11}b_{12} + a_{12}b_{22}$$

Similarly, the elements in the second row of AB are obtained using the second row of A and the first and second columns of B.

Examples

1. $\begin{bmatrix} 1 & 0 \\ 1 & 1 \end{bmatrix}\begin{bmatrix} 2 & 1 \\ 3 & 0 \end{bmatrix} = \begin{bmatrix} 1\cdot 2 + 0\cdot 3 & 1\cdot 1 + 0\cdot 0 \\ 1\cdot 2 + 1\cdot 3 & 1\cdot 1 + 1\cdot 0 \end{bmatrix} = \begin{bmatrix} 2 & 1 \\ 5 & 1 \end{bmatrix}$
2. $\begin{bmatrix} 0 & 1 \\ 1 & 0 \end{bmatrix}\begin{bmatrix} 2 & 3 \\ 4 & 5 \end{bmatrix} = \begin{bmatrix} 0\cdot 2 + 1\cdot 4 & 0\cdot 3 + 1\cdot 5 \\ 1\cdot 2 + 0\cdot 4 & 1\cdot 3 + 0\cdot 5 \end{bmatrix} = \begin{bmatrix} 4 & 5 \\ 2 & 3 \end{bmatrix}$
3. $\begin{bmatrix} 2 & 0 \\ -2 & 3 \end{bmatrix}\begin{bmatrix} 1 & 4 \\ 2 & 2 \end{bmatrix} = \begin{bmatrix} 2\cdot 1 + 0\cdot 2 & 2\cdot 4 + 0\cdot 2 \\ -2\cdot 1 + 3\cdot 2 & -2\cdot 4 + 3\cdot 2 \end{bmatrix} = \begin{bmatrix} 2 & 8 \\ 4 & -2 \end{bmatrix}$
4. $\begin{bmatrix} -1 & 3 \\ 2 & 7 \end{bmatrix}\begin{bmatrix} 4 & 2 \\ 1 & 6 \end{bmatrix} = \begin{bmatrix} -1 & 16 \\ 15 & 46 \end{bmatrix}$

We shall extend this product to larger matrices in the next section.

Exercises

1. Determine the composite $S \circ T$ of the following linear transformations.
   (a) $S(x_1, x_2) = (x_1 - x_2, x_2)$
   $T(y_1, y_2) = (2y_2, y_1)$
   (b) $S(x_1, x_2) = (2x_1 + x_2, x_1 - x_2, x_1 + 2x_2)$
   $T(y_1, y_2, y_3) = (y_1 - y_2, y_2 - y_3)$
   (c) $S(x_1, x_2, x_3) = (3x_3 - x_1, x_1 + x_3, 2x_2)$
   $T(y_1, y_2, y_3) = (y_2 - y_3, y_1 - y_2, y_1 + y_2 + y_3)$
   (d) $S(x_1, x_2, x_3) = (x_2 + x_1, x_3 - 2x_1 + x_2, 3x_3 - 2x_1)$
   $T(y_1, y_2, y_3) = (y_1 - y_3, y_2 - y_1, y_3 - y_2)$
2. Determine the following products.

   (a) $\begin{bmatrix} 1 & 0 \\ 1 & 1 \end{bmatrix}\begin{bmatrix} 2 & 3 \\ 4 & 5 \end{bmatrix}$ (b) $\begin{bmatrix} 2 & 1 \\ 1 & -1 \end{bmatrix}\begin{bmatrix} 3 & 0 \\ 1 & 1 \end{bmatrix}$

   (c) $\begin{bmatrix} 7 & 1 \\ -6 & 4 \end{bmatrix}\begin{bmatrix} 2 & 7 \\ 2 & 7 \end{bmatrix}$ (d) $\begin{bmatrix} 4 & -1 \\ 2 & 5 \end{bmatrix}\begin{bmatrix} 7 & 2 \\ 1 & -1 \end{bmatrix}$

   (e) $\begin{bmatrix} a & b \\ c & d \end{bmatrix}\begin{bmatrix} 0 & 1 \\ 1 & 0 \end{bmatrix}$ (f) $\begin{bmatrix} 0 & 1 \\ 1 & 0 \end{bmatrix}\begin{bmatrix} a & b \\ c & d \end{bmatrix}$
3. Find two 2×2 matrices A, B with $AB \neq BA$.

4. Find two 2×2 matrices, neither of which is the zero matrix but whose product is the zero matrix.
5. Let A, B, C be 2×2 matrices and r a real number. Prove the following:
   (a) $r(AB) = (rA)B = A(rB)$.
   (b) $(A + B)C = AC + BC$.
   (c) $A(B + C) = AB + AC$.
   (d) If A is the 2×2 identity matrix,
   $$A = \begin{bmatrix} 1 & 0 \\ 0 & 1 \end{bmatrix}$$
   then $AB = BA = B$.
   (e) $(AB)C = A(BC)$.
6. State and prove the analogues of the statements in Exercise 5 for linear transformations $Q, S, T : \mathbf{R}^2 \to \mathbf{R}^2$.
7. A 2×2 matrix A is said to be *invertible* if there is a matrix B such that $AB = I$, the 2×2 identity matrix. The matrix B is called an *inverse* for A. Find an inverse for the matrix
   $$A = \begin{bmatrix} 2 & 1 \\ 1 & 1 \end{bmatrix}$$

*8. Prove that $AB = I$ implies $BA = I$.

*9. Prove that a 2×2 matrix
   $$A = \begin{bmatrix} a & b \\ c & d \end{bmatrix}$$
   is invertible if and only if $ad - bc \neq 0$ and find an inverse in this case. (See Exercise 7.)

*10. Determine which of the following matrices are invertible and find the inverses of those that are.

   (a) $\begin{bmatrix} 2 & 3 \\ 3 & 4 \end{bmatrix}$ (b) $\begin{bmatrix} 2 & 1 \\ 4 & 2 \end{bmatrix}$

   (c) $\begin{bmatrix} 2 & 3 \\ 2 & 2 \end{bmatrix}$ (d) $\begin{bmatrix} 5 & 2 \\ 6 & 3 \end{bmatrix}$

9. Matrix Multiplication

In the previous section, we defined the product of 2×2 matrices. The entries of the product were obtained by multiplying each component of a row of the first matrix by the corresponding component of a column of the second matrix and adding the results. As we see now, this definition extends to larger matrices.

Let $A = (a_{ij})$ and $B = (b_{ij})$ be two matrices. The *product* AB is defined to be the matrix (c_{ij}), where the number c_{ij} is obtained by multiplying each component of the ith row of A by the corresponding component of the jth column of B and then adding the result. Note that, in order to carry this out, the number of elements in each row of A (that is, the number of columns of A) must equal the number of elements in each column of B (that is, the number of rows of B). Thus, *the product AB is defined only when the number of columns of A equals the number of rows of B.*

We can make this prescription more explicit. If $A = (a_{ij})$ is a $q \times p$ matrix and $B = (b_{ij})$ is a $p \times n$ matrix, then the product AB is the $q \times n$ matrix (c_{ij}) with

$$c_{ij} = a_{i1}b_{1j} + a_{i2}b_{2j} + \cdots + a_{ip}b_{pj} = \sum_{k=1}^{p} a_{ik}b_{kj} \tag{9.1}$$

Examples

1. $$\begin{bmatrix} 1 & 2 \\ 2 & 3 \end{bmatrix}\begin{bmatrix} 2 & 0 & 7 \\ 1 & -1 & 4 \end{bmatrix} = \begin{bmatrix} 1\cdot 2 + 2\cdot 1 & 1\cdot 0 + 2\cdot(-1) & 1\cdot 7 + 2\cdot 4 \\ 2\cdot 2 + 3\cdot 1 & 2\cdot 0 + 3\cdot(-1) & 2\cdot 7 + 3\cdot 4 \end{bmatrix}$$
$$= \begin{bmatrix} 4 & -2 & 15 \\ 7 & -3 & 26 \end{bmatrix}$$

   Note that the product

$$\begin{bmatrix} 2 & 0 & 7 \\ 1 & -1 & 4 \end{bmatrix}\begin{bmatrix} 1 & 2 \\ 2 & 3 \end{bmatrix}$$

   is not defined, since the first matrix has three columns and the second has only two rows.

2. $$\begin{bmatrix} 3 & -1 & 1 \\ 1 & 0 & 2 \end{bmatrix}\begin{bmatrix} 1 & 7 \\ 6 & -2 \\ 2 & 3 \end{bmatrix}$$
$$= \begin{bmatrix} 3\cdot 1 + (-1)\cdot 6 + 1\cdot 2 & 3\cdot 7 + (-1)\cdot(-2) + 1\cdot 3 \\ 1\cdot 1 + 0\cdot 6 + 2\cdot 2 & 1\cdot 7 + 0\cdot(-2) + 2\cdot 3 \end{bmatrix}$$
$$= \begin{bmatrix} -1 & 26 \\ 5 & 13 \end{bmatrix}$$

3. $$\begin{bmatrix} 2 & 2 & 1 \\ 0 & 1 & 7 \end{bmatrix}\begin{bmatrix} 6 & 0 & -1 \\ -1 & 4 & 1 \\ 3 & 2 & 3 \end{bmatrix} = \begin{bmatrix} 13 & 10 & 3 \\ 20 & 18 & 22 \end{bmatrix}$$

4. $$\begin{bmatrix} 1 & 0 & 2 \\ -2 & 1 & 1 \\ 0 & 3 & 7 \end{bmatrix}\begin{bmatrix} 2 & 0 & 7 & 1 \\ 1 & 5 & -1 & 0 \\ 2 & -2 & 0 & 4 \end{bmatrix} = \begin{bmatrix} 6 & -4 & 7 & 9 \\ -1 & 3 & -15 & 2 \\ 17 & 1 & -3 & 28 \end{bmatrix}$$

We saw in the previous section that, for linear transformations $S, T : \mathbf{R}^2 \to \mathbf{R}^2$, the matrix of the composite $S \circ T$ is the product of the matrix of S with the matrix of T. Indeed, it was this relationship which led us to the definition of matrix multiplication. We now see that this fact holds in general.

Proposition 9.1 *Let $T : \mathbf{R}^n \to \mathbf{R}^p$ be a linear transformation with matrix B and $S : \mathbf{R}^p \to \mathbf{R}^q$ a linear transformation with matrix A. Then the linear transformation $S \circ T : \mathbf{R}^n \to \mathbf{R}^q$ has matrix AB.*

Proof. We have already demonstrated this result when $n = p = q = 2$. To prove it in general, we must show that, for each j, $1 \le j \le n$, the jth row of AB coincides with the components of the vector $(S \circ T)(\mathbf{e}_j)$.

Suppose $A = (a_{ij})$ and $B = (b_{ij})$; then

$$T(\mathbf{e}_j) = (b_{1j}, \ldots, b_{pj}) = \sum_{i=1}^{p} b_{ij}\mathbf{e}_i$$

and

$$S(\mathbf{e}_i) = (a_{1i}, \ldots, a_{qi})$$

It follows that

$$\begin{aligned}(S \circ T)(\mathbf{e}_j) &= S(T(\mathbf{e}_j)) \\ &= S\left(\sum_{i=1}^{p} b_{ij}\mathbf{e}_i\right) \\ &= \sum_{i=1}^{p} b_{ij} S(\mathbf{e}_i) \\ &= \sum_{i=1}^{p} b_{ij}(a_{1i}, \ldots, a_{qi}) \\ &= \left(\sum_{i=1}^{p} b_{ij}a_{1i}, \ldots, \sum_{i=1}^{p} b_{ij}a_{qi}\right) \\ &= \left(\sum_{i=1}^{p} a_{1i}b_{ij}, \ldots, \sum_{i=1}^{p} a_{qi}b_{ij}\right)\end{aligned}$$

which is exactly the jth column of AB.

We note that the operations of addition, scalar multiplication, and matrix multiplication satisfy a number of natural properties. For example, if A is a $q \times p$ matrix and B, C are $p \times n$ matrices, then

$$A(B + C) = AB + AC$$

These properties are given in Exercise 6.

Remark. The product of matrices is not generally commutative. In fact, if AB is defined, BA need not be. Even if both are defined, they may be of different sizes; even if AB and BA have the same size, they need not be equal. Examples occur in Exercises 1 and 2.

We close this section by showing how matrix multiplication can be used to determine the value of a linear transformation on a particular vector.

Let $T:\mathbf{R}^2 \to \mathbf{R}^2$ be a linear transformation with matrix $A = (a_{ij})$ and let $\mathbf{v} = (x_1, x_2)$ be a vector in $\mathbf{R}^2$. Then,

$$\begin{aligned} T(\mathbf{v}) &= T(x_1\mathbf{e}_1 + x_2\mathbf{e}_2) \\ &= x_1 T(\mathbf{e}_1) + x_2 T(\mathbf{e}_2) \\ &= x_1(a_{11}, a_{21}) + x_2(a_{12}, a_{22}) \\ &= (a_{11}x_1 + a_{12}x_2, a_{21}x_1 + a_{22}x_2) \end{aligned}$$

Now, the components of this vector $T(\mathbf{v})$ resemble the expressions that arise when we multiply matrices. In fact, if we write $\mathbf{v}$ as a column vector instead of a row vector, that is a 2×1 matrix

$$V = \begin{bmatrix} x_1 \\ x_2 \end{bmatrix}$$

then the product AV is the 2×1 matrix obtained by considering $T(\mathbf{v})$ as a column vector.

For example, suppose $T:\mathbf{R}^2 \to \mathbf{R}^2$ has matrix

$$\begin{bmatrix} 1 & 2 \\ 3 & 4 \end{bmatrix}$$

and we wish to determine $T(1, -2)$. Following the recipe above, we compute the product

$$\begin{bmatrix} 1 & 2 \\ 3 & 4 \end{bmatrix} \begin{bmatrix} 1 \\ -2 \end{bmatrix} = \begin{bmatrix} -3 \\ -5 \end{bmatrix}$$

and conclude that $T(1, -2) = (-3, -5)$. (Check this.)

More generally, if $T:\mathbf{R}^n \to \mathbf{R}^p$ is a linear transformation with matrix $A = (a_{ij})$ and $\mathbf{v} = (x_1, \ldots, x_n) \in \mathbf{R}^n$, then $T(\mathbf{v})$ is obtained by multiplying A by the $n \times 1$ matrix whose entries are the components of $\mathbf{v}$ and considering the resulting $p \times 1$ matrix

$$\begin{bmatrix} a_{11} & \cdots & a_{1n} \\ \vdots & & \vdots \\ a_{p1} & \cdots & a_{pn} \end{bmatrix} \begin{bmatrix} x_1 \\ \vdots \\ x_n \end{bmatrix}$$

as a row vector. For instance, suppose $T:\mathbf{R}^5 \to \mathbf{R}^3$ is a linear transformation with matrix

$$A = \begin{bmatrix} 1 & 0 & 2 & 1 & 3 \\ -1 & 4 & 2 & 0 & -1 \\ 0 & -2 & 1 & 3 & 0 \end{bmatrix}$$

Then $T(1, 2, -1, 0, 3) = (8, 2, -5)$ since

$$\begin{bmatrix} 1 & 0 & 2 & 1 & 3 \\ -1 & 4 & 2 & 0 & -1 \\ 0 & -2 & 1 & 3 & 0 \end{bmatrix} \begin{bmatrix} 1 \\ 2 \\ -1 \\ 0 \\ 3 \end{bmatrix} = \begin{bmatrix} 8 \\ 2 \\ -5 \end{bmatrix}$$

and $T(2, -1, 4, -3, 1) = (10, 1, -3)$ since

$$\begin{bmatrix} 1 & 0 & 2 & 1 & 3 \\ -1 & 4 & 2 & 0 & -1 \\ 0 & -2 & 1 & 3 & 0 \end{bmatrix} \begin{bmatrix} 2 \\ -1 \\ 4 \\ -3 \\ 1 \end{bmatrix} = \begin{bmatrix} 10 \\ 1 \\ -3 \end{bmatrix}$$

Exercises

1. Compute the following:

(a) $\begin{bmatrix} 1 & 8 \\ 4 & 3 \end{bmatrix}\begin{bmatrix} 2 & 7 \\ 3 & 4 \end{bmatrix}$

(b) $\begin{bmatrix} -2 & 3 & 5 \\ 6 & 1 & 7 \end{bmatrix}\begin{bmatrix} 3 & 2 & 5 \\ 8 & -9 & 4 \\ -6 & 7 & 1 \end{bmatrix}$

(c) $\begin{bmatrix} 3 & 4 \\ -2 & 4 \\ 1 & 4 \end{bmatrix}\begin{bmatrix} 0 & 3 \\ 1 & 2 \end{bmatrix}$

(d) $\begin{bmatrix} 4 & 5 & 6 & 8 \\ 1 & 3 & 2 & 4 \end{bmatrix}\begin{bmatrix} -9 & -1 & 4 \\ 3 & 2 & 3 \\ 2 & 6 & 0 \\ -7 & 1 & 4 \end{bmatrix}$

(e) $\begin{bmatrix} 2 & 0 & 0 \\ 0 & 3 & 0 \\ 0 & 0 & 7 \end{bmatrix}\begin{bmatrix} 3 & 2 & 1 & 4 \\ -6 & 4 & 5 & -2 \\ 3 & 1 & 6 & -8 \end{bmatrix}$

(f) $\begin{bmatrix} 1 & 3 & 7 \end{bmatrix}\begin{bmatrix} 7 \\ 2 \\ 1 \end{bmatrix}$

(g) $\begin{bmatrix} 7 \\ 2 \\ 1 \end{bmatrix}\begin{bmatrix} 1 & 3 & 7 \end{bmatrix}$

(h) $\begin{bmatrix} 2 & 4 & 8 \\ -6 & 3 & -5 \end{bmatrix}\begin{bmatrix} 3 & -6 \\ -2 & 1 \\ 7 & 5 \end{bmatrix}$

(i) $\begin{bmatrix} 3 & 6 \\ -2 & 1 \\ 7 & 5 \end{bmatrix}\begin{bmatrix} 2 & 4 & 8 \\ -6 & 3 & -5 \end{bmatrix}$

(j) $\begin{bmatrix} 6 & 4 & 3 \\ 1 & 8 & 7 \\ 2 & 2 & 5 \end{bmatrix}\begin{bmatrix} 1 & 6 \\ -4 & 3 \\ 7 & 2 \end{bmatrix}$

2. Let $A = \begin{bmatrix} 1 & 3 \\ 2 & 4 \end{bmatrix}$, $B = \begin{bmatrix} 1 & 4 \\ 3 & -6 \end{bmatrix}$, and $C = \begin{bmatrix} 3 & -4 & 5 \\ 2 & 1 & 6 \end{bmatrix}$.

Determine which of the following make sense and compute those that do.

(a) $A^2(=AA)$ (b) B^2
(c) C^2 (d) AB
(e) BA (f) AC
(g) CA (h) BC
(i) CB (j) $(CB)A$

3. Let I be the $n \times n$ identity matrix.
(a) Show that $AI = A$ for any $p \times n$ matrix A.
(b) Show that $IA = A$ for any $n \times p$ matrix A.

4. Let $T: \mathbf{R}^3 \to \mathbf{R}^3$ be the linear transformation with matrix

$$\begin{bmatrix} 2 & 1 & 3 \\ 5 & -2 & 1 \\ 0 & 3 & -3 \end{bmatrix}$$

Use the method described at the end of this section to evaluate the following:
(a) $T(1, 1, 2)$ (b) $T(-1, 3, 0)$
(c) $T(2, 5, -4)$ (d) $T(6, 1, 2)$

5. Let $T: \mathbf{R}^4 \to \mathbf{R}^5$ be the linear transformation with matrix

$$\begin{bmatrix} 1 & 0 & 2 & 1 \\ -3 & 4 & 1 & 1 \\ 2 & 0 & -1 & 0 \\ 5 & 1 & 1 & 1 \\ 3 & 7 & 0 & -2 \end{bmatrix}$$

Use the method described at the end of this section to evaluate the following:
(a) $T(1, 0, 2, 1)$ (b) $T(2, 1, -2, 3)$
(c) $T(7, 4, -4, 2)$ (d) $T(3, 6, -3, 4)$

6. Let A be a $q \times p$ matrix, B, C $p \times n$ matrices, D an $n \times m$ matrix, and r a real number. Prove the following:
(a) $r(AB) = (rA)B = A(rB)$
(b) $A(B + C) = AB + AC$
(c) $(B + C)D = BD + CD$
(d) $(AB)D = A(BD)$

7. Let I_n be the $n \times n$ identity matrix, 0_n the $n \times n$ zero matrix, A any $n \times p$ matrix, and B any $p \times n$ matrix.
(a) Prove that $I_nA = A$ and $BI_n = B$.
(b) Prove that $0_nA = B0_n = 0_n$.

8. Let $S: \mathbf{R}^p \to \mathbf{R}^q$, $T: \mathbf{R}^n \to \mathbf{R}^p$, and $Q: \mathbf{R}^m \to \mathbf{R}^n$ be linear transformations and prove that

$$S \circ (T \circ Q) = (S \circ T) \circ Q$$

Use this fact and Proposition 9.1 to prove part (d) of Exercise 6.

9. (a) A 3×3 matrix of the form

$$\begin{bmatrix} a_{11} & a_{12} & a_{13} \\ 0 & a_{22} & a_{23} \\ 0 & 0 & a_{33} \end{bmatrix}$$

is called *upper triangular*. Show that if A and B are upper triangular matrices, so is AB.

*(b) State and prove a similar result for $n \times n$ matrices.

10. (a) A 3×3 matrix of the form

$$\begin{bmatrix} 0 & a_{12} & a_{13} \\ 0 & 0 & a_{23} \\ 0 & 0 & 0 \end{bmatrix}$$

is called *strictly upper triangular*. Show that if A, B, and C are strictly upper triangular 3×3 matrix, then $ABC = 0_3$, the 3×3 zero matrix.

*(b) State and prove a similar result for $n \times n$ matrices.

*11. An $n \times n$ matrix A is said to be *invertible* if there is an $n \times n$ matrix B such that $AB = BA = I$, where I is the $n \times n$ identity matrix. The matrix B is called the *inverse* of A. Let A and B be given by

$$A = \begin{bmatrix} 1 & 3 & 2 \\ 0 & 4 & 3 \\ 6 & 7 & 4 \end{bmatrix} \qquad B = \begin{bmatrix} -5 & 2 & 1 \\ 18 & -8 & -3 \\ -24 & 11 & 4 \end{bmatrix}$$

and show that B is an inverse of A.

12. Find the inverse of the matrix

$$\begin{bmatrix} 1 & 0 & a \\ 0 & 1 & 0 \\ 0 & 0 & 1 \end{bmatrix}$$

where a is some real number.

13. Prove that the upper triangular matrix of Exercise 8 is invertible if $a_{11}a_{22}a_{33} \neq 0$ and determine the inverse in this case.

*14. Generalize Exercise 13 to $n \times n$ matrices.

3

Continuous Functions

The functions that we are primarily interested in are the differentiable functions defined in the next chapter. However, in order to study these functions, we need the notion of continuity. In this chapter, we define continuous function, give some examples, and study some of their elementary properties.

Many of the results of this chapter, although interesting, are technical and not explicitly needed in most of what follows. As a result, this chapter need not be read in detail; the reader can instead refer back to it when necessary.

1. Continuous Functions of One Variable

Before considering the notion of continuity for functions between Euclidean spaces, we review this notion for functions $f: \mathbf{R} \to \mathbf{R}$.

There are many ways of describing the concept of continuity. We can say, for example, that a function $f: \mathbf{R} \to \mathbf{R}$ is continuous if its graph has no "breaks." Thus, the function whose graph is given in Figure 1.1 is continuous whereas the function whose graph is given in Figure 1.2 is not continuous. This description has a great deal of intuitive value but, unfortunately, is not precise enough to be usable.

Another way of describing continuity is to say that $f: \mathbf{R} \to \mathbf{R}$ is continuous at $a \in \mathbf{R}$ if we can make $f(x)$ as close as we like to $f(a)$ by choosing x close enough to a. This description, too, needs to be made more precise, as follows. We say that the function $f: \mathbf{R} \to \mathbf{R}$ is *continuous at a* if, for any $\varepsilon > 0$, there is a $\delta > 0$ such that $|f(x) - f(a)| < \varepsilon$ wherever $|x - a| < \delta$.

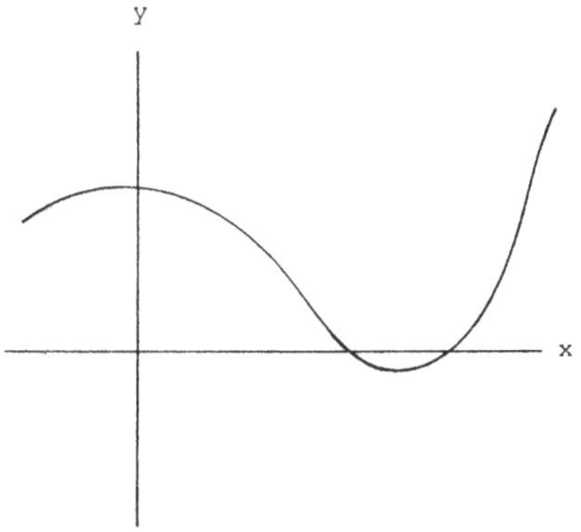

Figure 1.1 The graph of a continuous function.

In terms of the graph of f, this means that the piece of the graph of f between the vertical lines through the points $a - \delta$ and $a + \delta$ must be entirely between the horizontal lines through the points $f(a) - \varepsilon$ and $f(a) + \varepsilon$. (See Figure 1.3.)

We now look at some examples.

1. The function $f(x) = 3x + 2$ is continuous at all $a \in \mathbf{R}$. To prove this, we must show that for any $\varepsilon > 0$, we can find a $\delta > 0$ such that

$$\begin{aligned} |f(x) - f(a)| &= |3x + 2 - (3a - 2)| \\ &= |3x - 3a| \\ &= 3|x - a| \end{aligned}$$

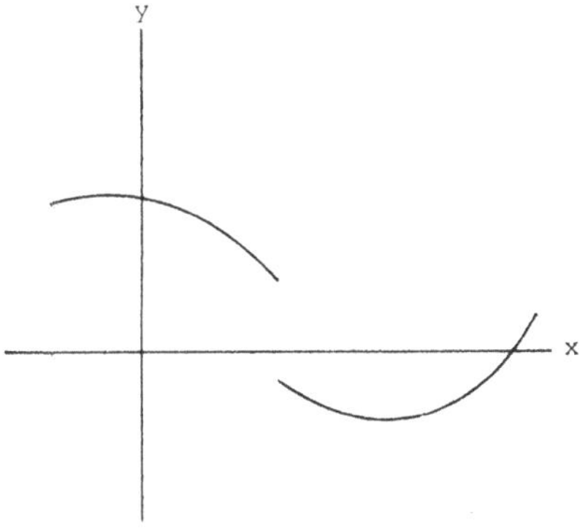

Figure 1.2 The graph of a discontinuous function.

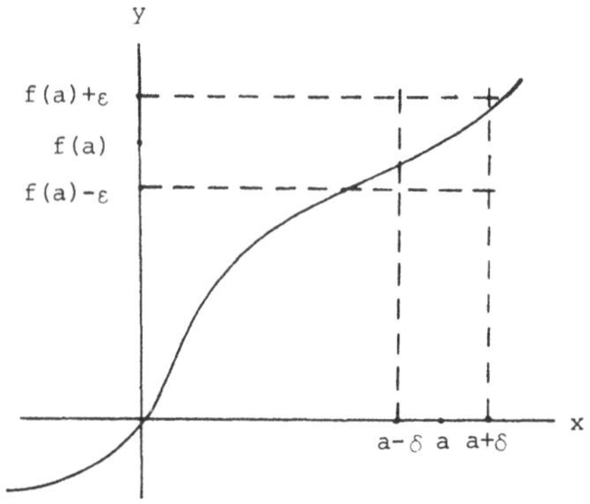

Figure 1.3 Continuity of f at a.

is less than ε whenever $|x - a| < \delta$. In other words, if we imagine that someone gives us a small number ε, we must find a number δ so that the above holds. Here the choice of δ is clear; we simply set $\delta = \varepsilon/3$. Then

$$|f(x) - f(a)| = 3|x - a|$$

is certainly less than ε when $|x - a| < \delta = \varepsilon/3$.

2. The function $f(x) = x^2$ is continuous at all $a \in \mathbf{R}$. To see this, let ε be any positive number and note that

$$|f(x) - f(a)| = |x^2 - a^2| = |x - a| \cdot |x + a|$$

Furthermore, if $|x - a| < 1$, then

$$|x + a| = |x - a + 2a| \leq 1 + 2|a|$$

Thus $$|f(x) - f(a)| < (1 + 2|a|) \cdot |x - a|$$

if $|x - a| < 1$. If we choose δ to be the smaller of the two numbers, 1, $\varepsilon/(1 + 2|a|)$, we see that

$$|f(x) - f(a)| < \varepsilon$$

whenever $|x - a| < \delta$. Notice that here we need to work harder to find δ. One reason is that δ depends on a as well as ε.

3. The function $f: \mathbf{R} \to \mathbf{R}$ defined by

$$f(x) = \begin{cases} x & \text{if } x \leq 0 \\ x + 1 & \text{if } x > 0 \end{cases}$$

is *not* continuous at 0. (See Figure 1.4.)

To prove this, we must show the following (see Section 1.2): there is a

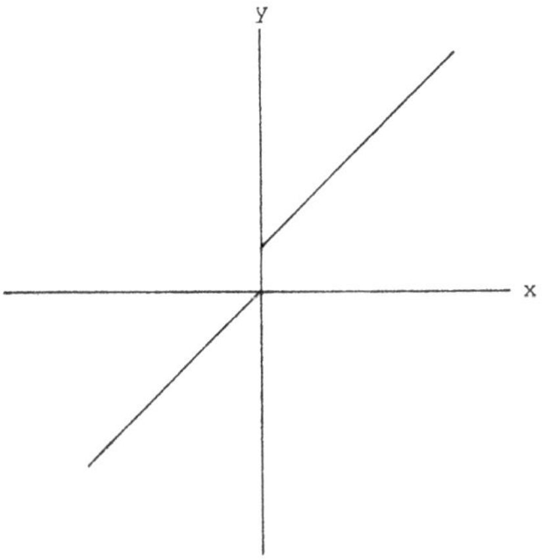

Figure 1.4 A function not continuous at 0.

number $\varepsilon > 0$ such that, for all $\delta > 0$, there is an x such that $|x - 0| < \delta$ and $|f(x) - f(0)| \geq \varepsilon$. If we choose $\varepsilon = 1/2$, we see that, no matter how small we choose δ, we can always find $x > 0$ with $|x - 0| < \delta$. However, for $x > 0$, $|f(x) - f(0)| = f(x) > 1 > \varepsilon$.

4. The function $f: \mathbf{R} \to \mathbf{R}$ defined by

$$f(x) = \begin{cases} \dfrac{1}{x} & \text{if } x \neq 0 \\ 0 & \text{if } x = 0 \end{cases}$$

is not continuous at $x = 0$.
For instance, if $\varepsilon = 1$ and δ is any positive number, we can clearly find x with $|x| < \delta$ and $f(x) \geq 1$.

Thus far, we have spoken only of continuity at a point. We define $f: \mathbf{R} \to \mathbf{R}$ to be *continuous* if it is continuous at every point in $\mathbf{R}$. The functions in Examples 1 and 2 are therefore continuous.

We note that continuous functions can be combined in various ways to obtain new continuous functions. For example, the sum and product of continuous functions are again continuous. We shall deal with this in a more general setting later in the chapter.

Exercises

1. Prove the following functions continuous at the values of x specified:
   (a) $f(x) = 3x + 3$, all $x \in \mathbf{R}$.
   (b) $f(x) = 3x^2$, all $x \in \mathbf{R}$.

(c) $f(x) = x^2 - x - 3$, all $x \in \mathbf{R}$.
(d) $f(x) = 1/x$, $x \neq 0$.
(e) $f(x) = |x|$, all $x \in \mathbf{R}$.

2. Show that for all values of a and b, $f(x) = ax + b$ and $g(x) = ax^2$ are continuous for all $x \in \mathbf{R}$.
3. Let $f: \mathbf{R} \to \mathbf{R}$ be continuous, let c be a constant, and let $g(x) = f(x) + c$. Show that g is continuous.
4. Let $f(x) = e^x$, and assume that f is continuous at 0. Prove that f is continuous.

*5. Suppose that $f: \mathbf{R} \to \mathbf{R}$ and $g: \mathbf{R} \to \mathbf{R}$ are continuous. Prove that $f + g$ is continuous.

*6. Suppose that $f: \mathbf{R} \to \mathbf{R}$ and $g: \mathbf{R} \to \mathbf{R}$ are continuous. Prove that fg is continuous.

*7. Define

$$f(x) = \begin{cases} 0 & \text{if } x \text{ is irrational} \\ 1 & \text{if } x \text{ is rational} \end{cases}$$

Prove that f is not continuous at any point.

*8. Set

$$f(x) = \begin{cases} 0 & \text{if } x \text{ is irrational} \\ \dfrac{1}{q} & \text{if } x = \dfrac{p}{q} \text{ as a fraction in lowest terms} \end{cases}$$

Show that f is continuous at all irrational points and discontinuous at all rational ones.

2. Continuous Functions between Euclidean Spaces

We now define continuity for functions between Euclidean spaces. We obtain the definition of continuity for functions $f: \mathbf{R}^n \to \mathbf{R}^p$ by simply replacing the absolute value by the norm in the definition of continuity given in Section 1. Since the norm and absolute value coincide in $\mathbf{R}^1 = \mathbf{R}$, this definition agrees with the other one in the case $n = p = 1$.

A function $f: \mathbf{R}^n \to \mathbf{R}^p$ is defined to be *continuous at* $\mathbf{v}_0 \in \mathbf{R}^n$ if, for any $\varepsilon > 0$, there is a $\delta > 0$ such that $\| f(\mathbf{v}) - f(\mathbf{v}_0) \| < \varepsilon$ whenever $\|\mathbf{v} - \mathbf{v}_0\| < \delta$.

Of course, f need not be defined on all of $\mathbf{R}^n$, only on some subset X containing $\mathbf{v}_0$. In this case, we require that the vectors $\mathbf{v}$ in the definition of continuity also lie in X.

If $f: \mathbf{R}^n \to \mathbf{R}^p$ and S is any subset of $\mathbf{R}^n$, we say that f is *continuous on* S if f is continuous at each point of S. We say that f is *continuous* if f is continuous at each $\mathbf{v} \in \mathbf{R}^n$.

Examples

1. Let $f: \mathbf{R}^2 \to \mathbf{R}$ be defined by $f(x, y) = 2x + y$. Then f is continuous at any point (x_0, y_0) in $\mathbf{R}^2$. To prove this, we let ε be any positive number and note

that

$$\begin{aligned}\| f(x, y) - f(x_0, y_0)\| &= |2x + y - (2x_0 + y_0)| \\ &= |2(x - x_0) + (y - y_0)| \\ &\le 2|x - x_0| + |y - y_0|\end{aligned}$$

Now
$$\begin{aligned}|x - x_0| &= [(x - x_0)^2]^{1/2} \\ &\le [(x - x_0)^2 + (y - y_0)^2]^{1/2} \\ &= \|(x, y) - (x_0, y_0)\|\end{aligned}$$

and
$$|y - y_0| = [(y - y_0)^2]^{1/2} \le \|(x, y) - (x_0, y_0)\|$$

Thus, if we choose $\delta = \varepsilon/3$, we have

$$|x - x_0| < \frac{\varepsilon}{3}$$

and
$$|y - y_0| < \frac{\varepsilon}{3}$$

whenever $\|(x, y) - (x_0, y_0)\| < \delta$. Therefore

$$\begin{aligned}\| f(x, y) - f(x_0, y_0)\| &\le 2|x - x_0| + |y - y_0| \\ &< \frac{2\varepsilon}{3} + \frac{\varepsilon}{3} = \varepsilon\end{aligned}$$

whenever $\|(x, y) - (x_0, y_0)\| < \delta$, and f is continuous at (x_0, y_0).

The following two examples will be useful later in this chapter.

2. Define $g : \mathbf{R}^n \to \mathbf{R}$ by

$$g(x_1, \ldots, x_n) = x_1$$

This function is continuous. To prove this, we need to show that for any $\mathbf{v}_0 = (a_1, \ldots, a_n) \in \mathbf{R}^n$ and $\delta > 0$, there is an $\delta > 0$ such that

$$\|g(\mathbf{v}) - g(\mathbf{v}_0)\| = \|g(x_1, \ldots, x_n) - g(a_1, \ldots, a_n)\| = |x_1 - a_1|$$

is less than ε whenever $\|\mathbf{v} - \mathbf{v}_0\|$ is less than δ. However,

$$\begin{aligned}|x_1 - a_1| &= [(x_1 - a_1)^2]^{1/2} \\ &\le [(x_1 - a_1)^2 + \cdots + (x_n - a_n)^2]^{1/2} \\ &= \|\mathbf{v} - \mathbf{v}_0\|\end{aligned}$$

Therefore, if we set $\delta = \varepsilon$, we have

$$\|g(\mathbf{v}) - g(\mathbf{v}_0)\| = |x_1 - a_1| \le \|\mathbf{v} - \mathbf{v}_0\| < \varepsilon$$

wherever $\|\mathbf{v} - \mathbf{v}_0\| < \delta$.

Of course, the same argument proves that each of the functions $g_j : \mathbf{R}^n \to \mathbf{R}$, $1 \le j \le n$, defined by

$$g_j(x_1, \ldots, x_n) = x_j$$

is continuous. (See Exercise 3.)

The next example is more difficult.

3. Let $h : \mathbf{R}^2 \to \mathbf{R}$ be defined by $h(x, y) = xy$. Then h is continuous. To demonstrate this, let ε be any positive number, (a, b) a vector in $\mathbf{R}^2$, and consider the following.

$$\begin{aligned} |h(x, y) - h(a, b)| &= |xy - ab| \\ &= |xy - ay + ay - ab| \\ &\le |xy - ay| + |ay - ab| \\ &= |x - a| \cdot |y| + |a| \cdot |y - b| \end{aligned}$$

Now suppose that $|y - b| < 1$. Then

$$|y| = |y - b + b| \le |y - b| + |b| < 1 + |b|$$

so that

$$(2.1) \qquad |h(x, y) - h(a, b)| < |x - a|(1 + |b|) + |a| \cdot |y - b|$$

when $|y - b| < 1$. Furthermore,

$$(2.2) \qquad \begin{aligned} |x - a| &= \sqrt{(x - a)^2} \\ &\le \sqrt{(x - a)^2 + (y - b)^2} \\ &\le \| (x, y) - (a, b) \| \end{aligned}$$

and, in the same way,

$$(2.3) \qquad |y - b| \le \| (x, y) - (a, b) \|$$

Thus, if we let δ be the smallest of the three numbers

$$1 \qquad \frac{\varepsilon}{2(1 + |b|)} \qquad \frac{\varepsilon}{2|a|}$$

we see from (2.1), (2.2), and (2.3) that

$$\begin{aligned} |h(x, y) - h(a, b)| &< \frac{\varepsilon}{2(1 + |b|)}(1 + |b|) + |a|\frac{\varepsilon}{2|a|} \\ &= \frac{\varepsilon}{2} + \frac{\varepsilon}{2} = \varepsilon \end{aligned}$$

whenever $\| (x, y) - (a, b) \| < \delta$. Thus h is continuous.

Exercises

1. Prove the following functions continuous:
   (a) $f: \mathbf{R}^2 \to \mathbf{R}$, $f(x, y) = 7x + 4y$
   (b) $g: \mathbf{R}^2 \to \mathbf{R}$, $g(x, y) = 2x - 9y + 3$
   (c) $h: \mathbf{R}^2 \to \mathbf{R}$, $h(x, y) = 5x - 6y$
   (d) $F: \mathbf{R}^2 \to \mathbf{R}$, $F(x, y) = 6x$
   (e) $G: \mathbf{R}^3 \to \mathbf{R}$, $G(x, y, z) = 2x + 3y - 4z$
   (f) $H: \mathbf{R}^3 \to \mathbf{R}$, $H(x, y, z) = x + y/2 + z/3 + 1/4$
2. Prove the following functions continuous:
   (a) $f: \mathbf{R}^2 \to \mathbf{R}$, $f(x, y) = 3xy + 2$
   (b) $g: \mathbf{R}^2 \to \mathbf{R}$, $g(x, y) = x^2 + 2y^2$
   (c) $h: \mathbf{R}^3 \to \mathbf{R}$, $h(x, y, z) = xy + yz$
   (d) $F: \mathbf{R}^3 \to \mathbf{R}$, $F(x, y, z) = xyz$
   (e) $G: \mathbf{R}^3 \to \mathbf{R}^2$, $G(x, y, z) = (x + 2y - z, y + z)$
   (f) $H: \mathbf{R}^3 \to \mathbf{R}^2$, $H(x, y, z) = (x^2 + y, y^2 + yz)$
3. Define $g_j: \mathbf{R}^k \to \mathbf{R}$ by $g_j(\mathbf{v}) = x_j$ [where $\mathbf{v} = (x_1, \ldots, x_k)$]. Prove that g_j is continuous. (See Example 2.)
4. Let $F: \mathbf{R}^3 \to \mathbf{R}^2$ be given by

   $$F(x, y, z) = (f_1(x, y, z), f_2(x, y, z))$$

   That is, f_1 and f_2 are the *coordinate functions* of F. Show that F is continuous if and only if f_1 and f_2 are continuous.
5. Let $f: \mathbf{R}^k \to \mathbf{R}$ be continuous at $\mathbf{v}_0$, and suppose that $f(\mathbf{v}_0) > a > 0$. Show that there is a number $\delta > 0$ such that if $\|\mathbf{v} - \mathbf{v}_0\| < \delta$, then $f(\mathbf{v}) > a$.
6. Let B be a subset of $\mathbf{R}^n$. A function $f: B \to \mathbf{R}^m$ is said to be *bounded* if there is a number $M > 0$ such that $\|f(\mathbf{v})\| \leq M$ for all $\mathbf{v} \in B$. Give an example of a continuous function $f: (0, 1) \to \mathbf{R}$ which is not bounded.

*7. A function $f: \mathbf{R}^n \to \mathbf{R}^p$ is said to be *locally bounded* if, for each $\mathbf{v} \in \mathbf{R}^n$, there is an $r > 0$ such that f is bounded on $B_r(\mathbf{v})$. Prove that f is locally bounded whenever f is continuous.

*8. State and prove the generalization of Exercise 4 to functions $F: \mathbf{R}^n \to \mathbf{R}^p$.

3. Sums and Scalar Products of Continuous Functions

In this section and the next, we develop some of the elementary properties of continuous functions. We begin here by showing that the sum of two continuous functions and the product of a continuous function by a real number are continuous. We also prove that linear transformations are continuous.

Let $f, g: \mathbf{R}^n \to \mathbf{R}^p$ be functions and let a be a real number. We define functions

$$f+g:\mathbf{R}^n\to\mathbf{R}^p$$
$$af:\mathbf{R}^n\to\mathbf{R}^p$$

by

$$(f+g)(\mathbf{v})=f(\mathbf{v})+g(\mathbf{v})$$
$$(af)(\mathbf{v})=af(\mathbf{v})$$

The following two propositions are analogues of well-known results for continuous functions $f\colon\mathbf{R}\to\mathbf{R}$.

Proposition 3.1 *Suppose f, $g:\mathbf{R}^n\to\mathbf{R}^p$ are continuous at $\mathbf{v}_0\in\mathbf{R}^n$. Then $f+g:\mathbf{R}^n\to\mathbf{R}^p$ is also continuous at $\mathbf{v}_0$.*

Proof. We need to show that for any $\varepsilon>0$ there is a $\delta>0$ such that $\|(f+g)(\mathbf{v})-(f+g)(\mathbf{v}_0)\|<\varepsilon$ whenever $\|\mathbf{v}-\mathbf{v}_0\|<\delta$. Now

$$\begin{aligned}\|(f+g)(\mathbf{v})-(f+g)(\mathbf{v}_0)\| &= \|f(\mathbf{v})+g(\mathbf{v})-f(\mathbf{v}_0)-g(\mathbf{v}_0)\|\\ &= \|[f(\mathbf{v})-f(\mathbf{v}_0)]+[g(\mathbf{v})-g(\mathbf{v}_0)]\|\\ &\le \|f(\mathbf{v})-f(\mathbf{v}_0)\|+\|g(\mathbf{v})-g(\mathbf{v}_0)\|\end{aligned}$$

The idea now is to use the continuity of f and g at $\mathbf{v}_0$. Since both $\|f(\mathbf{v})-f(\mathbf{v}_0)\|$ and $\|g(\mathbf{v})-g(\mathbf{v}_0)\|$ are small when $\|\mathbf{v}-\mathbf{v}_0\|$ is small enough, $\|(f+g)(\mathbf{v})-(f+g)(\mathbf{v}_0)\|$ will be small when $\|\mathbf{v}-\mathbf{v}_0\|$ is small enough. More precisely, since f is continuous at $\mathbf{v}_0$, we can find $\delta_1>0$ so that $\|f(\mathbf{v})-f(\mathbf{v}_0)\|<\varepsilon/2$ when $\|\mathbf{v}-\mathbf{v}_0\|<\delta_1$. Since g is continuous at $\mathbf{v}_0$, we can find $\delta_2>0$ so that $\|g(\mathbf{v})-g(\mathbf{v}_0)\|<\varepsilon/2$ when $\|\mathbf{v}-\mathbf{v}_0\|<\delta_2$.
Thus, if $\delta=\min\{\delta_1,\delta_2\}$,

$$\begin{aligned}\|(f+g)(\mathbf{v})-(f+g)(\mathbf{v}_0)\| &\le \|f(\mathbf{v})-f(\mathbf{v}_0)\|+\|g(\mathbf{v})-g(\mathbf{v}_0)\|\\ &<\frac{\varepsilon}{2}+\frac{\varepsilon}{2}=\varepsilon\end{aligned}$$

whenever $\|\mathbf{v}-\mathbf{v}_0\|<\delta$, so $f+g$ is continuous at $\mathbf{v}_0$.

Proposition 3.2 *If $f\colon\mathbf{R}^n\to\mathbf{R}^p$ is continuous at $\mathbf{v}_0\in\mathbf{R}^n$ and a is a real number, then $af\colon\mathbf{R}^n\to\mathbf{R}^p$ is also continuous at $\mathbf{v}_0$.*

Proof. We need to show that given any $\varepsilon>0$, we can find a $\delta>0$ such that

$$\|(af)(\mathbf{v})-(af)(\mathbf{v}_0)\|<\varepsilon$$

whenever $\|\mathbf{v}-\mathbf{v}_0\|<\delta$. Now

$$\begin{aligned}\|(af)(\mathbf{v})-(af)(\mathbf{v}_0)\| &= \|af(\mathbf{v})-af(\mathbf{v}_0)\|\\ &= |a|\cdot\|f(\mathbf{v})-f(\mathbf{v}_0)\|\end{aligned}$$

so the assertion is trivial if $a = 0$. Assume $a \neq 0$. Since f is continuous at $\mathbf{v}_0$, we can find $\delta > 0$ so that

$$\|f(\mathbf{v}) - f(\mathbf{v}_0)\| < \frac{\varepsilon}{|a|}$$

when $\|\mathbf{v} - \mathbf{v}_0\| < \delta$. It follows immediately that

$$\|(af)(\mathbf{v}) - (af)(\mathbf{v}_0)\| = |a| \cdot \|f(\mathbf{v}) - f(\mathbf{v}_0)\| < \varepsilon$$

when $\|\mathbf{v} - \mathbf{v}_0\| < \delta$, so af is continuous at $\mathbf{v}_0$.

An easy induction argument gives the following generalization of Propositions 3.1 and 3.2.

Proposition 3.3 *Let $g_1, \ldots, g_k : \mathbf{R}^n \to \mathbf{R}^p$ be continuous at $\mathbf{v}_0$, and let $a_1, \ldots, a_k$ be real numbers. Then the function*

$$g = \sum_{j=1}^{k} a_j g_j$$

is also continuous at $\mathbf{v}_0$.

The proof is left as Exercise 1.

For example, suppose $T: \mathbf{R}^n \to \mathbf{R}$ is a linear transformation. Then, according to the results of Section 2.6, there are real numbers $a_1, \ldots, a_n$ such that

$$T(x_1, \ldots, x_n) = \sum_{j=1}^{n} a_j x_j$$

Let $g_j : \mathbf{R}^n \to \mathbf{R}$, $1 \leq j \leq n$, be defined by

$$g_j(x_1, \ldots, x_n) = x_j$$

These functions are continuous (see Example 2 of Section 2) and T can be written

$$T(x_1, \ldots, x_n) = \sum_{j=1}^{n} a_j g_j(x_1, \ldots, x_n)$$

It follows from Proposition 3.3 that T is continuous. Thus any linear transformation $T: \mathbf{R}^n \to \mathbf{R}$ is continuous.

The next proposition will allow us to generalize this result.

Proposition 3.4 *Let $f: \mathbf{R}^n \to \mathbf{R}^p$ be given by*

$$f(\mathbf{v}) = (f_1(\mathbf{v}), f_2(\mathbf{v}), \ldots, f_p(\mathbf{v}))$$

where $f_j : \mathbf{R}^n \to \mathbf{R}$. *Then* f *is continuous at* $\mathbf{v}_0$ *in* $\mathbf{R}^n$ *if and only if each* f_j *is continuous at* $\mathbf{v}_0$, $1 \le j \le p$.

Proof. We first prove that f is continuous at $\mathbf{v}_0$ if each f_j is continuous at $\mathbf{v}_0$.

Suppose we are given $\varepsilon > 0$. We must find a $\delta > 0$ such that $\| f(\mathbf{v}) - f(\mathbf{v}_0)\| < \varepsilon$ whenever $\|\mathbf{v} - \mathbf{v}_0\| < \delta$. Since each f_j is continuous, we can find $\delta_j > 0$ so that

$$\|\mathbf{v} - \mathbf{v}_j\| < \delta_j \Rightarrow |f_j(\mathbf{v}) - f_j(\mathbf{v}_0)| < \frac{\varepsilon}{\sqrt{p}}$$

Thus, if $\delta = \min\{\delta_1, \ldots, \delta_p\}$, we have

$$\|\mathbf{v} - \mathbf{v}_0\| < \delta \Rightarrow |f_j(\mathbf{v}) - f_j(\mathbf{v}_0)| < \frac{\varepsilon}{\sqrt{p}} \qquad \text{for all } j = 1, \ldots, m$$

so that

$$\left[\sum_{j=1}^{p} [f_j(\mathbf{v}) - f_j(\mathbf{v}_0)]^2\right]^{1/2} < \left(\sum_{j=1}^{p} \frac{\varepsilon^2}{p}\right)^{1/2} = \varepsilon$$

It follows that f is continuous at $\mathbf{v}_0$.

Assume now that f is continuous at $\mathbf{v}_0$. For any $\varepsilon > 0$, choose $\delta > 0$ so that

$$\|\mathbf{v} - \mathbf{v}_0\| < \delta \Rightarrow \| f(\mathbf{v}) - f(\mathbf{v}_0)\| < \varepsilon$$

However, for any $j = 1, \ldots, p$,

$$\begin{aligned} |f_j(\mathbf{v}) - f_j(\mathbf{v}_0)| &= \{[f_j(\mathbf{v}) - f_j(\mathbf{v}_0)]^2\}^{1/2} \\ &\le \left\{\sum_{i=1}^{p} [f_i(\mathbf{v}) - f_i(\mathbf{v}_0)]^2\right\}^{1/2} \\ &= \| f(\mathbf{v}) - f(\mathbf{v}_0)\| \end{aligned}$$

so
$$\|\mathbf{v} - \mathbf{v}_0\| < \delta \Rightarrow |f_j(\mathbf{v}) - f_j(\mathbf{v}_0)| < \varepsilon$$

Therefore f_j is continuous at $\mathbf{v}_0$ for all $j = 1, \ldots, p$.

We can now prove that any linear transformation $T: \mathbf{R}^n \to \mathbf{R}^p$ is continuous.

Proposition 3.5 *Let* $T: \mathbf{R}^n \to \mathbf{R}^p$ *be a linear transformation. Then* T *is continuous.*

Proof. According to the results of Section 2.6, we have

$$T(x_1, \ldots, x_n) = (T_1(x_1, \ldots, x_n), \ldots, T_p(x_1, \ldots, x_n))$$

where
$$T_i(x_1, \ldots, x_n) = \sum_{j=1}^{n} a_{ij}x_j$$

$1 \le i \le p$, and (a_{ij}) is the matrix of T. We know from Proposition 3.3 that each of the functions T_i is continuous. (See the example following Proposition 3.3.) Thus T is continuous by Proposition 3.4.

Exercises

1. Write out the proof of Proposition 3.3, giving the inductive argument.
2. Let $T: \mathbf{R}^n \to \mathbf{R}^p$ be a linear transformation. Show that there is a positive number K such that $\|T(\mathbf{v})\| \le K\|\mathbf{v}\|$ for all vectors $\mathbf{v} \in \mathbf{R}^n$. [*Hint*: Use the continuity of T at $\mathbf{0}$ to show that $\|T(\mathbf{v})\| \le \varepsilon\|\mathbf{v}\|$ for all vectors $\mathbf{v}$ with $\|\mathbf{v}\| = \delta/2$ for some $\delta > 0$.]

4. Composites and Products of Continuous Functions

We continue with our study of continuous functions, dealing here with composites and products.

Suppose $f\colon \mathbf{R}^n \to \mathbf{R}^p$ and $g: \mathbf{R}^p \to \mathbf{R}^q$ are functions. We define the *composite* of f with g to be the function $g \circ f\colon \mathbf{R}^n \to \mathbf{R}^q$ given by
$$(g \circ f)(\mathbf{v}) = g(f(\mathbf{v}))$$
For example, if $n = p = q = 1$ and $f(x) = x^2 + 1$, $g(y) = \sin y$, then
$$(g \circ f)(x) = g(f(x)) = g(x^2 + 1) = \sin(x^2 + 1)$$

Our first result here deals with the composite of continuous functions.

Proposition 4.1 *Suppose $f: \mathbf{R}^n \to \mathbf{R}^p$ is continuous at $\mathbf{v}_0 \in \mathbf{R}^n$ and $g: \mathbf{R}^p \to \mathbf{R}^q$ is continuous at $\mathbf{w}_0 = f(\mathbf{v}_0) \in \mathbf{R}^p$. Then $g \circ f\colon \mathbf{R}^n \to \mathbf{R}^q$ is continuous at $\mathbf{v}_0$.*

Proof. We need to show that, given $\varepsilon > 0$, there is a $\delta > 0$ such that
$$\|(g \circ f)(\mathbf{v}) - (g \circ f)(\mathbf{v}_0)\| = \|g(f(\mathbf{v})) - g(f(\mathbf{v}_0))\| < \varepsilon$$
when $\|\mathbf{v} - \mathbf{v}_0\| < \delta$. The idea is to use the continuity of g to assert that $\|g(f(\mathbf{v})) - g(f(\mathbf{v}_0))\|$ will be small when $\|f(\mathbf{v}) - f(\mathbf{v}_0)\|$ is small. Since f is continuous, we can make $\|f(\mathbf{v}) - f(\mathbf{v}_0)\|$ small be choosing $\mathbf{v}$ close enough to $\mathbf{v}_0$.

Explicitly, since g is continuous at $\mathbf{w}_0$, we can find $\delta_1 > 0$ such that
$$\|g(\mathbf{w}) - g(\mathbf{w}_0)\| < \varepsilon$$
whenever $\|\mathbf{w} - \mathbf{w}_0\| < \delta_1$. Now, using the continuity of f, we choose $\delta > 0$ so that

$$\| f(\mathbf{v}) - f(\mathbf{v}_0)\| < \delta_1$$

whenever $\|\mathbf{v} - \mathbf{v}_0\| < \delta$. We then have

$$\begin{aligned}\|\mathbf{v} - \mathbf{v}_0\| < \delta &\Rightarrow \| f(\mathbf{v}) - f(\mathbf{v}_0)\| < \delta_1 \\ &\Rightarrow \|g(f(\mathbf{v})) - g(f(\mathbf{v}_0))\| < \varepsilon\end{aligned}$$

so that $g\circ f$ is continuous at $\mathbf{v}_0$.

Suppose now that f: $\mathbf{R}^n \to \mathbf{R}^p$, g: $\mathbf{R}^p \to \mathbf{R}^q$, and h: $\mathbf{R}^q \to \mathbf{R}^m$ are functions. We can then define the composite

$$h\circ g\circ f\colon \mathbf{R}^n \to \mathbf{R}^m$$

by

$$(h\circ g\circ f)(\mathbf{v}) = h(g(f(\mathbf{v})))$$

It is immediate that

$$h\circ g\circ f = h\circ (g\circ f) = (h\circ g)\circ f$$

Now, if f is continuous at $\mathbf{v}_0$, g continuous at $\mathbf{w}_0 = f(\mathbf{v}_0)$, and h continuous at $\mathbf{u}_0 = g(\mathbf{w}_0) = (g\circ f)(\mathbf{v}_0)$, then $g\circ f$ is continuous at $\mathbf{v}_0$ by Proposition 4.1 and $h\circ(g\circ f) = h\circ g\circ f$ is continuous at $\mathbf{v}_0$, again by Proposition 4.1. Thus we see that the composite of three continuous functions is continuous. The statement and proof of the general result of this kind is left as Exercise 3.

Let $f, g : \mathbf{R}^n \to \mathbf{R}$ be two functions. We define their product $fg : \mathbf{R}^n \to \mathbf{R}$ in the obvious way:

$$(fg)(\mathbf{v}) = f(\mathbf{v})g(\mathbf{v})$$

Proposition 4.2 *Let $f, g : \mathbf{R}^n \to \mathbf{R}$ be functions, each of which is continuous at $\mathbf{v}_0 \in \mathbf{R}^n$. Then the product $fg : \mathbf{R}^n \to \mathbf{R}$ is also continuous at $\mathbf{v}_0$.*

Proof. This result follows easily from earlier results as follows. Let F : $\mathbf{R}^n \to \mathbf{R}^2$ and $h : \mathbf{R}^2 \to \mathbf{R}$ be defined by

$$F(\mathbf{v}) = (f(\mathbf{v}), g(\mathbf{v}))$$

$$h(x, y) = xy$$

Then F is continuous at $\mathbf{v}_0$ by Proposition 3.4 and h was proved continuous as Example 3 in Section 2. Thus $h\circ F$ is continuous at $\mathbf{v}_0$ by Proposition 4.1. However,

$$\begin{aligned}(h\circ F)(\mathbf{v}) &= h(F(\mathbf{v})) \\ &= h(f(\mathbf{v}), g(\mathbf{v})) \\ &= f(\mathbf{v})g(\mathbf{v}) \\ &= (fg)(\mathbf{v})\end{aligned}$$

so that fg is continuous at $\mathbf{v}_0$.

A straightforward induction argument establishes the following.

Corollary *Let* $f_1, \ldots, f_k : \mathbf{R}^n \to \mathbf{R}$ *be functions, each of which is continuous at* $\mathbf{v}_0 \in \mathbf{R}^n$. *Then the product* $f_1 f_2 \cdots f_k : \mathbf{R}^n \to \mathbf{R}$ *defined by*

$$f_1 f_2 \cdots f_k(\mathbf{v}) = f_1(\mathbf{v}) f_2(\mathbf{v}) \cdots f_k(\mathbf{v})$$

is continuous at $\mathbf{v}_0$.

Here are some examples.

1. A *polynomial function* $P : \mathbf{R} \to \mathbf{R}$ is a function of the form

$$P(x) = a_k x^k + a_{k-1} x^{k-1} + \cdots + a_1 x + a_0$$

where $a_k, a_{k-1}, \ldots, a_0 \in \mathbf{R}$. For instance,

$$P(x) = 6x^3 - 3x^2 + 2x - 1$$

is a polynomial function.

To see that a polynomial function is continuous, we note that the function $x \to x$ is continuous (see Example 3 of Section 2) so that, using the corollary to Proposition 4.2, the function

$$x \to x^j$$

is continuous for any integer $j > 0$. Therefore P is continuous by Proposition 3.3.

2. A function $f: \mathbf{R}^n \to \mathbf{R}$ is called a *monomial function* if

$$f(x_1, \ldots, x_n) = a x_{i_1} x_{i_2} \cdots x_{i_k}$$

where $a \in \mathbf{R}$ and $1 \leq i_1, i_2, \ldots, i_k \leq n$. For example,

$$f(x_1, x_2, x_3, x_4) = x_1 x_2 x_1 x_4 = x_1^2 x_2 x_4$$

is a monomial function on $\mathbf{R}^4$.

A monomial function is constructed by multiplying a real number by the product of functions of the form $g_j(x_1, \ldots, x_n) = x_j$, $1 \leq j \leq n$. Since the functions g_j are continuous (see Example 2 of Section 2), Proposition 3.2 and the corollary to Proposition 4.2 tell us the same is true for any monomial function.

3. A function $f: \mathbf{R}^n \to \mathbf{R}^p$, $f(\mathbf{v}) = (f_1(\mathbf{v}), \ldots, f_p(\mathbf{v}))$, is called a *polynomial function* if each of the functions $f_i : \mathbf{R}^n \to \mathbf{R}$ is the sum of monomial functions. For example, the function $f: \mathbf{R}^3 \to \mathbf{R}^2$ defined by

$$f(x_1, x_2, x_3) = (x_1^2 + 3x_1x_3,\ x_2x_3 - 7x_1x_3^2 + 4x_1x_2x_3)$$

is a polynomial function.

The continuity of polynomial functions follows from Propositions 3.1 and 3.4 and the continuity of monomial functions.

Exercises

1. (a) Prove that $f(x) = x^{1/2}$ is continuous for $x > 0$.
   (b) Prove that $g(x) = x^{1/3}$ is continuous for all x.
   (c) Prove that $h(x) = (3 + x^2)^{1/2}$ is continuous for all x.
   (d) Prove that $f(x, y) = (x + y)^{1/2}$ is continuous for $x, y > 0$.
   (e) Prove that $g(x, y) = (x^2 + 2xy + 3y^2 + 1)^{1/2}$ is continuous for all x, y.
2. Let $f: \mathbf{R}^n \to \mathbf{R}$ be a function. Prove that $g: \mathbf{R}^n \to \mathbf{R}$ defined by $g(\mathbf{v}) = [f(\mathbf{v})]^2$ is continuous when f is. Can g be continuous when f is not continuous?
3. State and prove a result concerning the continuity of the composite of k continuous functions.
4. Prove the corollary to Proposition 4.2.
5. Show that if $f: \mathbf{R}^n \to \mathbf{R}$ is continuous at $\mathbf{v}_0$ and $f(\mathbf{v}_0) \neq 0$, then $g(\mathbf{v}) = 1/f(\mathbf{v})$ is continuous at $\mathbf{v}_0$.
6. A function $f: \mathbf{R}^n \to \mathbf{R}$ is called *rational* if it is the quotient of two polynomial functions. Show that a rational function is continuous at any point where the denominator is nonzero.

5. Open and Closed Sets

Two important kinds of subsets of a Euclidean space are the open and closed sets. In this section, we give their definitions and some of their elementary properties.

Let $\mathbf{v}_0$ be a vector in $\mathbf{R}^n$ and r any positive real number. The *ball* (or *open ball*) *of radius r about* $\mathbf{v}_0$ is the set

$$B_r(\mathbf{v}_0) = \{\mathbf{v} \in \mathbf{R}^n : \|\mathbf{v} - \mathbf{v}_0\| < r\}$$

For example, if $n = 2$, then $B_r(\mathbf{v}_0)$ is a disk of radius r with center $\mathbf{v}_0$. (See Figure 5.1.) If $n = 3$, then $B_r(\mathbf{v}_0)$ is in fact a ball of radius r with center $\mathbf{v}_0$. (See Figure 5.2.)

A subset $U \subset \mathbf{R}^n$ is said to be *open* if for each $\mathbf{v}_0$ in U, there is an $r > 0$ so that $B_r(\mathbf{v}_0) \subset U$. That is, all points sufficiently close to $\mathbf{v}_0$ are also in U. This, in fact, is what makes open sets so useful to us. When we deal with a notion like the continuity of a function at a point $\mathbf{v}_0$, we need to examine the behavior of the function at points near $\mathbf{v}_0$, and we need, therefore, to know that the function is defined everywhere near $\mathbf{v}_0$. One way of insuring this is to deal with functions defined on open sets.

Examples

1. $\mathbf{R}^n$ itself is an open subset of $\mathbf{R}^n$.
2. The empty set $\varnothing$ is an open subset of $\mathbf{R}^n$. (Since there are no points $\mathbf{v}_0 \in \varnothing$, there is no point at which the criterion could fail to be met.)

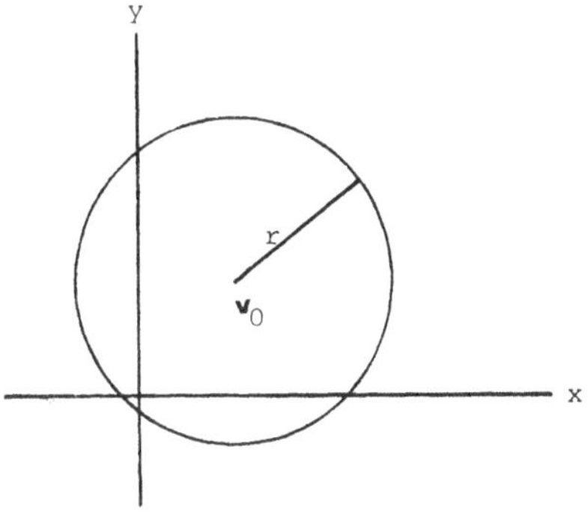

Figure 5.1 An open ball in $\mathbf{R}^2$.

3. The open interval (a, b) (see Section 1.4) is an open subset of $\mathbf{R} = \mathbf{R}^1$; if $x \in (a, b)$ and $r = \min\{x - a, b - x\}$, then $B_r(x) \subset (a, b)$.
4. Any open ball is an open set.

Proof of 4. Let $\mathbf{v}_1$ be a vector in $B_r(\mathbf{v}_0)$ and set $s = \|\mathbf{v}_1 - \mathbf{v}_0\|$. Then $r - s > 0$, and if $\mathbf{v} \in B_{r-s}(\mathbf{v}_1)$,

$$\|\mathbf{v} - \mathbf{v}_0\| \leq \|\mathbf{v} - \mathbf{v}_1\| + \|\mathbf{v}_1 - \mathbf{v}_0\| < (r - s) + s = r$$

(See Figure 5.3.) This proves that $B_{r-s}(\mathbf{v}_1) \subset B_r(\mathbf{v}_0)$, so $B_r(\mathbf{v}_0)$ is open.

5. The set $X = \{(x_1, x_2) : x_2 = 0\}$ is *not* an open subset of $\mathbf{R}^2$. To see this, we note that any ball about a point in X must contain points outside of X, since it contains points (x_1, x_2) with $x_2 \neq 0$.

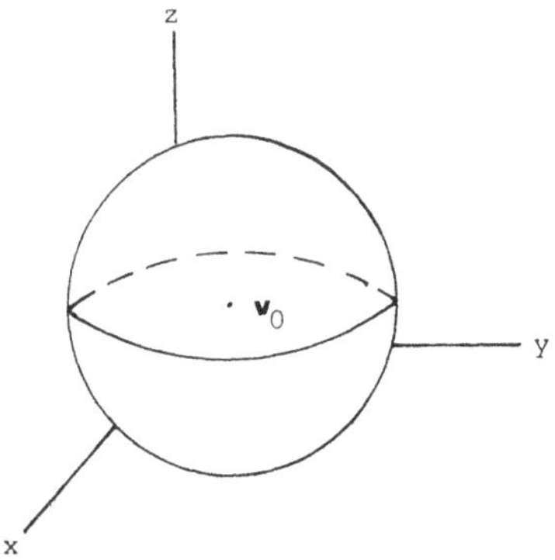

Figure 5.2 An open ball in $\mathbf{R}^3$.

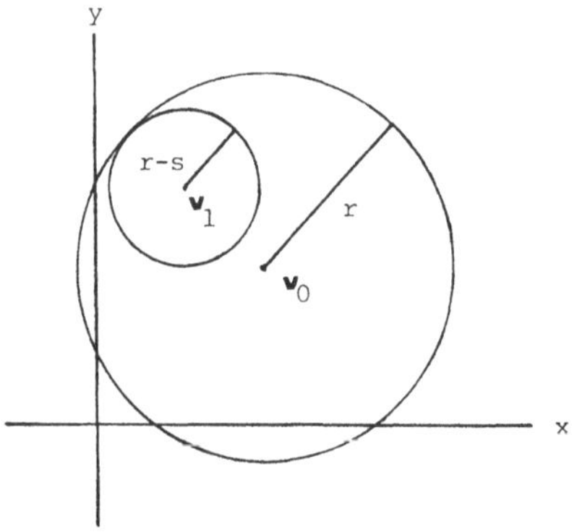

Figure 5.3 An open ball is an open set.

We now derive two important properties of open sets.

Proposition 5.1 *The union of an arbitrary collection of open subsets of* $\mathbf{R}^n$ *is open. The intersection of a finite collection of open subsets of* $\mathbf{R}^n$ *is open.*

Proof. Let $\{U_j\}$ be any collection of open subsets of $\mathbf{R}^n$ and let $\mathbf{v}_0$ be a point in $U = \cup U_j$. Then $\mathbf{v}_0$ is an element of some U_j; so since U_j is open, we can find an $r > 0$ with $B_r(\mathbf{v}_0) \subset U_j$. However, $U_j \subset U$, and so $B_r(\mathbf{v}_0) \subset U$. This proves that U is open.

Suppose $U_1, \ldots, U_k$ is a finite collection of open subsets of $\mathbf{R}^n$ and let $\mathbf{v}_0$ be a point of $U = U_1 \cap U_2 \cap \cdots \cap U_k$. Then $\mathbf{v}_0 \in U_j$ for all $j = 1, \ldots, k$, so we can find positive numbers $r_1, \ldots, r_k$ with

$$B_{r_j}(\mathbf{v}_0) \subset U_j \qquad 1 \le j \le k$$

Set $r = \min_j r_j$, so that $r \le r_j$ for all $j = 1, \ldots, k$. Then

$$B_r(\mathbf{v}_0) \subset B_{r_j}(\mathbf{v}_0) \subset U_j$$

for all $j = 1, \ldots, k$, which means that $B_r(\mathbf{v}_0) \subset U$.

A subset C of $\mathbf{R}^n$ is said to be *closed* if its complement $\mathbf{R}^n - C$ is open.

Examples

6. $\mathbf{R}^n$ itself is a closed subset of $\mathbf{R}^n$.
7. The empty set is a closed subset of $\mathbf{R}^n$.
8. Any finite subset of points in $\mathbf{R}^n$ is closed.

Proof. Let $\mathbf{C} = \{\mathbf{v}_1, \ldots, \mathbf{v}_k\}$ and $\mathbf{v}$ a point of $\mathbf{R}^n - \mathbf{C}$. Set $r = \min_j \|\mathbf{v} - \mathbf{v}_j\|$. Then $B_r(\mathbf{v}) \subset \mathbf{R}^n - C$; so $\mathbf{R}^n - C$ is open and C is closed.

9. A *closed ball of radius r about* $\mathbf{v}_0$ in $\mathbf{R}^n$ is defined to be the set

$$\bar{B}_r(\mathbf{v}_0) = \{\mathbf{v} \in \mathbf{R}^n : \|\mathbf{v} - \mathbf{v}_0\| \leq r\}$$

(Note the $\leq$ sign!) It is not difficult to show that $\bar{B}_r(\mathbf{v}_0)$ is a closed set. (See Exercise 4.)

The following result is an analogue of Proposition 5.2 for closed sets. It follows from Proposition 5.1 using the De Morgan laws. (See Section 1.1.) We leave the proof as Exercise 6.

Proposition 5.2 *The intersection of an arbitrary collection of closed subsets of* $\mathbf{R}^n$ *is closed. The union of a finite collection of closed subsets of* $\mathbf{R}^n$ *is closed.*

We note that a set can be neither open nor closed. For example, the set

$$(0, 1] = \{x \in \mathbf{R}^1 : 0 < x \leq 1\}$$

is neither open nor closed. (See Exercise 8.)

Exercises

1. Which of the following subsets of $\mathbf{R}^2$ are open?
   (a) $\{(x, y) : x > y\}$
   (b) $\{(x, y) : x^2 > -7\}$
   (c) $\{(x, y) : xy \neq 0\}$
   (d) $\{(x, y) : 2x + 3y - 5 < 1\}$
   (e) $\{(x, y) : y = 3\}$
   (f) $\{(x, y) : \|(x, y)\| < 3\}$
   (g) $\{(x, y) : |x| > 2\}$
   (h) $\{(x, y) : y \geq |x|\}$
   (i) $\{(x, y) : x^2 + y^2 = 9\}$
   (j) $\{(x, y) : y - x \leq 6\}$
2. Which of the following subsets of $\mathbf{R}^2$ are closed?
   (a) $\{(x, y) : 2x + 3y - 7 \geq 0\}$
   (b) $\{(x, y) : xy = 1\}$
   (c) $\{(x, y) : 0 < x^2 + y^2 \leq 1\}$
   (d) $\{(x, y) : |x - y| \neq 1\}$
   (e) $\{(x, y) : x^2 = y^2\}$
   (f) $\{(x, y) : x \geq y\}$
   (g) $\{(x, y) : 2x - y < 4\}$
   (h) $\{(x, y) : x \geq 3 \text{ or } y < 2\}$
   (i) $\{(x, y) : \|(x, y)\| < 7\}$
   (j) $\{(x, y) : y \geq |x|\}$
3. Determine which of the following subsets of $\mathbf{R}^3$ are open, which are closed, and which are neither.
   (a) $\{(x, y, z) : x + y + z = 2\}$
   (b) $\{(x, y, z) : x^2 + y^2 + z^2 < 1\}$
   (c) $\{(x, y, z) : x > 0, y > 0, z > 0\}$
   (d) $\{(x, y, z) : x \geq 0, y \geq 0, z \geq 0\}$
   (e) $\{(x, y, z) : x \geq 0, y > 0, z > 0\}$
   (f) $\{(x, y, z) : |x| + |y| + |z| \neq 0\}$
   (g) $\{(x, y, z) : x \leq y \leq z\}$
   (h) $\{(x, y, z) : x \leq y < z\}$
4. Prove that a closed interval

$$[a, b] = \{x \in \mathbf{R} : a \leq x \leq b\}$$

   is a closed set.
5. Prove that a closed ball is a closed set.
6. Prove Proposition 5.2.
7. Let U be an open subset of $\mathbf{R}^n$ and $\mathbf{v}_1, \ldots, \mathbf{v}_k \in U$. Prove that $V = U - \{\mathbf{v}_1, \ldots, \mathbf{v}_k\}$ is open.

8. Prove that the set $\{x \in \mathbf{R}^1 : 0 < x \leq 1\}$ is neither open nor closed.
9. (a) Give an example to show that the intersection of an infinite collection of open sets need not be open.
   (b) Give an example to show that the union of an infinite collection of closed sets need not be closed.
10. Give examples to show that the intersection of a closed set and an open set can be open, closed, or neither.
11. Give examples to show that the union of an open set and a closed set can be open, closed, or neither.

*12. Let S be any subset of $\mathbf{R}^n$. Prove that there is a largest open set, $\operatorname{Int}(S)$, with $\operatorname{Int}(S) \subset S$. [That is, if U is any open set with $U \subset S$, then $U \subset \operatorname{Int}(S)$.] $\operatorname{Int}(S)$ is called the *interior* of S.

*13. Determine the interiors of each of the sets of Exercises 1 and 2.

*14. Let S be any subset of $\mathbf{R}^n$. Prove that there is a smallest closed set, $\operatorname{Cl}(S)$, with $S \subset \operatorname{Cl}(S)$. $\operatorname{Cl}(S)$ is called the *closure* of S.

*15. Determine the closures of each of the sets in Exercises 1 and 2.

*16. If S is any subset of $\mathbf{R}^n$, prove that $\operatorname{Cl}(\mathbf{R}^n - S) = \mathbf{R}^n - \operatorname{Int}(S)$.

*17. Prove that a set S is open if and only if $\operatorname{Int}(S) = S$.

*18. Prove that a set S is closed if and only if $\operatorname{Cl}(S) = S$.

*19. Let S be any subset of $\mathbf{R}^n$. Define the *boundary* of S, $\partial(S)$, by $\partial(S) = \operatorname{Cl}(S) \cap \operatorname{Cl}(\mathbf{R}^n - S)$. Prove that $\mathbf{v} \in \partial(S)$ if and only if, for any $r > 0$, $B_r(\mathbf{v}) \cap S$ and $B_r(\mathbf{v}) \cap (\mathbf{R}^n - S)$ are both nonempty.

*20. Describe the boundaries of the sets in Exercises 1 and 2.

*21. Let $f : \mathbf{R}^n \to \mathbf{R}^m$ be a function. The *graph* of f, $G(f)$, is the subset of $\mathbf{R}^{n+m}$ defined by

$$G(f) = \{(\mathbf{v}, \mathbf{w}) : \mathbf{w} = f(\mathbf{v})\}$$

(We identify $\mathbf{R}^{n+m}$ and $\mathbf{R}^n \times \mathbf{R}^m$ in the obvious way.) Prove that if f is continuous, then $G(f)$ is a closed subset of $\mathbf{R}^{n+m}$. Is the converse true? (Give a proof or a counterexample.)

6. Convergent Sequences

In this section, we introduce the notion of a convergent sequence in a Euclidean space. Our definition is the obvious extension of the corresponding notion for real numbers.

Recall that a sequence $a_1, a_2, \ldots, a_n, \ldots$ of real numbers is said to *converge* to the number a if, for any $\varepsilon > 0$, there is an integer N such that $|a_n - a| < \varepsilon$ whenever $n > N$. That is, no matter how small an ε we choose, all of the terms of the sequence from some point on are within ε of a.

For example, the sequence $1, \frac{1}{2}, \frac{1}{3}, \ldots, 1/n, \ldots$ converges to 0. For if ε is any positive number, we pick an integer $N > 1/\varepsilon$. Then, wherever $n > N$,

$$\left|\frac{1}{n} - 0\right| = \frac{1}{n} < \frac{1}{N} < \varepsilon$$

which proves that the sequence converges to 0.

On the other hand, the sequence $1, -1, 1, -1, \ldots$ does not converge at all. Intuitively, this is clear; the terms of the sequence are not getting close to any single number. To prove it, we must show that, no matter what a we choose, we can find an $\varepsilon > 0$ such that, for all N, there is a term a_n of this sequence with $n > N$ and $|a_n - a| \geq \varepsilon$.

If $a \neq \pm 1$, this is clear; we simply let ε be the smaller of the two numbers $|a - 1|$, $|a + 1|$. Then $|a_n - a| \geq \varepsilon$ for all n.

If $a = 1$, we see that, no matter how large N is, we can find $n > N$ such that $a_n = -1$ so that $|a_n - a| = 2$. Thus, we can set $\varepsilon = 2$ in this case.

The same argument works if we set $a = -1$.

Since $\|x\| = |x|$ for $x \in \mathbf{R}$, we extend this idea to an arbitrary Euclidean space as follows: let $\mathbf{v}_1, \mathbf{v}_2, \ldots, \mathbf{v}_n, \ldots$ be a sequence of vectors in $\mathbf{R}^n$. We say that this sequence *converges to a vector* $\mathbf{v} \in \mathbf{R}^n$ if for any $\varepsilon > 0$, there is an integer N such that $\|\mathbf{v}_n - \mathbf{v}\| < \varepsilon$ whenever $n > N$.

We often write a sequence as $\{\mathbf{v}_n\}_{n=1}^{\infty}$ or simply as $\{\mathbf{v}_n\}$. If a sequence $\{\mathbf{v}_n\}$ converges to $\mathbf{v}$, we call $\mathbf{v}$ a *limit point* of the sequence and write

$$\lim_{n \to \infty} \mathbf{v}_n = \mathbf{v}$$

We now give generalizations of some of the standard results about sequences of real numbers.

Proposition 6.1 *A convergent sequence in* $\mathbf{R}^n$ *cannot have two distinct limit points.*

Proof. Suppose on the contrary that $\{\mathbf{v}_n\}$ is a sequence in $\mathbf{R}^n$ with

$$\lim_{n \to \infty} \mathbf{v}_n = \mathbf{v} \qquad \lim_{n \to \infty} \mathbf{v}_n = \mathbf{v}'$$

with $\mathbf{v} \neq \mathbf{v}'$. Then $r = \|\mathbf{v} - \mathbf{v}'\| > 0$ and we set $\varepsilon = r/2$. By assumption, we can find integers N and N' such that

$$\|\mathbf{v}_n - \mathbf{v}\| < \frac{r}{2} \qquad \text{when } n > N$$

$$\|\mathbf{v}_m - \mathbf{v}'\| < \frac{r}{2} \qquad \text{when } m > N'$$

Pick some integer p larger than both N and N'. Then

$$\begin{aligned} r = \|\mathbf{v} - \mathbf{v}'\| &= \|\mathbf{v} - \mathbf{v}_p - (\mathbf{v}' - \mathbf{v}_p)\| \\ &\leq \|\mathbf{v} - \mathbf{v}_p\| + \|\mathbf{v}' - \mathbf{v}_p\| \\ &< \frac{r}{2} + \frac{r}{2} = r \end{aligned}$$

This contradiction means that r cannot be positive; so $\mathbf{v} = \mathbf{v}'$.

If $\{\mathbf{v}_n\}$ and $\{\mathbf{w}_n\}$ are sequences, we can define the *sum* of the sequences to be the sequence $\{\mathbf{v}_n + \mathbf{w}_n\}$.

Proposition 6.2 *Suppose $\{\mathbf{v}_n\}$ and $\{\mathbf{w}_n\}$ are convergent sequences in $\mathbf{R}^n$. Then the sequence $\{\mathbf{v}_n + \mathbf{w}_n\}$ is convergent and*

$$\lim_{n\to\infty} (\mathbf{v}_n + \mathbf{w}_n) = \lim_{n\to\infty} \mathbf{v}_n + \lim_{n\to\infty} \mathbf{w}_n$$

Similarly, if $\{\mathbf{v}_n\}$ is a sequence in $\mathbf{R}^n$ and b is a real number, we define a sequence $\{b\mathbf{v}_n\}$ whose nth term is $b\mathbf{v}_n$.

Proposition 6.3 *Let $\{\mathbf{v}_n\}$ be a convergent sequence and b any real number. Then the sequence $\{b\mathbf{v}_n\}$ is convergent and*

$$\lim_{n\to\infty} b\mathbf{v}_n = b \lim_{n\to\infty} \mathbf{v}_n$$

The proofs of these two propositions are similar to the proofs of Propositions 3.1 and 3.2. We give only the proof of Proposition 6.2.

Let ε be any positive real number and let $\lim_{n\to\infty} \mathbf{v}_n = \mathbf{v}$, $\lim_{n\to\infty} \mathbf{w}_n = \mathbf{w}$. We need to show that there is an N such that

$$\|(\mathbf{v}_n + \mathbf{w}_n) - (\mathbf{v} + \mathbf{w})\| < \varepsilon \qquad \text{whenever } n > N$$

Since $\{\mathbf{v}_n\}$ and $\{\mathbf{w}_n\}$ are convergent, we can find N_1 and N_2 so that

$$\|\mathbf{v}_n - \mathbf{v}\| < \frac{\varepsilon}{2}$$

when $n > N_1$ and

$$\|\mathbf{w}_n - \mathbf{w}\| < \frac{\varepsilon}{2}$$

when $n > N_2$. If we set $N = \max\{N_1, N_2\}$, we have

$$\begin{aligned} \|(\mathbf{v}_n + \mathbf{w}_n) - (\mathbf{v} + \mathbf{w})\| &= \|(\mathbf{v}_n - \mathbf{v}) + (\mathbf{w}_n - \mathbf{w})\| \\ &\le \|\mathbf{v}_n - \mathbf{v}\| + \|\mathbf{w}_n - \mathbf{w}\| \\ &< \frac{\varepsilon}{2} + \frac{\varepsilon}{2} = \varepsilon \end{aligned}$$

when $n > N$. Thus

$$\lim_{n\to\infty} (\mathbf{v}_n + \mathbf{w}_n) = \mathbf{v} + \mathbf{w}$$

Let $\{\mathbf{v}_n\}$ be a sequence in $\mathbf{R}^n$ and $n_1, n_2, \ldots$ positive integers with

$$n_1 < n_2 < \cdots < n_m < \cdots$$

The sequence $\{\mathbf{w}_m\}$ defined by

$$\mathbf{w}_m = \mathbf{v}_{n_m}$$

is called a *subsequence* of the sequence $\{\mathbf{v}_n\}$. This terminology is reasonable since $\{\mathbf{w}_m\}$ is obtained by omitting some of the terms of the sequence $\{\mathbf{v}_n\}$. For example, the sequence

$$\frac{1}{2}, \frac{1}{4}, \frac{1}{6}, \ldots, \frac{1}{2m}, \ldots$$

is the subsequence of the sequence

$$1, \frac{1}{2}, \frac{1}{3}, \frac{1}{4}, \ldots, \frac{1}{n}, \ldots$$

obtained by omitting the "odd" terms. More explicitly, the sequence $b_m = 1/2m$ is a subsequence of the sequence $a_n = 1/n$ corresponding to the positive integers $n_m = 2m$:

$$2 < 4 < 6 < \cdots$$

Proposition 6.4 *Any subsequence of a convergent sequence is convergent and has the same limit point.*

Proof. Let $\{\mathbf{w}_m\}$ be a subsequence of the sequence $\{\mathbf{v}_n\}$ corresponding to the positive integers

$$n_1 < n_2 < \cdots < n_m < \cdots$$

Suppose $\lim_{n \to \infty} \mathbf{v}_n = \mathbf{v}$, let ε be any positive number, and choose N so that

$$\|\mathbf{v}_n - \mathbf{v}\| < \varepsilon$$

whenever $n > N$. Since $\mathbf{w}_m = \mathbf{v}_{n_m}$ and $n_m \geq n$, we see that

$$\|\mathbf{w}_m - \mathbf{v}\| < \varepsilon$$

whenever $m > N$. Thus $\lim_{m \to \infty} \mathbf{w}_m = \mathbf{v}$.

We say a subset S of $\mathbf{R}^n$ is *bounded* if there is a real number M such that $\|\mathbf{v}\| \leq M$ for all $\mathbf{v} \in S$. For example, the set $\{1, \frac{1}{2}, \frac{1}{3}, \ldots\}$ is bounded whereas the set $\{1, 2, 3, \ldots\}$ is not bounded.

Proposition 6.5 *Let $\{\mathbf{v}_n\}$ be a convergent sequence in $\mathbf{R}^n$. Then the set $\{\mathbf{v}_1, \mathbf{v}_2, \ldots, \mathbf{v}_n, \ldots\}$ is bounded.*

Proof. The idea of the proof is to use the fact that the sequence $\{\mathbf{v}_n\}$ is convergent to take care of all but a finite number of terms of the sequence. Since any finite set is clearly bounded, this will be sufficient.

Suppose $\lim_{n\to\infty} \mathbf{v}_n = \mathbf{v}_0$, and pick N so that $\|\mathbf{v}_n - \mathbf{v}_0\| < 1$ when $n > N$. It follows that

$$\|\mathbf{v}_n\| = \|\mathbf{v}_n - \mathbf{v}_0 + \mathbf{v}_0\| \leq \|\mathbf{v}_n - \mathbf{v}_0\| + \|\mathbf{v}_0\| \leq 1 + \|\mathbf{v}_0\|$$

for $n > N$. Let M be defined by

$$M = \max\{\|\mathbf{v}_1\|, \|\mathbf{v}_2\|, \ldots, \|\mathbf{v}_N\|, 1 + \|\mathbf{v}_0\|\}$$

It follows that $\|\mathbf{v}_n\| \leq M$ for all n.

The following proposition is quite useful. Its proof is straightforward and is left to the reader.

Proposition 6.6 *Let $\{\mathbf{v}_n\}$ be a sequence in $\mathbf{R}^n$. Then $\{\mathbf{v}_n\}$ converges to $\mathbf{v}_0$ if and only if the sequence $\{\|\mathbf{v}_n - \mathbf{v}_0\|\}$ converges to 0 in $\mathbf{R}$.*

Exercises

1. Which of the following sequences converge? Prove your assertions.

   (a) $1, \frac{1}{2}, -1, \frac{1}{3}, 1, \frac{1}{4}, -1, \frac{1}{5}, \ldots$

   (b) $2, \frac{1}{2}, 1\frac{1}{3}, \frac{3}{4}, 1\frac{1}{5}, \frac{5}{6}, 1\frac{1}{7}, \ldots$

   (c) $\frac{1}{3}, \frac{2}{5}, \frac{3}{7}, \frac{4}{9}, \frac{5}{11}, \ldots$

   (d) $1, -2, 3, -4, 5, -6, \ldots$

   (e) $1 + \frac{1}{1}, 1 + \frac{1}{4}, 1 + \frac{1}{9}, 1 + \frac{1}{16}, \ldots$

   (f) $1\frac{1}{2}, 2\frac{1}{4}, 1\frac{1}{8}, 2\frac{1}{16}, 1\frac{1}{32}, \ldots$

   (g) $1, 2^2, 3^3, 4^4, \ldots$

   (h) $\frac{1^{100}}{3^1}, \frac{2^{100}}{3^2}, \frac{3^{100}}{3^3}, \frac{4^{100}}{3^4}, \frac{5^{100}}{3^5}, \ldots$

2. Prove Proposition 6.3.
3. Prove Proposition 6.6.
4. Let a_n be a convergent sequence in $\mathbf{R}$ and $\mathbf{v}$ a fixed vector in $\mathbf{R}^n$. Prove that the sequence $\{a_n\mathbf{v}\}$ converges and that

$$\lim_{n\to\infty} (a_n\mathbf{v}) = (\lim_{n\to\infty} a_n)\mathbf{v}$$

5. Define a sequence of real numbers by $a_n = r^n$. Prove that this sequence converges to 0 if $|r| < 1$, converges to 1 if $r = 1$, and diverges otherwise.

6. Let $\{\mathbf{v}_n\}$ and $\{\mathbf{w}_n\}$ be sequences in $\mathbf{R}^n$ with

$$\lim_{n\to\infty} \mathbf{v}_n = \mathbf{v}_0 \qquad \lim_{n\to\infty} \mathbf{w}_n = \mathbf{w}_0$$

Prove that the sequence

$$\mathbf{v}_1, \mathbf{w}_1, \mathbf{v}_2, \mathbf{w}_2, \mathbf{v}_3, \mathbf{w}_3, \ldots$$

diverges if $\mathbf{v}_0 \neq \mathbf{w}_0$ and converges if $\mathbf{v}_0 = \mathbf{w}_0$.

7. Let $\mathbf{v}_n = (x_n, y_n, z_n)$ be a sequence in $\mathbf{R}^3$. Prove that

$$\lim_{n\to\infty} \mathbf{v}_n = \mathbf{v} = (x, y, z)$$

if and only if

$$\lim_{n\to\infty} x_n = x \qquad \lim_{n\to\infty} y_n = y \qquad \lim_{n\to\infty} z_n = z$$

8. A sequence $\{\mathbf{v}_n\}$ in $\mathbf{R}^n$ is called a *Cauchy sequence* if for any $\varepsilon > 0$, there is an integer N such that $\|\mathbf{v}_n - \mathbf{v}_m\| < \varepsilon$ whenever $n, m > N$. Prove that any convergent sequence is a Cauchy sequence.

9. If $\{\mathbf{v}_n\}$ is a Cauchy sequence, prove that the set

$$\{\mathbf{v}_1, \mathbf{v}_2, \ldots\}$$

is bounded.

*10. State and prove the analogue of Exercise 7 for sequences in $\mathbf{R}^n$.

11. Let $\{x_n\}$ be a convergent sequence of real numbers converging to a number x. Show that if $a \geq x_n$ for all n, then $a \geq x$; and that if $b \leq x_n$ for all n, then $b \leq x$. Is it always true that if $a > x_n$ for all n, then $a > x$?

*12. Suppose that $\{\mathbf{v}_n\}$ is a sequence of vectors in $\mathbf{R}^n$ such that no subsequence of $\{\mathbf{v}_n\}$ converges. Show that
   (a) No vector is repeated infinitely many times in the sequence.
   (b) $\{\mathbf{v}_n\}$ (as a subset of $\mathbf{R}^n$) is closed.

7. Further Properties of Continuous Functions

We begin with a useful criterion for continuity in terms of convergent sequences.

Proposition 7.1 *A function $f: \mathbf{R}^n \to \mathbf{R}^p$ is continuous at $\mathbf{v}_0 \subset \mathbf{R}^n$ if and only if whenever $\{\mathbf{v}_n\}$ is a convergent sequence in $\mathbf{R}^n$ with limit point $\mathbf{v}_0$, then $\{f(\mathbf{v}_n)\}$ is a convergent sequence in $\mathbf{R}^p$ with limit point $f(\mathbf{v}_0)$.*

Proof. Suppose first of all that f is continuous at $\mathbf{v}_0$ and let $\{\mathbf{v}_n\}$ be a convergent sequence with limit point $\mathbf{v}_0$. We must show that

$$\lim_{n\to\infty} f(\mathbf{v}_n) = f(\mathbf{v}_0)$$

Equivalently, we must show that, for any $\varepsilon > 0$, there is an integer N such that

$$\| f(\mathbf{v}_n) - f(\mathbf{v}_0)\| < \varepsilon$$

whenever $n > N$.

Suppose then that we are given $\varepsilon > 0$. Since f is continuous at $\mathbf{v}_0$, we can find δ so that

$$\| f(\mathbf{v}) - f(\mathbf{v}_0)\| < \varepsilon$$

when $\|\mathbf{v} - \mathbf{v}_0\| < \delta$. Furthermore, $\lim_{n\to\infty} \mathbf{v}_n = \mathbf{v}_0$; so we can find an integer N such that

$$\|\mathbf{v}_n - \mathbf{v}_0\| < \delta$$

whenever $n > N$. Thus

$$\begin{aligned} n > N &\Rightarrow \|\mathbf{v}_n - \mathbf{v}_0\| < \delta \\ &\Rightarrow \| f(\mathbf{v}_n) - f(\mathbf{v}_0)\| < \varepsilon \end{aligned}$$

so $f(\mathbf{v}_n)$ converges to $f(\mathbf{v}_0)$.

We now need to prove the converse, that if $f\colon \mathbf{R}^n \to \mathbf{R}^p$ takes sequences converging to $\mathbf{v}_0$ into sequences converging to $f(\mathbf{v}_0)$, then f is continuous at $\mathbf{v}_0$. We begin by transforming this statement into a logically equivalent statement. (Use Section 1.2 as a reference for these logical manipulations.) First of all, we may instead prove the contrapositive of the converse; that is, we shall prove that if f is not continuous at $\mathbf{v}_0$, then there is a sequence $\{\mathbf{v}_n\}$ in $\mathbf{R}^n$ converging to $\mathbf{v}_0$ such that $\{f(\mathbf{v}_n)\}$ does *not* converge to $f(\mathbf{v}_0)$ in $\mathbf{R}^p$.

Next, we need to interpret the statement that f is not continuous at $\mathbf{v}_0$. We know (from the definition) that it means that the statement $(\forall\varepsilon > 0)(\exists\delta > 0)(\forall\mathbf{v} \in \mathbf{R}^n)$ if $\|\mathbf{v} - \mathbf{v}_0\| < \delta$, then $\| f(\mathbf{v}) - f(\mathbf{v}_0)\| < \varepsilon$ is *false*. We gave a rule in Section 1.2 for negating such statements: change $\forall$ to $\exists$ and $\exists$ to $\forall$, and deny the final statement. Thus "f is not continuous at $\mathbf{v}_0$" means "$(\exists\varepsilon > 0)(\forall\delta > 0)(\exists\mathbf{v} \in \mathbf{R}^n)$: it is false that if $\|\mathbf{v} - \mathbf{v}_0\| < \delta$, then $\| f(\mathbf{v}) - f(\mathbf{v}_0)\| < \varepsilon$," or

"There is an $\varepsilon > 0$ such that for any $\delta > 0$, we can pick $\mathbf{v} \in \mathbf{R}^n$ such that $\|\mathbf{v} - \mathbf{v}_0\| < \delta$, but $\| f(\mathbf{v}) - f(\mathbf{v}_0)\| \geq \varepsilon$."

In particular, for any integer n, we can find a vector $\mathbf{v}_n$ in $\mathbf{R}^n$ with

$$\|\mathbf{v}_0 - \mathbf{v}_n\| < \frac{1}{n} \qquad \text{and} \qquad \| f(\mathbf{v}_0) - f(\mathbf{v}_n)\| > \varepsilon$$

It follows from Proposition 6.6 that the sequence $\{\mathbf{v}_n\}$ converges to $\mathbf{v}_0$. However, $\{f(\mathbf{v}_n)\}$ cannot converge to $f(\mathbf{v}_0)$ since $\| f(\mathbf{v}_n) - f(\mathbf{v}_0)\| > \varepsilon > 0$ for all n. This concludes the proof of Proposition 7.1.

Armed with this result, we go back to deriving properties of continuous functions.

We say a function $f\colon \mathbf{R}^n \to \mathbf{R}^m$ is *locally bounded* at $\mathbf{v}_0 \in \mathbf{R}^n$ if there are positive numbers M and δ such that $\| f(\mathbf{v})\| \leq M$ whenever $\|\mathbf{v} - \mathbf{v}_0\| < \delta$. For instance, if $m = 1$, this says that values of $f(\mathbf{v})$ are no greater than M in absolute value when $\mathbf{v}$ is near $\mathbf{v}_0$ which is what locally bounded ought to mean.

Proposition 7.2 *Let $f\colon \mathbf{R}^n \to \mathbf{R}^p$ be continuous at $\mathbf{v}_0 \in \mathbf{R}^n$. Then f is locally bounded at $\mathbf{v}_0$.*

Proof. Suppose f is not locally bounded at $\mathbf{v}_0$. Then, for every $\varepsilon > 0$ and $M > 0$, there is a $\mathbf{v} \in \mathbf{R}^n$ with $\|\mathbf{v} - \mathbf{v}_0\| < \varepsilon$ and $\| f(\mathbf{v})\| > M$. In particular, for every positive integer n, we can find $\mathbf{v}_n \in \mathbf{R}^n$ with $\|\mathbf{v}_n - \mathbf{v}_0\| < 1/n$ and $\| f(\mathbf{v}_n)\| > n$. It follows from Proposition 6.6 that the sequence $\{\mathbf{v}_n\}$ converges to $\mathbf{v}_0$. Furthermore, since the sequence $\{ f(\mathbf{v}_n)\}$ is not bounded, it follows from Proposition 6.5 that it cannot be convergent. Thus f cannot be continuous at $\mathbf{v}_0$ by Proposition 7.1.

We close this section with two additional characterizations of continuity involving open and closed sets.

Proposition 7.3 *A function $f\colon \mathbf{R}^n \to \mathbf{R}^p$ is continuous if and only if $f^{-1}(U)$ is an open subset of $\mathbf{R}^n$ whenever U is an open subset of $\mathbf{R}^p$.*

Proof. Recall that $f^{-1}(U)$ is defined by

$$f^{-1}(U) = \{\mathbf{v} \in \mathbf{R}^n : f(\mathbf{v}) \in U\}$$

Suppose, first of all, that f is continuous and U is an open subset of $\mathbf{R}^p$. Let $\mathbf{v}_0$ be a vector in $f^{-1}(U)$ and $\mathbf{w}_0 = f(\mathbf{v}_0) \in U$. Since U is open, we can find an $\varepsilon > 0$ so that $B_\varepsilon(\mathbf{w}_0) \subset U$. Since f is continuous at $\mathbf{v}_0$, we can find a $\delta > 0$ so that

$$\| f(\mathbf{v}) - f(\mathbf{v}_0)\| < \varepsilon \qquad \text{whenever } \|\mathbf{v} - \mathbf{v}_0\| < \delta$$

or, equivalently, $\quad f(\mathbf{v}) \in B_\varepsilon(\mathbf{w}_0) \qquad$ whenever $\mathbf{v} \in B_\delta(\mathbf{v}_0)$

Thus

$$B_\delta(\mathbf{v}_0) \subset f^{-1}(B_\varepsilon(\mathbf{w}_0)) \subset f^{-1}(U)$$

and $f^{-1}(U)$ is open.

Conversely, suppose $f^{-1}(U)$ is an open subset of $\mathbf{R}^n$ whenever U is an open subset of $\mathbf{R}^p$. Let $\mathbf{v}_0$ be a vector in $\mathbf{R}^n$ with $\mathbf{w}_0 = f(\mathbf{v}_0)$. Then for any $\varepsilon > 0$, $B_\varepsilon(\mathbf{w}_0)$ is open in $\mathbf{R}^p$, so $f^{-1}(B_\varepsilon(\mathbf{w}_0))$ is open in $\mathbf{R}^n$. Since $\mathbf{v}_0 \in f^{-1}(B_\varepsilon(\mathbf{w}_0))$, we can find a $\delta > 0$ so that

$$B_\delta(\mathbf{v}_0) \subset f^{-1}(B_\varepsilon(\mathbf{w}_0))$$

This is the same as saying that $\|\mathbf{v} - \mathbf{v}_0\| < \delta$ implies that $\| f(\mathbf{v}) - f(\mathbf{v}_0)\| < \varepsilon$;

so f is continuous at $\mathbf{v}_0$. Since $\mathbf{v}_0$ was arbitrarily chosen in $\mathbf{R}^n$, f is continuous on $\mathbf{R}^n$ and the proposition is proved.

Proposition 7.4 *A function $f : \mathbf{R}^n \to \mathbf{R}^p$ is continuous if and only if $f^{-1}(C)$ is a closed subset of $\mathbf{R}^n$ whenever C is a closed subset of $\mathbf{R}^p$.*

This follows easily from Proposition 7.3 and the fact that $f^{-1}(\mathbf{R}^p - U) = \mathbf{R}^n - f^{-1}(U)$. We leave the details to the reader.

Exercises

1. Use Proposition 7.1 to prove that the composite of continuous functions is continuous.
2. Give an example to show that a function locally bounded at $\mathbf{v}_0$ need not be continuous at $\mathbf{v}_0$.
3. If $f : \mathbf{R}^n \to \mathbf{R}^p$ is continuous and U is open in $\mathbf{R}^n$, is $f(U)$ necessarily open in $\mathbf{R}^p$?
4. If $f : \mathbf{R}^n \to \mathbf{R}^p$ is continuous and C is closed in $\mathbf{R}^n$, is $f(C)$ necessarily closed in $\mathbf{R}^p$?
5. Prove Proposition 7.4.
6. Prove that $f : \mathbf{R}^n \to \mathbf{R}^p$ is continuous at $\mathbf{v}_0 \in \mathbf{R}^n$ if and only if for any open set U_1 of $\mathbf{R}^p$ containing $f(\mathbf{v}_0)$, there is an open set U_0 in $\mathbf{R}^n$ containing $\mathbf{v}_0$ such that $f(U_0) \subset U_1$.
7. Prove the following analogue of Proposition 7.3: Let V be an open subset of $\mathbf{R}^n$ and $f : V \to \mathbf{R}^p$ a function. Prove that f is continuous on V if and only if $f^{-1}(U)$ is an open subset of $\mathbf{R}^n$ whenever U is an open subset of $\mathbf{R}^p$.
8. Prove the following analogue of Proposition 7.4: Let X be a closed subset of $\mathbf{R}^n$ and $f : X \to \mathbf{R}^p$ a function. Prove that f is continuous on X if and only if $f^{-1}(C)$ is a closed subset of $\mathbf{R}^n$ whenever C is a closed subset of $\mathbf{R}^p$.

4

The Derivative

In elementary calculus, the main mathematical tool for studying the behavior of functions was the derivative. This state of affairs persists when we examine functions from one Euclidean space to another. When we work in this more general setting, however, we need to change our notion of the derivative. If $f:\mathbf{R}\to\mathbf{R}$ is a function, the derivative of f at a point x_0 is a number $f'(x_0)$ giving the slope of the graph of f at x_0. However, we can also view $f'(x_0)$ as defining a linear transformation $L:\mathbf{R}\to\mathbf{R}$, given by

$$L(x) = f'(x_0)x$$

This point of view may seem artificial when dealing with functions of one variable, but it turns out to be very useful when one does multivariable calculus. For one thing, it simplifies some formulas for computing derivatives. For another, it emphasizes the fundamental idea underlying the calculus: to study the behavior of nonlinear functions, we approximate them by linear functions.

In this chapter, we define the derivative of a function between euclidean spaces, develop some of its properties, and show how to compute it.

1. The Derivative

In this section, we give the definition and some examples of the derivative of a function between Euclidean spaces. As the definition may seem somewhat mysterious at first, we begin with a review of the one variable case.

Recall that if U is an open subset of $\mathbf{R}$, then the function $f: U \to \mathbf{R}$ is said to be *differentiable* at $x_0 \in U$ if

$$\lim_{x \to x_0} \frac{f(x) - f(x_0)}{x - x_0}$$

exists. The value of this limit is called the *derivative of f at x_0* and is denoted by $f'(x_0)$.

As we saw in Section 2.1, this definition can be restated as follows: the function $f: U \to \mathbf{R}$ is differentiable at $x_0 \in U$ if there is a number $f'(x_0)$ such that

$$\text{(1.1)} \qquad \lim_{x \to x_0} \frac{|f(x) - f(x_0) - f'(x_0)(x - x_0)|}{|x - x_0|} = 0$$

Now, any real number a determines a linear transformation $L: \mathbf{R} \to \mathbf{R}$ defined by

$$L(x) = ax$$

In particular, $f'(x_0)$ determines a linear transformation $L: \mathbf{R} \to \mathbf{R}$. Using L, we may rewrite (1.1) as

$$\text{(1.2)} \qquad \lim_{x \to x_0} \frac{|f(x) - f(x_0) - L(x - x_0)|}{|x - x_0|} = 0$$

We can now use (1.2) to define differentiability of a function $f: \mathbf{R}^n \to \mathbf{R}^p$. Let $U \subset \mathbf{R}^n$ be an open subset (see Section 3.5) and $f: U \to \mathbf{R}^p$ a function. We say that f is *differentiable* at $\mathbf{v}_0 \in U$ if there is a linear transformation $L: \mathbf{R}^n \to \mathbf{R}^p$ such that

$$\text{(1.3)} \qquad \lim_{\mathbf{v} \to \mathbf{v}_0} \frac{\| f(\mathbf{v}) - f(\mathbf{v}_0) - L(\mathbf{v} - \mathbf{v}_0)\|}{\|\mathbf{v} - \mathbf{v}_0\|} = 0$$

This means that, for any $\varepsilon > 0$, there is a $\delta > 0$ such that

$$\| f(\mathbf{v}) - f(\mathbf{v}_0) - L(\mathbf{v} - \mathbf{v}_0)\| < \varepsilon \|\mathbf{v} - \mathbf{v}_0\|$$

whenever $\|\mathbf{v} - \mathbf{v}_0\| < \delta$.

Note that we define differentiability only for functions defined on open sets so that, in the limit (1.3), the vector $\mathbf{v}$ can approach $\mathbf{v}_0$ from all directions. For example, if $n = 1$, $\mathbf{v}$ can approach $\mathbf{v}_0$ both from the right and from the left.

The basic feature that a function f differentiable at $\mathbf{v}_0$ possesses is that it can be approximated near $\mathbf{v}_0$ by the relatively simple function

$$\text{(1.4)} \qquad h(\mathbf{v}) = L(\mathbf{v}) + [f(\mathbf{v}_0) - L(\mathbf{v}_0)]$$

In fact, this approximation is so good that, not only does $\| f(\mathbf{v}) - h(\mathbf{v})\|$ tend to zero as $\mathbf{v}$ approaches $\mathbf{v}_0$, but $\| f(\mathbf{v}) - h(\mathbf{v})\|$ multiplied by the large number $1/\|\mathbf{v} - \mathbf{v}_0\|$ also tends to zero as $\mathbf{v}$ approaches $\mathbf{v}_0$.

The linear transformation L of equation (1.3) is called the *derivative of f at* $\mathbf{v}_0$. We denote it by $D_{\mathbf{v}_0} f$ or by $f'(\mathbf{v}_0)$. As we shall see in the next section, L is uniquely defined by equation (1.3), so that a function differentiable at $\mathbf{v}_0 \in U$ has exactly one derivative there.

We have seen that if U is an open subset of $\mathbf{R}^1$ and $f: U \to \mathbf{R}$ has a derivative $f'(x_0)$ at x_0 in the usual calculus sense, then $f'(x_0)$ is also the derivative in this new sense, once we interpret the number $f'(x_0)$ as the linear transformation

$$L(x) = f'(x_0)x$$

Here are some other examples of derivatives.

1. If $f: \mathbf{R}^n \to \mathbf{R}^p$ is a constant function and $L : \mathbf{R}^n \to \mathbf{R}^p$ is the zero linear transformation, then

$$f(\mathbf{v}) - f(\mathbf{v}_0) - L(\mathbf{v} - \mathbf{v}_0) = \mathbf{0}$$

for all $\mathbf{v}$, $\mathbf{v}_0 \in \mathbf{R}^n$. Thus $D_{\mathbf{v}} f$ is the zero transformation for all $\mathbf{v} \in \mathbf{R}^n$.

2. If $f: \mathbf{R}^n \to \mathbf{R}^p$ is a linear transformation and $L = f: \mathbf{R}^n \to \mathbf{R}^p$, then

$$f(\mathbf{v}) - f(\mathbf{v}_0) - L(\mathbf{v} - \mathbf{v}_0) = \mathbf{0}$$

for all $\mathbf{v}_0, \mathbf{v} \in \mathbf{R}^n$. It follows that $D_{\mathbf{v}} f = f$ for all $\mathbf{v} \in \mathbf{R}^n$ whenever f is a linear transformation. That is, a linear transformation is its own derivative. [Warning: This means, in the case of functions from $\mathbf{R}$ to $\mathbf{R}$, that the derivative of $f(x) = ax$ is the linear function given by $y = ax$. It does *not* mean that the derivative of ax is ax.]

3. Suppose $h : \mathbf{R}^2 \to \mathbf{R}^1$ is defined by

$$h(x, y) = xy$$

We shall compute $D_{\mathbf{v}_0} h$.

We know that the most general linear transformation $L : \mathbf{R}^2 \to \mathbf{R}^1$ is given by

$$L(x, y) = ax + by$$

(See Section 2.6.) Now let $\mathbf{v}_0 = (x_0, y_0)$ and $\mathbf{v} = (x, y)$. To find the derivative of h, we need to produce numbers a and b such that

$$\lim_{\mathbf{v} \to \mathbf{v}_0} \frac{| h(\mathbf{v}) - h(\mathbf{v}_0) - a(x - x_0) - b(y - y_0)|}{\|\mathbf{v} - \mathbf{v}_0\|} = 0$$

or

$$\text{(1.5)} \qquad \lim_{(x, y) \to (x_0, y_0)} \frac{|xy - x_0 y_0 - a(x - x_0) - b(y - y_0)|}{[(x - x_0)^2 + (y - y_0)^2]^{1/2}} = 0$$

We can rewrite the numerator of the left side of equation (1.5) as

$$\begin{aligned}&|x(y-y_0)+y_0(x-x_0)-a(x-x_0)-b(y-y_0)|\\&=|(x-b)(y-y_0)+(y_0-a)(x-x_0)|\\&\le|x-b|\,|y-y_0|+|y_0-a|\,|x-x_0|\end{aligned}$$

Therefore

$$\begin{aligned}&\frac{|xy-x_0y_0-a(x-x_0)-b(y-y_0)|}{[(x-x_0)^2+(y-y_0)^2]^{1/2}}\\&\le\frac{|x-b|\,|y-y_0|}{[(x-x_0)^2+(y-y_0)^2]^{1/2}}\\&+\frac{|y_0-a|\,|x-x_0|}{[(x-x_0)^2+(y-y_0)^2]^{1/2}}\\&\le|x-b|+|y_0-a|\end{aligned}$$

since $|y-y_0|=[(y-y_0)^2]^{1/2}\le[(x-x_0)^2+(y-y_0)^2]^{1/2}$

and $|x-x_0|=[(x-x_0)^2]^{1/2}\le[(x-x_0)^2+(y-y_0)^2]^{1/2}$

It follows that

$$\lim_{(x,y)\to(x_0,y_0)}\frac{|xy-x_0y_0-a(x-x_0)-b(y-y_0)|}{[(x-x_0)^2+(y-y_0)^2]^{1/2}}\le\lim_{(x,y)\to(x_0,y_0)}(|x-b|+|y_0-a|)$$

However, $\lim_{(x,y)\to(x_0,y_0)}(|x-b|+|y_0-a|)=0$

if and only if $b=x_0$ and $a=y_0$. Thus $D_{\mathbf{v}_0}h$ is the linear transformation taking (x,y) into y_0x+x_0y.

As this calculation makes clear, we need a better way of computing derivatives. We already know good methods for functions $f\colon \mathbf{R}\to\mathbf{R}$ from elementary calculus. In Section 3, we shall see how to reduce the problem of computing the derivative of a function $f\colon \mathbf{R}^n\to\mathbf{R}^p$ to the one variable case.

We say that f is *differentiable on* U if f is differentiable at each $\mathbf{v}\in U$. If S is any subset of $\mathbf{R}^n$ (not necessarily open) and $g: S\to\mathbf{R}^p$ is a function, we say that g is *differentiable on* S if there is an open set U containing S and a function $f\colon U\to\mathbf{R}^p$ such that $f(\mathbf{v})=g(\mathbf{v})$ for $\mathbf{v}\in S$ and f is differentiable on U. In this case, we define the derivative of g at $\mathbf{v}_0\in S$ to be $D_{\mathbf{v}_0}f$ and denote it by $D_{\mathbf{v}_0}g$. (The definition may depend on our choice of g, but it will not in the cases where we use it.)

Exercises

1. Determine, directly from the definition, the derivative of the function $f\colon\mathbf{R}^2\to\mathbf{R}^2$ defined by

$$f(x_1, x_2) = (x_1^2 + x_1x_2, x_1 - x_2^2)$$

Express the derivative of f as T_A for some 2×2 matrix A. (See Section 2.6 and Example 3 of this section.)

2. Determine, directly from the definition, the derivative of the function $f: \mathbf{R}^3 \to \mathbf{R}^2$ defined by

$$f(x_1, x_2, x_3) = (x_1^2 - 2x_1x_2 + x_3, x_2^2 + x_3^2)$$

3. Let $f: \mathbf{R} \to \mathbf{R}^n$ be given by the formula

$$f(x) = (f_1(x), f_2(x), \ldots, f_n(x))$$

where each $f_j: \mathbf{R} \to \mathbf{R}$ is differentiable, $1 \le j \le n$. Show that

$$(D_{x_0} f)(x) = x\mathbf{v}_0$$

where $\mathbf{v}_0 = (f_1'(x_0), f_2'(x_0), \ldots, f_n'(x_0))$.

*4. Let $f: \mathbf{R}^n \to \mathbf{R}$ and $g: \mathbf{R}^n \to \mathbf{R}^p$ be differentiable at $\mathbf{v}_0$. Show that $fg: \mathbf{R}^n \to \mathbf{R}^p$ [defined by $(fg)(\mathbf{v}) = f(\mathbf{v})g(\mathbf{v})$] is differentiable at $\mathbf{v}_0$, and that

$$D_{\mathbf{v}_0}(fg)(\mathbf{v}) = f(\mathbf{v}_0)(D_{\mathbf{v}_0}g)(\mathbf{v}) + ((D_{\mathbf{v}_0}f)(\mathbf{v}))g(\mathbf{v}_0)$$

5. Prove that the function $f: \mathbf{R} \to \mathbf{R}$ defined by $f(x) = |x|$ is not differentiable at $x = 0$.

6. Let $f: (a, b) \to \mathbf{R}$ be differentiable with $f'(c) > 0$, $c \in (a, b)$. Prove that there is a $\delta > 0$ such that the quotient

$$\frac{f(x) - f(c)}{x - c} > 0$$

for $x \in (c - \delta, c + \delta)$, $x \neq c$. (A similar result holds if $f'(c) < 0$.) Note that (a, b) and $(c - \delta, c + \delta)$ denote intervals in $\mathbf{R}$.

7. Let $f: (a, b) \to \mathbf{R}$ be a differentiable function and suppose that f has a local

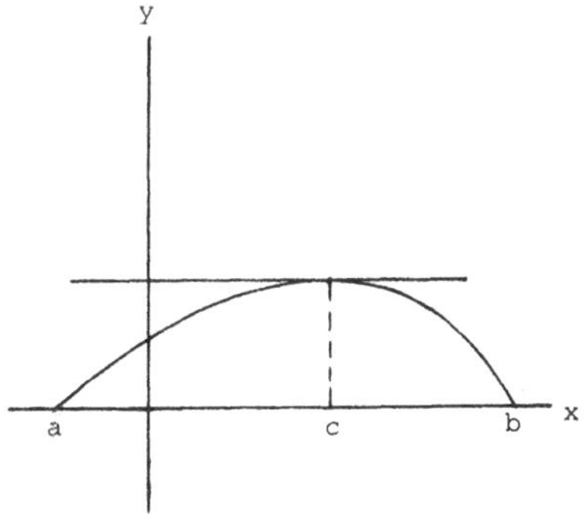

Figure 1.1 Rolle's theorem.

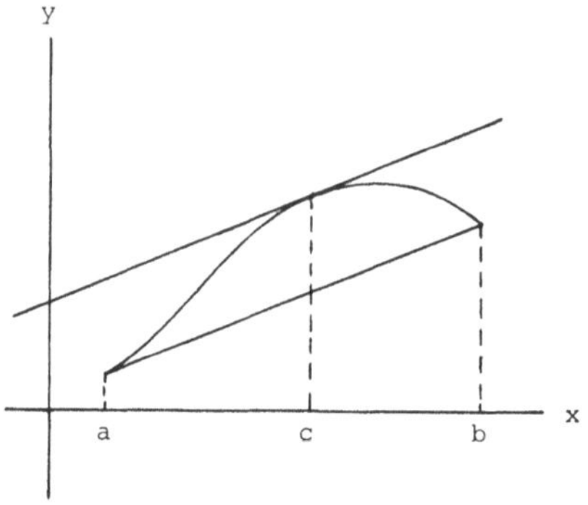

Figure 1.2 The mean value theorem.

maximum (or a local minimum) at $c \in (a, b)$. Prove that $f'(c) = 0$. (*Hint*: Use Exercise 6.)

8. (*Rolle's Theorem*) Let f: $[a, b] \to \mathbf{R}$ be a continuous function which is differentiable on (a, b) and suppose $f(a) = f(b) = 0$. Prove that there is a number $c \in (a, b)$ such that $f'(c) = 0$. (See Figure 1.1.) (You may assume that any continuous function on a closed interval attains its maximum and minimum values. This fact is proved in Chapter 7.)

*9. (*The Mean Value Theorem*) Let f: $[a, b] \to \mathbf{R}$ be a continuous function which is differentiable on (a, b). Prove that there is a number $c \in (a, b)$ such that

$$f'(c) = \frac{f(b) - f(a)}{b - a}$$

[This equation states that the tangent line to the graph to f at the point c is parallel to the chord joining the points $(a, f(a))$ and $(b, f(b))$ in $\mathbf{R}^2$; see Figure 1.2.]

10. Let f: $(a, b) \to \mathbf{R}$ be a differentiable function with $f'(x) = 0$ for all $x \in (a, b)$. Prove that f is a constant function. (*Hint*: Use the mean value theorem.)

2. Elementary Properties of the Derivative*

We now derive some of the elementary properties of the derivative. We begin by showing that a differentiable function has a unique derivative.

Proposition 2.1 *Let U be an open subset of $\mathbf{R}^n$ and f: $U \to \mathbf{R}^p$ a function differentiable at $\mathbf{v}_0 \in U$. Suppose that L_1, L_2 : $\mathbf{R}^n \to \mathbf{R}^p$ are derivatives of f at $\mathbf{v}_0$. Then $L_1 = L_2$.*

*Some readers may prefer to skip this section temporarily, going directly to Section 3, where we show how to compute the derivative.

Proof. Note first of all that

$$\begin{aligned}\|L_1(\mathbf{v}-\mathbf{v}_0)-L_2(\mathbf{v}-\mathbf{v}_0)\| &= \|f(\mathbf{v})-f(\mathbf{v}_0)-L_2(\mathbf{v}-\mathbf{v}_0)\\ &\quad -[f(\mathbf{v})-f(\mathbf{v}_0)-L_1(\mathbf{v}-\mathbf{v}_0)]\|\\ &\le \|f(\mathbf{v})-f(\mathbf{v}_0)-L_2(\mathbf{v}-\mathbf{v}_0)\|\\ &\quad +\|f(\mathbf{v})-f(\mathbf{v}_0)-L_1(\mathbf{v}-\mathbf{v}_0)\|\end{aligned}$$

Setting $L = L_1 - L_2$ and $\mathbf{u} = \mathbf{v} - \mathbf{v}_0$, it follows that

$$\lim_{\mathbf{u}\to 0}\frac{\|L(\mathbf{u})\|}{\|\mathbf{u}\|} = 0$$

since both L_1 and L_2 are derivatives for f at $\mathbf{v}_0$. Thus Proposition 2.1 will be proved if we can show that L is the zero transformation. We proceed by contradiction.

Suppose $L(\mathbf{u}) \neq \mathbf{0}$ and set $\varepsilon = \|L(\mathbf{u})\|/\|\mathbf{u}\|$. [We know that $\mathbf{u} \neq \mathbf{0}$ since $L(\mathbf{0}) = \mathbf{0}$.] Then, for all $\alpha > 0$, $\|L(\alpha\mathbf{u})\|/\|\alpha\mathbf{u}\| = \varepsilon$. Now we certainly can pick α so that $\|\alpha\mathbf{u}\|$ is less than any $\delta > 0$. This contradicts the fact that $\lim_{\mathbf{u}\to 0}(\|L(\mathbf{u})\|/\|\mathbf{u}\|) = 0$. Therefore $L(\mathbf{u}) = \mathbf{0}$ for all $\mathbf{u}$ in $\mathbf{R}^n$ and this proves the proposition.

We next prove an analogue of a well-known result for functions of one variable.

Recall that, if $S, T : \mathbf{R}^n \to \mathbf{R}^p$ are linear transformations and c is a real number, then $S + T$ and cT are the linear transformations from $\mathbf{R}^n$ to $\mathbf{R}^p$ defined by

$$(S + T)(\mathbf{v}) = S(\mathbf{v}) + T(\mathbf{v})$$
$$(cT)(\mathbf{v}) = cT(\mathbf{v})$$

Proposition 2.2 *Let a be a real number, U an open subset of $\mathbf{R}^n$, and suppose that $f, g : U \to \mathbf{R}^p$ are differentiable at $\mathbf{v}_0$. Then af and $f + g$ are differentiable at $\mathbf{v}_0$ with*

$$D_{\mathbf{v}_0}(af) = aD_{\mathbf{v}_0}f$$
$$D_{\mathbf{v}_0}(f+g) = D_{\mathbf{v}_0}f + D_{\mathbf{v}_0}g$$

Proof. We prove only that af is differentiable at $\mathbf{v}_0$ with $D_{\mathbf{v}_0}(af) = aD_{\mathbf{v}_0}f$, leaving the statement about $f + g$ as Exercise 1.

Set $L = D_{\mathbf{v}_0}f$. Then

$$\begin{aligned}\|(af)(\mathbf{v})-(af)(\mathbf{v}_0)-(aL)(\mathbf{v}-\mathbf{v}_0)\| &= \|af(\mathbf{v})-af(\mathbf{v}_0)-aL(\mathbf{v}-\mathbf{v}_0)\|\\ &= |a|\,\|f(\mathbf{v})-f(\mathbf{v}_0)-L(\mathbf{v}-\mathbf{v}_0)\|\end{aligned}$$

It follows easily that

$$\lim_{\mathbf{v}\to\mathbf{v}_0} \frac{\|(af)(\mathbf{v}) - (af)(\mathbf{v}_0) - (aL)(\mathbf{v} - \mathbf{v}_0)\|}{\|\mathbf{v} - \mathbf{v}_0\|}$$

$$= |a| \lim_{\mathbf{v}\to\mathbf{v}_0} \frac{\| f(\mathbf{v}) - f(\mathbf{v}_0) - L(\mathbf{v} - \mathbf{v}_0)\|}{\|\mathbf{v} - \mathbf{v}_0\|} = 0$$

since $L = D_{\mathbf{v}_0} f$. This completes the proof of the first part of Proposition 2.2.

We know from elementary calculus that a differentiable function $f : \mathbf{R} \to \mathbf{R}$ is continuous. The next result proves this true for functions between Euclidean spaces.

Proposition 2.3 *Let U be an open subset of $\mathbf{R}^n$. If $f : U \to \mathbf{R}^p$ is differentiable at a point $\mathbf{v}_0 \in U$, then f is continuous at $\mathbf{v}_0$.*

Proof. Given $\varepsilon > 0$, we know that there is a number $\delta_0 > 0$ such that

$$\frac{\| f(\mathbf{v}) - f(\mathbf{v}_0) - D_{\mathbf{v}_0} f(\mathbf{v} - \mathbf{v}_0)\|}{\|\mathbf{v} - \mathbf{v}_0\|} < \frac{\varepsilon}{2}$$

if $\|\mathbf{v} - \mathbf{x}_0| < \delta_0$. If $\|\mathbf{v} - \mathbf{v}_0\|$ is also less than 1, then

$$\| f(\mathbf{v}) - f(\mathbf{v}_0) - D_{\mathbf{v}_0} f(\mathbf{v} - \mathbf{v}_0)\| < \frac{\varepsilon}{2}$$

Since $D_{\mathbf{v}_0} f$ is linear, Proposition 3–3.4 implies that $D_{\mathbf{v}_0} f$ is continuous; thus we can choose δ_1 so that

$$\| D_{\mathbf{v}_0} f(\mathbf{v}) - D_{\mathbf{v}_0} f(\mathbf{v}_0)\| < \frac{\varepsilon}{2}$$

whenever $\|\mathbf{v} - \mathbf{v}_0\| < \delta_1$.

Now let δ be the smallest of 1, δ_0, δ_1. If $\|\mathbf{v} - \mathbf{v}_0\| < \delta$, then

$$\begin{aligned} \| f(\mathbf{v}) - f(\mathbf{v}_0)\| &\le \| f(\mathbf{v}) - f(\mathbf{v}_0) - D_{\mathbf{v}_0} f(\mathbf{v} - \mathbf{v}_0)\| + \| D_{\mathbf{v}_0} f(\mathbf{v} - \mathbf{v}_0)\| \\ &< \frac{\varepsilon}{2} + \| D_{\mathbf{v}_0} f(\mathbf{v}) - D_{\mathbf{v}_0} f(\mathbf{v}_0)\| \\ &< \frac{\varepsilon}{2} + \frac{\varepsilon}{2} = \varepsilon \end{aligned}$$

Therefore f is continuous at $\mathbf{v}_0$.

We note that a function continuous at $\mathbf{v}_0$ need not be differentiable at $\mathbf{v}_0$. For example, the function $f : \mathbf{R} \to \mathbf{R}$ defined by $f(x) = |x|$ is everywhere continuous but not differentiable at $x = 0$. (See Exercise 5, Section 1.)

Exercises

1. Prove the second statement of Proposition 2.2.
2. (a) Let $f, g : \mathbf{R}^n \to \mathbf{R}^p$ be continuous functions such that $f + g$ is differentiable. Are f and g differentiable? Give a proof or counterexample.
   (b) Suppose that af is differentiable (for $a \neq 0$). Is f? Again, give a proof or counterexample.

3. Partial Derivatives and the Jacobian Matrix

As we saw in the first section, the computation of the derivative of a function $f : \mathbf{R}^n \to \mathbf{R}^p$ directly from the definition is a complicated matter. We simplify it here by reducing the computation of the derivative of such a function to the case of functions from $\mathbf{R}$ to $\mathbf{R}$, a problem studied in elementary calculus.

Suppose that U is an open subset of $\mathbf{R}^n$ and $f : U \to \mathbf{R}$ is a function. The *partial derivative of f with respect to x_j at the* $\mathbf{v} = (x_1, \ldots, x_n)$ is defined by

$$\text{(3.1)} \quad \frac{\partial f}{\partial x_j} = \lim_{h \to 0} \frac{f(x_1, \ldots, x_{j-1}, x_j + h, x_{j+1}, \ldots, x_n) - f(x_1, \ldots, x_n)}{h}$$

when the limit exists. If we wish to indicate explicitly the point at which the partial derivative is taken, we shall write

$$\left.\frac{\partial f}{\partial x_j}\right|_{\mathbf{v}} \quad \text{or} \quad \frac{\partial f}{\partial x_j}(\mathbf{v})$$

Note that if $n = 1$, then the partial derivative defined above is just the ordinary derivative of a function of one variable. In general, if we fix the point $\mathbf{v}_0 = (a_1, \ldots, a_n)$ and define a function g of one variable x by

$$g(x) = f(a_1, \ldots, a_{j-1}, x, a_{j+1}, \ldots, a_n)$$

then the partial derivative of f with respect to x_j at the point $\mathbf{v}_0$ is the ordinary derivative of g with respect to x evaluated at $x = a_j$. This observation reduces the computation of partial derivatives to that of ordinary differentiation. Thus, to compute $\partial f/\partial x_j$, we treat all variables except x_j as constants and differentiate the resulting function of the single variable x_j in the usual way.

Here are some examples of partial derivatives.

1. If $f : \mathbf{R}^2 \to \mathbf{R}$ is given by $f(x_1, x_2) = x_1^2 + x_2$, then $\partial f/\partial x_2$ is obtained by treating the variable x_1 as a constant and differentiating f in the usual way with respect to x_2. Thus

$$\frac{\partial f}{\partial x_2} = 1$$

since $(\partial/\partial x_2)(x_1^2) = 0$ (x_1 is treated as a constant). Similarly, $\partial f/\partial x_1 = 2x_1$.

2. Let $f: \mathbf{R}^3 \to \mathbf{R}$ be defined by

$$f(x_1, x_2, x_3) = x_1 x_3^2 + x_1^2 x_2 x_3 + x_2^3$$

Then $\partial f/\partial x_1$ is obtained by treating the variables x_2 and x_3 as constants and differentiating with respect to x_1. Thus

$$\frac{\partial f}{\partial x_1} = x_3^2 + 2x_1 x_2 x_3$$

In the same way, we have

$$\frac{\partial f}{\partial x_2} = x_1^2 x_3 + 3x_2^2 \qquad \frac{\partial f}{\partial x_3} = 2x_1 x_3 + x_1^2 x_2$$

If $n = 2$, the graph of the function g defined above can be obtained by intersecting the graph of f with a plane perpendicular to one of the axes. For instance, if $f(x_1, x_2) = 1 - x_1^2 - x_2^2$ and $g: \mathbf{R} \to \mathbf{R}$ is defined by

$$g(x) = f(a, x) = 1 - a^2 - x^2$$

then the graph of g is the intersection of the graph of f and the plane perpendicular to the x_1 axis through the point $(a, 0, 0)$. (See Figure 3.1.)

Suppose now that $f: U \to \mathbf{R}^p$ is defined by

$$f(\mathbf{v}) = (f_1(\mathbf{v}), \ldots, f_p(\mathbf{v}))$$

where $f_i: U \to \mathbf{R}$, $1 \le i \le p$. For each $\mathbf{v} \in U$, we define the *Jacobian matrix of f at $\mathbf{v}$ by*

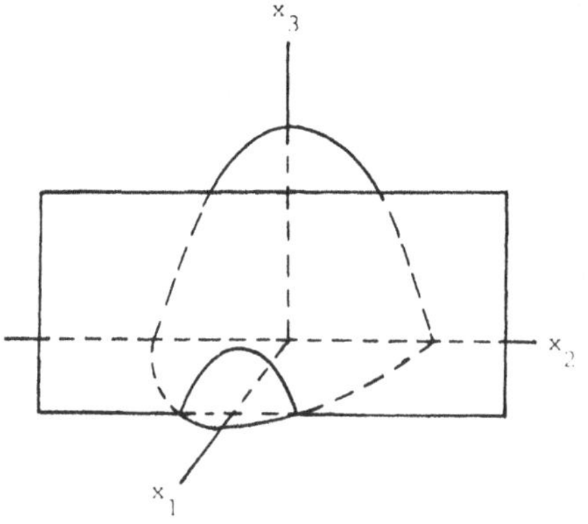

Figure 3.1 The graph of $g(x) = f(a, x)$.

$$J_f(\mathbf{v}) = (a_{ij})$$

where $a_{ij} = (\partial f_i/\partial x_j)(\mathbf{v})$, $1 \le i \le p$, $1 \le j \le n$.

Here are some examples of Jacobian matrices.

3. Let $f: \mathbf{R}^2 \to \mathbf{R}^3$ be given by

$$f(x_1, x_2) = (x_1^2, x_1x_2, x_2^2)$$

so that $f_1(x_1, x_2) = x_1^2$, $f_2(x_1, x_2) = x_1x_2$, and $f_3(x_1, x_2) = x_2^2$. Then we have

$$\frac{\partial f_1}{\partial x_1} = 2x_1 \qquad \frac{\partial f_2}{\partial x_1} = x_2 \qquad \frac{\partial f_3}{\partial x_1} = 0$$

$$\frac{\partial f_1}{\partial x_2} = 0 \qquad \frac{\partial f_2}{\partial x_2} = x_1 \qquad \frac{\partial f_3}{\partial x_2} = 2x_2$$

Thus

$$J_f(x_1, x_2) = \begin{bmatrix} 2x_1 & 0 \\ x_2 & x_1 \\ 0 & 2x_2 \end{bmatrix}$$

4. Let $f: \mathbf{R}^3 \to \mathbf{R}^3$ be given by

$$g(x_1, x_2, x_3) = (x_1 \cos x_2, x_2 e^{x_3}, \sin x_1 x_3)$$

Then

$$J_g(x_1, x_2, x_3) = \begin{bmatrix} \cos x_2 & -x_1 \sin x_2 & 0 \\ 0 & e^{x_3} & x_2 e^{x_3} \\ x_3 \cos x_1 x_3 & 0 & x_1 \cos x_1 x_3 \end{bmatrix}$$

Proposition 3.1 *Let U be an open subset of $\mathbf{R}^n$ and $f: U \to \mathbf{R}^p$ a function defined by $f(\mathbf{v}) = (f_1(\mathbf{v}), \ldots, f_p(\mathbf{v}))$, where each $f_i: U \to \mathbf{R}$, $1 \le i \le p$. Then if f is differentiable at a point $\mathbf{v}$ in U, each of the partial derivatives $(\partial f_i/\partial x_j)(\mathbf{v})$ exists, $1 \le i \le p$, $1 \le j \le n$. Furthermore, $D_\mathbf{v} f: \mathbf{R}^n \to \mathbf{R}^p$ is the linear transformation defined by the Jacobian matrix of f at $\mathbf{v}$. (See Section 2.6.)*

Proof. Let $L = D_\mathbf{v} f: \mathbf{R}^n \to \mathbf{R}^p$. Let $\mathbf{e}_1, \ldots, \mathbf{e}_n$ be the standard basis for $\mathbf{R}^n$ and $\mathbf{e}_1, \ldots, \mathbf{e}_p$, the standard basis for $\mathbf{R}^p$. Suppose that (a_{ij}) is the matrix of L, so that

$$L\mathbf{e}_j = \sum_{i=1}^{p} a_{ij}\mathbf{e}_i \qquad 1 \le j \le n$$

Since f is differentiable at $\mathbf{v}$, it follows that

$$\lim_{h \to 0} \frac{\| f(\mathbf{v} + h\mathbf{e}_j) - f(\mathbf{v}) - L(h\mathbf{e}_j)\|}{\| h\mathbf{e}_j \|} = 0$$

for each $j = 1, \ldots, n$. Equivalently, we have

$$\lim_{h\to 0} \frac{\| f(\mathbf{v} + h\mathbf{e}_j) - f(\mathbf{v}) - hL(\mathbf{e}_j)\|}{h} = 0$$

Now

$$\frac{\| f(\mathbf{v} + h\mathbf{e}_j) - f(\mathbf{v}) - hL(\mathbf{e}_j)\|}{h} = \left\{ \sum_{i=1}^{p} \left[\frac{f_i(\mathbf{v} + h\mathbf{e}_j) - f_i(\mathbf{v})}{h} - a_{ij} \right]^2 \right\}^{1/2}$$

For the right-hand side to tend to 0, each term in the sum must tend to 0. Therefore

$$\frac{\partial f_i}{\partial x_j}(\mathbf{v}) = \lim_{h\to 0} \frac{f_i(\mathbf{v} + h\mathbf{e}_j) - f_i(\mathbf{v})}{h} = a_{ij} \qquad 1 \le i \le p,\ 1 \le j \le n$$

This proves that $(\partial f_i/\partial x_j)(\mathbf{v})$ exists and equals a_{ij}, completing the proof of Proposition 3.1.

Proposition 3.1 reduces the problem of computing the derivative of a differentiable function $f: \mathbf{R}^n \to \mathbf{R}^p$ to that of computing partial derivatives for the functions $f_1, \ldots, f_n$. As we have seen, a partial derivative is simply an ordinary derivative with respect to one of the variables (the others regarded as constant). Computing the derivative of f thus becomes an exercise in elementary calculus.

We note that the existence of the partial derivatives of a function does not necessarily imply that the function is differentiable or, for that matter, continuous. For example, let $f: \mathbf{R}^2 \to \mathbf{R}$ be defined by

$$f(x_1, x_2) = \begin{cases} \dfrac{x_1 x_2}{x_1^2 + x_2^2} & \text{if } (x_1, x_2) \ne (0, 0) \\ 0 & \text{if } (x_1, x_2) = (0, 0) \end{cases}$$

This function is not continuous at (0, 0), since $f(x, x) = \frac{1}{2}$ for $x \ne 0$ and $f(0, 0) = 0$. Thus, f is not differentiable at (0, 0) by Proposition 2.3. However, $f(0, 0) = f(h, 0) = f(0, h) = 0$ so that both partial derivatives of f exist at (0, 0) (and are equal to zero). In Section 6 we show that if the partial derivatives of a function exist *and are continuous*, then the function is differentiable.

Exercises

1. Compute the partial derivatives of the following functions.
   (a) $f(x_1, x_2) = x_1^2 x_2 + 2x_2^2 - 3$
   (b) $f(x_1, x_2) = x_1 \sin(x_1 x_2)$
   (c) $f(x_1, x_2) = (x_1 + x_2)/(x_1^2 + x_2^2)$, $(x_1, x_2) \ne (0, 0)$
   (d) $f(x_1, x_2, x_3) = x_1 \cos x_2 + x_2 \sin x_3$
   (e) $f(x_1, x_2, x_3) = \sqrt{x_1}\, e^{x_2 + x_1 x_2}$

2. Compute the Jacobians of the following functions.
   (a) $f: \mathbf{R} \to \mathbf{R}^2$, $f(x_1) = (\cos x_1, \sin x_1)$
   (b) $f: \mathbf{R}^2 \to \mathbf{R}^2$, $f(x_1, x_2) = (x_1^2 + x_2, 2x_1x_2 - x_2^2)$
   (c) $f: \mathbf{R}^3 \to \mathbf{R}^3$, $f(x_1, x_2, x_3) = (x_1x_2, x_1^2 + x_2x_3, x_2x_3^2)$
   (d) $f: \mathbf{R}^3 \to \mathbf{R}^2$, $f(x_1, x_2, x_3) = (x_1e^{x_2}, x_2^2 + x_3 \sin x_1)$
   (e) $f: \mathbf{R}^3 \to \mathbf{R}^3$, $f(x_1, x_2, x_3) = (x_1^2 + x_2x_3, \arctan(x_1 + x_3), x_3 - x_2e^{x_1})$
3. Let $f: \mathbf{R}^2 \to \mathbf{R}$ be defined by

$$f(x_1, x_2) = \begin{cases} \dfrac{x_1x_2}{x_1^2 + x_2^2} & \text{if } (x_1, x_2) \neq (0, 0) \\ 0 & \text{if } (x_1, x_2) = (0, 0) \end{cases}$$

   (a) Prove that f is not continuous at (0, 0).
   (b) Prove that both partial derivatives of f exist and equal 0 at (0, 0).
   (c) Prove that the partial derivatives of f are not continuous at (0, 0).
4. (a) Let $f, g: \mathbf{R}^n \to \mathbf{R}^p$ be differentiable at $\mathbf{v}_0$. Show that

$$\frac{\partial}{\partial x_j}(f_i + g_i)\bigg|_{\mathbf{v}_0} = \frac{\partial f_i}{\partial x_j}\bigg|_{\mathbf{v}_0} + \frac{\partial g_i}{\partial x_i}\bigg|_{\mathbf{v}_0}$$

   and

$$\frac{\partial}{\partial x_j}(a f_i)\bigg|_{\mathbf{v}_0} = a\frac{\partial}{\partial x_j} f_i\bigg|_{\mathbf{v}_0} \qquad \text{(where } a \text{ is a constant)}$$

   *(b) Why do part (a) and Proposition 3.1 not imply Proposition 2.2?
5. Let $f, g: \mathbf{R}^n \to \mathbf{R}$ be differentiable. Find an expression for $D_{\mathbf{v}}(fg)$. (You may assume that fg is differentiable.)

4. The Chain Rule

In this section, we state the chain rule for determining the derivative of the composite of two differentiable functions and give some examples to show how it is used. The proof is given in the next section.

Let U be an open subset of $\mathbf{R}^n$ and U' an open subset of $\mathbf{R}^p$.

The Chain Rule *Let $f: U \to \mathbf{R}^p$ be differentiable at $\mathbf{v}_0$, with $f(U) \subset U'$, and let $g: U' \to \mathbf{R}^k$ be differentiable at $\mathbf{w}_0 = f(\mathbf{v}_0)$. Then the composite function $g \circ f: U \to \mathbf{R}^k$ is differentiable at $\mathbf{v}_0$ and $D_{\mathbf{v}_0}(g \circ f)$ is the composite of $D_{\mathbf{v}_0} f$ with $D_{\mathbf{w}_0} g$:*

$$D_{\mathbf{v}_0}(g \circ f) = (D_{\mathbf{w}_0} g) \circ (D_{\mathbf{v}_0} f) \tag{4.1}$$

As stated above, we prove this assertion in Section 5.

We know from Proposition 3.1 that, relative to the standard bases, the matrix of $D_{\mathbf{w}_0}g$ is the Jacobian $J_g(\mathbf{w}_0)$ and the matrix of $D_{\mathbf{v}_0}f$ is the Jacobian $J_f(\mathbf{v}_0)$. Since the matrix of the composite of two linear transformations is the product of the matrices (see Section 2.8), the chain rule can be written

$$J_{g\circ f}(\mathbf{v}_0) = J_g(f(\mathbf{v}_0))J_f(\mathbf{v}_0) \tag{4.2}$$

More explicitly, let $(x_1, \ldots, x_n)$ be coordinates in $\mathbf{R}^n$, $(y_1, \ldots, y_p)$ coordinates in $\mathbf{R}^p$, and suppose that

$$f(x_1, \ldots, x_n) = (f_1(x_1, \ldots, x_n), \ldots, f_p(x_1, \ldots, x_n))$$
$$g(y_1, \ldots, y_p) = (g_1(y_1, \ldots, y_p), \ldots, g_k(y_1, \ldots, y_p))$$

Then

$$J_{g\circ f}(\mathbf{v}_0) = \begin{bmatrix} \dfrac{\partial g_1}{\partial y_1} & \cdots & \dfrac{\partial g_1}{\partial y_p} \\ \vdots & & \vdots \\ \dfrac{\partial g_k}{\partial y_1} & \cdots & \dfrac{\partial g_k}{\partial y_p} \end{bmatrix} \begin{bmatrix} \dfrac{\partial f_1}{\partial x_1} & \cdots & \dfrac{\partial f_1}{\partial x_n} \\ \vdots & & \vdots \\ \dfrac{\partial f_p}{\partial x_1} & \cdots & \dfrac{\partial f_p}{\partial x_n} \end{bmatrix}$$

where the entries in the first matrix are evaluated at $f(\mathbf{v}_0)$ and the entries in the second matrix are evaluated at $\mathbf{v}_0$.

Examples

1. If $n = p = k = 1$, then each of the matrices involved in a one-by-one matrix—that is, a number. In this case, we have

$$\frac{d(g\circ f)}{dx}(x_0) = \frac{dg}{dy}(f(x_0))\frac{df}{dx}(x_0)$$

or, with different notation,

$$(g\circ f)'(x_0) = g'(f(x_0))f'(x_0)$$

This is the standard one-variable chain rule.

2. If $n = p = 2$ and $k = 1$, we have

$$J_{g\circ f} = \begin{bmatrix} \dfrac{\partial g}{\partial y_1} & \dfrac{\partial g}{\partial y_2} \end{bmatrix} \begin{bmatrix} \dfrac{\partial f_1}{\partial x_1} & \dfrac{\partial f_1}{\partial x_2} \\ \dfrac{\partial f_2}{\partial x_1} & \dfrac{\partial f_2}{\partial x_2} \end{bmatrix}$$

$$= \begin{bmatrix} \dfrac{\partial g}{\partial y_1}\dfrac{\partial f_1}{\partial x_1} + \dfrac{\partial g}{\partial y_2}\dfrac{\partial f_2}{\partial x_1} & \dfrac{\partial g}{\partial y_1}\dfrac{\partial f_1}{\partial x_2} + \dfrac{\partial g}{\partial y_2}\dfrac{\partial f_2}{\partial x_2} \end{bmatrix}$$

where all partial derivatives of g are evaluated at $f(\mathbf{v}_0)$ and all partial derivatives of f_1 and f_2 are evaluated at $\mathbf{v}_0$. Since

$$J_{g\circ f}(\mathbf{v}_0) = \left[\frac{\partial(g\circ f)}{\partial x_1}(\mathbf{v}_0) \quad \frac{\partial(g\circ f)}{\partial x_2}(\mathbf{v}_0)\right]$$

by definition of the Jacobian, we have

$$\frac{\partial(g\circ f)}{\partial x_1}(\mathbf{v}_0) = \left[\frac{\partial g}{\partial y_1}(f(\mathbf{v}_0))\right]\left[\frac{\partial f_1}{\partial x_1}(\mathbf{v}_0)\right] + \left[\frac{\partial g}{\partial y_2}(f(\mathbf{v}_0))\right]\left[\frac{\partial f_2}{\partial x_1}(\mathbf{v}_0)\right]$$

and

$$\frac{\partial(g\circ f)}{\partial x_2}(\mathbf{v}_0) = \left[\frac{\partial g}{\partial y_1}(f(\mathbf{v}_0))\right]\left[\frac{\partial f_1}{\partial x_2}(\mathbf{v}_0)\right] + \left[\frac{\partial g}{\partial y_2}(f(\mathbf{v}_0))\right]\left[\frac{\partial f_2}{\partial x_2}(\mathbf{v}_0)\right]$$

For example, if $f(x_1, x_2) = (x_1 \sin x_2, x_1 x_2^2)$ and $g(y_1, y_2) = 2y_1 y_2 + 4$, then

$$\frac{\partial f_1}{\partial x_1} = \sin x_2 \qquad \frac{\partial f_1}{\partial x_2} = x_1 \cos x_2$$

$$\frac{\partial f_2}{\partial x_1} = x_2^2 \qquad \frac{\partial f_2}{\partial x_2} = 2x_1 x_2$$

$$\frac{\partial g}{\partial y_1} = 2y_2 \qquad \frac{\partial g}{\partial y_2} = 2y_1$$

It follows that

$$\frac{\partial g}{\partial y_1}(f(x_1, x_2)) = 2x_1 x_2^2 \qquad \frac{\partial g}{\partial y_2}(f(x_1, x_2)) = 2x_1 \sin x_2$$

so that

$$\frac{\partial(g\circ f)}{\partial x_1} = 2x_1 x_2^2 \sin x_2 + (2x_1 \sin x_2)x_2^2$$

$$\frac{\partial(g\circ f)}{\partial x_2} = 2x_1 x_2^2 x_1 \cos x_2 + (2x_1 \sin x_2)\, 2x_1 x_2$$

This can be checked by noting that

$$g\circ f(x_1, x_2) = 2(x_1 \sin x_2)(x_1 x_2^2) + 4$$

and computing the partial derivatives directly.

3. Let $n = 2$, $m = k = 3$ and define f and g by

$$f(x_1, x_2) = (x_2^2, x_1 x_2, x_1^3)$$
$$g(y_1, y_2, y_3) = (y_1 - y_3^2, 2y_1^2 - y_2, y_1 y_2 y_3)$$

Then
$$J_f(x, x_2) = \begin{bmatrix} 0 & 2x_2 \\ x_2 & x_1 \\ 3x_1^2 & 0 \end{bmatrix}$$

and
$$J_g(y_1, y_2, y_3) = \begin{bmatrix} 1 & 0 & -2y_3 \\ 4y_1 & -1 & 0 \\ y_2y_3 & y_1y_3 & y_1y_2 \end{bmatrix}$$

Thus
$$\begin{aligned} J_{g\circ f}(x_1, x_2) &= J_g(f(x_1, x_2))J_f(x_1, x_2) \\ &= \begin{bmatrix} 1 & 0 & -2x_1^3 \\ 4x_2^2 & -1 & 0 \\ x_1^4x_2 & x_1^3x_2^2 & x_1x_2^3 \end{bmatrix} \begin{bmatrix} 0 & 2x_2 \\ x_2 & x_1 \\ 3x_1^2 & 0 \end{bmatrix} \\ &= \begin{bmatrix} -6x_1^5 & 2x_2 \\ -x_2 & 8x_2^3 - x_1 \\ 4x_1^3x_2^3 & 3x_1^4x_2^2 \end{bmatrix} \end{aligned}$$

Again, this can be checked by writing out $g\circ f$ explicitly and computing the partial derivatives.

Exercises

1. Define functions f, g, h, F, G, H, K by
$f(x, y) = x^2 + xy - y^2$
$g(x, y) = x \sin y$
$h(x, y, z) = xyz^2$
$F(x, y) = (xy^2, xy)$
$G(x, y) = (x + y, x - y, xy)$
$H(x, y, z) = (xyz, ze^{xy})$
$K(x, y, z) = (xy, yz, zx)$
Compute the following derivatives, using the chain rule:
(a) $(f\circ F)'$ (b) $(g\circ H)'$
(c) $(h\circ G)'$ (d) $(h\circ K)'$
(e) $(F\circ H)'$ (f) $(K\circ G)'$
(g) $(f\circ H)'$ (h) $(G\circ H)'$
(i) $(H\circ G)'$ (j) $(F\circ F)'$
Check the answers by computing the composite functions and differentiating.
2. Let $f:\mathbf{R}\to\mathbf{R}$ be differentiable and define $g:\mathbf{R}^2\to\mathbf{R}$ by $g(x, y) = f(x + y)$. Prove that $\partial g/\partial x - \partial g/\partial y = 0$.
3. Let V be an open subset of $\mathbf{R}^n$ and let $f: V\to\mathbf{R}$ be a function differentiable at $\mathbf{v}_0$. Define $g: V\to\mathbf{R}$ by $g(\mathbf{v}) = [f(\mathbf{v})]^n$ where n is a positive integer. Prove that g is differentiable at $\mathbf{v}_0$ and that
$$D_{\mathbf{v}_0}g = n[f(\mathbf{v}_0)^{n-1}]D_{\mathbf{v}_0}f$$

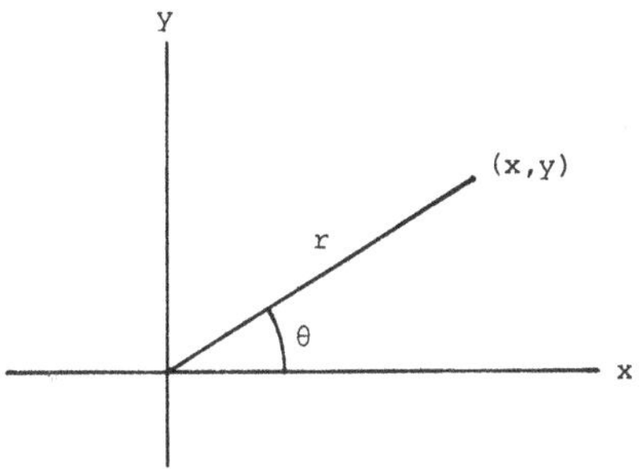

Figure 4.1 Polar coordinates.

4. Let $f: \mathbf{R}^2 \to \mathbf{R}$ be a function and define

$$F(r, \theta) = f(r \cos \theta, r \sin \theta)$$

This is the function f in *polar coordinates*. (See Figure 4.1.) Compute $\partial F/\partial r$ and $\partial F/\partial \theta$ in terms of the partial derivatives of f.

5. Let $f: \mathbf{R}^3 \to \mathbf{R}$ be a function and define

$$F(\rho, \theta, \varphi) = f(\rho \sin \varphi \cos \theta, \rho \sin \varphi \sin \theta, \rho \cos \varphi)$$

This is the function f in *spherical coordinates*. (See Figure 4.2.) Compute the partial derivatives of F in terms of those of f.

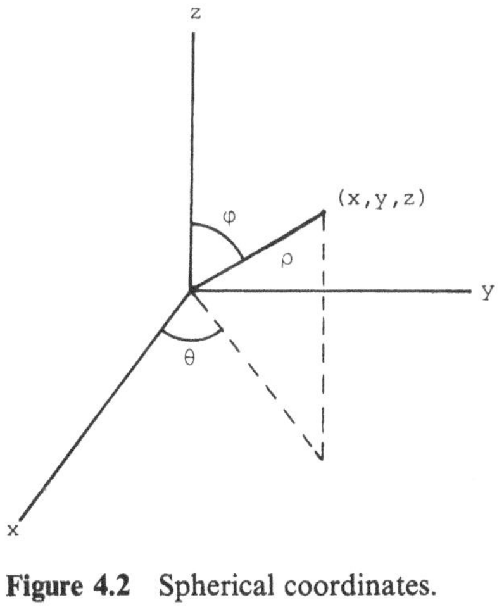

Figure 4.2 Spherical coordinates.

6. Suppose $f: \mathbf{R}^1 \to \mathbf{R}^4$ and $g: \mathbf{R}^4 \to \mathbf{R}$ are differentiable. Derive a formula for $(g \circ f)'$ in terms of the derivatives of f and g.
7. A function $f: V \to \mathbf{R}$ is said to be *homogeneous of degree k* if $f(t\mathbf{v}) = t^k f(\mathbf{v})$ for any real number $t > 0$. Prove that, if $f: \mathbf{R}^n \to \mathbf{R}$ is differentiable and homogeneous of degree k, then

$$\sum_{i=1}^{n} x_i \frac{\partial f}{\partial x_i}(x_1, \ldots, x_n) = k f(x_1, \ldots, x_n)$$

8. (a) Let u, v be functions from $\mathbf{R}^1$ to $\mathbf{R}^1$. Define functions $f: \mathbf{R}^1 \to \mathbf{R}^2$ and $g: \mathbf{R}^2 \to \mathbf{R}^1$ by

$$f(x) = (u(x), v(x))$$
$$g(x_1, x_2) = x_1 x_2$$

Compute $(g \circ f)'$ by the chain rule and thus derive the product rule for functions of one variable.

(b) Let U, $V: \mathbf{R}^n \to \mathbf{R}$ be differentiable functions and define $f: \mathbf{R}^n \to \mathbf{R}^2$ by $f(\mathbf{v}) = (U(\mathbf{v}), V(\mathbf{v}))$. Compute $(g \circ f)'$ by the chain rule; conclude that UV is differentiable, and find its derivative. (Compare Exercise 5 of Section 3.)

5. Proof of the Chain Rule

We begin with some preliminary results which are of independent interest.

We say that a linear transformation $T: \mathbf{R}^n \to \mathbf{R}^p$ is *bounded* if there is a number $M > 0$ such that

$$\| T(\mathbf{v}) \| \leq M \|\mathbf{v}\|$$

for all $\mathbf{v} \in \mathbf{R}^n$. The number M is called a *bound* for T.

Proposition 5.1 *Let $T: \mathbf{R}^n \to \mathbf{R}^p$ be a linear transformation. Then T is bounded.*

Proof. We know by Proposition 3.4 of Chapter 3 that T is continuous. Thus we can find a $\delta > 0$ so that $\| T(\mathbf{v})\| < 1$ whenever $\|\mathbf{v}\| < \delta$. Set $M = 2/\delta$. Then, for any nonzero $\mathbf{v} \in V$,

$$\left\| \frac{\delta \mathbf{v}}{2\|\mathbf{v}\|} \right\| = \frac{\delta}{2} < \delta$$

so that

$$\left\| T\left(\frac{\delta \mathbf{v}}{2\|\mathbf{v}\|}\right) \right\| < 1$$

However,

$$\left\| T\left(\frac{\delta \mathbf{v}}{2\|\mathbf{v}\|}\right) \right\| = \frac{\delta}{2\|\mathbf{v}\|} \| T(\mathbf{v})\| = \frac{\| T(\mathbf{v})\|}{M \|\mathbf{v}\|}$$

Since $\| T(\delta \mathbf{v}/2\|\mathbf{v}\|)\| < 1$, it follows that

$$\| T(\mathbf{v})\| \le \frac{2}{\delta}\|\mathbf{v}\| = M\|\mathbf{v}\|$$

for all $\mathbf{v} \in \mathbf{R}^n$.

Now suppose that U is an open subset of $\mathbf{R}^n$ and $f: U \to \mathbf{R}^p$ is differentiable at $\mathbf{v}_0$ in U. Then, by the definition of differentiability, f can be approximated near $\mathbf{v}_0$ (up to a constant)* by the linear transformation $D_{\mathbf{v}_0} f: V \to W$. As a result, we might expect that an analogue of Proposition 5.1 would hold for f near $\mathbf{v}_0$.

Proposition 5.2 *Let U be an open subset of $\mathbf{R}^n$ and $f: U \to \mathbf{R}^p$ a function differentiable at $\mathbf{v}_0$. Then there are positive real numbers N and δ so that*

$$\| f(\mathbf{v}) - f(\mathbf{v}_0)\| \le N\|\mathbf{v} - \mathbf{v}_0\| \tag{5.1}$$

whenever $\|\mathbf{v} - \mathbf{v}_0\| < \delta$.

A function satisfying equation (5.1) is said to satisfy a *local Lipschitz condition* at $\mathbf{v}_0$.

Proof of Proposition 5.2 Let $L = D_{\mathbf{v}_0} f$. Then

$$\begin{aligned}\| f(\mathbf{v}) - f(\mathbf{v}_0)\| &= \| f(\mathbf{v}) - f(\mathbf{v}_0) - L(\mathbf{v} - \mathbf{v}_0) + L(\mathbf{v} - \mathbf{v}_0)\| \\ &\le \| f(\mathbf{v}) - f(\mathbf{v}_0) - L(\mathbf{v} - \mathbf{v}_0)\| + \| L(\mathbf{v} - \mathbf{v}_0)\|\end{aligned}$$

Since f is differentiable at $\mathbf{v}_0$, we can choose δ so that

$$\| f(\mathbf{v}) - f(\mathbf{v}_0) - L(\mathbf{v} - \mathbf{v}_0)\| < \|\mathbf{v} - \mathbf{v}_0\|$$

for $\|\mathbf{v} - \mathbf{v}_0\| < \delta$. (We chose $\varepsilon = 1$ in the definition of differentiability.) Furthermore, we can use Proposition 5.1 to find an $M > 0$ such that

$$\| L(\mathbf{v} - \mathbf{v}_0)\| < M\|\mathbf{v} - \mathbf{v}_0\|$$

for all $\mathbf{v}$, $\mathbf{v}_0$ in $\mathbf{R}^n$. Now, defining $N = 1 + M$, we have

$$\|\mathbf{v} - \mathbf{v}_0\| < \delta \Rightarrow \| f(\mathbf{v}) - f(\mathbf{v}_0)\| \le N\|\mathbf{v} - \mathbf{v}_0\|$$

completing the proof of Proposition 5.2.

We can now prove the chain rule. Let $T = D_{\mathbf{v}_0} f$ and $S = D_{\mathbf{w}_0} g$. We must show that, for any $\varepsilon > 0$, we can find $\delta > 0$ with

$$\|(g \circ f)(\mathbf{v}) - (g \circ f)(\mathbf{v}_0) - (S \circ T)(\mathbf{v} - \mathbf{v}_0)\| < \varepsilon\|\mathbf{v} - \mathbf{v}_0\|$$

*We say "up to a constant" because we really approximate $f(\mathbf{v}) - f(\mathbf{v}_0)$.

whenever $\|\mathbf{v} - \mathbf{v}_0\| < \delta$. To begin with,

$$\begin{aligned}
\|(g\circ f)(\mathbf{v}) - (g\circ f)(\mathbf{v}_0) - (S\circ T)(\mathbf{v} - \mathbf{v}_0)\| &= \|(g\circ f)(\mathbf{v}) - (g\circ f)(\mathbf{v}_0) \\
&\quad - S(f(\mathbf{v}) - f(\mathbf{v}_0)) \\
&\quad + S(f(\mathbf{v}) - f(\mathbf{v}_0) - T(\mathbf{v} - \mathbf{v}_0))\| \\
&\leq \|(g\circ f)(\mathbf{v}) - (g\circ f)(\mathbf{v}_0) \\
&\quad - S(f(\mathbf{v}) - f(\mathbf{v}_0))\| \\
&\quad + \|S(f(\mathbf{v}) - f(\mathbf{v}_0) - T(\mathbf{v} - \mathbf{v}_0))\| \qquad (5.2)
\end{aligned}$$

We deal with these two terms separately.

Since f is differentiable at $\mathbf{v}_0$, we know by Proposition 5.2 that we can find real numbers k and $\delta_1 > 0$ so that

$$\|f(\mathbf{v}) - f(\mathbf{v}_0)\| < k\|\mathbf{v} - \mathbf{v}_0\| \tag{5.3}$$

when $\|\mathbf{v} - \mathbf{v}_0\| < \delta_1$. Since g is differentiable at $\mathbf{w}_0 = f(\mathbf{v}_0)$ with $S = D_{\mathbf{w}_0}g$, we can find $\delta_2 > 0$ with

$$\|g(\mathbf{w}) - g(\mathbf{w}_0) - S(\mathbf{w} - \mathbf{w}_0)\| < \frac{\varepsilon}{2k}\|\mathbf{w} - \mathbf{w}_0\| \tag{5.4}$$

when $\|\mathbf{w} - \mathbf{w}_0\| < \delta_2$. Finally, we know that f is continuous at $\mathbf{v}_0$, since it is differentiable there (by Proposition 2.3); thus there is a $\delta_3 > 0$ such that

$$\|\mathbf{v} - \mathbf{v}_0\| < \delta_3 \Rightarrow \|f(\mathbf{v}) - f(\mathbf{v}_0)\| < \delta_2 \tag{5.5}$$

Now, if $\|\mathbf{v} - \mathbf{v}_0\| < \min\{\delta_1, \delta_3\}$, we have [from equations (5.4) and (5.5)]

$$\begin{aligned}
\|g(f(\mathbf{v})) - g(f(\mathbf{v}_0)) - S(f(\mathbf{v}) - f(\mathbf{v}_0))\| &< \frac{\varepsilon}{2k}\|f(\mathbf{v}) - f(\mathbf{v}_0)\| \\
&< \frac{\varepsilon}{2k}(k\|\mathbf{v} - \mathbf{v}_0\|) \qquad \text{[from (5.3)]} \\
&= \frac{\varepsilon}{2}\|\mathbf{v} - \mathbf{v}_0\|
\end{aligned}$$

This takes care of the first term in the expression (5.2) above. To deal with the second term, we first use Proposition 5.1 to find an $M > 0$ with $\|S\mathbf{w}\| < M\|\mathbf{w}\|$. Then

$$\|S(f(\mathbf{v}) - f(\mathbf{v}_0) - T(\mathbf{v} - \mathbf{v}_0)) < M\|f(\mathbf{v}) - f(\mathbf{v}_0) - T(\mathbf{v} - \mathbf{v}_0)\|$$

Furthermore, since f is differentiable at $\mathbf{v}_0$, we can find $\delta_4 > 0$ so that

$$\|f(\mathbf{v}) - f(\mathbf{v}_0) - T(\mathbf{v} - \mathbf{v}_0)\| < \frac{\varepsilon}{2M}\|\mathbf{v} - \mathbf{v}_0\|$$

whenever $\|\mathbf{v} - \mathbf{v}_0\| < \delta_4$. It follows that

$$\| S(f(\mathbf{v}) - f(\mathbf{v}_0) - T(\mathbf{v} - \mathbf{v}_0))\| < \frac{\varepsilon}{2} \|\mathbf{v} - \mathbf{v}_0\|$$

whenever $\|\mathbf{v} - \mathbf{v}_0\| < \delta_4$.

If we now set $\delta = \min\{\delta_1, \delta_3, \delta_4\}$, we see from the inequality (5.2) above that

$$\|(g \circ f)(\mathbf{v}) - (g \circ f)(\mathbf{v}_0) - (S \circ T)(\mathbf{v} - \mathbf{v}_0)\| < \varepsilon \|\mathbf{v} - \mathbf{v}_0\|$$

whenever $\|\mathbf{v} - \mathbf{v}_0\| < \delta$. This finishes the proof of the chain rule.

Exercises

1. Show that any function $f: \mathbf{R}^n \to \mathbf{R}^p$ satisfying the local Lipschitz condition (5.1) is continuous at $\mathbf{v}_0$.
2. Give an example of a continuous function $f: (-1, 1) \to \mathbf{R}$ satisfying (5.1) at $x_0 = 0$ but not differentiable there.

*3. Let $f: \mathbf{R} \to \mathbf{R}$ be a function satisfying the condition

$$|f(x) - f(y)| < M|x - y|^n$$

for some $n > 1$. Prove that f is constant.

6. A Sufficient Condition for Differentiability

Let U be an open subset of $\mathbf{R}^n$ and $f: U \to \mathbf{R}^p$ a function, $f(\mathbf{v}) = (f_1(\mathbf{v}), \ldots, f_p(\mathbf{v}))$. We proved in Section 3 that, if f is differentiable at $\mathbf{v}_0$, then the first-order partial derivatives of the component functions $f_1, \ldots, f_p$ exist at $\mathbf{v}_0$. In addition, an example was given to show that the converse is not true in general. We now show that, with an additional hypothesis, a converse to the result above can be proved.

Proposition 6.1 *Let U be an open subset of $\mathbf{R}^n$ and $f: U \to \mathbf{R}^p$ a function with $f(\mathbf{v}) = (f_1(\mathbf{v}), \ldots, f_p(\mathbf{v}))$. Suppose that each of the functions $\partial f_i/\partial x_j$ exists and is continuous on U, $1 \le i \le p$, $1 \le j \le n$. Then f is differentiable on U.*

Proof. We prove this result for $n = 2$ and $p = 1$; the case $p > 1$ follows easily by considering the component functions and the proof for $n > 2$ is the same as the proof given below except that the Cauchy inequality (Theorem 2.1 of Chapter 5) is needed.

Let $\mathbf{v}_0 = (a_1, a_2)$ be any vector in U and $L: \mathbf{R}^2 \to \mathbf{R}$ the linear transformation whose matrix is $J_f(\mathbf{v}_0)$. Then

$$L(x_1, x_2) = \sum_{i=1}^{2} \frac{\partial f}{\partial x_i}(\mathbf{v}_0) x_i \tag{6.1}$$

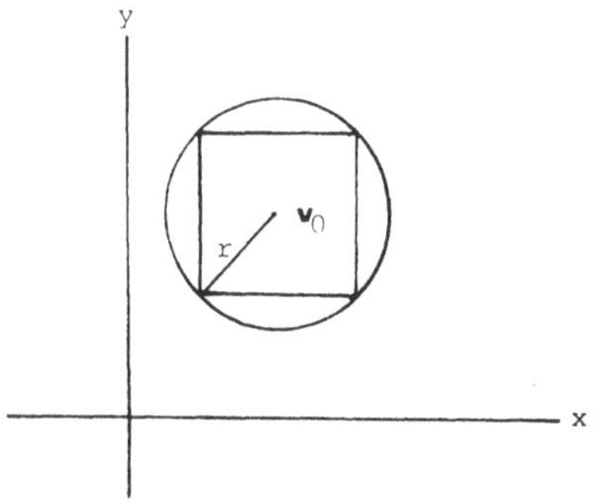

Figure 6.1 A square in a ball.

We must show that, for any $\varepsilon > 0$, there is a $\delta > 0$ so that

$$|f(\mathbf{v}) - f(\mathbf{v}_0) - L(\mathbf{v} - \mathbf{v}_0)| < \varepsilon \|\mathbf{v} - \mathbf{v}_0\|$$

when $\|\mathbf{v} - \mathbf{v}_0\| < \delta$.

Pick $r > 0$ so that $B_r(\mathbf{v}_0) \subset U$ (which we can do since U is open). It then follows that the "square" S with center $\mathbf{v}_0$ and side length $r\sqrt{2}$,

$$S = \left\{(x_1, x_2) \in \mathbf{R}^2 : |x_1 - a_1| < \frac{r}{\sqrt{2}} \text{ and } |x_2 - a_2| < \frac{r}{\sqrt{2}}\right\}$$

is contained entirely within $B_r(\mathbf{v}_0)$, so is contained entirely within U. (See Figure 6.1.) For the remainder of the section, we assume that all vectors that we deal with are in the square S.

Let $\mathbf{v} = (x_1, x_2)$ be in S and let $\mathbf{v}' = (x_1, a_2)$. Then $\mathbf{v}'$ is in S and we define $g : [a_1, x_1] \to \mathbf{R}$ by $g(x) = f(x_1, a_2)$. (See Figure 6.2.) Then g is differentiable on (a_1, x_1) and continuous on $[a_1, x_1]$ (since $\partial f/\partial x_1 = g'$ exists on U). Thus,

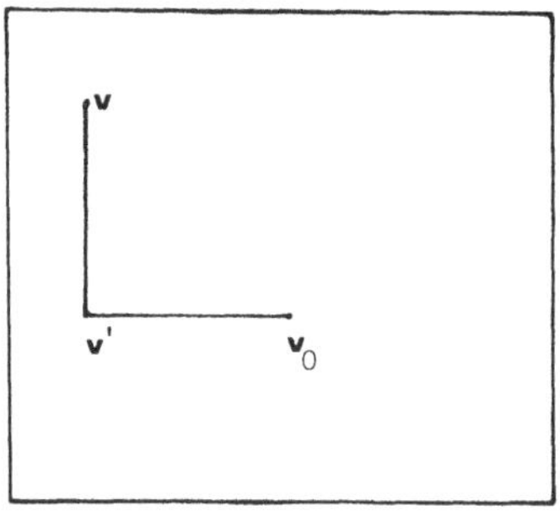

Figure 6.2

by the mean value theorem (see Exercise 9 of Section 1), there is a number $c_1 \in (a_1, x_1)$ such that

$$g(x_1) - g(a_1) = g'(c_1)(x_1 - a_1)$$

or, in terms of f,

$$f(x_1, a_2) - f(a_1, a_2) = \frac{\partial f}{\partial x_1}(\mathbf{v}_1)(x_1 - a_1)$$

where $\mathbf{v}_1 = (c_1, a_2)$.

Similarly, we apply the mean value theorem to the function $h : [a_2, x_2] \to \mathbf{R}$ defined by $h(x) = f(x_1, x)$ to obtain $c_2 \in (a_2, x_2)$ such that

$$f(x_1, x_2) - f(x_1, a_2) = \frac{\partial f}{\partial x_2}(\mathbf{v}_2)(x_2 - a_2)$$

where $\mathbf{v}_2 = (x_1, c_2)$.
Combining these two equations, we find that

$$f(\mathbf{v}) - f(\mathbf{v}_0) = \frac{\partial f}{\partial x_1}(\mathbf{v}_1)(x_1 - a_1) + \frac{\partial f}{\partial x_2}(\mathbf{v}_2)(x_2 - a_2) \tag{6.2}$$

Now, we use (6.1) and (6.2):

(6.3)

$$\begin{aligned} |f(\mathbf{v}) - f(\mathbf{v}_0) - L(\mathbf{v} - \mathbf{v}_0)| &= \left| \sum_{i=1}^{2} \frac{\partial f}{\partial x_i}(\mathbf{v}_i)(x_i - a_i) - \sum_{i=1}^{2} \frac{\partial f}{\partial x_i}(\mathbf{v}_0)(x_i - a_i) \right| \\ &= \left| \sum_{i=1}^{2} \left[\frac{\partial f}{\partial x_i}(\mathbf{v}_i) - \frac{\partial f}{\partial x_i}(\mathbf{v}_0) \right](x_i - a_i) \right| \\ &\le \left\{ \sum_{i=1}^{2} \left[\frac{\partial f}{\partial x_i}(\mathbf{v}_i) - \frac{\partial f}{\partial x_i}(\mathbf{v}_0) \right]^2 \right\}^{1/2} \|\mathbf{v} - \mathbf{v}_0\| \end{aligned}$$

by Lemma 4.2 of Chapter 2. (For $n > 2$, we need the Cauchy inequality here.)

Since the functions $\partial f/\partial x_j$ are continuous at $\mathbf{v}_0$, we find a $\delta > 0$ such that, for any $j = 1, 2$,

$$\left| \frac{\partial f}{\partial x_j}(\mathbf{v}) - \frac{\partial f}{\partial x_j}(\mathbf{v}_0) \right| < \frac{\varepsilon}{\sqrt{2}}$$

whenever $\|\mathbf{v} - \mathbf{v}_0\| < \delta$. Furthermore, we have

$$\|\mathbf{v}_j - \mathbf{v}_0\| \le \|\mathbf{v} - \mathbf{v}_0\| < \delta \qquad j = 1, 2$$

by the definition of $\mathbf{v}_1$ and $\mathbf{v}_2$, so that

$$\left| \frac{\partial f}{\partial x_j}(\mathbf{v}_j) - \frac{\partial f}{\partial x_j}(\mathbf{v}_0) \right| < \frac{\varepsilon}{\sqrt{2}} \qquad j = 1, 2$$

Therefore

$$\sum_{j=1}^{2}\left[\frac{\partial f}{\partial x_j}(\mathbf{v}_j)-\frac{\partial f}{\partial x_j}(\mathbf{v}_0)\right]^2<\varepsilon^2$$

whenever $\|\mathbf{v}-\mathbf{v}_0\|<\delta$. It now follows from (6.3) that

$$|f(\mathbf{v})-f(\mathbf{v}_0)-L(\mathbf{v}-\mathbf{v}_0)|<\varepsilon\,\|\mathbf{v}-\mathbf{v}_0\|$$

when $\|\mathbf{v}-\mathbf{v}_0\|<\varepsilon$. Thus f is differentiable at $\mathbf{v}_0$.

Exercises

1. Verify that the function $f\colon \mathbf{R}^2\to\mathbf{R}$ defined by

$$f(x, y)=\begin{cases}\dfrac{xy}{x^2+y^2} & \text{if } (x, y)\neq(0, 0)\\ 0 & \text{if } (x, y)=(0, 0)\end{cases}$$

does not have continuous partial derivatives at (0, 0).

2. Define $f\colon \mathbf{R}^2\to\mathbf{R}$ by

$$f(x_1, x_2)=\begin{cases}\dfrac{x_1^2x_2}{x_1^4+x_2^2} & \text{if } (x_1, x_2)\neq(0, 0)\\ 0 & \text{if } (x_1, x_2)=(0, 0)\end{cases}$$

(a) Show that f has partial derivatives everywhere.
(b) Show that f is discontinuous at 0 (and hence not differentiable there).

3. Define $f\colon \mathbf{R}\to\mathbf{R}$ by

$$f(x)=\begin{cases}x^2\sin\dfrac{1}{x^2} & \text{if } x\neq 0\\ 0 & \text{if } x=0\end{cases}$$

Show that f is differentiable everywhere on $\mathbf{R}$, but that f' is not continuous.

4. (a) Let $g, h\colon \mathbf{R}\to\mathbf{R}$ be differentiable, and define $f\colon \mathbf{R}^2\to\mathbf{R}$ by

$$f(x_1, x_2)=g(x_1)+h(x_2)$$

Show that f is differentiable.
(b) Find a function $f\colon \mathbf{R}^2\to\mathbf{R}$ such that f is differentiable, but the partial derivatives of f are discontinuous.

5. Prove Proposition 6.1 for $p=1$ and arbitrary n.
6. Prove the following:

Proposition Let U be an open subset of $\mathbf{R}^n$, and let $f: U \to \mathbf{R}^p$ be a function with $f(\mathbf{v}) = (f_1(\mathbf{v}), \ldots, f_p(\mathbf{v}))$. Suppose that the functions $f_1, \ldots, f_p$ all are differentiable (as functions from U to $\mathbf{R}$). Then f is differentiable on U.

(This result together with Exercise 5 proves the general case of Proposition 6.1.)

5

The Geometry of Euclidean Spaces

In this chapter, we introduce the notion of an inner product on a Euclidean space. This allows us to define the angle between two vectors and to describe some of the geometrical objects in Euclidean space such as planes and hyperplanes. We then show how these ideas can be used to obtain greater insight into the meaning of the derivative.

1. The Inner Product

When we discussed vectors in Chapter 2, we spoke of their "length" and "direction." Whereas the norm of a vector in $\mathbf{R}^n$ can be thought of as its length, we have as yet no notion of direction for vectors. We shall remedy this situation here by defining the inner product of two vectors in $\mathbf{R}^n$. This will allow us to define the angle between two vectors.

We begin with vectors in $\mathbf{R}^2$. Let $\mathbf{v} = (x_1, x_2)$ and $\mathbf{w} = (y_1, y_2)$ be vectors in $\mathbf{R}^2$. Using the law of cosines, we showed in the proof of Lemma 4.2 of Chapter 2 that

$$\|\mathbf{v}\| \cdot \|\mathbf{w}\| \cos\theta = x_1 y_1 + x_2 y_2$$

where θ is the angle between $\mathbf{v}$ and $\mathbf{w}$. (See Figure 1.1.) It follows that the angle θ between $\mathbf{v}$ and $\mathbf{w}$ is determined by their lengths and the quantity $x_1 y_1 + x_2 y_2$. This quantity is called the *inner product* of $\mathbf{v}$ and $\mathbf{w}$ and is denoted by $\langle \mathbf{v}, \mathbf{w} \rangle$:

(1.1) $$\langle \mathbf{v}, \mathbf{w} \rangle = x_1 y_1 + x_2 y_2$$

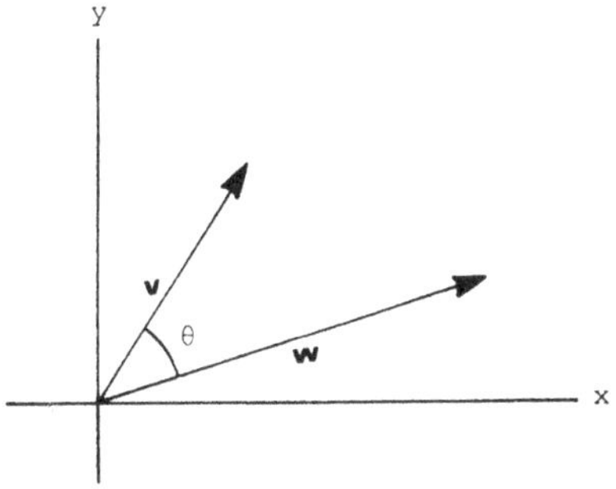

Figure 1.1 The angle between two vectors.

With this notation, the *angle* between **v** and **w** is defined to be the angle θ, $0 \leq \theta \leq 180°$, with

$$\cos\theta = \frac{\langle \mathbf{v}, \mathbf{w} \rangle}{\|\mathbf{v}\| \cdot \|\mathbf{w}\|} \tag{1.2}$$

(We assume here that neither **v** or **w** is the zero vector.)

Remark. The inner product $\langle \mathbf{v}, \mathbf{w} \rangle$ is sometimes written $\mathbf{v} \cdot \mathbf{w}$, and is therefore sometimes called the *dot product* of **v** and **w**. We shall not use this notation, however.

For example, if $\mathbf{v} = (1, \sqrt{3})$ and $\mathbf{w} = (\sqrt{3}, 1)$, then

$$\|\mathbf{v}\| = [(\sqrt{3})^2 + 1^2]^{1/2} = 2 = \|\mathbf{w}\|$$

and

$$\langle \mathbf{v}, \mathbf{w} \rangle = \sqrt{3} + \sqrt{3} = 2\sqrt{3}$$

Thus, if θ is the angle between **v** and **w**, we have

$$\cos\theta = \frac{2\sqrt{3}}{4} = \frac{\sqrt{3}}{2}$$

so $\theta = 30°$, which is what we expect. (See Figure 1.2.)

We can now follow the example of (1.1) to define the inner product of vectors in any Euclidean space. Let $\mathbf{v} = (x_1, \ldots, x_n)$, $\mathbf{w} = (y_1, \ldots, y_n)$ be vectors in $\mathbf{R}^n$. The *inner product* $\langle \mathbf{v}, \mathbf{w} \rangle$ of **v** with **w** is defined to be the number

$$\langle \mathbf{v}, \mathbf{w} \rangle = \sum_{j=1}^{n} x_j y_j \tag{1.3}$$

We define the angle between two vectors $\mathbf{v}, \mathbf{w} \in \mathbf{R}^n$ using equation (1.2).

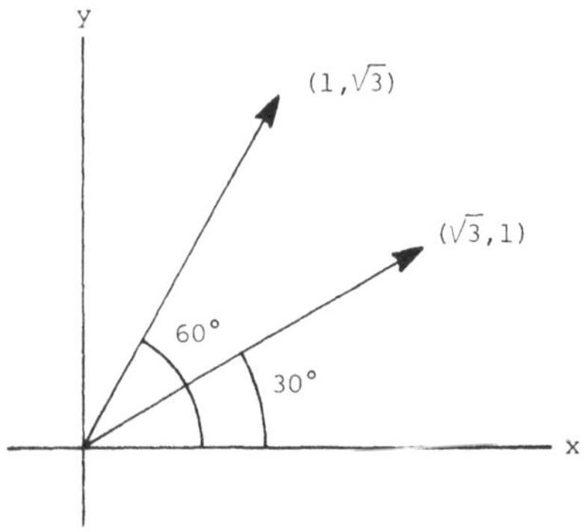

Figure 1.2

For example, if $\mathbf{v} = (1, 0, 3, 1)$ and $\mathbf{w} = (0, 2, 1, -1)$ are vectors in $\mathbf{R}^4$, then

$$\langle \mathbf{v}, \mathbf{w} \rangle = 1 \cdot 0 + 0 \cdot 2 + 3 \cdot 1 + 1 \cdot (-1) = 2$$

and the angle θ between $\mathbf{v}$ and $\mathbf{w}$ satisfies

$$\cos\theta = \frac{\langle \mathbf{v}, \mathbf{w} \rangle}{\|\mathbf{v}\| \cdot \|\mathbf{w}\|} = \frac{2}{\sqrt{11}\sqrt{6}} = \frac{2}{\sqrt{66}}$$

Thus, θ is approximately 76 degrees or 1.3 radians.

The next result gives some of the elementary properties of the inner product.

Proposition 1.1 *The inner product defined above has the following properties:*

(a) $\langle \mathbf{v}, \mathbf{w} \rangle = \langle \mathbf{w}, \mathbf{v} \rangle$ *for all* $\mathbf{v}, \mathbf{w} \in \mathbf{R}^n$.
(b) $\langle a\mathbf{v} + b\mathbf{v}', \mathbf{w} \rangle = a\langle \mathbf{v}, \mathbf{w} \rangle + b\langle \mathbf{v}', \mathbf{w} \rangle$ *for all* $\mathbf{v}, \mathbf{v}', \mathbf{w} \in \mathbf{R}^n$ *and all* $a, b \in \mathbf{R}$.
(c) $\langle \mathbf{v}, \mathbf{v} \rangle \geq 0$ *for all* $\mathbf{v} \in \mathbf{R}^n$ *and* $\langle \mathbf{v}, \mathbf{v} \rangle = 0$ *if and only if* $\mathbf{v} = \mathbf{0}$.

Proof. Statement (a) follows immediately from the commutativity of multiplication of real numbers. To prove (b), suppose $\mathbf{v} = (x_i)$, $\mathbf{w} = (y_i)$, and $\mathbf{v}' = (x_i')$. Then

$$\begin{aligned}
\langle a\mathbf{v} + b\mathbf{v}', \mathbf{w} \rangle &= \langle a(x_1, \ldots, x_n) + b(x_1', \ldots, x_n'), (y_1, \ldots, y_n) \rangle \\
&= \langle (ax_1 + bx_1', ax_2 + bx_2', \ldots, ax_n + bx_n'), \\
&\quad (y_1, \ldots, y_n) \rangle \\
&= (ax_1 + bx_1')y_1 + \cdots + (ax_n + bx_n')y_n \\
&= ax_1y_1 + bx_1'y_1 + \cdots + ax_ny_n + bx_n'y_n \\
&= a(x_1y_1 + \cdots + x_ny_n) + b(x_1'y_1 + \cdots + x_n'y_n) \\
&= a\langle \mathbf{v}, \mathbf{w} \rangle + b\langle \mathbf{v}', \mathbf{w} \rangle
\end{aligned}$$

Finally, $\langle \mathbf{v}, \mathbf{v}\rangle = x_1^2 + \cdots + x_n^2$, which is always nonnegative and is zero if and only if $x_1 = x_2 = \cdots = x_n = 0$.

Note that, combining (a) and (b), we have

$$\begin{aligned}\langle \mathbf{v}, a\mathbf{w} + b\mathbf{w}'\rangle &= \langle a\mathbf{w} + b\mathbf{w}', \mathbf{v}\rangle \\ &= a\langle \mathbf{w}, \mathbf{v}\rangle + b\langle \mathbf{w}', \mathbf{v}\rangle \\ &= a\langle \mathbf{v}, \mathbf{w}\rangle + b\langle \mathbf{v}, \mathbf{w}'\rangle\end{aligned}$$

The notion of the angle between two vectors defined above will not be of much use to us in what follows. There is one important exception, however. If θ is a right angle, then $\cos\theta = 0$, or $\langle \mathbf{v}, \mathbf{w}\rangle = 0$. We therefore define two vectors $\mathbf{v}$ and $\mathbf{w}$ to be *orthogonal* (and write $\mathbf{v} \perp \mathbf{w}$) if their inner product is 0:

$$\mathbf{v} \perp \mathbf{w} \qquad \text{if} \quad \langle \mathbf{v}, \mathbf{w}\rangle = 0$$

Notice that $\mathbf{0}$ is orthogonal to every vector $\mathbf{v}$, according to this definition.

Exercises

1. Let $\mathbf{v} = (1, 3, 1)$, $\mathbf{w} = (-2, 1, 0)$, and $\mathbf{u} = (1, -1, 2)$, and compute the following:
   (a) $\langle \mathbf{v}, \mathbf{w}\rangle$ (b) $\langle \mathbf{u}, \mathbf{v}\rangle$
   (c) $\langle 2\mathbf{u} + \mathbf{v}, \mathbf{w}\rangle$ (d) $\langle \mathbf{u} - \mathbf{w}, \mathbf{v}\rangle$
   (e) $\langle \mathbf{v} + \mathbf{w}, \mathbf{v} - \mathbf{w}\rangle$ (f) $\langle \mathbf{u} - 2\mathbf{w}, 3\mathbf{v} + \mathbf{u}\rangle$
2. Let $\mathbf{v} = (2, 0, 2, 1)$, $\mathbf{w} = (-1, 2, 1, 1)$, and $\mathbf{u} = (3, 1, -1, -1)$, and compute the following:
   (a) $\langle \mathbf{v}, \mathbf{u}\rangle$ (b) $\langle \mathbf{v}, \mathbf{w}\rangle$
   (c) $\langle \mathbf{u} - \mathbf{v}, \mathbf{w}\rangle$ (d) $\langle 2\mathbf{w} + \mathbf{u}, \mathbf{u}\rangle$
   (e) $\langle 3\mathbf{u} + 2\mathbf{w}, \mathbf{v} + \mathbf{u}\rangle$ (f) $\langle 2\mathbf{v} + \mathbf{u}, 3\mathbf{v} - 2\mathbf{w}\rangle$
3. Let $\mathbf{v}$, $\mathbf{w}$, and $\mathbf{u}$ be the vectors defined in Exercise 1 and compute the angle between
   (a) $\mathbf{v}$ and $\mathbf{w}$ (b) $\mathbf{v}$ and $\mathbf{u} - \mathbf{w}$
   (c) $2\mathbf{u} + \mathbf{v}$ and $\mathbf{w}$ (d) $\mathbf{v} + \mathbf{w}$ and $\mathbf{v} - \mathbf{w}$
4. Let $\mathbf{v}$, $\mathbf{w}$, and $\mathbf{u}$ be the vectors defined in Exercise 2 and compute the angle between
   (a) $\mathbf{v}$ and $\mathbf{u}$ (b) $\mathbf{u} - \mathbf{v}$ and $\mathbf{w}$
   (c) $2\mathbf{w} + \mathbf{u}$ and $\mathbf{u}$ (d) $\mathbf{v}$ and $\mathbf{w}$
5. Suppose that $\mathbf{u}$, $\mathbf{v}$, and $\mathbf{w}$ are vectors in $\mathbf{R}^n$ and that both $\mathbf{u}$ and $\mathbf{v}$ are orthogonal to $\mathbf{w}$. Prove that for any $a \in \mathbf{R}$, the vectors $a\mathbf{u}$ and $\mathbf{u} + \mathbf{v}$ are orthogonal to $\mathbf{w}$.
6. Let $\mathbf{v}_0$ be a fixed vector in $\mathbf{R}^n$ and define $f: \mathbf{R}^n \to \mathbf{R}$ by $f(\mathbf{v}) = \langle \mathbf{v}, \mathbf{v}_0\rangle$. Prove that f is continuous.
7. Let $\mathbf{v}$ and $\mathbf{w}$ be nonzero vectors in $\mathbf{R}^n$. Prove that
$$\|\mathbf{u} + \mathbf{v}\|^2 = \|\mathbf{u}\|^2 + \|\mathbf{v}\|^2$$
   if and only if $\mathbf{u}$ and $\mathbf{v}$ are orthogonal.
8. Let $\mathbf{v}$, $\mathbf{w}$ be vectors in $\mathbf{R}^n$. Prove the *parallelogram law*:
$$\|\mathbf{v} + \mathbf{w}\|^2 + \|\mathbf{v} - \mathbf{w}\|^2 = 2\|\mathbf{v}\|^2 + 2\|\mathbf{w}\|^2$$
   (Draw a picture to see where the name comes from.)

9. Prove the *polarization identity*: if $\mathbf{v}$, $\mathbf{w}$ are vectors in $\mathbf{R}^n$, then

$$\langle \mathbf{v}, \mathbf{w} \rangle = \tfrac{1}{2}(\|\mathbf{v} + \mathbf{w}\|^2 - \|\mathbf{v}\|^2 - \|\mathbf{w}\|^2)$$

(Thus one can recover the inner product from the norm.)

*10. Show that if $\mathbf{w}_1$ and $\mathbf{w}_2$ are vectors in $\mathbf{R}^n$ such that for every vector $\mathbf{v}$ in $\mathbf{R}^n$,

$$\langle \mathbf{v}, \mathbf{w}_1 \rangle = \langle \mathbf{v}, \mathbf{w}_2 \rangle$$

then

$$\mathbf{w}_1 = \mathbf{w}_2$$

*11. Show that if $T: \mathbf{R}^n \to \mathbf{R}^1$ is a linear transformation, then there is a unique vector $\mathbf{w}$ in $\mathbf{R}^n$ such that

$$T(\mathbf{v}) = \langle \mathbf{v}, \mathbf{w} \rangle$$

for all vectors $\mathbf{v}$ in $\mathbf{R}^n$. (This result is often called the *Riesz lemma.*)

12. Let $\mathbf{v}_0 = (a_1, \ldots, a_n)$ and define $f: \mathbf{R}^n \to \mathbf{R}$ by $f(\mathbf{v}) = \langle \mathbf{v}, \mathbf{v}_0 \rangle$. Prove that f is differentiable and that the Jacobian of f is the $1 \times n$ matrix $[a_1 \ \cdots \ a_n]$.

2. The Cauchy Inequality

The norm of a vector $\mathbf{v} = (x_1, \ldots, x_n) \in \mathbf{R}^n$ was defined in Section 2.5 to be the number

$$\|\mathbf{v}\| = (x_1^2 + \cdots + x_n^2)^{1/2}$$

In terms of the inner product of the last section, this becomes

$$\|\mathbf{v}\| = \sqrt{\langle \mathbf{v}, \mathbf{v} \rangle} \tag{2.1}$$

In fact, Lemma 4.2 of Chapter 2 can be interpreted as a relation between the norm and inner product in $\mathbf{R}^2$:

$$|\langle \mathbf{v}, \mathbf{w} \rangle| \leq \|\mathbf{v}\| \cdot \|\mathbf{w}\|$$

for all $\mathbf{v}, \mathbf{w} \in \mathbf{R}^2$. Our next result, the *Cauchy inequality*, asserts that this inequality holds in $\mathbf{R}^n$.

Theorem 2.1 *Let* $\mathbf{u}$, $\mathbf{v}$ *be vectors in* $\mathbf{R}^n$. *Then*

$$|\langle \mathbf{u}, \mathbf{v} \rangle| \leq \|\mathbf{u}\| \cdot \|\mathbf{v}\|$$

Proof. This proof is not difficult but is somewhat artificial. Note first of all that if either $\mathbf{u}$ or $\mathbf{v}$ is the zero vector, the inequality above is trivially satisfied. Next observe that, for any real numbers a and b, we have

$$0 \leq \|a\mathbf{u} + b\mathbf{v}\|^2 = a^2\|\mathbf{u}\|^2 + 2ab\langle \mathbf{u}, \mathbf{v} \rangle + b^2\|\mathbf{v}\|^2$$

Setting $a = \|\mathbf{v}\|$, $b = -\|\mathbf{u}\|$, we obtain

$$0 \leq 2\|\mathbf{u}\|^2 \cdot \|\mathbf{v}\|^2 - 2\|\mathbf{u}\| \cdot \|\mathbf{v}\| \langle \mathbf{u}, \mathbf{v} \rangle$$

Since we assume $\|\mathbf{u}\| \neq 0$ and $\|\mathbf{v}\| \neq 0$, we can divide this inequality through by $2\|\mathbf{u}\| \cdot \|\mathbf{v}\|$ and obtain

$$\langle \mathbf{u}, \mathbf{v} \rangle \leq \|\mathbf{u}\| \cdot \|\mathbf{v}\|$$

Similarly, if we set $a = \|\mathbf{v}\|$ and $b = \|\mathbf{u}\|$, we obtain the inequality

$$-\langle \mathbf{u}, \mathbf{v} \rangle \leq \|\mathbf{u}\| \cdot \|\mathbf{v}\|$$

Since $|\langle \mathbf{u}, \mathbf{v} \rangle|$ is either $\langle \mathbf{u}, \mathbf{v} \rangle$ or $-\langle \mathbf{u}, \mathbf{v} \rangle$, the Cauchy inequality follows from the two inequalities above. (See Proposition 4.1, Chapter 1.)

We note that the Cauchy inequality is what is needed to complete the proof of Proposition 6.1 of Chapter 4. (See Exercise 1.)

As we saw in Proposition 4.1 of Chapter 2, the Cauchy inequality is what is needed to prove the important *triangle inequality*.

Proposition 2.2 *For any two vectors* $\mathbf{v}_1$ *and* $\mathbf{v}_2$ *in* $\mathbf{R}^n$, *we have*

$$\|\mathbf{v}_1 + \mathbf{v}_2\| \leq \|\mathbf{v}_1\| + \|\mathbf{v}_2\|$$

Proof. We proceed directly:

$$\begin{aligned}
\|\mathbf{v}_1 + \mathbf{v}_2\|^2 &= \|\mathbf{v}_1\|^2 + \|\mathbf{v}_2\|^2 + 2\langle \mathbf{v}_1, \mathbf{v}_2 \rangle && \text{(by Exercise 9 of Section 1)} \\
&\leq \|\mathbf{v}_1\|^2 + \|\mathbf{v}_2\|^2 + 2\|\mathbf{v}_1\| \cdot \|\mathbf{v}_2\| && \text{(by Theorem 2.1)} \\
&= (\|\mathbf{v}_1\| + \|\mathbf{v}_2\|)^2
\end{aligned}$$

Since $\|\mathbf{v}_1 + \mathbf{v}_2\|$ and $\|\mathbf{v}_1\| + \|\mathbf{v}_2\|$ are nonnegative, we can take square roots to obtain the triangle inequality.

We note that the proofs of these last two results used only the properties of the inner product given in Proposition 1.1 and not the explicit form of the inner product.

Another useful inequality is the following.

Proposition 2.3 *For any vectors* $\mathbf{v}$, $\mathbf{w}$ *in* $\mathbf{R}^n$ *we have*

$$|(\|\mathbf{v}\| - \|\mathbf{w}\|)| \leq \|\mathbf{v} - \mathbf{w}\|$$

Proof. There are two cases to consider, depending on whether $\|\mathbf{v}\| - \|\mathbf{w}\|$ is positive or negative.

(a) If $\|\mathbf{v}\| - \|\mathbf{w}\| \geq 0$, we need to show that

$$\|\mathbf{v}\| - \|\mathbf{w}\| \leq \|\mathbf{v} - \mathbf{w}\|$$

This is easy: the triangle inequality says that

$$\|\mathbf{v}\| \leq \|\mathbf{v} - \mathbf{w}\| + \|\mathbf{w}\|$$

and we simply subtract $\|\mathbf{w}\|$ from both sides.

(b) If $\|\mathbf{v}\| - \|\mathbf{w}\| < 0$, we need to show that

$$\|\mathbf{w}\| - \|\mathbf{v}\| \leq \|\mathbf{v} - \mathbf{w}\|$$

However, $\|\mathbf{v} - \mathbf{w}\| = \|\mathbf{w} - \mathbf{v}\|$ and we can prove this inequality exactly as in case (a).

Exercise

1. Use the Cauchy inequality to prove Proposition 6.1 of Chapter 4 for arbitrary n and p.

3. Line, Planes, and Hyperplanes

We stated in Section 2.3 that the linear transformations between Euclidean spaces were the generalizations of the functions $f: \mathbf{R} \to \mathbf{R}$ whose graphs were straight lines through the origin. In order to demonstrate this, we need to do some geometry. We begin in this section with the notion of a hyperplane in $\mathbf{R}^m$. Hyperplanes are the generalizations of lines in $\mathbf{R}^2$ and planes in $\mathbf{R}^3$.

We first consider lines in $\mathbf{R}^2$. Suppose that we wish to describe the line in $\mathbf{R}^2$ which is orthogonal to the vector $\mathbf{n} = (a_1, a_2)$ and contains the endpoint of the vector $\mathbf{b} = (b_1, b_2)$. Then, if the endpoint of the vector $\mathbf{v} = (x_1, x_2)$ is on this line, it follows that the vector $\mathbf{v} - \mathbf{b}$ lies along the line (see Figure 3.1) and so it must be orthogonal to $\mathbf{n}$:

$$\langle \mathbf{n}, \mathbf{v} - \mathbf{b} \rangle = 0 \tag{3.1}$$

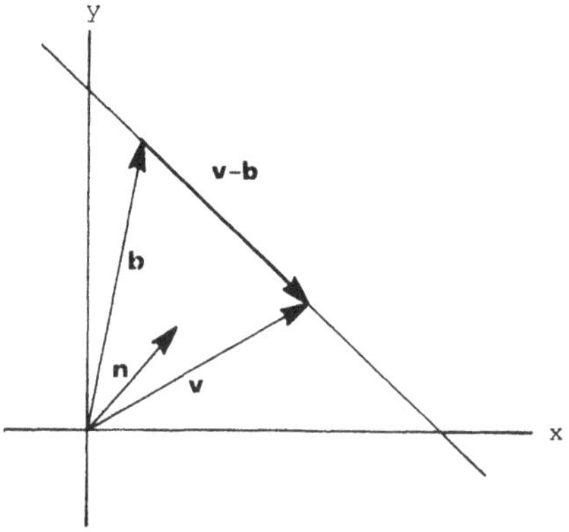

Figure 3.1 The line containing **b** with normal **n**.

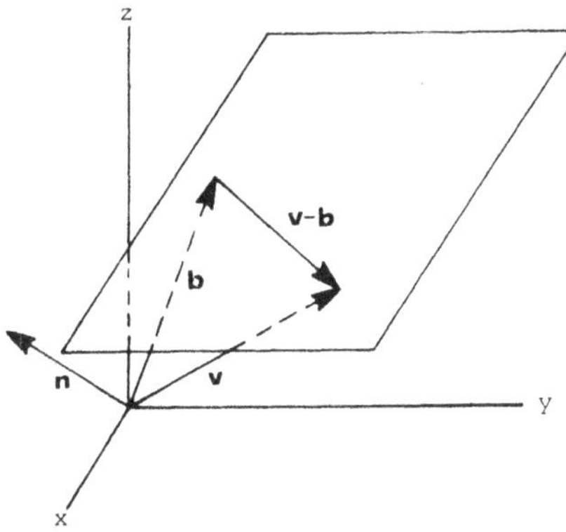

Figure 3.2 The plane containing **b** with normal **n**.

Conversely, if **v** is any vector satisfying (3.1), it is not difficult to show that the endpoint of **v** must lie on the line orthogonal to **n** containing the endpoint of **b**.

Equation (3.1) is called the *vector equation* of the line and the vector **n** is called a *normal* to the line. If we write (3.1) out in terms of components, we obtain the equation

$$a_1(x_1 - b_1) + a_2(x_2 - b_2) = 0$$

For example, the line orthogonal to the vector $\mathbf{n} = (2, -4)$ containing the point $\mathbf{b} = (7, -3)$ has the equation

$$2(x_1 - 7) - 4(x_2 + 3) = 0$$

or

$$2x_1 - 4x_2 - 26 = 0$$

Next, suppose we wish to describe the plane in $\mathbf{R}^3$ orthogonal to the vector $\mathbf{n} = (a_1, a_2, a_3)$ and containing the endpoint of the vector $\mathbf{b} = (b_1, b_2, b_3)$. Then, if $\mathbf{v} = (x_1, x_2, x_3)$ is any vector in $\mathbf{R}^3$ whose endpoint is on this plane, the vector $\mathbf{v} - \mathbf{b}$ must lie in the plane (see Figure 3.2) and thus must be orthogonal to **n**:

(3.2) $$\langle \mathbf{n}, \mathbf{v} - \mathbf{b} \rangle = 0$$

Conversely, if **v** is any vector satisfying (3.2), the endpoint of **v** must lie in the plane orthogonal to **n** containing **b**.

Equation (3.1) is called the *vector equation* of the plane and the vector **n** is called the *normal* to the plane.

For instance, the plane orthogonal to the vector $\mathbf{n} = (2, 1, -3)$ containing the point $\mathbf{b} = (2, 4, 7)$ has the equation

$$2(x_1 - 2) + (x_2 - 4) - 3(x_3 - 7) = 0$$

or

$$2x_1 + x_2 - 3x_3 + 13 = 0$$

Armed with the two cases above, we define the *hyperplane* in $\mathbf{R}^m$ with normal $\mathbf{n} \in \mathbf{R}^m$ containing the point $\mathbf{b} \in \mathbf{R}^m$ to be the set of all vectors $\mathbf{v} \in \mathbf{R}^m$ satisfying the equation

$$\langle \mathbf{n}, \mathbf{v} - \mathbf{b} \rangle = 0 \tag{3.3}$$

If $\mathbf{n} = (a_1, \ldots, a_m)$, $\mathbf{b} = (b_1, \ldots, b_m)$, and $\mathbf{v} = (x_1, \ldots, x_m)$, equation (3.3) becomes

$$\sum_{j=1}^{m} a_j(x_j - b_j) = 0 \tag{3.4}$$

For example, the hyperplane in $\mathbf{R}^7$ with normal $\mathbf{n} = (1, 0, -2, 0, 0, 12, 0)$ containing the vector $\mathbf{b} = (2, -7, 3, 3, 1, 4, 6)$ has equation

$$(x_1 - 2) - 2(x_3 - 3) + 12(x_6 - 4) = 0$$

or

$$x_1 - 2x_3 + 12x_6 - 44 = 0$$

Remark. Our point of view above is based upon the geometrical "fact" that a plane in $\mathbf{R}^3$ is determined by its normal direction together with one of its points. Of course, there are other ways of determining a plane in $\mathbf{R}^3$; we can give two lines in the plane or a line and a point in the plane. However, these descriptions do not generalize as easily to higher dimensional Euclidean spaces.

Let $f: \mathbf{R}^n \to \mathbf{R}^p$ be a function. The *graph of f*, $G(f)$, is the subset of $\mathbf{R}^{n+p}$ defined by

$$G(f) = \{(x_1, \ldots, x_{n+p}) \in \mathbf{R}^{n+p} : f(x_1, \ldots, x_n) = (x_{n+1}, \ldots, x_{n+p})\}$$

For example, if $f(x) = x^2$, the graph of f consists of all $(x_1, x_2) \in \mathbf{R}^2$ such that $x_1^2 = x_2$. We recognize this as a parabola. (See Figure 3.3.)

Suppose now that $T: \mathbf{R}^n \to \mathbf{R}$ is the linear transformation given by

$$T(x_1, \ldots, x_n) = a_{11}x_1 + \cdots + a_{1n}x_n \tag{3.5}$$

The graph of T consists of all points $(x_1, \ldots, x_{n+1}) \in \mathbf{R}^{n+1}$ satisfying the equation

$$x_{n+1} = T(x_1, \ldots, x_n) = a_{11}x_1 + \cdots + a_{1n}x_n$$

or, equivalently, the equation

$$a_{11}x_1 + \cdots + a_{1n}x_n - x_{n+1} = 0$$

This is the equation of a hyperplane in $\mathbf{R}^{n+1}$ through $\mathbf{0}$ with normal $(a_{11}, \ldots, a_{1n}, -1)$.

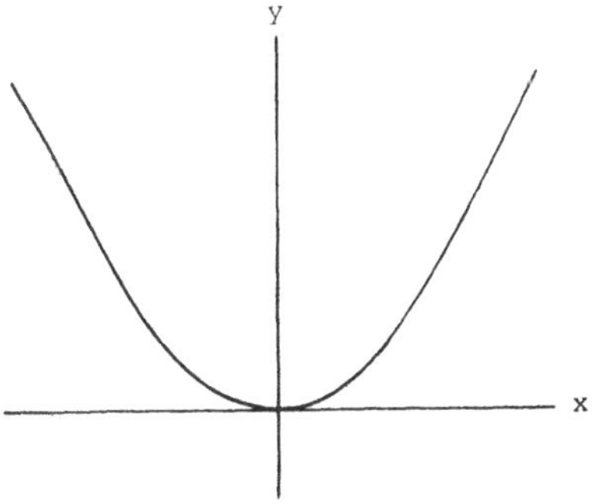

Figure 3.3 The parabola.

More generally, suppose $h : \mathbf{R}^n \to \mathbf{R}$ is defined by

$$h(\mathbf{v}) = T(\mathbf{v}) + c \tag{3.6}$$

where $T : \mathbf{R}^n \to \mathbf{R}$ is the linear transformation defined by (3.5) and c is a fixed real number. The graph of h is the set of all $(x_1, \ldots, x_{n+1}) \in \mathbf{R}^{n+1}$ satisfying the equation

$$a_{11}x_1 + \cdots + a_{1n}x_n - x_{n+1} + c = 0 \tag{3.7}$$

To see that this is a hyperplane, suppose that $(b_1, \ldots, b_{n+1})$ is any point in the graph of h, so that

$$a_{11}b_1 + \cdots + a_{1n}b_n - b_{n+1} + c = 0 \tag{3.8}$$

Subtracting equation (3.8) from (3.7), we obtain

$$a_{11}(x_1 - b_1) + \cdots + a_{1n}(x_n - b_n) - (x_{n+1} - b_{n+1}) = 0 \tag{3.9}$$

It follows that the graph of h is the set of points $(x_1, \ldots, x_{n+1}) \in \mathbf{R}^{n+1}$ satisfying equation (3.9), so is indeed a hyperplane.

We can now further justify our definition of differentiability for functions between Euclidean spaces. We observed in Section 2.1 that a function $f : \mathbf{R} \to \mathbf{R}$ is differentiable at x_0 if, near x_0, f can be approximated in a strong sense by a function whose graph is a straight line. The definition of differentiability for functions $f : \mathbf{R}^n \to \mathbf{R}$ given in Section 4.1 involved approximating the function by a function of the form (3.6) [where $b = f(\mathbf{v}_0) - T(\mathbf{v}_0)$]. As we demonstrated above, the graphs of these functions are hyperplanes in $\mathbf{R}^{n+1}$. Since hyperplanes in $\mathbf{R}^{n+1}$ are generalizations of lines in $\mathbf{R}^2$, our definition of differentiability is a reasonable generalization of the standard definition, at least for functions $f : \mathbf{R}^n \to \mathbf{R}$. To deal with functions $f : \mathbf{R}^n \to \mathbf{R}^p, p > 1$, we need the notion of k planes in $\mathbf{R}^m$. This is developed in the next section.

Exercises

1. Find the equation of the hyperplane in $\mathbf{R}^m$ with normal $\mathbf{n}$ containing the point $\mathbf{b}$, where
   (a) $\mathbf{n} = (0, 1, 1)$, $\mathbf{b} = (2, -1, 3)$
   (b) $\mathbf{n} = (2, 1, -3)$, $\mathbf{b} = (1, 0, -5)$
   (c) $\mathbf{n} = (-4, 0, 2, 2)$, $\mathbf{b} = (10, 7, -7, 14)$
   (d) $\mathbf{n} = (1, 0, 4, -1, 2)$, $\mathbf{b} = (0, 0, 2, 1, 0)$
   (e) $\mathbf{n} = (2, -3, 2, 1, 6)$, $\mathbf{b} = (5, 1, 0, 1, -1)$
2. Find the equation of the hyperplane in $\mathbf{R}^3$ through the origin containing the vectors $(1, 1, 0)$ and $(0, 1, 1)$.
3. Find the equation of the hyperplane in $\mathbf{R}^4$ through the origin containing the vectors $(1, 0, 1, 0)$, $(1, 0, 0, 1)$, and $(0, 1, 0, 1)$.
4. Show that any function $f: \mathbf{R}^n \to \mathbf{R}$ whose graph is a hyperplane must have the form (3.6) for some linear transformation $T: \mathbf{R}^n \to \mathbf{R}$ and $b \in \mathbf{R}$.
5. Determine exactly which hyperplanes in $\mathbf{R}^{n+1}$ are the graphs of functions $f: \mathbf{R}^n \to \mathbf{R}$.

4. *k* Planes in Euclidean Space

In the last section, we introduced the notion of a hyperplane in $\mathbf{R}^n$. These were analogues of lines in $\mathbf{R}^2$ and planes in $\mathbf{R}^3$. We now discuss the higher dimensional analogues of lines in $\mathbf{R}^3$.

Suppose L is a line in $\mathbf{R}^3$. Then, if P_1 and P_2 are distinct planes containing L, the intersection $P_1 \cap P_2$ is exactly the line L. (See Figure 4.1.) Thus we can think of L as determined by the two planes P_1 and P_2.

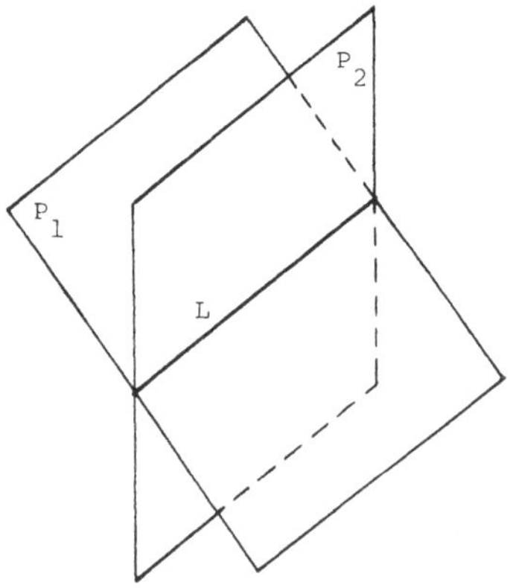

Figure 4.1 The intersection of two planes.

In fact, if P_1 and P_2 are any two *nonparallel* planes in $\mathbf{R}^3$, their intersection is a line. Of course, different planes can determine the same line. Nevertheless, we will define a line in $\mathbf{R}^3$ to be the intersection of two nonparallel planes.

For example, the x_1 axis is determined by the planes $x_2 = 0$ and $x_3 = 0$. More generally, if P_1 has the equation

$$a_{11}(x_1 - b_{11}) + a_{12}(x_2 - b_{12}) + a_{13}(x_3 - b_{13}) = 0 \tag{4.1}$$

and P_2 has the equation

$$a_{21}(x_1 - b_{21}) + a_{22}(x_2 - b_{22}) + a_{23}(x_3 - b_{23}) = 0 \tag{4.2}$$

then their intersection consists of all points (x_1, x_2, x_3) that satisfy both (4.1) and (4.2).

Of course, we can describe lines in $\mathbf{R}^3$ in other ways. For example, suppose we wish to define the line L through the origin in the direction of the vector $\mathbf{v}_1 = (1, 1, -1)$ as the intersection of two nonparallel planes. Two such planes will contain the origin so they are determined by their normals $\mathbf{n}_1$ and $\mathbf{n}_2$. Since the vector $\mathbf{v}_1$ is to lie in both planes, $\mathbf{v}_1$ will be orthogonal to both $\mathbf{n}_1$ and $\mathbf{n}_2$. Thus, all we need do is find two vectors $\mathbf{n}_1 = (a_{11}, a_{12}, a_{13})$, $\mathbf{n}_2 = (a_{21}, a_{22}, a_{23})$ orthogonal to $\mathbf{v}_1$:

$$\langle \mathbf{v}_1, \mathbf{n}_1 \rangle = a_{11} + a_{12} - a_{13} = 0$$

$$\langle \mathbf{v}_1, \mathbf{n}_2 \rangle = a_{21} + a_{22} - a_{23} = 0$$

We see by inspection that the vectors $\mathbf{n}_1 = (1, -1, 0)$ and $\mathbf{n}_2 = (0, 1, 1)$ will do. Therefore L is the set of all points $(x_1, x_2, x_3) \in \mathbf{R}^3$ which satisfy the two equations

$$x_1 - x_2 = 0 \qquad x_2 + x_3 = 0$$

We would now like to define a k plane in $\mathbf{R}^m$ to be the set of points common to hyperplanes $P_1, \ldots, P_{m-k}$ in $\mathbf{R}^m$. To do this we need a condition on these planes which is analogous to planes not being parallel. The condition that the two planes defining a line in $\mathbf{R}^3$ be nonparallel is equivalent to requiring that neither of the normals of the two planes be a multiple of the other. In general, we say that vectors $\mathbf{v}_1, \ldots, \mathbf{v}_r$ in $\mathbf{R}^m$ are *linearly independent* if no one of the vectors can be expressed as a linear combination of the others. (See Exercise 3, Section 2.2.)

For example, the vectors (1, 0, 0), (0, 1, 0) and (0, 0, 1) are obviously linearly independent, whereas $\mathbf{v}_1 = (1, 2, -1)$, $\mathbf{v}_2 = (3, 0, 5)$, $\mathbf{v}_3 = (2, 1, 2)$ are not linearly independent (thus *linearly dependent*) since $\mathbf{v}_2 = 2\mathbf{v}_3 - \mathbf{v}_1$.

We can now define a *k plane in* $\mathbf{R}^m$ to consist of all points on the intersection of hyperplanes, $P_1, \ldots, P_{m-k}$, whose normal vectors are linearly independent. We call a 1 plane in $\mathbf{R}^m$ a *line* in $\mathbf{R}^m$.

To describe a k plane in terms of equations, we first suppose that $\mathbf{b} = (b_1, \ldots, b_m)$ is a point on the k plane defined by the hyperplanes $P_1, \ldots, P_{m-k}$. In particular, $\mathbf{b}$ lies on each of the hyperplanes $P_1, \ldots, P_{m-k}$. Suppose further that $\mathbf{n}_j$ is a normal for P_j, $1 \leq j \leq m-k$, where $\mathbf{n}_j = (a_{j1}, \ldots, a_{jm})$. The equation of the hyperplane P_j is then

$$\sum_{i=1}^{m} a_{ji}(x_i - b_i) = 0$$

It follows that the k plane defined by $P_1, \ldots, P_{m-k}$ is the set of all points satisfying the $m-k$ equations

$$\begin{aligned}
a_{11}(x_1 - b_1) + \cdots + a_{1m}(x_m - b_m) &= 0 \\
a_{21}(x_1 - b_1) + \cdots + a_{2m}(x_m - b_m) &= 0 \\
\cdots\cdots\cdots\cdots\cdots\cdots\cdots\cdots\cdots\cdots \\
a_{m-k}(x_1 - b_1) + \cdots + a_{m-k}(x_m - b_m) &= 0
\end{aligned}$$

In order better to understand the condition that the normals of the hyperplanes defining a k plane be independent, we consider an example. Let P_1, P_2, P_3 be the hyperplanes in $\mathbf{R}^5$ each containing the point $\mathbf{b} = (1, 3, 1, -2, 2)$ with normals

$$\begin{aligned}
\mathbf{n}_1 &= (3,\ 1,\ -1,\ 0,\ 5) \\
\mathbf{n}_2 &= (2,\ 4,\ 1,\ 0,\ 1) \\
\mathbf{n}_3 &= (4,\ -2,\ -3,\ 0,\ 9)
\end{aligned}$$

The intersection P of these three hyperplanes is the set of all $(x_1, \ldots, x_5) \in \mathbf{R}^5$ satisfying the three equations

$$\begin{aligned}
3(x_1 - 1) + (x_2 - 3) - (x_3 - 1) + 5(x_5 - 2) &= 0 \\
2(x_1 - 1) + 4(x_2 - 3) + (x_3 - 1) + (x_5 - 2) &= 0 \\
4(x_1 - 1) - 2(x_2 - 3) - 3(x_3 - 1) + 9(x_5 - 2) &= 0
\end{aligned}$$

We would expect P to be a 2 plane in $\mathbf{R}^5$. However, the normals $\mathbf{n}_1$, $\mathbf{n}_2$, and $\mathbf{n}_3$ are not linearly independent: $2\mathbf{n}_1 - \mathbf{n}_2 = \mathbf{n}_3$. It follows that any point in $\mathbf{R}^5$ satisfying the first two equations above *automatically* satisfies the third equation. Thus,

$$P = P_1 \cap P_2 \cap P_3 = P_1 \cap P_2$$

so that P is actually the 3 plane in $\mathbf{R}^5$ defined by P_1 and P_2.

More generally, if $P_1, \ldots, P_r$ are hyperplanes with normal vectors $\mathbf{n}_1, \ldots, \mathbf{n}_r$ and if $\mathbf{n}_j$ is a linear combination of $\mathbf{n}_1, \ldots, \mathbf{n}_{j-1}, \mathbf{n}_{j+1}, \ldots, \mathbf{n}_r$, then

$$P_1 \cap \cdots \cap P_r = P_1 \cap \cdots \cap P_{j-1} \cap P_{j+1} \cap \cdots \cap P_r$$

Suppose now that $T: \mathbf{R}^n \to \mathbf{R}^p$ is a linear transformation with matrix (a_{ij}). Then $G(T)$, the graph of T, is the set of all points $(x_1, \ldots, x_{n+p}) \in \mathbf{R}^{n+p}$ satisfying $T(x_1, \ldots, x_n) = (x_{n+1}, \ldots, x_{n+p})$ or, in terms of the matrix of T,

$$(x_{n+1}, \ldots, x_{n+p}) = \left(\sum_{j=1}^{n} a_{j1} x_j, \ldots, \sum_{j=1}^{n} a_{jp} x_j \right)$$

This equation is equivalent to the p equations

$$\begin{aligned} x_{n+1} &= a_{11}x_1 + \cdots + a_{n1}x_n \\ &\cdots\cdots\cdots\cdots\cdots\cdots \\ x_{n+p} &= a_{1p}x_1 + \cdots + a_{np}x_n \end{aligned}$$

which, in turn, are equivalent to the p equations

$$\begin{aligned} a_{11}x_1 + \cdots + a_{n1}x_n - x_{n+1} &= 0 \\ \cdots\cdots\cdots\cdots\cdots\cdots\cdots\cdots & \\ a_{1p}x_1 + \cdots + a_{np}x_n - x_{n+p} &= 0 \end{aligned} \tag{4.3}$$

Now, each of the equations in (4.3) is the equation of a plane and the normals to these planes are the vectors

$$\begin{aligned} \mathbf{n}_1 &= (a_{11}, \ldots, a_{n1}, 1, 0, \ldots, 0) \\ \mathbf{n}_2 &= (a_{12}, \ldots, a_{n2}, 0, 1, 0, \ldots, 0) \\ &\cdots\cdots\cdots\cdots\cdots\cdots\cdots\cdots \\ \mathbf{n}_p &= (a_{1p}, \ldots, a_{np}, 0, \ldots, 0, 1) \end{aligned} \tag{4.4}$$

These vectors are linearly independent; $\mathbf{n}_j$ cannot be expressed as a linear combination of the other $\mathbf{n}_i$ since $\mathbf{n}_j$ has $(n+j)$th component equal to 1 whereas each $\mathbf{n}_i$, $i \neq j$, has $(n+j)$th component equal to zero. Thus the equations (4.3) define an $(n-p)$ plane in $\mathbf{R}^{n+p}$ and we see that the graph of a linear transformation is an $(n-p)$ plane through the origin in $\mathbf{R}^{n+p}$.

If $h: \mathbf{R}^n \to \mathbf{R}^p$ is defined by

$$h(\mathbf{v}) = T(\mathbf{v}) + \mathbf{c}$$

where $T: \mathbf{R}^n \to \mathbf{R}^p$ is linear and $\mathbf{c} \in \mathbf{R}^p$ is a fixed vector, we see, just as we did in the previous section when $p = 1$, that the graph of $\mathbf{h}$ is an $(n-p)$ plane in $\mathbf{R}^{n+p}$. (The proof is left as an exercise.) Thus we have shown that a function $f: \mathbf{R}^n \to \mathbf{R}^p$ is differentiable at $\mathbf{v}_0$ if near $\mathbf{v}_0$, f can be approximated in a strong sense by a function whose graph is an $(n-p)$ plane, namely the function $h: \mathbf{R}^n \to \mathbf{R}^p$ defined by

$$h(\mathbf{v}) = L(\mathbf{v}) + \mathbf{c}$$

where $L = D_{\mathbf{v}_0} f$ and $\mathbf{c} = f(\mathbf{v}_0) - L(\mathbf{v}_0)$.

Exercises

1. Find two planes in $\mathbf{R}^3$ whose intersection defines the line through the origin containing the vector
   (a) $(1, 0, 1)$ (b) $(2, 1, 0)$
   (c) $(1, 2, 1)$ (d) $(2, -3, 1)$
2. Find three hyperplanes in $\mathbf{R}^4$ whose intersection defines the line through the origin containing the vector
   (a) $(1, 0, 1, 0)$ (b) $(1, 0, 1, 1)$
   (c) $(1, 2, -1, 0)$ (d) $(1, 2, -1, 3)$
3. Find two hyperplanes in $\mathbf{R}^4$ whose intersection defines the 2 plane through the origin containing
   (a) $(1, 0, 1, 0), (0, 1, 0, 1)$ (b) $(1, 0, 1, 0), (1, 0, 0, 1)$
   (c) $(1, 0, 2, 1), (0, 1, -1, 1)$ (d) $(1, 2, 0, 1), (1, -1, 3, 2)$
4. Determine which of the following sets of vectors are linearly independent.
   (a) $(1, 0, 2), (0, 1, 1)$
   (b) $(1, 1, 2), (0, 2, 1), (3, -1, 4)$
   (c) $(1, 1, 0, 1), (0, 1, 1, 0), (0, -1, 2, 1)$
   (d) $(2, 1, 0, 1), (-1, 3, 1, 0), (1, 1, 2, 1), (1, -2, 1, 1)$
5. Find the equation of the line through the point $\mathbf{v}_0$ orthogonal to the hyperplane P, where
   (a) $\mathbf{v}_0 = (0, 0, 0)$ and P has the equation $x + y + z = 1$.
   (b) $\mathbf{v}_0 = (0, 2, -1)$ and P has the equation $2x - 3y + z - 5 = 0$.
   (c) $\mathbf{v}_0 = (1, 0, 1, 0)$ and P has the equation $x_1 + 2x_2 + 3x_3 + 4x_4 = 5$.
   (d) $\mathbf{v}_0 = (2, -2, 1, 3)$ and P has the equation $2x_1 - x_3 + 7x_4 = 11$.
   (e) $\mathbf{v}_0 = (1, 0, -1, 2, 0)$ and P has the equation $2x_1 - x_4 + 4x_5 = 0$.
6. Find the equation of the hyperplane through $(1, 2, 3)$ perpendicular to the line given by the two planes $x + y - z = 0$ and $x - 2y + z = 1$.
7. Let $T : \mathbf{R}^n \to \mathbf{R}^p$ be a linear transformation and $\mathbf{c} \in \mathbf{R}^p$ a fixed vector. Prove that the graph of the function $h : \mathbf{R}^n \to \mathbf{R}^p$ defined by $h(\mathbf{v}) = T(\mathbf{v}) + \mathbf{c}$ is an $(n - p)$ plane in $\mathbf{R}^{n+p}$.
8. Let $h : \mathbf{R}^n \to \mathbf{R}^p$ be a function whose graph is an $(n - p)$ plane in $\mathbf{R}^{n+p}$. Prove that $h(\mathbf{v}) = T(\mathbf{v}) + \mathbf{c}$ for some linear transformation $T: \mathbf{R}^n \to \mathbf{R}^p$ and fixed vector $\mathbf{c} \in \mathbf{R}^p$.
9. Determine exactly which $(n - p)$ planes in $\mathbf{R}^{n+p}$ are the graphs of functions $f: \mathbf{R}^n \to \mathbf{R}^p$.
10. Prove that the vectors (4.4) are linearly independent.

5. The Directional Derivative

A notion closely related to the partial derivative is that of the directional derivative. Let U be an open subset of $\mathbf{R}^n$, and let $f: U \to \mathbf{R}$ be a function. Pick vectors $\mathbf{v} \in U$ and $\mathbf{v}_0 \in \mathbf{R}^n$, with $\mathbf{v}_0$ a *unit vector* (that is, a vector with $\|\mathbf{v}_0\| = 1$). *The directional derivative of f at* $\mathbf{v}$ in the direction $\mathbf{v}_0$ is defined to be the number

$$\lim_{h\to 0} \frac{f(\mathbf{v} + h\mathbf{v}_0) - f(\mathbf{v})}{h}$$

if the limit exists.

The reason for the name is the following: one can specify a direction by a unit vector. In $\mathbf{R}^2$, for instance, $(1, 0)$ specifies the direction along the x axis, $(0, -1)$ the direction along the negative y axis, and $(\sqrt{2}/2, \sqrt{2}/2)$ the direction making a 45° angle with the positive x axis and with the positive y axis. Now consider the function f, but restricted to the line L through $\mathbf{v}$ and parallel to $\mathbf{v}_0$. (See Figure 5.1.) This line consists of all points of the form $\mathbf{v} + h\mathbf{v}_0$, $h \in \mathbf{R}$. Thus if one looks only at this line, f can be regarded as a function of one variable h and its derivative at $h = 0$ is the directional derivative of f at $\mathbf{v}$ in the direction $\mathbf{v}_0$.

Some examples follow.

1. If $\mathbf{v}_0 = \mathbf{e}_j$, the directional derivative of f at $\mathbf{v}$ in the direction $\mathbf{v}_0$ is just $(\partial f/\partial x_j)(\mathbf{v})$. To see why, write out the definition of the directional derivative in this case (for simplicity of notation, we set $j = 1$):

$$\text{Directional derivative} = \lim_{h\to 0} \frac{f(\mathbf{v} + h\mathbf{e}_1) - f(\mathbf{v})}{h}$$

$$= \lim_{h\to 0} \frac{f(x_1 + h_1, x_2, \ldots, x_n) - f(x_1, x_2, \ldots, x_n)}{h} = \frac{\partial f}{\partial x_1}(\mathbf{v})$$

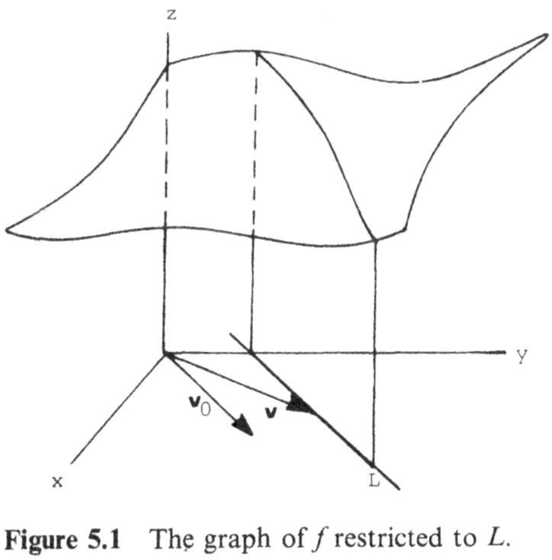

Figure 5.1 The graph of f restricted to L.

2. If $f: \mathbf{R}^2 \to \mathbf{R}$ is given by

$$f(x, y) = x^2 y$$

and $\mathbf{v}_0 = (\sqrt{2}/2, \sqrt{2}/2)$, then the directional derivative of f at $\mathbf{v} = (x, y)$ in the direction $\mathbf{v}_0$ is

$$\lim_{h \to 0} \frac{f(x + h\sqrt{2}/2,\, y + h\sqrt{2}/2) - f(x, y)}{h}$$

$$= \lim_{h \to 0} \frac{(x + h\sqrt{2}/2)^2 (y + h\sqrt{2}/2) - x^2 y}{h}$$

$$= \lim_{h \to 0} \frac{x^2 h\sqrt{2}/2 + xyh\sqrt{2} + 2xh^2 \frac{1}{2} + yh^2 \frac{1}{2} + h^3\sqrt{2}/4}{h}$$

$$= x^2 \frac{\sqrt{2}}{2} + xy\sqrt{2}$$

3. If $n = 1$, then there are only two unit vectors, (1) and (-1). If $\mathbf{v}_0 = (1)$, then, as one can check, the directional derivative is $f'(\mathbf{v})$. [For the case $\mathbf{v}_0 = (-1)$, see Exercise 5.]

The phrase "directional derivative of f at $\mathbf{v}$ in the direction $\mathbf{v}_0$" is fairly cumbersome, and it may seem strange that we have not introduced a special symbol for it. The reason is given by the following proposition.

Proposition 5.1 *If $f: U \to \mathbf{R}$ is differentiable at the point $\mathbf{v} \in U$, then the directional derivative of f at $\mathbf{v}$ in the direction $\mathbf{v}_0$ is $D_{\mathbf{v}} f(\mathbf{v}_0)$.*

Proof. We know (since f is differentiable) that

$$\lim_{t \to 0} \frac{|f(\mathbf{v} + t\mathbf{v}_0) - f(\mathbf{v}) - D_{\mathbf{v}} f(t\mathbf{v}_0)|}{\|t\mathbf{v}_0\|} = 0$$

Since $D_{\mathbf{v}} f$ is linear and $\mathbf{v}_0$ is a unit vector, we have

$$\lim_{t \to 0} \left| \frac{f(\mathbf{v} + t\mathbf{v}_0) - f(\mathbf{v}) - tD_{\mathbf{v}} f(\mathbf{v}_0)}{t} \right| = 0$$

or

$$\lim_{t \to 0} \left| \frac{f(\mathbf{v} + t\mathbf{v}_0) - f(\mathbf{v})}{t} - D_{\mathbf{v}} f(\mathbf{v}_0) \right| = 0$$

This last equation can be restated as

$$\lim_{t \to 0} \frac{f(\mathbf{v} + t\mathbf{v}_0) - f(\mathbf{v})}{t} = D_{\mathbf{v}} f(\mathbf{v}_0)$$

which proves the proposition.

Proposition 5.1 takes care of directional derivatives for differentiable functions. What about others? We saw in Section 4.3 that a function can have all its partial derivatives at every point and yet fail to be continuous. That function, however, does not have directional derivatives in every direction. Is it true that a function with directional derivatives in every direction at every point is differentiable, or at least continuous? The answer is no. The standard example is given in Exercise 7. Of course, if the directional derivatives are continuous, the situation is considerably different; we saw in Section 4.6 that the function has a continuous derivative if all the partial derivatives are continuous. For most of the rest of this text, we shall be assuming that our functions have continuous partial derivatives (a reasonable enough assumption in practice), and Proposition 5.1 will automatically apply.

Exercises

1. Let $f: \mathbf{R}^2 \to \mathbf{R}$ be defined by $f(x_1, x_2) = x_1^2 x_2$. Compute the directional derivative of f at $\mathbf{v}$ in the direction $\mathbf{v}_0$, where
   (a) $\mathbf{v} = (1, 0)$, $\mathbf{v}_0 = (0, 1)$.
   (b) $\mathbf{v} = (1, 1)$, $\mathbf{v}_0 = (\sqrt{2}/2, \sqrt{2}/2)$.
   (c) $\mathbf{v} = (-1, 2)$, $\mathbf{v}_0 = (1/2, \sqrt{3}/2)$.
   (d) $\mathbf{v} = (2, 1)$, $\mathbf{v}_0 = (-\sqrt{3}/2, -1/2)$.
2. Let $f: \mathbf{R}^3 \to \mathbf{R}$ be defined by $f(x_1, x_2, x_3) = x_1 x_2 x_3$. Compute the directional derivative of f at $\mathbf{v}$ in the direction $\mathbf{v}_0$, where
   (a) $\mathbf{v} = (1, 0, 0)$, $\mathbf{v}_0 = (\sqrt{2}/2, 0, \sqrt{2}/2)$.
   (b) $\mathbf{v} = (1, 1, -1)$, $\mathbf{v}_0 = (\sqrt{3}/2, -1/2, 0)$.
   (c) $\mathbf{v} = (2, 0, 1)$, $\mathbf{v}_0 = (\frac{2}{3}, \frac{-1}{3}, \frac{2}{3})$.
   (d) $\mathbf{v} = (1, 2, -1)$, $\mathbf{v}_0 = (-\sqrt{3}/4, 3/4, -1/2)$
3. Let $g: \mathbf{R}^2 \to \mathbf{R}$ be defined by $f(x_1, x_2) = \cos x_1 x_2$. Compute the directional derivative of f at $\mathbf{v}$ in the direction $\mathbf{v}_0$ for the vectors $\mathbf{v}$ and $\mathbf{v}_0$ of Exercise 1.
4. Let $g: \mathbf{R}^3 \to \mathbf{R}$ be defined by $g(x_1, x_2, x_3) = \sin(x_1 x_3 - 2x_2^2)$. Compute the directional derivative of f at $\mathbf{v}$ in the direction $\mathbf{v}_0$ for the vectors $\mathbf{v}$ and $\mathbf{v}_0$ of Exercise 2.
5. Let $f: \mathbf{R} \to \mathbf{R}$ be a differentiable function. Compute the directional derivative of f at $\mathbf{v} = x$ in the direction $\mathbf{v}_0 = -1$.
6. Suppose $f: \mathbf{R}^2 \to \mathbf{R}$ is a differentiable function and $\mathbf{v}_0 = (\cos\theta, \sin\theta)$ is a unit vector in $\mathbf{R}^2$. Prove that the directional derivative of f at $\mathbf{v}$ in the direction $\mathbf{v}_0$ is

$$\left[\frac{\partial f}{\partial x}(\mathbf{v})\right]\cos\theta + \left[\frac{\partial f}{\partial y}(\mathbf{v})\right]\sin\theta$$

7. Let $f: \mathbf{R}^2 \to \mathbf{R}$ be defined by

$$f(x_1, x_2) = \begin{cases} \dfrac{x_1^2 x_2}{x_1^4 + x_2^2} & \text{if } (x_1, x_2) \neq (0, 0) \\ 0 & \text{if } (x_1, x_2) = (0, 0) \end{cases}$$

(a) Show that f has a directional derivative at (0, 0) in every direction. [*Hint*: If $\mathbf{v} = (a_1, a_2)$ is the direction, consider the cases $a_1 = 0$ and $a_2 = 0$ separately.]
(b) Show that if $x_2 = x_1^2$, then $f(x_1, x_2) = \frac{1}{2}$, so that f is not continuous at 0.

8. What directional derivatives does the function

$$f(x, y) = \begin{cases} \dfrac{xy}{x^2 + y^2} & \text{if } (x, y) \neq (0, 0) \\ 0 & \text{if } (x, y) = (0, 0) \end{cases}$$

have at $(x, y) = (0, 0)$?

6. The Gradient

There is another way of computing directional derivatives, and one which is interesting geometrically. It requires another definition. Let U be an open subset of $\mathbf{R}^n$ and $f: U \to \mathbf{R}$ a differentiable function. We define the *gradient of* f, grad f, at a point $\mathbf{v} \in U$ by the formula

$$(\text{grad } f)(\mathbf{v}) = \left(\frac{\partial f}{\partial x_1}(\mathbf{v}), \ldots, \frac{\partial f}{\partial x_n}(\mathbf{v})\right)$$

It is common to use the symbol ∇ for gradients and to write ∇f for grad f. We shall use both notations.

Notice that $(\text{grad } f)(\mathbf{v})$ is a vector in $\mathbf{R}^n$; therefore grad f may be regarded as a function from U to $\mathbf{R}^n$. Notice also that grad f is closely related to the Jacobian matrix for f at $\mathbf{v}$: the Jacobian matrix is the $1 \times n$ matrix

$$\left[\frac{\partial f}{\partial x_1}(\mathbf{v}) \cdots \frac{\partial f}{\partial x_n}(\mathbf{v})\right]$$

and the gradient is the vector with the same entries.

Proposition 6.1 *If $f: U \to R$ is differentiable at the point $\mathbf{v} \in U$, then the directional derivative of f at $\mathbf{v}$ in the direction $\mathbf{v}_0$ is $\langle(\text{grad } f)(\mathbf{v}), \mathbf{v}_0\rangle$.*

Proof. This follows immediately by writing both $D_{\mathbf{v}} f(\mathbf{v}_0)$ and $\langle(\text{grad } f)(\mathbf{v}), \mathbf{v}_0\rangle$ in terms of the components of $\mathbf{v}$ and $\mathbf{v}_0$. We leave the details as Exercise 2.

Corollary *Suppose $f: U \to \mathbf{R}$ is differentiable at the point $\mathbf{v} \in U$ with $(\text{grad } f)(\mathbf{v}) \neq \mathbf{0}$. Let $\mathbf{v}_1$ be the unit vector*

$$\mathbf{v}_1 = \frac{(\text{grad } f)(\mathbf{v})}{\|(\text{grad } f)(\mathbf{v})\|}$$

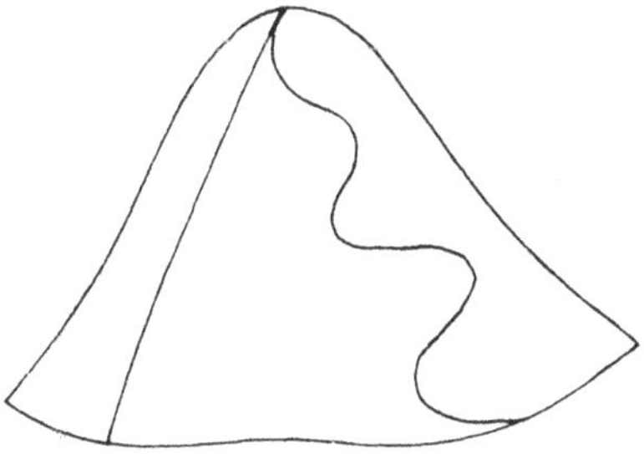

Figure 6.1 Climbing a hill.

Then $\mathbf{v}_1$ *is the direction in which the directional derivative is largest.*

Proof. We know from the Cauchy inequality (Theorem 2.1) that

$$\langle (\text{grad } f)(\mathbf{v}), \mathbf{v}_0 \rangle \le \|(\text{grad } f)(\mathbf{v})\| \cdot \|\mathbf{v}_0\| = \|(\text{grad } f)(\mathbf{v})\|$$

for any unit vector $\mathbf{v}_0$. On the other hand,

$$\begin{aligned}\langle (\text{grad } f)(\mathbf{v}), \mathbf{v}_1 \rangle &= \left\langle (\text{grad } f)(\mathbf{v}), \frac{(\text{grad } f)(\mathbf{v})}{\|(\text{grad } f)(\mathbf{v})\|} \right\rangle \\ &= \frac{1}{\|(\text{grad } f)(\mathbf{v})\|} \langle (\text{grad } f)(\mathbf{v}), (\text{grad } f)(\mathbf{v}) \rangle \\ &= \|(\text{grad } f)(\mathbf{v})\|\end{aligned}$$

Anyone who has ridden a bicycle up a steep hill has experienced Proposition 6.1 and its corollary. Let $f: U \to \mathbf{R}$ be the function representing the height above sea level; then the directional derivative of f at a point $\mathbf{v}$ in the direction $\mathbf{v}_0$ is the slope of the ground at $\mathbf{v}$ in the direction $\mathbf{v}_0$. The higher the value of this directional derivative, the harder the bicycle rider must work. A rider heading straight up the hill is always traveling in the direction of the gradient, and therefore always making the directional derivative as large as possible. Thus he is giving himself as hard a time as he can. (On the other hand, he also reaches the top rapidly). A rider who weaves back and forth across the street, on the other hand, is usually meeting the gradient almost at a right angle. (See Figure 6.1.) As a result, the directional derivative is small (except at the switchbacks), and the ride is comparatively easy, though longer.

Exercises

1. Compute the gradients and directions of steepest ascent of the following functions at the points indicated.
   (a) $f(x_1, x_2) = x_1^2 x_2$ at $\mathbf{v} = (1, 2)$.

(b) $f(x_1, x_2) = x_1 \sin x_1x_2$ at $\mathbf{v} = (\pi, \frac{1}{3})$.
(c) $f(x_1, x_2, x_3) = x_1x_2^2x_3$ at $\mathbf{v} = (1, -1, 2)$.
(d) $f(x_1, x_2, x_3) = \cos x_1x_2x_3$ at $\mathbf{v} = (\frac{1}{2}, -\frac{1}{2}, \pi)$.
(e) $f(x_1, \ldots, x_4) = x_1x_2(x_1^2 - x_3x_4)$ at $\mathbf{v} = (1, 2, -2, 0)$.
(f) $f(x_1, \ldots, x_4) = x_1 \cos(x_2 + x_3 + x_4)$ at $\mathbf{v} = (2, \pi/3, 0, \pi/3)$.
(g) $f(x_1, \ldots, x_n) = x_1^2 + x_2^2 + \cdots + x_n^2$ at $\mathbf{v} = (x_1, \ldots, x_n)$.

2. Prove Proposition 6.1.
3. Let $f, g : \mathbf{R}^n \to \mathbf{R}$ be differentiable functions and a, b real numbers. Prove that

$$\operatorname{grad}(af + bg) = a \operatorname{grad} f + b \operatorname{grad} g$$

4. Let $g : \mathbf{R}^2 \to \mathbf{R}^2$ and $f : \mathbf{R}^2 \to \mathbf{R}^1$ be differentiable functions. Use the chain rule to derive a formula for $\operatorname{grad}(f \circ g)$.
5. Let $g : \mathbf{R}^2 \to \mathbf{R}$ and $f : \mathbf{R} \to \mathbf{R}$ be differentiable functions. Use the chain rule to derive a formula for $\operatorname{grad}(f \circ g)$.
6. Do Exercise 4 supposing that $g : \mathbf{R}^n \to \mathbf{R}^n$ and $f : \mathbf{R}^n \to \mathbf{R}^1$.
7. Do Exercise 5 supposing that $g : \mathbf{R}^n \to \mathbf{R}$.

7. The Tangent Plane

One interpretation given to the derivative in elementary calculus is that it gives the slope of the tangent line to the graph of the function. Explicitly, if $f : \mathbf{R} \to \mathbf{R}$ is a differentiable function, then the slope of the graph of f at the point $(a_1, f(a_1))$ is $f'(a_1)$ and the equation of the tangent line is

$$x_2 - f(a_1) = f'(a_1)(x_1 - a_1) \tag{7.1}$$

The observation that this line is close to the graph of f near $(a_1, f(a_1))$ is a geometric analogue of the fact that f can be approximated by a linear function near x_0. (See Figure 7.1.)

Similarly, if $f : \mathbf{R}^2 \to \mathbf{R}$ is differentiable at $\mathbf{v}_0 \in \mathbf{R}^2$, we should be able to define a tangent plane to the graph of f which is close to the graph of f near $(\mathbf{v}_0, f(\mathbf{v}_0))$. (See Figure 7.2.) To do this, we need to reinterpret the equation of the tangent line (7.1).

Let $h : \mathbf{R} \to \mathbf{R}$ be the function defined by

$$h(x) = f'(a_1)(x - a_1) + f(a_1)$$

As we saw in Section 2.1, the graph of h is exactly the line with equation (7.1). Since h is a good approximation for f near a_1, the graph of h should be a good approximation to the graph of f near $(a_1, f(a_1))$.

Motivated by this, we can now define the tangent plane. Suppose $f : \mathbf{R}^n \to \mathbf{R}$ is a function differentiable at $\mathbf{v}_0 \in \mathbf{R}^n$. Define $h : \mathbf{R}^n \to \mathbf{R}$ by

$$h(\mathbf{v}) = (D_{\mathbf{v}_0} f)(\mathbf{v} - \mathbf{v}_0) + f(\mathbf{v}_0)$$

We know from Section 3 that the graph of h is a hyperplane and that h is a

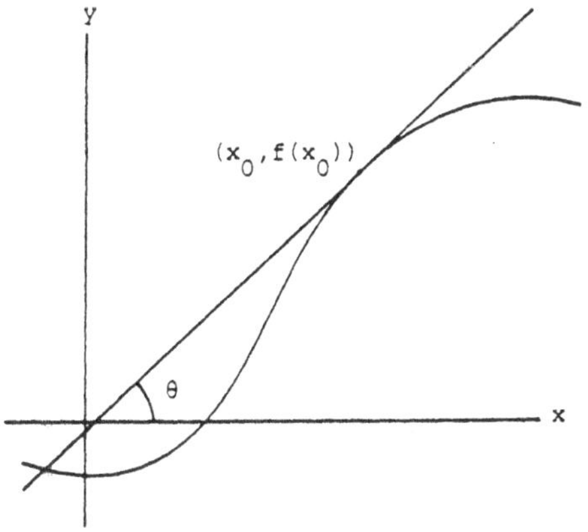

Figure 7.1 The tangent line.

good approximation for f near $\mathbf{v}_0$. It follows that the graph of h should be close to the graph of f near $(\mathbf{v}_0, f(\mathbf{v}_0))$. We define the *tangent plane to the graph of f at the point* $(\mathbf{v}_0, f(\mathbf{v}_0))$ to be the graph of h. By definition, this is the set of all $(x_1, \ldots, x_{n+1}) \in \mathbf{R}^{n+1}$ satisfying the equation

$$x_{n+1} = (D_{\mathbf{v}_0} f)(\mathbf{v} - \mathbf{v}_0) + f(\mathbf{v}_0) \tag{7.2}$$

where $\mathbf{v} = (x_1, \ldots, x_n)$ and $\mathbf{v}_0 = (a_1, \ldots, a_n)$. We can rewrite (7.2) in terms of the gradient as

$$x_{n+1} - f(\mathbf{v}_0) = \langle (\text{grad } f)(\mathbf{v}_0), \mathbf{v} - \mathbf{v}_0 \rangle \tag{7.3}$$

or, in terms of coordinates, as

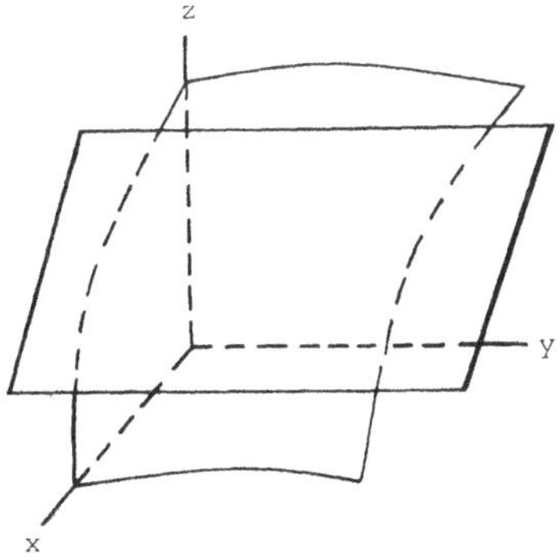

Figure 7.2 The tangent plane.

(7.4) $$x_{n+1} - f(\mathbf{v}_0) = \sum_{i=1}^{n} \frac{\partial f}{\partial x_i}(\mathbf{v}_0)(x_i - a_i)$$

where $\mathbf{v}_0 = (a_1, \ldots, a_n)$.

Here are some examples.

1. Let $f(x_1, x_2) = 4 - 2x_1 - x_2$. The graph of f is a plane in $\mathbf{R}^3$ (Figure 7.3), and so we might expect that the graph of f will be its own tangent plane. Indeed, using equation (7.4) we see that the equation of the tangent plane through the point $(a_1, a_2, f(a_1, a_2))$ is

$$\begin{aligned} x_3 - f(a_1, a_2) &= -2(x_1 - a_1) - (x_2 - a_2) \\ &= -2x_1 + 2a_1 - x_2 + a_2 \end{aligned}$$

   Since $f(a_1, a_2) = 4 - 2a_1 - a_2$, this equation becomes

$$x_3 = 4 - 2x_1 - x_2$$

   which is exactly the equation of the graph of f.

   For a generalization of this situation, see Exercise 4.

2. Suppose $f(x_1, x_2) = x_1^2 + x_2^2$. Then the graph of f is the *circular paraboloid* pictured in Figure 7.4. In this case, the graph of f at the point (1, 2, 5) is given by

$$x_3 - 5 = 2(x_1 - 1) + 4(x_2 - 2)$$

   More generally, the tangent plane to the graph of f at the point $(a_1, a_2, f(a_1, a_2))$ is given by

$$x_3 - f(a_1, a_2) = 2a_1(x_1 - a) + 2a_2(x_2 - a_2)$$

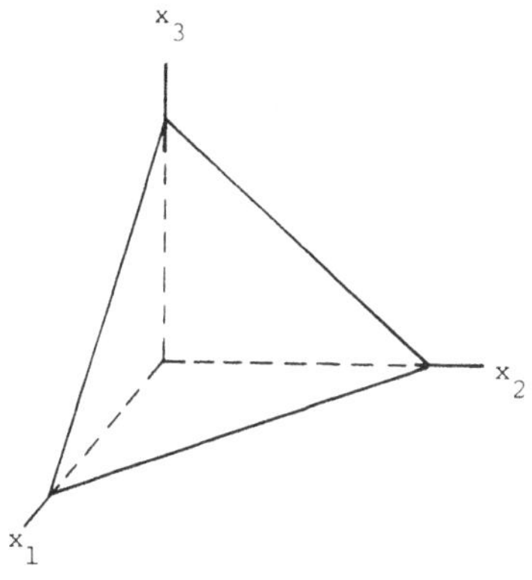

Figure 7.3 A plane in $\mathbf{R}^3$.

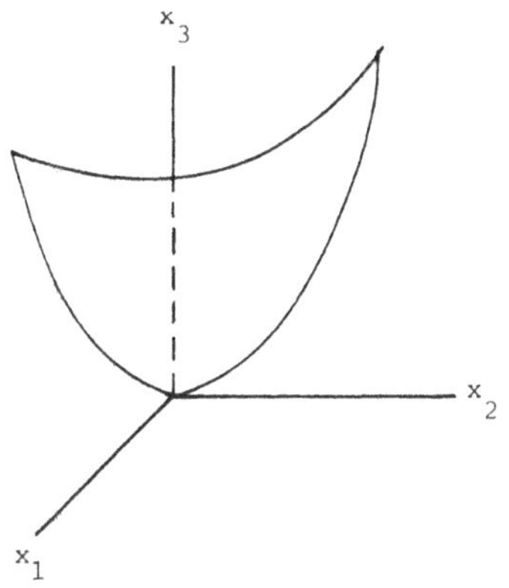

Figure 7.4 The circular paraboloid.

3. Let $g : \mathbf{R}^4 \to \mathbf{R}$ be given by $g(x_1, x_2, x_3, x_4) = x_1 x_3 - x_2 x_4$. The tangent plane to the graph of g at the point $(1, 2, 2, 1, 0)$ [note that $f(1, 2, 2, 1) = 0$] has the equation

$$x_5 = 2(x_1 - 1) - (x_2 - 2) + (x_3 - 2) - 2(x_4 - 1)$$

More generally, the tangent plane to the graph of g at the point $(a_1, a_2, a_3, a_4, f(a_1, a_2, a_3, a_4))$ has the equation

$$\begin{aligned} x_5 - f(a_1, a_2, a_3, a_4) &= a_3(x_1 - a_1) - a_4(x_2 - a_2) \\ &\quad + a_1(x_3 - a_3) - a_2(x_4 - a_4) \end{aligned}$$

In order to motivate the definition of a tangent plane for functions $f: \mathbf{R}^n \to \mathbf{R}^p$, $p > 1$, we first consider a special case.

Let $f: \mathbf{R} \to \mathbf{R}^2$, $f(x_1) = (f_1(x_1), f_2(x_1))$, be a function differentiable at the point a_1. The graph G of this function consists of all points (x_1, x_2, x_3) with

$$f(x_1) = (x_2, x_3)$$

Equivalently, G consists of all (x_1, x_2, x_3) satisfying the two equations

$$x_2 = f_1(x_1) \qquad x_3 = f_2(x_1)$$

Thus, G is the intersection $G_1 \cap G_2$ of the two sets

$$G_1 = \{(x_1, x_2, x_3) : x_2 = f_1(x_1)\}$$
$$G_2 = \{(x_1, x_2, x_3) : x_3 = f_2(x_1)\}$$

Now each of these sets is a surface in $\mathbf{R}^3$, indeed, a surface of a very special kind. For example, the condition $x_2 = f_1(x_1)$ defining G_1 does not involve the coordinate x_3. Thus, G_1 has the property that, if $(x_1, x_2, x_3) \in G_1$, then

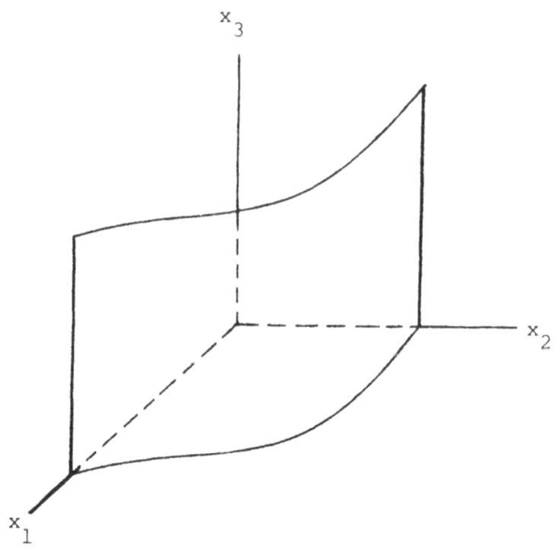

Figure 7.5 A cylinder along the x_3 axis.

$(x_1, x_2, x'_3) \in G_1$ for any $x'_3 \in \mathbf{R}$. Such a surface is called a *cylinder along the* x_3 *axis*. (See Figure 7.5.)

Similarly, G_2 has the property that $(x_1, x_2, x_3) \in G_2$ implies that $(x_1, x'_2, x_3) \in G_2$ for any $x'_2 \in \mathbf{R}$. Thus G_2 is a *cylinder along the* x_2 *axis*. (See Figure 7.6.) It follows that G is the intersection of two cylindrical surfaces in

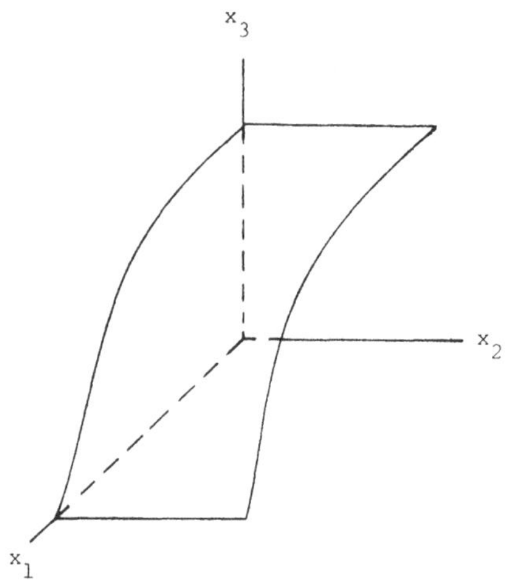

Figure 7.6 A cylinder along the x_2 axis.

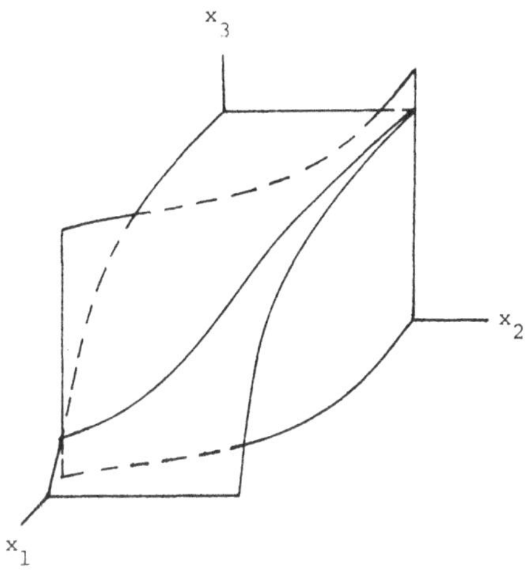

Figure 7.7 The curve of intersection.

$\mathbf{R}^3$ and is therefore a curve. (See Figure 7.7.) We would expect then that the graph G of f would have a tangent line. In fact, if $\mathbf{v}_0 = (a_1, a_2, a_3) \in G = G_1 \cap G_2$, this tangent line should be the intersection of the tangent plane of G_1 at $\mathbf{v}_0$ with the tangent plane of G_2 at $\mathbf{v}_0$. The equation of the tangent plane to G_1 at $\mathbf{v}_0 = (a_1, a_2, a_3)$ is

$$(7.5) \qquad x_2 - a_2 = f_1'(a_1)(x_1 - a_1)$$

and the equation of the tangent plane to G_2 at $\mathbf{v}_0$ is

$$(7.6) \qquad x_3 - a_3 = f_2'(a_1)(x_1 - a_1)$$

Therefore, the tangent line to the graph of f at $\mathbf{v}_0 = (a_1, a_2, a_3)$ is the line defined by equations (7.5) and (7.6).

Another way of describing this line is as follows. Let $h : \mathbf{R}^1 \to \mathbf{R}^2$ be defined by

$$\begin{aligned} h(x_1) &= (D_{a_1} f)(x_1 - a_1) + f(a_1) \\ &= (f_1'(a_1)(x_1 - a_1) + f_1(a_1), f_2'(a_1)(x_1 - a_1) + f_2(a_2)) \end{aligned}$$

We know from the definition of differentiability that h is a very good approximation to f near a_1. Furthermore, the graph of h is exactly the line defined by equations (7.5) and (7.6). This observation motivates the following definition.

Let $f : \mathbf{R}^n \to \mathbf{R}^p$ *be a function differentiable at* $\mathbf{v}_0 \in \mathbf{R}^n$. The *tangent plane to the graph of f at the point* $(\mathbf{v}_0, f(\mathbf{v}_0))$ is defined to be the graph of the function $h : \mathbf{R}^n \to \mathbf{R}^p$ given by

$$h(\mathbf{v}) = (D_{\mathbf{v}_0} f)(\mathbf{v} - \mathbf{v}_0) + f(\mathbf{v}_0) \tag{7.7}$$

As we saw in Section 4, the graph of h is an n plane in $\mathbf{R}^{n+p}$.

Here are some examples.

1. Let $f: \mathbf{R} \to \mathbf{R}^2$ be defined by

$$f(x_1) = (2x_1, 3x_1)$$

Since f is linear, it is equal to its derivative, so that $h(x_1) = f(x_1)$ in this case. The graph of f is the intersection of the two planes

$$x_2 = 2x_1 \qquad x_3 = 3x_2$$

and is therefore a line. (See Figure 7.8.) In this case, the tangent line to the graph of f is the graph itself.

2. Let $f: \mathbf{R}^2 \to \mathbf{R}^3$ be defined by

$$f(x_1, x_2) = (x_1 x_2^2, x_2^3, x_1^2 x_2)$$

Then $D_{\mathbf{v}} f$ has matrix

$$\begin{bmatrix} x_2^2 & 2x_1x_2 \\ 0 & 3x_2^2 \\ 2x_1x_2 & x_1^2 \end{bmatrix}$$

where $\mathbf{v} = (x_1, x_2)$. If $\mathbf{v}_0 = (2, 1)$, then $D_{\mathbf{v}_0} f$ has matrix

$$\begin{bmatrix} 1 & 4 \\ 0 & 3 \\ 4 & 4 \end{bmatrix}$$

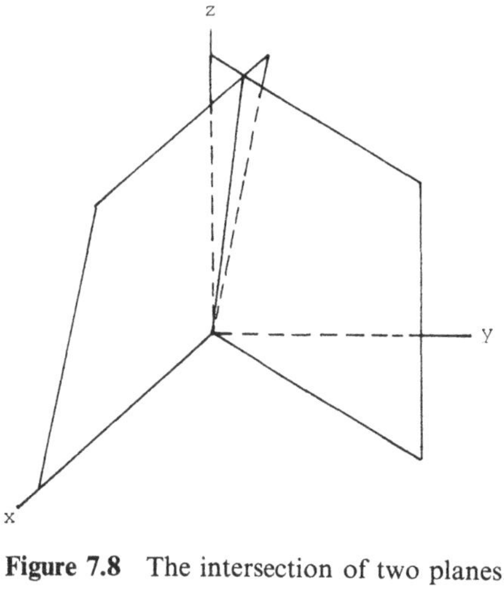

Figure 7.8 The intersection of two planes.

so that

$$\begin{aligned} h(x_1, x_2) &= (D_{\mathbf{v}_0} f)(\mathbf{v} - \mathbf{v}_0) + f(\mathbf{v}_0) \\ &= ((x_1 - 2) + 4(x_2 - 1),\ 3(x_2 - 1),\ 4(x_1 - 2) + 4(x_2 - 1)) \\ &\quad + (2, 1, 4) \\ &= ((x_1 - 2) + 4(x_2 - 1) + 2,\ 3(x_2 - 1) + 1, \\ &\quad 4(x_1 - 2) + 4(x_2 - 1) + 4) \end{aligned}$$

Thus, the tangent plane to the graph of f at the point (2, 1, 2, 1, 4) is the 2 plane in $\mathbf{R}^5$ defined by the three equations

$$\begin{aligned} x_3 &= (x_1 - 2) + 4(x_2 - 1) + 2 \\ x_4 &= 3(x_2 - 1) + 1 \\ x_5 &= 4(x_1 - 2) + 4(x_2 - 1) + 4 \end{aligned}$$

Exercises

1. Find the equations of the tangent plane to the graph of the function f at the point $(\mathbf{v}_0, f(\mathbf{v}_0))$, where
   (a) $f: \mathbf{R}^2 \to \mathbf{R}$, $f(x_1, x_2) = 2x_1 + 5x_2 - 7$, $\mathbf{v}_0 = (1, -1)$.
   (b) $f: \mathbf{R}^2 \to \mathbf{R}$, $f(x_1, x_2) = x_1^2 x_2^5$, $\mathbf{v}_0 = (2, -1)$.
   (c) $f: \mathbf{R}^3 \to \mathbf{R}$, $f(x_1, x_2, x_3) = x_1^2 + x_2 x_3$, $\mathbf{v}_0 = (1, 0, -2)$.
   (d) $f: \mathbf{R}^3 \to \mathbf{R}$, $f(x_1, x_2, x_3) = x_2 \sin x_1 x_3$, $\mathbf{v}_0 = (\pi/3, 7, -2)$.
   (e) $f: \mathbf{R}^4 \to \mathbf{R}$, $f(x_1, x_2, x_3, x_4) = 2x_1^2 x_3 - x_4$, $\mathbf{v}_0 = (0, 1, 1, -1)$.
   (f) $f: \mathbf{R}^4 \to \mathbf{R}$, $f(x_1, x_2, x_3, x_4) = x_2 e^{x_3} - x_1^2 x_4$, $\mathbf{v}_0 = (2, 7, 0, -3)$.
   (g) $f: \mathbf{R} \to \mathbf{R}^2$, $f(x_1) = (2x_1 - 3, x_1 + 5)$, $\mathbf{v}_0 = -7$.
   (h) $f: \mathbf{R} \to \mathbf{R}^2$, $f(x_1) = (\cos x_1, \sin x_1)$, $\mathbf{v}_0 = \pi/6$.
   (i) $f: \mathbf{R} \to \mathbf{R}^3$, $f(x_1) = (2\cos x_1, 2\sin x_1, x_1)$, $\mathbf{v}_0 = \pi/4$.
   (j) $f: \mathbf{R}^2 \to \mathbf{R}^2$, $f(x_1, x_2) = (2x_2, x_1 + x_2 - 7)$, $\mathbf{v}_0 = (2, 0)$.
   (k) $f: \mathbf{R}^2 \to \mathbf{R}^3$, $f(x_1, x_2) = (x_1 \cos x_2, x_2 \sin x_1, x_1 x_2)$, $\mathbf{v}_0 = (\pi/3, \pi/4)$.
   (l) $f: \mathbf{R}^2 \to \mathbf{R}^4$, $f(x_1, x_2) = (x_1^2, -x_1, x_1^2 + x_2, x_1 x_2^2 + 3)$, $\mathbf{v}_0 = (1, -3)$.

2. Let f: $\mathbf{R}^2 \to \mathbf{R}^3$ be a function differentiable at $\mathbf{v}_0 \in \mathbf{R}^2$, $f(\mathbf{v}) = (f_1(\mathbf{v}), f_2(\mathbf{v}), f_3(\mathbf{v}))$. Prove that the tangent plane to the graph of f at the point $(\mathbf{v}_0, f(\mathbf{v}_0))$ is defined by the three equations

$$\begin{aligned} x_3 &= \langle (\operatorname{grad} f_1)(\mathbf{v}_0), \mathbf{v} - \mathbf{v}_0 \rangle + f_1(\mathbf{v}_0) \\ x_4 &= \langle (\operatorname{grad} f_2)(\mathbf{v}_0), \mathbf{v} - \mathbf{v}_0 \rangle + f_2(\mathbf{v}_0) \\ x_5 &= \langle (\operatorname{grad} f_3)(\mathbf{v}_0). \mathbf{v} - \mathbf{v}_0 \rangle + f_3(\mathbf{v}_0) \end{aligned}$$

3. State and prove the analogue of Exercise 2 for functions $f: \mathbf{R}^n \to \mathbf{R}^p$.
4. Let f: $\mathbf{R}^n \to \mathbf{R}^p$ be defined by $f(\mathbf{v}) = T(\mathbf{v}) + \mathbf{b}$, where T: $\mathbf{R}^n \to \mathbf{R}^p$ is a linear transformation and $\mathbf{b}$ a fixed vector in $\mathbf{R}^p$. Prove that the tangent plane to the graph of f at any point $(\mathbf{v}_0, f(\mathbf{v}_0))$ coincides with the graph of f.

6

Higher Order Derivatives and Taylor's Theorem

As in the case of real-valued functions of one variable, higher derivatives play an important role in the study of functions between Euclidean spaces. The higher derivatives can again be considered as linear transformations; however, our emphasis will be on the higher order partial derivatives.

The basic result of this chapter is Taylor's theorem. This result gives conditions under which a function can be approximated near a point by a polynomial and will be useful to us when we study maxima and minima in Chapter 8.

1. Higher Order Partial Derivatives

Let U be an open subset of $\mathbf{R}^n$ and $f\colon U \to \mathbf{R}$ a differentiable function. The derivative of f at $\mathbf{v} \in \mathbf{R}^n$ is a linear transformation $D_{\mathbf{v}} f \colon \mathbf{R}^n \to \mathbf{R}$. As we saw in Section 2.6, this linear transformation determines (and is determined by) a $1 \times n$ matrix. If we use the entries of this matrix to define a vector in $\mathbf{R}^n$, we obtain the gradient of f at $\mathbf{v}$,

$$(\operatorname{grad} f)(\mathbf{v}) = \left(\frac{\partial f}{\partial x_1}(\mathbf{v}), \ldots, \frac{\partial f}{\partial x_n}(\mathbf{v})\right)$$

which we defined in Section 5.6. Therefore, we can consider the derivative of f on U to be the function $\operatorname{grad} f\colon U \to \mathbf{R}^n$.

It makes sense now to ask whether or not this function is differentiable. Indeed, since $\operatorname{grad} f$ corresponds to the first derivative of f, the derivative of grad f ought to correspond to the second derivative of f. Furthermore, if grad f is differentiable, then

$$D_{\mathbf{v}}(\text{grad } f) : \mathbf{R}^n \to \mathbf{R}^n$$

is a linear transformation whose matrix has entries which are the appropriate partial derivatives of the component functions of grad f, that is, of

$$\frac{\partial f}{\partial x_i} : \mathbf{R}^n \to \mathbf{R}$$

(See Proposition 3.1, Chapter 4.) This leads us to the notion of the higher order partial derivative.

Let U be an open subset of $\mathbf{R}^n$ and $f: U \to \mathbf{R}$ a function. Suppose each of the partial derivatives of f exist on U and, further, that each of the partial derivatives of the functions

$$\frac{\partial f}{\partial x_i} : U \to \mathbf{R}$$

also exist on U. We denote $\partial/\partial x_j(\partial f/\partial x_i)$ by

$$\frac{\partial^2 f}{\partial x_j \, \partial x_i}$$

and call these the *second-order partial derivatives of f.*

Of course, one can define *higher order partial derivatives* in the same way. For instance,

$$\frac{\partial^3 f}{\partial x_1 \, \partial x_2^2} = \frac{\partial}{\partial x_1}\left(\frac{\partial}{\partial x_2}\left(\frac{\partial f}{\partial x_2}\right)\right)$$

is a partial derivative of f of order 3. Here are some examples.

1. Let $f: \mathbf{R}^2 \to \mathbf{R}$ be defined by $f(x_1, x_2) = x_1 x_2$. Then

$$\frac{\partial f}{\partial x_1} = x_2 \qquad \frac{\partial f}{\partial x_2} = x_1$$

and $$\frac{\partial^2 f}{\partial x_1^2} = \frac{\partial^2 f}{\partial x_2^2} = 0 \qquad \frac{\partial^2 f}{\partial x_1 \partial x_2} = \frac{\partial^2 f}{\partial x_2 \partial x_1} = 1$$

In this case, all partial derivatives of order greater than 2 vanish.

2. Let $f: \mathbf{R}^3 \to \mathbf{R}$ be defined by $f(x_1, x_2, x_3) = x_1 \sin x_2 x_3$. Then

$$\frac{\partial^2 f}{\partial x_3 \partial x_1} = \frac{\partial}{\partial x_3} \sin x_2 x_3 = x_2 \cos x_2 x_3$$

and
$$\frac{\partial^3 f}{\partial x_2^2 \partial x_1} = \frac{\partial^2}{\partial x_2^2} \sin x_2 x_3$$
$$= \frac{\partial}{\partial x_2} x_3 \cos x_2 x_3$$
$$= - x_3^2 \sin x_2 x_3$$

We can now compute the matrix of the second derivative of a function $f: U \to \mathbf{R}$ in terms of the second-order partial derivatives of f. If we identify the derivative of f with the gradient of f as above,

$$\text{grad } f: U \to \mathbf{R}^n$$

then Proposition 3.1 of Chapter 4 tells that the matrix of $D_{\mathbf{v}}(\text{grad } f)$ is the Jacobian of grad f evaluated at $\mathbf{v}$:

$$\begin{bmatrix} \frac{\partial^2 f}{\partial x_1 \partial x_1}(\mathbf{v}) & \cdots & \frac{\partial^2 f}{\partial x_n \partial x_1}(\mathbf{v}) \\ \vdots & & \vdots \\ \frac{\partial^2 f}{\partial x_1 \partial x_n}(\mathbf{v}) & \cdots & \frac{\partial^2 f}{\partial x_n \partial x_n}(\mathbf{v}) \end{bmatrix} \tag{1.1}$$

The matrix (1.1) is called the *Hessian of f* and denoted by $H(f)$. Its value at $\mathbf{v}$ will be written as $H(f)(\mathbf{v})$ or $H_{\mathbf{v}}(f)$.

Examples

1. Let $f: \mathbf{R}^2 \to \mathbf{R}$ be defined by $f(x_1, x_2) = x_1^2 x_2$. Then

$$\frac{\partial f}{\partial x_1} = 2x_1 x_2 \qquad \frac{\partial f}{\partial x_2} = x_1^2$$

and
$$\frac{\partial^2 f}{\partial x_1^2} = 2x_2 \qquad \frac{\partial^2 f}{\partial x_2^2} = 0$$

$$\frac{\partial^2 f}{\partial x_1 \partial x_2} = \frac{\partial^2 f}{\partial x_2 \partial x_1} = 2x_1$$

Thus, the Hessian of f is the 2×2 matrix

$$H(f) = \begin{bmatrix} 2x_2 & 2x_1 \\ 2x_1 & 0 \end{bmatrix}$$

In particular, $H(f)(2, -3)$ is the 2×2 matrix

$$\begin{bmatrix} -6 & 4 \\ 4 & 0 \end{bmatrix}$$

2. Let $g : \mathbf{R}^3 \to \mathbf{R}$ be defined by $g(x_1, x_2, x_3) = x_1 \sin x_2 x_3$. We compute $H(g)(\mathbf{v})$ for $\mathbf{v} = (3, 2, \pi/6)$. To begin with,

$$\frac{\partial g}{\partial x_1} = \sin x_2 x_3$$

$$\frac{\partial g}{\partial x_2} = x_1 x_3 \cos x_2 x_3$$

$$\frac{\partial g}{\partial x_3} = x_1 x_2 \cos x_2 x_3$$

It follows that

$$\frac{\partial^2 g}{\partial x_1^2} = 0 \qquad \frac{\partial^2 g}{\partial x_2^2} = -\, x_1 x_3^2 \sin x_2 x_3 \qquad \frac{\partial^2 g}{\partial x_3^2} = -\, x_1 x_2^2 \sin x_2^3$$

$$\frac{\partial^2 g}{\partial x_1\, \partial x_2} = x_3 \cos x_2 x_3 = \frac{\partial^2 g}{\partial x_2\, \partial x_1}$$

$$\frac{\partial^2 g}{\partial x_1\, \partial x_3} = x_2 \cos x_2 x_3 = \frac{\partial^2 g}{\partial x_3 \partial x_1}$$

$$\frac{\partial^2 g}{\partial x_2\, \partial x_3} = x_1 \cos x_2 x_3 - x_1 x_2 x_3 \sin x_2 x_3 = \frac{\partial^2 g}{\partial x_3\, \partial x_2}$$

Evaluating these partial derivatives at $(3, 2, \pi/6)$, we see that

$$H(g)\left(3, 2, \frac{\pi}{6}\right) = \begin{bmatrix} 0 & \frac{\pi}{12} & 1 \\ \frac{\pi}{12} & \frac{-\pi^2\sqrt{3}}{24} & \frac{3 - \pi\sqrt{3}}{2} \\ 1 & \frac{3 - \pi\sqrt{3}}{2} & -6\sqrt{3} \end{bmatrix}$$

The reader may have noticed that

$$\frac{\partial^2 f}{\partial x_i \partial x_j} = \frac{\partial^2 f}{\partial x_j \partial x_i}$$

for the functions we have dealt with. This is not always the case; the standard counterexample is given as Exercise 11 at the end of the section. However, these "mixed partials" are equal if we assume that both are continuous.

Theorem 1.1 *Let U be an open subset of $\mathbf{R}^n$ and $f : U \to \mathbf{R}$ a function such that for some fixed i, j, $1 \le i, j \le n$,*

$$\frac{\partial^2 f}{\partial x_i \, \partial x_j} \qquad \frac{\partial^2 f}{\partial x_j \, \partial x_i}$$

exist on U and are continuous at a point $\mathbf{v}$ in U. Then

$$\frac{\partial^2 f}{\partial x_i \, \partial x_j}(\mathbf{v}) = \frac{\partial^2 f}{\partial x_j \, \partial x_i}(\mathbf{v})$$

We prove this result in Section 5.

Remark. The obvious analogue of Theorem 1.1 for partial derivatives of order greater than 2 is true. For example, if all partial derivatives of $f: \mathbf{R}^n \to \mathbf{R}$ of order 3 exist and are continuous at $\mathbf{v}$, then the order of differentiation of these third-order partial derivatives does not matter. For example,

$$\frac{\partial^3 f}{\partial x_1 \, \partial x_2 \, \partial x_3}(\mathbf{v}) = \frac{\partial^3 f}{\partial x_2 \, \partial x_3 \, \partial x_1}(\mathbf{v})$$

in this case. (See Exercise 12.)

We conclude this section with an example showing how the chain rule is used to compute higher order partial derivatives of composite functions. Although in principle this is a straightforward application of the chain rule, in practice it can be quite complicated.

Let $f: \mathbf{R}^1 \to \mathbf{R}^1$ and $g: \mathbf{R}^2 \to \mathbf{R}^1$ be functions whose second derivatives exist. We wish to compute the second-order partial derivative $\partial^2(f \circ g)/\partial x^2$ of the composite function $f \circ g : \mathbf{R}^2 \to \mathbf{R}^1$.

To begin with, let (x, y) be coordinates in the domain of g and z the coordinate in the domain of f. Then

$$\frac{\partial(f \circ g)}{\partial x}(x, y) = \frac{df}{dz}(g(x, y))\frac{\partial g}{\partial x}(x, y)$$

$$\frac{\partial(f \circ g)}{\partial y}(x, y) = \frac{df}{dz}(g(x, y))\frac{\partial g}{\partial y}(x, y)$$

by the chain rule. Now

$$(1.2) \qquad \frac{\partial^2(f \circ g)}{\partial x^2}(x, y) = \frac{\partial}{\partial x}\left(\frac{\partial(f \circ g)}{\partial x}\right)(x, y)$$

$$= \left[\frac{\partial}{\partial x}\left(\frac{df}{dz}(g(x, y))\right)\right]\left[\frac{\partial g}{\partial x}(x, y)\right]$$

$$+ \left[\frac{df}{dz}(g(x, y))\right]\left[\frac{\partial}{\partial x}\left(\frac{\partial g}{\partial x}(x, y)\right)\right]$$

by the product rule for differentiation. The second of the two terms on the right side of equation (1.2) is easily identified with

$$\frac{df}{dz}(g(x, y))\frac{\partial^2 g}{\partial x^2}(x, y)$$

We compute the first term.

Note that $(df/dz)(g(x, y))$ is the composite of two functions, namely $df/dz : \mathbf{R} \to \mathbf{R}$ and $g : \mathbf{R}^2 \to \mathbf{R}$. Therefore, we need to use the chain rule to differentiate it:

$$\begin{aligned}\frac{\partial}{\partial x}\left(\frac{df}{dz}(g(x, y))\right) &= \left(\frac{d}{dz}\left(\frac{df}{dz}\right)\right)(g(x, y))\frac{\partial g}{\partial x}(x, y)\\ &= \frac{d^2 f}{dz^2}(g(x, y))\frac{\partial g}{\partial x}(x, y)\end{aligned}$$

It follows that

$$\frac{\partial^2 (f \circ g)}{\partial x^2}(x, y) = \frac{d^2 f}{dz^2}(g(x, y))\left[\frac{\partial g}{\partial x}(x, y)\right]^2 + \frac{df}{dz}(g(x, y))\frac{\partial^2 g}{\partial x^2}(x, y)$$

For example, if $f(z) = z^3$ and $g(x, y) = x^2 + y$, then

$$\begin{aligned}\frac{\partial^2 (f \circ g)}{\partial x^2}(x, y) &= 6(x^2 + y)(2x)^2 + 3(x^2 + y)^2 2\\ &= 24(x^4 + yx^2) + 6(x^4 + 2x^2 y + y^2)\\ &= 30x^4 + 36\, x^2 y + 6y^2\end{aligned}$$

Of course, this can be verified by first computing the composite $f \circ g$ and then differentiating it.

Exercises

1. Compute $\partial^2 f/\partial x^2$, $\partial^2 f/\partial x\,\partial y$, and $\partial^3 f/\partial x\,\partial y^2$ for the following functions.
   (a) $f(x, y) = x^2 y - y^3$
   (b) $f(x, y) = x \sin y$
   (c) $f(x, y) = xy\,e^{x^2}$
   (d) $f(x, y) = \arctan(x + y)$
   (e) $f(x, y) = (xy + y^2)^{1/3}$
   (f) $f(x, y) = 2y \tan xy$
   (g) $f(x, y) = x \arcsin x^2 y$
2. Compute $H(f)(0, 0)$ for the functions in Exercise 1.
3. Compute $H(f)(\mathbf{v}_0)$, where
   (a) $f: \mathbf{R}^3 \to \mathbf{R}$, $f(x_1, x_2, x_3) = x_1 x_2 - x_3^3$, $\mathbf{v}_0 = (0, 0, 0)$.
   (b) f is as in (a), $\mathbf{v}_0 = (1, 3, -1)$.
   (c) $f: \mathbf{R}^3 \to \mathbf{R}$, $f(x_1, x_2, x_3) = x_1 \sin x_2 x_3$, $\mathbf{v}_0 = (0, 0, 0)$.

(d) f is as in (c), $\mathbf{v}_0 = (3, \pi/6, -2)$.
(e) $f: \mathbf{R}^4 \to \mathbf{R}$, $f(x_1, x_2, x_3, x_4) = x_1x_3 - x_2x_4$, $\mathbf{v}_0 = (0, 0, 0, 0)$
(f) f is as in (e), $\mathbf{v}_0 = (2, -1, 1, 3)$.

4. Let f, $g: \mathbf{R} \to \mathbf{R}$ be differentiable functions and define $h: \mathbf{R}^2 \to \mathbf{R}$ by $h(x, t) = f(x - at) + g(x + at)$, where a is a constant. Prove that h satisfies the *wave equation*

$$a^2 \frac{\partial^2 h}{\partial x^2} = \frac{\partial^2 h}{\partial t^2}$$

(This equation is satisfied by vibrating strings and by air in a pipe.)

5. Let $f: \mathbf{R} \to \mathbf{R}$ and $g: \mathbf{R}^2 \to \mathbf{R}$ be functions whose second derivatives exist. Compute all second-order partial derivatives of the composite $f \circ g$ in terms of the derivatives of f and g. (See the example at the end of the section.)
6. Let $f: \mathbf{R}^n \to \mathbf{R}$, $g: \mathbf{R} \to \mathbf{R}^n$ be functions whose second derivatives exist. Compute the second derivative of the composite $f \circ g: \mathbf{R} \to \mathbf{R}$ in terms of the derivatives of f and g.
7. If $f: \mathbf{R}^2 \to \mathbf{R}$ is a function, then $F(r, \theta) = f(r\cos\theta, r\sin\theta)$ is the function f in polar coordinates. (See Exercise 4, Section 4.4.) Suppose that the second derivative of the function f exists and compute all second-order partial derivatives of $F(r, \theta)$ in terms of the partial derivatives of f.
8. Let $f: \mathbf{R}^n \to \mathbf{R}$ be a twice differentiable function. The *Laplacian* of f, Δf, is defined by

$$\Delta f = \sum_{i=1}^{n} \frac{\partial^2 f}{\partial x_i^2}$$

If $n = 2$ and $F(r, \theta) = f(r\cos\theta, r\sin\theta)$, prove that

$$\Delta f = \frac{\partial^2 F}{\partial r^2} + \frac{1}{r^2}\frac{\partial^2 F}{\partial \theta^2} + \frac{1}{r}\frac{\partial F}{\partial r}$$

9. If $f: \mathbf{R}^3 \to \mathbf{R}$ is a function, then

$$F(\rho, \theta, \varphi) = f(\rho \sin\varphi \cos\theta, \rho \sin\varphi \sin\theta, \rho \cos\varphi)$$

is the function f in spherical coordinates. (See Exercise 5, Section 4.4.) Determine the second-order partial derivatives of F in terms of the partial derivatives of f.

10. With notation as in Exercise 9, derive a formula for the Laplacian of f in terms of the derivatives of F. (See Exercise 8.)
11. Consider the function $f: \mathbf{R}^2 \to \mathbf{R}$ defined by

$$f(x, y) = \begin{cases} xy\dfrac{x^2 - y^2}{x^2 + y^2} & \text{if } (x, y) \neq (0, 0) \\ 0 & \text{if } (x, y) = (0, 0) \end{cases}$$

Show that $\partial f/\partial x$ and $\partial f/\partial y$ exist and are continuous, that $\partial^2 f/\partial x\partial y$ and $\partial^2 f/\partial y\partial x$ exist at every point, and that

$$\left.\frac{\partial^2 f}{\partial x \partial y}\right|_{(0,0)} \neq \left.\frac{\partial^2 f}{\partial y \partial x}\right|_{(0,0)}$$

What conclusion can you draw about the functions $\partial^2 f/\partial x\partial y$ and $\partial^2 f/\partial y\partial x$?

12. Use induction to prove the following extension of Theorem 1.1. Let U be an open subset of $\mathbf{R}^n$ and $f: U \to \mathbf{R}$ a function. Suppose that all partial derivatives of f of order less than or equal to r exist and are continuous at $\mathbf{v} \in U$. Prove that the order of differentiation of these partial derivatives does not matter. For example, if $r = 3$, then

$$\frac{\partial^3 f}{\partial x_1 \partial x_2 \partial x_3}(\mathbf{v}) = \frac{\partial^3 f}{\partial x_2 \partial x_3 \partial x_1}(\mathbf{v})$$

2. Taylor's Theorem for Functions of One Variable

As we saw in Section 4.1, the derivative $D_{\mathbf{v}_0} f$ of a function $f: \mathbf{R}^n \to \mathbf{R}$ at $\mathbf{v}_0$ provides us with a good linear approximation to f near $\mathbf{v}_0$:

$$h(\mathbf{v}) = D_{\mathbf{v}_0}(\mathbf{v} - \mathbf{v}_0) + f(\mathbf{v}_0)$$

If f has higher derivatives, we might expect to get an even better approximation using polynomials instead of linear functions. We show now that this is indeed the case. We begin with $n = 1$, since the result for $n > 1$ can be reduced to this case.

Let U be an open interval in $\mathbf{R}^1$ and let $f: U \to \mathbf{R}$ be a function such that the jth derivative $f^{(j)}$ exists on U for $1 \leq j \leq m + 1$. [Of course, $f^{(j)}$ is continuous for $j \leq m$, but $f^{(m+1)}$ need not be.] Define the *mth degree Taylor polynomial of f about a* by

$$P_m(x) = f(a) + f'(a)(x - a) + \frac{f''(a)}{2}(x - a)^2 + \cdots + \frac{f^{(m)}(a)}{m!}(x - a)^m$$

$$= \sum_{j=0}^{m} \frac{f^{(j)}(a)}{j!}(x - a)^j$$

and the *remainder* by

$$R_m(x) = f(x) - P_m(x)$$

The following result is known as *Taylor's remainder theorem* for functions of one variable.

Theorem 2.1 *Let U be an open interval in $\mathbf{R}$. Pick $a \in U$, and let $f: U \to \mathbf{R}$ be a function having $m + 1$ derivatives on U. Then for any x in U, there is a $c \in U$ between x and a such that*

$$R_{m+1}(x) = \frac{f^{(m+1)}(c)}{(m+1)!}(x - a)^{m+1}$$

Proof. For definiteness, assume $a < x$. (If $a = x$, the theorem is trivial; if $a > x$ the proof is essentially the same as for $a < x$.) Define a number K by the equation

$$(2.1) \quad f(x) - f(a) - \sum_{j=1}^{m} \frac{f^{(j)}(a)}{j!}(x-a)^j - \frac{K}{(m+1)!}(x-a)^{m+1} = 0$$

We need to prove that $K = f^{(m+1)}(c)$ for some c between x and a.

Let g be a function defined by

$$g(t) = f(t) - f(a) - \sum_{j=1}^{m} \frac{f^{(j)}(a)}{j!}(t-a)^j - \frac{K}{(m+1)!}(t-a)^{m+1}$$

Then $$g'(t) = f'(t) - f'(a) - \sum_{j=1}^{m-1} \frac{f^{(j)}(a)}{j!}(t-a)^j - \frac{K}{m!}(t-a)^m$$

and, similarly,

$$(2.2) \quad g^{(p)}(t) = f^{(p)}(t) - f^{(p)}(a) - \sum_{j=1}^{m-p} \frac{f^{(j)}(a)}{j!}(t-a)^j - \frac{K}{(m+1-p)!}(t-a)^{m+1-p}$$

for $1 \leq p \leq m$. This is easily proved by induction; we omit the details.

In particular,

$$g(a) = g'(a) = \cdots = g^{(m)}(a) = 0$$

Furthermore, $g(x) = 0$, so by Rolle's theorem (Exercise 8, Section 4.1), there is a number c_1, $a < c_1 < x$, such that $g'(c_1) = 0$. Similarly, since $g'(c_1) = g'(a) = 0$, there is a number c_2, $a < c_2 < c_1$, with $g''(c_2) = 0$. Continuing inductively, we eventually find a number $c = c_{m+1}$ between a and x with $g^{(m+1)}(c) = 0$. However,

$$g^{(m+1)}(t) = f^{(m+1)}(t) - K$$

so $f^{(m+1)}(c) = K$ and we are finished.

Here are some examples.

1. Let $f(x) = e^x$ and $a = 0$. Then $f^{(k)}(x) = e^x$ and $f^{(k)}(0) = 1$ for all k. Therefore, the mth degree Taylor polynomial of the function e^x about $x = 0$ is

$$P_m(x) = \sum_{j=0}^{m} \frac{x^j}{j!}$$
$$= 1 + x + \frac{x^2}{2!} + \frac{x^3}{3!} + \cdots + \frac{x^m}{m!}$$

and the remainder is given by

$$R_m(x) = \frac{e^c}{(m+1)!} x^{m+1}$$

for some c between 0 and x.

This expression for the remainder is useful in determining how good an approximation $P_m(x)$ is for e^x. If, for example, $0 \le x \le 1$, then $c \le 1$ and $e^c \le 3$. Therefore

$$|e^x - P_m(x)| = |R_m(x)| \le \frac{3x^{m+1}}{(m+1)!}$$

In particular, if $x = 0.1$ and $n = 5$, we have

$$|e^{0.1} - P_5(0.1)| \le \frac{3 \cdot 10^{-6}}{6!} = \frac{10^{-7}}{24}$$

2. Let $f(x) = \sin x$ and $a = 0$. Then

$$f^{(k)}(x) = \begin{cases} \sin x & \text{if } k = 4l \\ \cos x & \text{if } k = 4l+1 \\ -\sin x & \text{if } k = 4l+2 \\ -\cos x & \text{if } k = 4l+3 \end{cases}$$

where l is a nonnegative integer. Thus

$$f^{(k)}(0) = \begin{cases} 0 & \text{if } k \text{ is even} \\ 1 & \text{if } k = 4l+1 \\ -1 & \text{if } k = 4l+3 \end{cases}$$

so that if m is odd, the mth degree Taylor polynomial of $\sin x$ about $a = 0$ is

$$\begin{aligned} P_m(x) &= x - \frac{x^3}{3!} + \frac{x^5}{5!} - \cdots \pm \frac{x^m}{m!} \\ &= \sum_{k=0}^{(m-1)/2} (-1)^k \frac{x^{2k+1}}{(2k+1)!} \end{aligned}$$

A similar expression holds if m is even. The remainder is given by

$$R_m(x) = \begin{cases} \pm \dfrac{\sin c}{(m+1)!} x^{m+1} & \text{if } m \text{ is odd} \\ \pm \dfrac{\cos c}{(m+1)!} x^{m+1} & \text{if } m \text{ is even} \end{cases}$$

where c is between 0 and x. In particular,

$$|\sin x - P_m(x)| = |R_m(x)| \le \frac{|x^{m+1}|}{(m+1)!}$$

for all x.

3. Let $f(x) = \cos x$ and $a = \pi/4$. Then

$$f^{(k)}(a) = \begin{cases} -\dfrac{\sqrt{2}}{2} & \text{if } k = 4l+1 \text{ or } 4l+2 \\ \dfrac{\sqrt{2}}{2} & \text{if } k = 4l \text{ or } 4l+3 \end{cases}$$

Thus the mth degree Taylor polynomial for $\cos x$ about $a = \pi/4$ is

$$P(x) = \frac{\sqrt{2}}{2}\left[1 - \left(x - \frac{\pi}{4}\right) - \frac{(x-\pi/4)^2}{2!} + \frac{(x-\pi/4)^3}{3!} + \cdots \pm \frac{(x-\pi/4)^m}{m!}\right]$$

and the remainder is

$$R_m(x) = \begin{cases} \pm\dfrac{(\cos c)(x-\pi/4)^{m+1}}{(m+1)!} & \text{if } m = 4l-1 \text{ or } 4l+1 \\ \pm\dfrac{(\sin c)(x-\pi/4)^{m+1}}{(m+1)!} & \text{if } m = 4l \text{ or } 4l+2 \end{cases}$$

where c is between x and a. In particular,

$$|\cos x - P_m(x)| \le \frac{|x-\pi/4|^{m+1}}{(m+1)!}$$

so that $P_m(x)$ provides a good approximation to $\cos x$ for values of x near $\pi/4$.

Exercises

1. Determine the mth degree Taylor polynomial of the function $f(x)$ about a and the remainder, where
   (a) $f(x) = \ln(1+x)$, $a = 0$, m arbitrary.
   (b) $f(x) = \cos x$, $a = 0$, m arbitrary.
   (c) $f(x) = \arctan x$, $a = 0$, m arbitrary.
   (d) $f(x) = \sin x$, $a = \pi/6$, $m = 5$.
   (e) $f(x) = e^x$, $a = 2$, $m = 5$.
   (f) $f(x) = (1+x)^r$, $r > 0$, $a = 0$, $m = 1$.
   (g) $f(x) = (1+x)^r$, $r > 0$, $a = 0$, $m = 3$.
2. Determine the following with error less than 10^{-3}.
   (a) $\sin 0.2$ (b) $e^{0.3}$
   (c) $\cos 43°$ (d) $\ln 1.2$
   (e) $\sqrt{2}$
3. (a) Let $f(x) = e^x, -1 < x < 1$. Show that if $P_m(x)$ is the mth degree Taylor

polynomial about 0 for f and $R_m(x)$ is the remainder term, then

$$\lim_{m\to\infty} R_m(x) = 0 \qquad \text{all } x$$

so that f can be approximated arbitrarily closely by the polynomials P_m. Prove similar results for

(b) $g(x) = \cos x$

(c) $h(x) = \sin x$

4. Let $f(x)$ be a polynomial of degree n. Prove that $f(x) = P_m(x)$ for all $m \geq n$.
5. There are various other forms of Taylor's theorem with remainder, in which one gives a different expression for the remainder term. We give one in this exercise. Show that if f is as in Theorem 2.1 and $f^{(n+1)}$ is also continuous, then

$$f(x) = f(a) + f'(a)(x-a) + \cdots + \frac{f^{(m)}(a)}{m!}(x-a)^m$$
$$+ \int_a^x \frac{f^{(n+1)}(t)(t-a)^n}{n!}\,dt$$

[*Hint*: Write $f(x) - f(a) = \int_a^x f'(t)\,dt$ and integrate by parts.]

6. (a) Prove one case of *L'Hôpital's rule*: if f, g have continuous derivatives in U and $f(a) = g(a) = 0$ $(a \in U)$, $g'(a) \neq 0$, then

$$\lim_{x\to a} \frac{f(x)}{g(x)} = \frac{f'(a)}{g'(a)}$$

If $g'(a) = 0$ and $f'(a) \neq 0$, then $\lim_{x\to a}[f(x)/g(x)]$ does not exist.

(b) Now suppose that f and g have n continuous derivatives, and suppose that

$$f(a) = f'(a) = \cdots = f^{(n-1)}(a) = 0$$
$$g(a) = g'(a) = \cdots = g^{(n-1)}(a) = 0$$

with $g^{(n)}(a) \neq 0$. Show that

$$\lim_{x\to a} \frac{f(x)}{g(x)} = \frac{f^{(n)}(a)}{g^{(n)}(a)}$$

7. (a) Prove the following variant of the mean value theorem: if f and g are differentiable functions on the open interval $U \subset \mathbf{R}^1$, and if $a, b \in U$, then there is a ξ between a and b such that

$$[f(b) - f(a)]g'(\xi) = [g(b) - g(a)]f'(\xi)$$

(b) Prove the following more precise form of L'Hôpital's rule: if f and g are as in (a) and if $f(a) = g(a) = 0$, then

$$\lim_{x\to a} \frac{f(x)}{g(x)} = \lim_{x\to a} \frac{f'(x)}{g'(x)}$$

in the sense that if the right-hand side has a limit, then so does the left-hand side, and they are equal.

3. Taylor's Theorem for Functions of Two Variables

The general form of Taylor's theorem is notationally quite complicated. We preview it here for functions of two variables, deferring the general case to the next section.

Let U be an open subset of $\mathbf{R}^2$ and $f: U \to \mathbf{R}$ be a function whose partial derivatives of order less than or equal to m exist and are continuous. Let $\mathbf{a} = (a_1, a_2)$ be a vector in U. The *Taylor polynomial of f of degree m* about $\mathbf{a}$ is defined by

$$P_m(x_1, x_2) = \sum_{i+j=m} \frac{1}{i!j!}\left(\frac{\partial^{i+j} f}{\partial x_1^i \, \partial x_2^j}(\mathbf{a})\right)(x_1 - a_1)^i (x_2 - a_2)^j$$

For example,

$$P_1(x_1, x_2) = f(\mathbf{a}) + \frac{\partial f}{\partial x_1}(\mathbf{a})(x_1 - a_1) + \frac{\partial f}{\partial x_2}(\mathbf{a})(x_2 - a_2)$$

and

$$\begin{aligned} P_2(x_1, x_2) &= f(\mathbf{a}) + \frac{\partial f}{\partial x_1}(\mathbf{a})(x_1 - a_1) + \frac{\partial f}{\partial x_2}(\mathbf{a})(x_2 - a_2) \\ &\quad + \frac{1}{2}\frac{\partial^2 f}{\partial x_1^2}(\mathbf{a})(x_1 - a_1)^2 \\ &\quad + \frac{\partial^2 f}{\partial x_1 \, \partial x_2}(\mathbf{a})(x_1 - a_1)(x_2 - a_2) \\ &\quad + \frac{1}{2}\frac{\partial^2 f}{\partial x_2^2}(\mathbf{a})(x_2 - a_2) \\ &= P_1(x_1, x_2) + \frac{1}{2}\left[\frac{\partial^2 f}{\partial x_1^2}(\mathbf{a})(x_1 - a_1)^2\right. \\ &\quad + 2\frac{\partial^2 f}{\partial x_1 \, \partial x_2}(\mathbf{a})(x_1 - a_1)(x_2 - a_2) \\ &\quad \left. + \frac{\partial^2 f}{\partial x_2^2}(\mathbf{a})(x_2 - a_2)^2\right] \end{aligned}$$

Similarly, $P_3(x_1, x_2)$ is obtained by adding the term

$$\begin{aligned} &\frac{1}{6}\frac{\partial^3 f}{\partial x_1^3}(\mathbf{a})(x_1 - a_1)^3 + \frac{1}{2}\frac{\partial^3 f}{\partial x_1^2 \, \partial x_2}(\mathbf{a})(x_1 - a_1)^2(x_2 - a_2) \\ &\qquad + \frac{1}{2}\frac{\partial^3 f}{\partial x_1 \, \partial x_2^2}(\mathbf{a})(x_1 - a_1)(x_2 - a_2)^2 + \frac{1}{6}\frac{\partial^3 f}{\partial x_2^3}(\mathbf{a})(x_2 - a_2)^3 \end{aligned}$$

to $P_2(x_1, x_2)$. (We use Exercise 12 of Section 1 here.)

Here are some examples.

1. Let $f(x_1x_2) = x_1^2x_2 + 1$ and $\mathbf{a} = (0, 0)$. Then

$$\frac{\partial f}{\partial x_1} = 2x_1x_2 \qquad \frac{\partial f}{\partial x_2} = x_1^2$$

$$\frac{\partial^2 f}{\partial x_1^2} = 2x_2 \qquad \frac{\partial^2 f}{\partial x_2^2} = 0$$

$$\frac{\partial^2 f}{\partial x_1\,\partial x_2} = 2x_1 = \frac{\partial^2 f}{\partial x_2\,\partial x_1}$$

and

$$\frac{\partial^3 f}{\partial x_1^2\,\partial x_2} = 2$$

whereas all other third-order derivatives vanish. In addition, all partial derivatives of order greater than or equal to 4 vanish.

Evaluating the partial derivatives above at $\mathbf{a} = (0, 0)$, we see that

$$P_1(x_1, x_2) = f(0, 0) = 1$$

$$P_2(x_1, x_2) = 1$$

$$\begin{aligned} P_3(x_1, x_3) &= 1 + \frac{1}{2}\left[\frac{\partial^3 f}{\partial x_1^2\,\partial x_2}(\mathbf{a})\right]x_1^2x_2 \\ &= 1 + x_1^2x_2 \\ &= f(x_1, x_2) \end{aligned}$$

and

$$P_m(x_1, x_2) = f(x_1, x_2)$$

for all $m \geq 4$.

We generalize the result of this example in Exercises 2 and 3.

2. Let $f(x_1, x_2) = \sin x_1x_2$ and $\mathbf{a} = (0, 0)$. Then

$$\frac{\partial f}{\partial x_1} = x_2 \cos x_1x_2$$

$$\frac{\partial f}{\partial x_2} = x_1 \cos x_1x_2$$

$$\frac{\partial^2 f}{\partial x_1^2} = -\,x_2^2 \sin x_1x_2$$

$$\frac{\partial^2 f}{\partial x_1\,\partial x_2} = \cos x_1x_2 - x_1x_2 \sin x_1x_2 = \frac{\partial^2 f}{\partial x_2\,\partial x_1}$$

$$\frac{\partial^2 f}{\partial x_2^2} = -\,x_1^2 \sin x_1x_2$$

All of these derivatives vanish at $\mathbf{a} = (0, 0)$ except for $\partial^2 f/\partial x_1\, \partial x_2$, which equals one at $\mathbf{a}$. Thus

$$P_2(x_1, x_2) = x_1 x_2$$

It is also easy to see that all of the third-order derivatives of f vanish at a. Thus $P_3(x_1, x_2)$ is also equal to $x_1 x_2$.

3. Let $f(x_1, x_2) = \sin x_1 x_2$ as in the previous example and $\mathbf{a} = (2, \pi/4)$. Then

$$f(\mathbf{a}) = \sin\frac{\pi}{2} = 1$$

$$\frac{\partial f}{\partial x_1}(\mathbf{a}) = \frac{\pi}{4}\cos\frac{\pi}{2} = 0$$

$$\frac{\partial f}{\partial x_2}(\mathbf{a}) = 0$$

$$\frac{\partial f^2}{\partial x_1^2}(\mathbf{a}) = -\frac{\pi^2}{16}\sin\frac{\pi}{2} = \frac{\pi^2}{16}$$

$$\frac{\partial^2 f}{\partial x_1\, \partial x_2}(\mathbf{a}) = \cos\frac{\pi}{2} - \frac{\pi}{2}\sin\frac{\pi}{2} = -\frac{\pi}{2}$$

$$= \frac{\partial^2 f}{\partial x_2\, \partial x_1}(\mathbf{a})$$

and

$$\frac{\partial^2 f}{\partial x_2^2}(\mathbf{a}) = -4\sin\frac{\pi}{2} = -4$$

Therefore

$$P_2(x_1, x_2) = 1 - \frac{\pi^2}{32}(x_1 - 2)^2 - \frac{\pi}{2}(x_1 - 2)\left(x_2 - \frac{\pi}{4}\right) - 4\left(x_2 - \frac{\pi}{4}\right)^2$$

Just as in the previous section, we define the *remainder* by

$$R_m(\mathbf{v}) = f(\mathbf{v}) - P_m(\mathbf{v})$$

We also have a Taylor's remainder theorem in this case.

Theorem 3.1 *Let U be an open subset of $\mathbf{R}^2$, let $\mathbf{a} = (a_1, a_2) \in U$, and let $f\colon U \to \mathbf{R}$ be a function whose partial derivatives of order $\leq m + 1$ exist. Let $\mathbf{v} = (x_1, x_2) \in U$ be such that all vectors on the line segment from $\mathbf{v}$ to $\mathbf{a}$ lie in U.* (See Figure 3.1.) *Then there is a vector $\mathbf{c}$ on this line segment such that*

$$R_m(\mathbf{v}) = \sum_{i+j=m+1} \frac{1}{i\,!\,j\,!}\left[\frac{\partial^{i+j} f}{\partial x_1^i\, \partial x_2^j}(\mathbf{c})\right](x_1 - a_1)^i (x_2 - a_2)^j$$

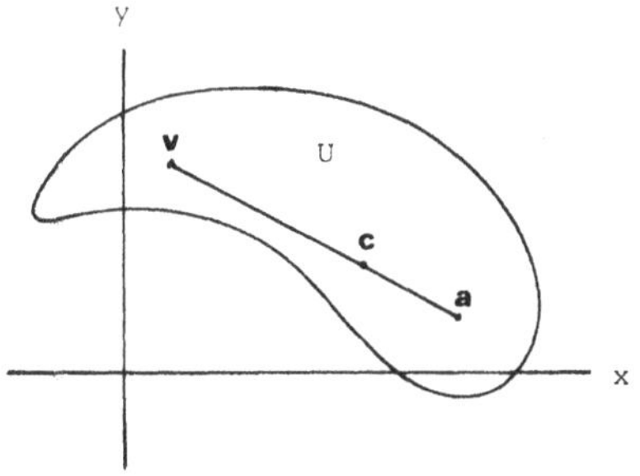

Figure 3.1 A line segment in U.

Since Theorem 4.1, proved in the next section, is a generalization of this result, we omit the proof of Theorem 3.1.

Exercises

1. Write down the Taylor polynomial of degree m about $\mathbf{a} \in \mathbf{R}^2$, where
   (a) $f(x_1, x_2) = 3 + x_1x_2 + x_2^3$, $\mathbf{a} = (0, 0)$, $m = 3$.
   (b) $f(x_1, x_2) = 3 + x_1x_2 + x_2^3$, $\mathbf{a} = (0, 0)$, $m = 57$.
   (c) $f(x_1, x_2) = 3 + x_1x_2 + x_2^3$, $\mathbf{a} = (1, -3)$, $m = 4$.
   (d) $f(x_1, x_2) = \cos x_1x_2$, $\mathbf{a} = (0, 0)$, $m = 2$.
   (e) $f(x_1, x_2) = \cos x_1x_2$, $\mathbf{a} = (0, 0)$, $m = 3$.
   (f) $f(x_1, x_2) = \cos x_1x_2$, $\mathbf{a} = (2, \pi/4)$, $m = 2$.
   (g) $f(x_1, x_2) = x_1 \tan x_1x_2$, $\mathbf{a} = (2, \pi/8)$, $m = 2$.
   (h) $f(x_1, x_2) = e^{x_1 \sin x_2}$, $\mathbf{a} = (0, \pi/3)$, $m = 2$.
2. Let $f(x_1, x_2)$ be a polynomial of degree 2,

$$f(x_1, x_2) = \sum_{i+j \le 2} a_{ij}x_1^i x_2^j$$
$$= a_{00} + a_{10}x_1 + a_{01}x_2 + a_{20}x_1^2 + a_{11}x_1x_2 + a_{02}x_2^2$$

   Prove that the Taylor polynomial of f about $\mathbf{a} = (0, 0)$ of degree m coincides with f whenever $m \ge 2$. What can be said when $m < 2$?
3. Let $f(x_1, x_2)$ be a polynomial of degree r,

$$f(x_1, x_2) = \sum_{i+j \le r} a_{ij}x_1^i x_2^j$$

   Prove that the Taylor polynomial of f about $\mathbf{a} = (0, 0)$ of degree m coincides with f whenever $m \ge r$. What can be said when $m < r$?
4. Let $f: \mathbf{R}^2 \to \mathbf{R}$ be a function having continuous partial derivatives of order ≤ 2 and let $g(x_1, x_2)$ be the polynomial of degree 2 such that

$$\frac{\partial^{i+j} f}{\partial x_1^i \, \partial x_2^j}(0, 0) = \frac{\partial^{i+j} g}{\partial x_1^i \, \partial x_2^j}(0, 0) \tag{3.1}$$

for $i + j \leq 2$. Prove that g is the Taylor polynomial of f of degree 2 about $a = (0, 0)$.

5. Let $f: \mathbf{R}^2 \to \mathbf{R}$ be a function having continuous partial derivatives of order $\leq m$ and let g be a polynomial of degree m such that (3.1) holds for $i + j \leq m$. Prove that g is the Taylor polynomial of f of degree m about $\mathbf{a} = (0, 0)$.
6. State and prove the analogue of Exercise 5 for arbitrary $\mathbf{a} = (a_1, a_2)$.

4. Taylor's Theorem for Functions of *n* Variables

We now state and prove the general form of Taylor's remainder theorem. We begin with some notation.

Fix an integer $n > 0$ and let $J = (j_1, j_2, \ldots, j_n)$ be a sequence of nonnegative integers. We define the *norm* of J by $|J| = j_1 + j_2 + \cdots + j_n$ and $J!$ by $J! = j_1! j_2! \cdots j_n!$. If $|J| = k$, we define the differential operator D_J by

$$D_J = \frac{\partial^k}{\partial x_1^{j_1} \partial x_2^{j_2} \cdots \partial x_n^{j_n}}$$

Thus, if f is a function of n variables,

$$D_J f = \frac{\partial^k f}{\partial x_1^{j_1} \partial x_2^{j_2} \cdots \partial x_n^{j_n}}$$

If $\mathbf{v} = (x_1, x_2, \ldots, x_n)$, $\mathbf{a} = (a_1, a_2, \ldots, a_n) \in \mathbf{R}^n$, we define $(\mathbf{v} - \mathbf{a})^J$ by

$$(\mathbf{v} - \mathbf{a})^J = (x_1 - a_1)^{j_1} (x_2 - a_2)^{j_2} \cdots (x_n - a_n)^{j_n}$$

In particular, if $n = 2$ and $J = (1, 3)$, then

$$|J| = 4 \qquad J! = 1! \cdot 3! = 6$$

$$D_J f = \frac{\partial^4 f}{\partial x_1 \partial x_2^3}$$

and

$$(\mathbf{v} - \mathbf{a})^J = (x_1 - a_1)(x_2 - a_2)^3$$

Let U be an open subset of $\mathbf{R}^n$, let $\mathbf{a} \in U$, and let $f: U \to \mathbf{R}$ be a function whose partial derivatives of order $\leq m + 1$ exist on U. *The Taylor polynomial of f of degree m about* $\mathbf{a}$ is defined by

$$P_m(\mathbf{v}) = \sum_{|J| \leq m} \frac{1}{J!} (D_J f)(\mathbf{a}) (\mathbf{x} - \mathbf{a})^J$$

Here,

$$\sum_{|J| \leq m}$$

means that we sum over all sequences $J = (j_1, j_2, \ldots, j_a)$ of nonnegative integers with $|J| = j_1 + j_2 + \cdots + j_n \leq m$.

It is easy to see that this definition agrees with that given in the previous section for $n = 2$.

As before, we define the remainder by

$$R_m(f)(\mathbf{v}) = f(\mathbf{v}) - P_m(\mathbf{v})$$

Theorem 4.1 *Let U be an open subset of $\mathbf{R}^n$, $\mathbf{a} \in U$, and $f : U \to \mathbf{R}$ a function whose partial derivatives of order $\leq m + 1$ exist. Let $\mathbf{v}$ be a vector in U with the property that each point on the line segment between $\mathbf{v}$ and $\mathbf{a}$ is also in U. Then there is a vector $\mathbf{c}$ on the line segment from $\mathbf{a}$ to $\mathbf{v}$ such that*

$$R_m(\mathbf{v}) = \sum_{|J| = m+1} \frac{1}{J!}(D_J f)(\mathbf{c})(\mathbf{v} - \mathbf{a})^J$$

This is the *Taylor's remainder theorem* for functions of n variables.

Proof. Define a function $g : [0, 1] \to \mathbf{R}^n$ by $g(t) = t\mathbf{v} + (1 - t)\mathbf{a}$. The image of g, $\mathrm{Im}(g)$, is the line segment between $\mathbf{v}$ and $\mathbf{a}$; so $\mathrm{Im}(g) \subset U$. In fact, since U is open,

$$g(-\varepsilon, 1 + \varepsilon) \subset U$$

for some $\varepsilon > 0$.

The function g has derivatives of arbitrary order; in fact, $D_t g = \mathbf{v} - \mathbf{a}$ and those of order greater than 1 vanish. Hence, the composite $F(t) = f(g(t))$ has derivatives of order $\leq m + 1$ on $(-\varepsilon, 1 + \varepsilon)$. Applying Theorem 2.1, we have

(4.1)
$$F(t) = \sum_{j=0}^{m} \frac{F^j(0)}{j!} t^j + \frac{F^{(m+1)}(b)}{(m + 1)!} t^{m+1}$$

for some b between 0 and t.

We now wish to express the $F^{(j)}(t)$ in terms of the partial derivatives of f. In principle, this is a matter of using the chain rule. In practice, it is a bit complicated.

To begin with,

(4.2)
$$\begin{aligned} F'(t) &= f'(g(t))g'(t) \\ &= \begin{bmatrix} \dfrac{\partial f}{\partial x_1} & \cdots & \dfrac{\partial f}{\partial x_n} \end{bmatrix} \begin{bmatrix} x_1 - a_1 \\ \vdots \\ x_n - a_n \end{bmatrix} \\ &= \sum_{j=1}^{n} \frac{\partial f}{\partial x_j}(g(t))(x_j - a_j) \end{aligned}$$

At $t = 0$, $g(t) = \mathbf{a}$, and we have

$$F'(0) = \sum_{j=1}^{k} \frac{\partial f}{\partial x_j}(\mathbf{a})(x_j - a_j)$$
$$= \sum_{|J|=1} \frac{1}{J!}(D_J f)(\mathbf{a})(\mathbf{v} - \mathbf{a})^J$$

since $J! = 1$ if $|J| = 1$.

To evaluate $F''(t)$, we differentiate $F'(t)$ as given in equation (4.2). The chain rule says that

$$\frac{d}{dt}\left(\frac{\partial f}{\partial x_j}(g(t))\right) = \left(\frac{\partial}{\partial x_i}\frac{\partial f}{\partial x_j}\right)(g(t))g'(t)$$
$$= \sum_{i=1}^{k} \frac{\partial^2 f}{\partial x_i \partial x_j}(g(t))(x_i - a_i)$$

Therefore, using equation (4.2), we have

$$\frac{1}{2!}F''(t) = \frac{1}{2!}\sum_{j=1}^{k}\sum_{i=1}^{k} \frac{\partial^2 f}{\partial x_i \partial x_j}(g(t))(x_i - a_i)(x_j - a_j)$$
$$= \sum_{|J|=2} \frac{1}{J!}(D_J f)(g(t))(\mathbf{v} - \mathbf{a})^J$$

since if, say, $J = (2, 0, 0, \ldots)$, then the sum contributes only one $D_J f$ term (the $\partial^2/\partial x_1^2$ term), while if, say, $J = (1, 1, 0, \ldots)$, then the sum contributes two $D_J f$ terms (the $\partial^2/\partial x_1 \partial x_2$ and $\partial^2/\partial x_2 \partial x_1$ terms). It follows that

$$\frac{1}{2!}F''(0) = \sum_{|J|=2} \frac{1}{J!}(D_J f)(\mathbf{a})(\mathbf{v} - \mathbf{a})^J$$

Continuing in this way,* we get

$$(4.3)\quad F^{(j)}(t) = \sum_{i_1=1}^{k} \cdots \sum_{i_j=1}^{k} \frac{\partial^j}{\partial x_{i_1} \cdots \partial x_{i_j}} f(g(t))(x_{i_1} - a_{i_1}) \cdots (x_{i_j} - a_{i_j})$$

We need now to know how many terms in this sum correspond to each sequence J. This is a standard combinatorial problem, and we shall simply state the result: one can write (4.3) as

$$(4.4)\qquad F^{(j)}(t) = \sum_{|J|=j} \frac{|J|!}{J!}(D_J f)(g(t))(\mathbf{v} - \mathbf{a})^J$$

Thus, if $\mathbf{v} = g(t)$, equations (4.2) and (4.4) imply that

*As is usually the case, this phrase conceals a use of mathematical induction; see Exercise 7.

$$
\begin{aligned}
R_m(\mathbf{v}) &= F(t) - \sum_{j=0}^{m} \frac{F^{(j)}(0)}{j!} t^j \\
&= \frac{F^{(m+1)}(b)}{(m+1)!} t^{m+1} \qquad \text{(for some } b \text{ between 0 and 1)} \\
&= \sum_{|J|=m+1} \frac{1}{J!} (D_J f)(\mathbf{c})(\mathbf{v} - \mathbf{a})^J
\end{aligned}
$$

where $\mathbf{c} = g(b)$.

Remark. We note that $F''(0)$ can be written in another way. Recall that the Hessian of f, $H(f)$, is the $n \times n$ matrix (a_{ij}), where each a_{ij} is the function

$$a_{ji} = \frac{\partial^2 f}{\partial x_i \partial x_j}$$

If f has continuous second partial derivative, $H(f)$ is a symmetric matrix (Theorem 1.1). Thus $H(f)(\mathbf{a})$ defines a linear transformation $T: \mathbf{R}^n \to \mathbf{R}^n$, and it is easy to check that $F''(0)$ is given by the formula

$$F''(0) = \langle T(\mathbf{v} - \mathbf{a}), \mathbf{v} - \mathbf{a} \rangle$$

Exercises

1. Find the Taylor polynomial of the function $f: \mathbf{R}^3 \to \mathbf{R}$ of degree m about $\mathbf{a}$, where
   (a) $f(x_1, x_2, x_3) = x_1 x_2 x_3$, $\mathbf{a} = (0, 0, 0)$, $m = 3$.
   (b) $f(x_1, x_2, x_3) = x_1 x_2 x_3$, $\mathbf{a} = (0, 0, 0)$, $m = 44$.
   (c) $f(x_1, x_2, x_3) = x_1 x_2 x_3$, $\mathbf{a} = (1, 2, -1)$, $m = 2$.
   (d) $f(x_1, x_2, x_3) = \sin x_1 x_2 x_3$, $\mathbf{a} = (0, 0, 0)$, $m = 2$.
   (e) $f(x_1, x_2, x_3) = \sin x_1 x_2 x_3$, $\mathbf{a} = (2, 0, 1)$, $m = 3$.
   (f) $f(x_1, x_2, x_3) = \sin x_1 x_2 x_3$, $\mathbf{a} = (1, 2, \pi/6)$, $m = 2$.
   (g) $f(x_1, x_2, x_3) = x_1 \cos(x_2 x_3 + x_1^2)$, $\mathbf{a} = (0, 0, 0)$, $m = 2$.
   (h) $f(x_1, x_2, x_3) = x_1 \cos(x_2 x_3 + x_1^2)$, $\mathbf{a} = (0, 2, \pi/8)$, $m = 2$.
2. Prove that $F''(0) = \langle T(\mathbf{v} - \mathbf{a}), \mathbf{v} - \mathbf{a} \rangle$, where $T: \mathbf{R}^n \to \mathbf{R}^n$ is the linear transformation defined by the matrix $H(f)(\mathbf{a})$. (This is the assertion of the last paragraph of this section.)
3. Let $f: \mathbf{R}^3 \to \mathbf{R}$ be a function having continuous partial derivatives of order ≤ 2 and let $g(x_1, x_2, x_3)$ be the polynomial of degree 2 such that

   $$\frac{\partial^{i+j} f}{\partial x_1^i \partial x_2^j \partial x_3^k}(0, 0, 0) = \frac{\partial^{i+j} g}{\partial x_1^i \partial x_2^j \partial x_3^k}(0, 0, 0) \tag{4.5}$$

   for $i + j + k \leq 2$. Prove that g is the Taylor polynomial of f of degree 2 about $\mathbf{a} = (0, 0, 0)$.
4. Let $f: \mathbf{R}^3 \to \mathbf{R}$ be a function having continuous partial derivatives of order $\leq m$ and let g be a polynomial of degree m such that (4.5) holds for $i + j \leq m$. Prove that g is the Taylor polynomial of f of degree m about $\mathbf{a} = (0, 0, 0)$.

5. State and prove the analogue of Exercise 5 for arbitrary $\mathbf{a} = (a_1, a_2, a_3)$.
6. State and prove the analogue of Exercise 5 for functions $f: \mathbf{R}^n \to \mathbf{R}$ for
   (a) $\mathbf{a} = (0, 0, \ldots, 0)$
   (b) $\mathbf{a} = (a_1, a_2, \ldots, a_n)$ arbitrary
7. Use induction to prove equation (4.3).

5. The Equality of Mixed Partial Derivatives

We now give the proof of Theorem 1.1.

We can clearly assume $i = 1$, $j = 2$ by reordering the variables if necessary. Since all other variables are held fixed, we can also assume $n = 2$. For convenience, we denote the points in $\mathbf{R}^2$ by (x, y).

Let $\mathbf{v} = (a, b)$ and let $\mathbf{w} = (c, d)$ be a point in $\mathbf{R}^2$ with $a < c, b < d$, and such that the rectangle

$$R(\mathbf{v}, \mathbf{w}) = \{(x, y) \in \mathbf{R}^2 : a \leq x \leq c,\ b \leq y \leq d\}$$

is entirely contained in U. Let $k(\mathbf{v}, \mathbf{w}) \in \mathbf{R}$ be defined by

$$k(\mathbf{v}, \mathbf{w}) = f(c, d) - f(c, b) - f(a, d) + f(a, b)$$

We evaluate $k(\mathbf{v}, \mathbf{w})$ in two different ways.

Define $g: \mathbf{R} \to \mathbf{R}$ by $g(x) = f(x, b)$. Then $k(\mathbf{v}, \mathbf{w}) = g(c) - g(a) = (c - a)g'(s_1)$ for some s_1 between a and c, by the mean value theorem for functions of one variable. (See Exercise 9, Section 4.1.) However,

$$\begin{aligned} g'(s_1) &= \frac{\partial f}{\partial x}(s_1, d) - \frac{\partial f}{\partial x}(s_1, b) \\ &= (d - b)\frac{\partial^2 f}{\partial y \partial x}(s_1, s_2) \end{aligned}$$

for s_2 between b and d, again by the mean value theorem. Thus

$$k(\mathbf{v}, \mathbf{w}) = (c - a)(d - b)\frac{\partial^2 f}{\partial y \partial x}(s_1, s_2)$$

Similarly, we define $h : \mathbf{R} \to \mathbf{R}$ by $h(y) = f(c, y) - f(a, y)$. Then, just as above,

$$k(\mathbf{v}, \mathbf{w}) = (c - a)(d - b)\frac{\partial^2 h}{\partial x \partial y}(t_1, t_2)$$

for t_1 between a and c, t_2 between b and d. It follows that

$$\frac{\partial^2 f}{\partial y \partial x}(s_1, s_2) = \frac{\partial^2 f}{\partial x \partial y}(t_1, t_2) \tag{5.1}$$

Let $\varphi, \psi : \mathbf{R}^2 \to \mathbf{R}$ be defined by

$$\varphi(\mathbf{z}) = \frac{\partial^2 f}{\partial y \partial x}(\mathbf{z}) \qquad \psi(\mathbf{z}) = \frac{\partial^2 f}{\partial x \partial y}(\mathbf{z})$$

Thus $\varphi(\mathbf{s}) = \psi(\mathbf{t})$ by equation (5.1), $\mathbf{s} = (s_1, s_2)$ and $\mathbf{t} = (t_1, t_2)$. We know that φ and ψ are continuous at $\mathbf{v}$; we need to prove that $\varphi(\mathbf{v}) = \psi(\mathbf{v})$. Let $\varepsilon > 0$ be any small number and choose $\delta > 0$ so that

$$\|\mathbf{v} - \mathbf{z}\| < \delta \Rightarrow |\varphi(\mathbf{v}) - \varphi(\mathbf{z})| < \frac{\varepsilon}{2} \qquad \text{and} \qquad |\psi(\mathbf{v}) - \psi(\mathbf{z})| < \frac{\varepsilon}{2}$$

Pick $\mathbf{w}$ above so that $\|\mathbf{v} - \mathbf{w}\| < \delta$. Then

$$\|\mathbf{v} - \mathbf{s}\| < \delta \qquad \text{and} \qquad \|\mathbf{v} - \mathbf{t}\| < \delta$$

so that

$$|\varphi(\mathbf{v}) - \varphi(\mathbf{s})| < \frac{\varepsilon}{2} \qquad \text{and} \qquad |\psi(\mathbf{v}) - \psi(\mathbf{t})| < \frac{\varepsilon}{2}$$

Then

$$\begin{aligned} |\varphi(\mathbf{v}) - \psi(\mathbf{v})| &= |\varphi(\mathbf{v}) - \varphi(\mathbf{s}) + \varphi(\mathbf{s}) - \psi(\mathbf{t}) + \psi(\mathbf{t}) - \psi(\mathbf{v})| \\ &\le |\varphi(\mathbf{v}) - \varphi(\mathbf{s})| + |\varphi(\mathbf{s}) - \psi(\mathbf{t})| + |\psi(\mathbf{t}) - \psi(\mathbf{v})| \\ &< \frac{\varepsilon}{2} + \frac{\varepsilon}{2} = \varepsilon \end{aligned}$$

since $\varphi(\mathbf{s}) = \psi(\mathbf{t})$. Thus $|\varphi(\mathbf{v}) - \psi(\mathbf{v})| < \varepsilon$ for any $\varepsilon > 0$, and it follows that $\varphi(\mathbf{v}) = \psi(\mathbf{v})$.

Remark. A somewhat stronger result than Theorem 1.1 is true: if f is differentiable and $\partial^2 f/\partial y \partial x$ exists in an open set about $\mathbf{v}$ and is continuous there, then $(\partial^2 f/\partial x \partial y)(\mathbf{v})$ exists and equal to $(\partial^2 f/\partial y \partial x)(\mathbf{v})$. We shall not prove this result here. (See Exercise 1.) In most cases arising in practice, both mixed partials are known to exist and be continuous, and then, of course, they are equal.

Exercises

1. Modify the proof of Theorem 2.1 to prove the result mentioned in the remark above: if f is continuously differentiable and $\partial^2 f/\partial y \partial x$ exists in an open set about $\mathbf{v}$, and if $\partial^2 f/\partial y \partial x$ is continuous at $\mathbf{v}$, then $(\partial^2 f/\partial x \partial y)(\mathbf{v})$ exists and equals $(\partial^2 f/\partial y \partial x)(\mathbf{v})$.
2. Let $f: \mathbf{R}^2 \to \mathbf{R}$ be defined by

$$f(x, y) = |x|$$

Show that $\partial^2 f/\partial x \partial y = 0$ for all $(x, y) \in \mathbf{R}^2$, but that $\partial^2 f/\partial y \partial x$ is not always defined. Does this contradict Exercise 1?

7

Compact and Connected Sets

The notion of a compact set, which we introduce in this chapter, is one of the most important in mathematics. We use it in order to state and prove some of the fundamental theorems in analysis, such as the theorem which states that a continuous real-valued function on a closed, bounded interval attains its maximum.

Besides compactness, the chapter is concerned with two other topics. We discuss upper and lower bounds of sets in **R** and prove various theorems about least upper bounds (lub) and greatest lower bounds (glb). We also define connectedness and prove another basic result, the intermediate value theorem.

The results of this chapter will be needed in a number of the proofs found later, and they are among the most fundamental in analysis. Many of them, however, are easy to accept on faith, and the reader who wishes to can skip this chapter on first reading, referring back when necessary.

1. Compact Sets

In Section 2.5, we singled out for study two classes of subsets of a Euclidean space, the open sets and the closed sets. Another important class of subsets of a normed vector space is the collection of compact sets.

A subset C of $\mathbf{R}^n$ is said to be *compact** if every sequence of points in C has a subsequence which converges to a limit point in C.

*One sometimes says that C is *sequentially* or *countably* compact.

For example, the open interval

$$(0, 1) = \{x \in \mathbf{R}^1 : 0 < x < 1\}$$

is *not* compact since the sequence $a_n = 1/n$ in (0, 1) has no subsequence converging to a point *in* (0, 1). This is true since the sequence converges to 0. Consequently, every subsequence of this sequence converges to 0, and $0 \notin (0, 1)$. (See Proposition 6.4, Chapter 3.)

Another set which is not compact is the set $C = \mathbf{R}^1 \subset \mathbf{R}^1$. This is because the sequence $a_n = n$ has no convergent subsequences.

To give examples of sets which are compact is more difficult. We give one (not very interesting) example here. Others will have to wait until Section 4, where we characterize the compact subsets of Euclidean spaces.

Proposition 1.1 *Any finite subset of* $\mathbf{R}^n$ *is compact.*

Proof. Suppose $C = \{\mathbf{v}_1, \ldots, \mathbf{v}_k\}$; let $\{\mathbf{a}_n\}$ be a sequence in C. Since C is finite, one of the elements of C must appear an infinite number of times in this sequence. More explicitly, there is a sequence of integers

$$0 < i_1 < i_2 < \cdots < i_r < \cdots$$

with $\mathbf{a}_{i_1} = \mathbf{a}_{i_2} = \cdots = \mathbf{a}_{i_r} = \cdots = \mathbf{v}_j$ for some j, $1 \leq j \leq k$. This gives us our convergent subsequence with limit point $\mathbf{v}_j$ in C.

Proposition 1.1 hardly provides evidence that compact sets are worth studying. We shall describe some more interesting compact sets in a moment. First, however, we prove two simple properties of compact sets.

Theorem 1.2 *Let C be a subset of* $\mathbf{R}^n$. *Then C is closed if and only if whenever* $\mathbf{v}_1, \mathbf{v}_2, \ldots$ *is a sequence of vectors in C with* $\lim_{j \to \infty} \mathbf{v}_j = \mathbf{v}$ *in* $\mathbf{R}^n$, *then* $\mathbf{v}$ *must be in C.*

Proof. Suppose C closed and $\mathbf{v}_1, \mathbf{v}_2, \ldots$ in C with $\lim_{n \to \infty} \mathbf{v}_n = \mathbf{v} \in \mathbf{R}^n - C$. Since $\mathbf{R}^n - C$ is open, there is an $r > 0$ with $B_r(\mathbf{v}) \subset \mathbf{R}^n - C$. But this implies that $\|\mathbf{v}_n - \mathbf{v}\| \geq r$ for all n, which contradicts the fact that $\lim_{n \to \infty} \mathbf{v}_n = \mathbf{v}$.

Now suppose $C \subset \mathbf{R}^n$ is such that whenever $\mathbf{v}_1, \ldots, \mathbf{v}_n, \ldots$ is a sequence in C with $\lim_{j \to \infty} \mathbf{v}_j = \mathbf{v}$ in $\mathbf{R}^n$, then $\mathbf{v} \in C$. If C is not closed, then $\mathbf{R}^n - C$ is not open; so we can find a vector $\mathbf{v} \in \mathbf{R}^n - C$ with the property that no ball about $\mathbf{v}$ is contained in $\mathbf{R}^n - C$. That is, every ball about $\mathbf{v}$ contains points of C. It follows that for each integer n, we can find a vector $\mathbf{v}_n$ in $B_{1/n}(\mathbf{v})$ which is also in C. Thus $\lim_{n \to \infty} \mathbf{v}_n = \mathbf{v}$, but $\mathbf{v} \notin C$. (See Figure 1.1.) This contradicts our hypothesis about C, and the result follows.

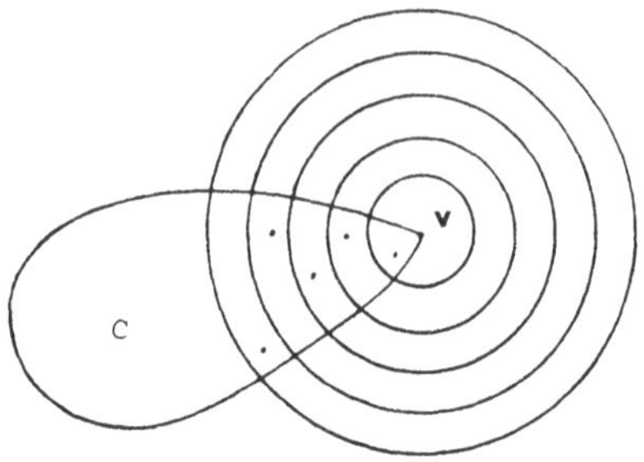

Figure 1.1 A sequence converging to $\mathbf{v}$.

We can now give one property of compact sets.

Proposition 1.3 *Any compact subset C of $\mathbf{R}^n$ is closed.*

Proof. According to Theorem 1.2, we must show that if $\{\mathbf{v}_n\}$ is a sequence in C which converges to a vector $\mathbf{v}$ in $\mathbf{R}^n$, then $\mathbf{v}$ is in C. Since C is compact, there is a subsequence of the sequence $\{\mathbf{v}_n\}$ converging to a vector $\mathbf{v}'$ in C. Since the sequence $\{\mathbf{v}_n\}$ is convergent, any subsequence must converge to $\mathbf{v}$ by Proposition 6.4 of Chapter 3. It follows that $\mathbf{v} = \mathbf{v}'$ is in C; so C is closed.

Recall that a subset C of $\mathbf{R}^n$ is said to be *bounded* if there is a real number $r > 0$ such that $\|\mathbf{v}\| < r$ for all $\mathbf{v}$ in C. Equivalently, C is bounded if there is an $r > 0$ with $C \subset B_r(\mathbf{0})$.

Proposition 1.4 *Any compact set is bounded.*

Proof. Suppose C is a subset of $\mathbf{R}^n$ which is not bounded. Then C is contained in no ball about $\mathbf{0}$ in $\mathbf{R}^n$. It follows that, for every integer n, we can find a vector $\mathbf{w}_n$ in C which is not in $B_n(\mathbf{0})$. Since no subsequence of the sequence $\{\mathbf{w}_n\}$ is bounded, it follows from Proposition 6.5 of Chapter 3 that no subsequence of this sequence can be convergent. Thus C cannot be compact if C is not bounded, and the proposition is proved.

The two results above tell us that a compact set is both closed and bounded. In Section 4, we will show that the converse is true; namely, that any closed and bounded subset of $\mathbf{R}^n$ is compact. The next result will be useful in proving this.

Proposition 1.5 *Suppose C is a compact subset of $\mathbf{R}^n$ and B is a closed subset of $\mathbf{R}^n$ contained in C. Then B is compact.*

Proof. Let $\{\mathbf{v}_n\}$ be a sequence in B. We must show that this sequence has a subsequence converging to a vector in B.

Since B is contained in C, $\{\mathbf{v}_n\}$ is a sequence in C, so we can find a subsequence $\mathbf{v}_{j_1}, \mathbf{v}_{j_2}, \ldots$ converging to a vector $\mathbf{v}$ in C. However, the subsequence $\mathbf{v}_{j_1}, \mathbf{v}_{j_2}, \ldots$ is a sequence in the closed set B which converges in $\mathbf{R}^n$, so its limit point $\mathbf{v}$ must be in B. Thus B is compact, and the proposition is proved.

Exercises

Note. Before doing these exercises, the reader may wish to read the next few sections to get some experience with the procedures used to prove statements about compact sets.

1. Prove that the union of a finite collection of compact sets is compact. Is the union of an infinite collection of compact sets necessarily compact?
2. Prove that the intersection of any collection of compact sets is compact.
3. Prove that if C_1 and C_2 are compact sets in $\mathbf{R}$, then $C_1 \times C_2$ is compact in $\mathbf{R} \times \mathbf{R} = \mathbf{R}^2$. (See Section 1.1 for the definition.)
4. State and prove the analogue of Exercise 3 for $\mathbf{R}^p \times \mathbf{R}^q$.

*5. Let $C_1, C_2, \ldots, C_n, \ldots$ be a decreasing collection of nonempty compact subsets of $\mathbf{R}^n$ (that is, $C_{n+1} \subset C_n$, all $n \geq 1$). Prove that the intersection $\cap_n C_n$ is nonempty. (*Hint*: Choose $\mathbf{v}_n \in C_n$, and show that some subsequence of the $\mathbf{v}_n$ converges to a point $\mathbf{v}$. Prove that $\mathbf{v}$ is in the intersection $\cap_n C_n$.)

*6. Give an example to show that the conclusion of Exercise 3 does not hold if we assume only that the sets C_n are closed sets.

*7. Prove that the set $C = \{0, 1, \frac{1}{2}, \frac{1}{3}, \frac{1}{4}, \ldots\} \subset \mathbf{R}$ is compact. (*Hint*: Let $a_1, a_2, \ldots$ be any sequence of elements of C. If there are elements of the sequence with arbitrarily large denominators, show that there is a subsequence converging to 0. If not, show that Proposition 1.1 applies.)

*8. Let C be a compact subset of $\mathbf{R}^n$. Prove that for any $\varepsilon > 0$, there is a finite set $F_\varepsilon \subset C$ such that

$$C \subset \bigcup_{\mathbf{v} \in F_\varepsilon} B_\varepsilon(\mathbf{v})$$

The remaining problems will involve the notion of an open covering of a set.

Let C be a subset of $\mathbf{R}^n$. An *open covering* of C is a collection $\mathscr{U} = \{U_\lambda : \lambda \in \Lambda\}$ of open subsets of $\mathbf{R}^n$ such that

$$C \subset \bigcup_{\lambda \in \Lambda} U_\lambda$$

We say that $\mathscr{U}$ is *finite* if Λ is finite and *countable* if Λ is countable (i.e., Λ is the natural numbers). An open covering $\mathscr{V}$ of C is a *subcovering* of the open covering $\mathscr{U}$ of C if each open set in $\mathscr{V}$ is also one of the open sets in $\mathscr{U}$.

For instance, if $C = (0, \frac{3}{2}) \subset \mathbf{R}$, one open covering of C is $\{B_{x/2}(x) : x \in$

$(0, \frac{3}{2})\}$, since $x \subset B_{x/2}(x)$. A countable subcovering of C is $\{B_{x/2}(x) : x = 1, \frac{1}{2}, \frac{1}{3}, \frac{1}{4}, \ldots\}$, as one can check. C has no finite subcovering: the set $\cup_{j=1}^{n} B_{x_j/2}(x_j)$, with $x_1 > x_2 > \cdots > x_n$, does not contain $x_n/2$. Another open covering of C is $\{B_{1/5}(x) : x \in (0, \frac{3}{2})\}$; $\{B_{1/5}(x) : x = \frac{1}{5}, \frac{2}{5}, \frac{3}{5}, \frac{4}{5}, 1, \frac{6}{5}, \frac{7}{5}\}$ is a finite subcovering.

*9. Prove that the subset $C \subset \mathbf{R}^n$ is compact if and only if every countable open covering of C has a finite subcovering. (That is, if $\mathscr{U} = \{U_j\}$ is any countable open covering of C, then there are indices $j_1, \ldots, j_m$ with $C \subset U_{j_1} \cup U_{j_2} \cup \cdots \cup U_{j_m}$.) (*Hint*: If $\{U_j\}$ has no such subcovering, then $\cup_{j=1}^{n} U_j \neq C$, all n. Pick $\mathbf{v}_n \in C - \cup_{j=1}^{n} U_j$. Let a subsequence $\{\mathbf{v}_{n_k}\}$ of $\{\mathbf{v}_n\}$ converge to $\mathbf{v}$; if $\mathbf{v} \in U_N$, show that $\mathbf{v}_{n_k} \in U_N$ for arbitrarily large numbers $\{n_k\}$. Deduce a contradiction.

(Conversely, suppose that $\{\mathbf{v}_n\}$ is a sequence such that no subsequence converges to any point in C. Show that if any subsequence of $\{\mathbf{v}_n\}$ converges, then C is not closed. If no subsequence converges in $\mathbf{R}^n$, then show that $U_j = \mathbf{R}^n - \{\mathbf{v}_j, \mathbf{v}_{j+1}, \ldots\}$ is open and that $\{U_j\}$ is a covering with no finite subcovering.) (See Exercise 12 of Section 3.6.)

*10. Let $\mathscr{U} = \{U_\lambda : \lambda \in \Lambda\}$ be an open covering of the compact set $C \subset \mathbf{R}^n$. Prove that there is an $r > 0$ such that, for any $\mathbf{v} \in C$, $B_r(\mathbf{v}) \subset U_\lambda$ for some $\lambda \in \Lambda$. (This result is called the *Lebesgue covering lemma.*) (*Hint*: If not, then for every n there is a $\mathbf{v}_n$ such that $B_{1/n}(\mathbf{v}_n) \notin U_\lambda$ for any λ. Pick a subsequence $\{\mathbf{v}_{n_k}\}$ converging to $\mathbf{v}$; since $\mathbf{v}$ is in some U_λ, $\exists N$ with $B_{1/N}(\mathbf{v}) \subset U_\lambda$. Show that for some n_k, $B_{1/n_k}(\mathbf{v}_{n_k}) \subset B_{1/N}(\mathbf{v})$, and get a contradiction.)

*11. Prove that $C \subset \mathbf{R}^n$ is compact if and only if every open covering of C has a finite subcovering. (This result is called the *Heine-Borel theorem.*) (*Hint*: Use Exercises 10 and 8 in one direction, and 9 in the other.)

*12. We say that a set C has the *finite intersection property* if whenever $C_1, C_2, \ldots, C_m, \ldots$ is a sequence of closed sets such that

$$C \cap \bigcap_{j=1}^{m} C_j \neq \varnothing$$

for all m, then

$$C \cap \bigcap_{j=1}^{\infty} C_j \neq \varnothing$$

Prove that C is compact if and only if it has the finite intersection property.

2. Upper and Lower Bounds

The material in this section will be used later in this chapter, but is somewhat different in content from the other sections. The reader may wish to read the basic definitions and then go ahead to Section 3, referring back as necessary.

Let X be a subset of the real numbers. An *upper bound* for X is a number M with the property that $x \leq M$ for all x in X. A *lower bound* for X is a number m

with $m \leq x$ for all x in X. We say that X is *bounded from above* if X has some upper bound and *bounded from below* if X has a lower bound.

Examples

1. Let $X = \{x \in \mathbf{R} : 0 \leq x < 1\}$. The number 2 is an upper bound for X, since every element of X is ≤ 2. The numbers 1 and 5 are also upper bounds for X. Similarly, -7 is a lower bound for X, since every element of X is ≥ -7. The number $\frac{1}{2}$ is neither an upper bound for X (since $\frac{3}{4} \in X$ and $\frac{3}{4} > \frac{1}{2}$) nor a lower bound (since $\frac{1}{4} \in X$ and $\frac{1}{4} < \frac{1}{2}$).
2. Let $X = \{x \in \mathbf{R} : x \text{ is an integer}\}$. Then X has no upper bound, since there is no real number greater than every integer. Similarly, X has no lower bound.

We say that M is a *least upper bound* for X if

(i) M is an upper bound for X.
(ii) If M' is any other upper bound for X, then $M \leq M'$.

Similarly, m is a *greatest lower bound* for X if

(i) m is a lower bound for X.
(ii) If m' is any other lower bound for X, then $m' \leq m$.

Examples

3. Let $X = \{x \in \mathbf{R} : 0 \leq x < 1\}$. The number 2 is an upper bound for X, since every element of X is ≤ 2. The number 1 is also an upper bound for X, and it is clear that 1 is the smallest upper bound we can choose. Therefore 1 is the least upper bound. Similarly, 0 is the greatest lower bound.
4. Let $X = \{x \in \mathbf{R} : x \text{ is rational and } x^3 < 2\}$. Then every rational negative number is in X; so no number is smaller than every number in X. Thus X has no lower bound. The number 2 is an upper bound for X, as is $\sqrt[3]{2}$. (The elements of X must be rational, but the upper bounds need not be.) In fact, $\sqrt[3]{2}$ is the least upper bound of X. (See Exercise 4.)

Our first result about upper bounds is fairly easy.

Proposition 2.1 *If $X \subset \mathbf{R}$ has a least upper bound, it is unique. If X has a greatest lower bound, it is unique.*

Proof. We prove the statement about least upper bounds, leaving the other half as an exercise. Suppose that x_1 and x_2 are two least upper bounds for X. Then x_1 is an upper bound, by (i) of the definition, so that, from (ii) of the definition, $x_1 \leq x_2$. Similarly, x_2 is an upper bound, and therefore $x_2 \leq x_1$. It follows that $x_1 = x_2$, as claimed.

We denote the least upper bound and greatest lower bound of X by lub X and glb X when they exist. Many books also write sup X (for *supremum* of X) instead of lub X, and inf X (*infimum* of X) for glb X.

It is often important to know that the glb and lub exist; the purpose of the next theorem is to state that they do.

Theorem 2.2 *If $X \subset \mathbf{R}$ is nonempty and bounded from above, then X has a least upper bound. If X is nonempty and bounded from below, then X has a greatest lower bound.*

To prove Theorem 2.2, we need to use special properties of the real numbers. For example, it is not necessarily true that if X is a nonempty set of *rational* numbers bounded from above, then X has a least upper bound which is *rational.* (See Example 4.) We need some sort of statement that the set of real numbers, unlike the rationals, does not have any "holes." Such a statement either must be taken as an axiom about the real numbers or must be proved in the course of providing a construction of the real numbers. Since we are not constructing the real numbers, we give an axiom.

We begin with a definition. A sequence $\{a_n\}$ of numbers in $\mathbf{R}$ is called a *Cauchy sequence* if for every $\varepsilon > 0$ there is an integer N such that $|a_m - a_n| < \varepsilon$ wherever $m, n > N$. (See Exercises 8 and 9 of Section 3.6.) Thus the points of a Cauchy sequence get closer and closer together as n gets large.

Our axiom can now be stated as follows:

Completeness of the Real Numbers Let $\{a_n\}$ be a Cauchy sequence of real numbers. Then there is a real number a_0 such that $\lim_{n\to\infty} a_n = a_0$.

Armed with this axiom, we can prove Theorem 2.2. We prove only the first half; the other half is similar.

To begin with, notice we can pick real numbers a_0, b_0 such that a_0 is not an upper bound and b_0 is an upper bound. We can find b_0 because we have assumed that X is bounded above. As for a_0, let x be any element of X (X is nonempty), and let $a_0 = x - 1$. Of course, $b_0 > a_0$.

We proceed to define two sequences, $\{a_n\}$ and $\{b_n\}$. Let $c_0 = (a_0 + b_0)/2$. If c_0 is an upper bound for X, let $b_1 = c_0$ and $a_1 = a_0$. If c_0 is not an upper bound for X, let $b_1 = b_0$ and $a_1 = c_0$. Either way, we have defined a_1 and b_1 so that $a_0 \le a_1$, $b_0 \ge b_1$, a_1 is not an upper bound for X, b_1 is an upper bound for X, $a_1 < b_1$, and $b_1 - a_1 = \frac{1}{2}(b_0 - a_0)$. Next, let $c_1 = (a_1 + b_1)/2$. If c_1 is an upper bound for X, let $b_2 = c_1$ and $a_2 = a_1$; if not, let $b_2 = b_1$ and $a_2 = c_1$. And so on.

In this way, we get sequences $\{a_n\}$ and $\{b_n\}$ such that

(2.1) $\{a_n\}$ is nondecreasing and $\{b_n\}$ is nonincreasing.

(2.2) The elements of $\{b_n\}$ are upper bounds of X, and the elements of $\{a_n\}$ are not.

(2.3) For all n, $a_n < b_n$, and $b_n - a_n = (1/2^n)(b_0 - a_0)$.

If $m > n$, then $|a_m - a_n| = a_m - a_n \leq b_m - a_n \leq b_n - a_n = 2^{-n}(b_0 - a_0)$. From this, it is easy to see that $\{a_n\}$ is a Cauchy sequence. Thus it converges to a limit, which we shall call a. Similarly, $\{b_n\}$ converges to a limit b. Since $b - a = \lim_{n\to\infty} b_n - a_n = \lim_{n\to\infty} 2^{-n}(b_0 - a_0) = 0$, $a = b$.

We are now ready to show that a is the least upper bound of X. First of all, we must show that it is an upper bound. If not, we can find $x \in X$ with $x > a$. Let $x - a = \varepsilon > 0$. Since $b_n \to a$, we can find an n with $|b_n - a| < \varepsilon$. Then $b_n < x$, which contradicts the fact that b_n is an upper bound.

To finish, we must show that if y is any upper bound, then $a \leq y$. This is not hard. Since a_n is not an upper bound, there is an $x_n \in X$ with $a_n < x_n$. But $y \geq x_n$; thus $y \geq a_n$, for all n. Taking limits (see Exercise 11 of Section 3.6), we see that $y \geq a$. This proves the theorem.

Remark. Our choice of a completeness axiom for the real numbers was somewhat arbitrary; there are various other axioms which work as well. One equivalent statement is that of Theorem 2.2: any nonempty set which is bounded from above has a least upper bound. Another one which is often used says that if $\{x_n\}$ is a nondecreasing sequence of real numbers which is bounded above, then $\{x_n\}$ converges. Each has some advantages. We chose the statement about Cauchy sequences in part because it will be the most convenient form for usc in Section 4 of this chapter.

The reader may want to prove each of the forms of the completeness axioms from the others. See Exercises 11 and 12 for examples.

Exercises

1. Determine glb X and lub X when they exist, where
   (a) $X = \{x \in \mathbf{R} : x^5 < 32\}$
   (b) $X = \{x \in \mathbf{R} : |x| \geq 1\}$
   (c) $X = \{x \in \mathbf{R} : x > 0 \text{ and } \sin x = \frac{1}{2}\}$
   (d) $X = \{x \in \mathbf{R} : |2x - 4| < 3\}$
   (e) $X = \{x \in \mathbf{R} : e^x \leq 3\}$
   (f) $X = \{x \in \mathbf{R} : x^2 + x < 2\}$
   (g) $X = \{x \in \mathbf{R} : x^3 - 3x^2 + 2x \geq 0\}$
2. Prove the second half of Proposition 2.1.
3. Let $\{a_n\}$ be a sequence of real numbers such that $a_j \leq a_{j+1}$ for all j and such that for some number M, $a_j \leq M$ for all j. Let $a = \text{lub}\{a_1, a_2, \ldots, a_n, \ldots\}$. Prove that $a = \lim_{n\to\infty} a_n$.

4. Prove that $\sqrt[3]{2}$ is the least upper bound for the set $\{x \in \mathbf{R} : x \text{ is rational and } x^3 < 2\}$.
5. Prove the second half of Theorem 2.2.
6. (a) Let X be a bounded subset of $\mathbf{R}$, and let a be positive. Set $aX = \{ax : x \in X\}$. Prove that

$$\text{lub } aX = a \text{ lub } X \qquad \text{glb } aX = a \text{ glb } X$$

(b) If a is negative, prove that

$$\text{lub } aX = a \text{ glb } X \qquad \text{glb } aX = a \text{ lub } X$$

7. Let A and B be nonempty subsets of $\mathbf{R}$ with $A \subset B$. Prove that

$$\text{lub } A \leq \text{lub } B \qquad \text{glb } A \geq \text{glb } B$$

*8. Prove *Archimedes' principle*: given positive real numbers a and b, then there exists an integer n such that $na > b$.
9. Let $T : \mathbf{R}^n \to \mathbf{R}^p$ be a linear transformation. Define the *norm* of T, $\|T\|$, by

$$\|T\| = \underset{\|\mathbf{v}\| = 1}{\text{lub}} \|T(\mathbf{v})\|$$

when this upper bound exists. In this case, T is said to be *bounded*. (See Section 4.5.)

(a) Prove that

$$\begin{aligned} \|T\| &= \underset{\|\mathbf{v}\| \leq 1}{\text{lub}} \|T(\mathbf{v})\| \\ &= \underset{\mathbf{v} \neq \mathbf{0}}{\text{lub}} \frac{\|T(\mathbf{v})\|}{\|\mathbf{v}\|} \end{aligned}$$

10. Let $T : \mathbf{R}^n \to \mathbf{R}^m$ have matrix (a_{ij}) relative to the standard bases.

(a) Suppose that both $\mathbf{R}^n$ and $\mathbf{R}^m$ have norms defined by

$$\|(x_i)\| = \sum_i |x_i|$$

Prove that if the norm of T is given as in Exercise 9,

$$\|T\| = \max_j \sum_i |a_{ij}|$$

(b) Suppose that both $\mathbf{R}^n$ and $\mathbf{R}^m$ have norms defined by

$$\|(x_i)\| = \max_i |x_i|$$

Prove that

$$\|T\| = \max_i \sum_j |a_{ij}|$$

11. Prove the completeness axiom, assuming the truth of Theorem 2.2.
12. Prove that if $\{x_n\}$ is a nondecreasing sequence of real numbers which is bounded above, then $\{x_n\}$ converges.

*13. Let $f : [a, b] \to \mathbf{R}$ be a bounded function and define the *oscillation of f at t*, $\omega(f, t)$, by

$$\omega(f, t) = \lim_{\delta \to 0} \text{lub}\{f(s) : |s - t| < \delta\} - \text{glb}\{f(s) : |s - t| < \delta\}$$

Prove this limit exists.

*14. Prove that the bounded function $f: [a, b] \to \mathbf{R}$ is continuous at t if and only if $\omega(f, t) = 0$.

*15. Let $\{a_n\}$ be an infinite sequence of real numbers and let L be the set of all points of $\mathbf{R}$ which are limit points of some convergent subsequence of $\{a_n\}$. Suppose L is bounded. Then the *upper limit* of the sequence a_n is defined by

$$\lim_{n \to \infty} \sup a_n = \text{lub } L$$

and the *lower limit* of the sequence $\{a_n\}$ is defined by

$$\lim_{n \to \infty} \inf a_n = \text{glb } L$$

(a) Prove that $\lim \sup_{n \to \infty} \{a_n\} = a$ if and only if $a \in L$ and whenever $x > a$, there is an integer N such that $a_n < x$ for $n \geq N$.

(b) State and prove the analogous characterization of the lower limit.

3. Continuous Functions on Compact Sets

It may not be clear yet what the point of defining compact sets is. To give some motivation, we prove an important theorem here: a continuous real-valued function defined on a compact set attains its maximum and minimum. This theorem generalizes one often quoted in calculus courses: a continuous function defined on a closed, bounded interval attains its maximum and minimum. To know that our theorem generalizes the other one, we need to know that closed, bounded intervals are compact, a fact that we shall prove in the next section.

We begin with a useful result about closed sets.

Proposition 3.1 *If $X \subset \mathbf{R}$ is closed and bounded, then lub X and glb X are elements of X.*

Proof. We prove that $M = \text{glb } X$ is in X, leaving the proof of the fact that lub X is in X to the reader.

For any integer $n > 0$, the number $M - 1/n$ cannot be an upper bound for X, since M is the least upper bound of X. Thus for each such n, we can find a real number $a_n \in X$ with

$$M - \frac{1}{n} \leq a_n \leq M$$

It follows easily that $\lim_{n \to \infty} a_n = M$. Since X is closed, M is in X by Theorem 1.2.

Our next result relates compact sets and continuous functions. It states that the notion of compactness is preserved by continuous functions.

Proposition 3.2 *Let $f: \mathbf{R}^n \to \mathbf{R}^m$ be a continuous function and C a compact subset of $\mathbf{R}^n$. Then the image of C under f,*

$$f(C) = \{ f(\mathbf{v}) : \mathbf{v} \in C \}$$

is a compact subset of $\mathbf{R}^m$.

Proof. Let $\{\mathbf{w}_n\}$ be a sequence of vectors in $f(C)$. We must show that there is a subsequence of this sequence converging in $f(C)$.

Since $\mathbf{w}_n$ is in $f(C)$, there is a vector $\mathbf{v}_n$ in C with $f(\mathbf{v}_n) = \mathbf{w}_n$. The vectors $\mathbf{v}_n$ form a sequence in the compact set C, so there is a subsequence $\{\mathbf{v}_{n_j}\}$ converging to a vector $\mathbf{v}$ in C. Applying Proposition 3.7.1, we see that the subsequence $\mathbf{w}_{n_j} = f(\mathbf{v}_{n_j})$ converges to the vector $f(\mathbf{v})$, which is clearly in $f(C)$. Therefore $f(C)$ is compact.

We can now state and prove the main theorem of this section.

Theorem 3.3 *Let C be a compact subset of $\mathbf{R}^n$ and $f: C \to \mathbf{R}$ a continuous function. Then f attains its maximum and minimum on C.*

More explicitly, there are vectors $\mathbf{v}_0$ and $\mathbf{v}_1$ in C such that

$$f(\mathbf{v}_0) \leq f(\mathbf{v}) \leq f(\mathbf{v}_1)$$

for all $\mathbf{v}$ in C.

Proof. According to Proposition 3.2, $f(C)$ is a compact subset of $\mathbf{R}$, and so by Propositions 1.3 and 1.4, is closed and bounded. It follows from Propositions 2.2 and 3.1 that $\operatorname{lub} f(C)$ and $\operatorname{glb} f(C)$ exist and are in $f(C)$. Pick $\mathbf{v}_0$, $\mathbf{v}_1$ such that

$$f(\mathbf{v}_0) = \operatorname{glb} f(C) \qquad f(\mathbf{v}_1) = \operatorname{lub} f(C)$$

Then

$$f(\mathbf{v}_0) \leq f(\mathbf{v}) \leq f(\mathbf{v}_1)$$

for all $\mathbf{v}$ in C, and the theorem is proved.

Corollary *Let $[a, b]$ be a closed interval and $f: [a, b] \to \mathbf{R}$ a continuous function. Then f attains its maximum and minimum on $[a, b]$.*

Strictly speaking, we should not call this result a corollary, since it depends on a fact we have not yet proved: the set $[a, b]$ is compact. This will be proved in the next section.

Theorem 3.3 states the existence of a maximum and a minimum for continuous functions on compact sets; it does not, however, give any information on how to find them. We shall examine this problem (for differentiable functions) in the next chapter.

Greatest lower bounds and least upper bounds often arise in connection with functions. Suppose that $f\colon S \to \mathbf{R}$ is a function; we may consider

$$M = \text{lub}\,\{f(\mathbf{v}) : \mathbf{v} \in S\}$$

and

$$m = \text{glb}\,\{f(\mathbf{v}) : \mathbf{v} \in S\}$$

Neither M nor m need exist, but if f is a bounded function (and S is not the empty set), then M and m both exist (by Theorem 1.2). If f takes on a maximum value, then M is that maximum; similarly, m is the minimum value if f has a minimum. Neither the maximum nor the minimum need exist. For example, if $f\colon(-1, 2) \to \mathbf{R}$ is defined by $f(x) = x^2$, then

$$f(S) = \{f(x) : x \in (-1, 2)\} = [0, 4)$$

Here, f has a minimum but no maximum. However, S has a least upper bound and a greatest lower bound; in fact, lub $S = 4$ and glb $S = 0$.

We conclude this section with a useful result.

Proposition 3.4 *Let $f, g\colon S \to \mathbf{R}$ be bounded functions, and let*

$$M = \text{lub}\,\{f(\mathbf{v}) : \mathbf{v} \in S\} \qquad m = \text{glb}\,\{f(\mathbf{v}) : \mathbf{v} \in S\}$$

$$M' = \text{lub}\,\{g(\mathbf{v}) : \mathbf{v} \in S\} \qquad m' = \text{glb}\,\{g(\mathbf{v}) : \mathbf{v} \in S\}$$

$$M'' = \text{lub}\,\{f(\mathbf{v}) + g(\mathbf{v}) : \mathbf{v} \in S\} \qquad m'' = \text{glb}\,\{f(\mathbf{v}) + g(\mathbf{v}) : \mathbf{v} \in S\}$$

Then $M'' \le M' + M$ and $m'' \ge m' + m$.

Proof. For every $\mathbf{v} \in S$, $M \ge f(\mathbf{v})$ and $M' \ge g(\mathbf{v})$. Therefore

$$M + M' \ge f(\mathbf{v}) + g(\mathbf{v})$$

which shows that $M + M'$ is an upper bound for $\{f(\mathbf{v}) + g(\mathbf{v}) : \mathbf{v} \in S\}$. It follows that $M'' \le M + M'$. The other part is similar.

The situation is more complicated for products; see Exercise 6.

Exercises

1. Let $T\colon \mathbf{R}^n \to \mathbf{R}^m$ be a linear transformation. Prove that T is bounded. (The definition of boundedness is given in Exercise 9 of Section 2.) Use this fact to prove that T is continuous.

2. Find a continuous function f on the set $S = \{\mathbf{v} \in \mathbf{R}^3 : \|\mathbf{v}\| < 1\}$ which has neither a maximum nor a minimum. (Note that S is not compact.)

*3. Let S be a set in $\mathbf{R}^n$ with the property that every continuous function on S has a maximum and a minimum. Show that S is compact. You may use the fact (to be proved in the next section) that every closed and bounded set is compact. [*Hint*: First prove that S is bounded. If S is not closed, use Theorem 1.2 to find a sequence $\mathbf{v}_1, \mathbf{v}_2, \ldots$ of points in S with $\lim_{n\to\infty} \mathbf{v}_n = \mathbf{v}_0 \notin S$. Let $f(\mathbf{v}) = \|\mathbf{v} - \mathbf{v}_0\|$.]

*4. Let $C \subset \mathbf{R}^n$ be compact, and let $f: C \to \mathbf{R}$ be a continuous function such that $f(\mathbf{v}) > 0$ for all $\mathbf{v} \in C$. Show that there is a number $\alpha > 0$ such that $f(\mathbf{v}) > \alpha$ for all $\mathbf{v} \in C$. Give an example to show that the conclusion need not hold if C is not compact.

*5. The following is a sketch of a different proof of Theorem 3.3. Fill in the details. In what follows, C is compact and $f: C \to \mathbf{R}$ is continuous. Use *no* properties of compactness except the definition.
   (a) Show that $f(C)$ is bounded above. [*Hint*: If not, then there are points $\mathbf{v}_n \in C$ with $f(\mathbf{v}_n) > n$. Choose a subsequence of the $\mathbf{v}_n$ which converges to $\mathbf{v}_0 \in C$. Find a contradiction from the fact that f is continuous at $\mathbf{v}_0$.]
   (b) Since $f(C)$ is bounded above, it has a lub, M. Show that there is a point $\mathbf{v} \in C$ with $f(\mathbf{v}) = M$. {*Hint*: If not, show that $g(\mathbf{v}) = 1/[M - f(\mathbf{v})]$ is not bounded above, and use (a) to get a contradiction.}
   (c) Parts (a) and (b) prove that f takes on a maximum value on C. Prove that f has a minimum.

6. Let $f, g: S \to \mathbf{R}$ be bounded *nonnegative* functions. Show that

$$\text{lub}\,\{f(\mathbf{v})g(\mathbf{v}) : \mathbf{v} \in S\} \leq \text{lub}\,\{f(\mathbf{v}) : \mathbf{v} \in S\} \cdot \text{lub}\,\{g(\mathbf{v}) : \mathbf{v} \in S\}$$

and

$$\text{glb}\,\{f(\mathbf{v})g(\mathbf{v}) : \mathbf{v} \in S\} \geq \text{glb}\,\{f(\mathbf{v}) : \mathbf{v} \in S\} \cdot \text{glb}\,\{g(\mathbf{v}) : \mathbf{v} \in S\}$$

(See Exercise 6 of Section 2.)

4. A Characterization of Compact Sets

As promised earlier, we now characterize the compact subsets of $\mathbf{R}^n$.

Theorem 4.1 *A subset C of $\mathbf{R}^n$ is compact if and only if it is closed and bounded.*

We have already proved half of this theorem, namely that a compact set is closed and bounded. What needs to be shown is that any closed and bounded subset of $\mathbf{R}^n$ is compact. We begin with the one-dimensional case.

Theorem 4.2 (The Bolzano-Weierstrass Theorem) *Any closed interval*

$$[a, b] = \{x \in \mathbf{R} : a \leq x \leq b\}$$

is compact.

Proof. This theorem, like Theorem 2.2, needs the completeness property of the real numbers: every Cauchy sequence converges. (See Section 2.) We need to prove that if $a_1, a_2, \ldots$ is a sequence in $[a, b]$, then some subsequence converges. We do this by finding a subsequence which is a Cauchy sequence.

Let $c = (a + b)/2$, and consider the two intervals $I_1 = [a, c]$ and $I_1' = [c, b]$ formed by dividing $[a, b]$ in half. Each of these intervals has length $(b - a)/2$. Furthermore, either I_1 or I_1' contains infinitely many of the a's. Choose one with infinitely many of the a's, and choose a_{i_1} to be a point of the sequence lying in this subinterval.

Now divide the interval just chosen into two subintervals of equal length, which we call I_2, I_2'. Each has length $(b - a)/4$. Again, choose one of these subintervals with infinitely many a's, and choose a term a_{i_2} in this interval with $i_2 > i_1$. (Since we have infinitely many terms to choose from, we can surely choose one with a subscript $> i_1$.) Repeat the procedure; that is, split this subinterval into two equal subintervals, choose one containing an infinite number of terms, and choose a_{i_3} in this new subinterval with $i_3 > i_2 > i_1$. Continue in this way.

An example may help. Let the interval be $[0, 2]$, and let the sequence be $\frac{1}{2}$, $\frac{3}{2}, \frac{1}{3}, \frac{4}{3}, \frac{1}{4}, \frac{5}{4}, \ldots$. We begin by considering the intervals $[0, 1]$ and $[1, 2]$. Each of these has infinitely many terms of the sequence (the odd-numbered terms are in $[0, 1]$ and the even-numbered terms in $[1, 2]$). We may therefore choose either interval; say we choose $[1, 2]$. We also need to pick some point of the sequence in $[1, 2]$, and we pick $\frac{4}{3}$. (We could equally well pick $\frac{3}{2}$.) Next, we divide $[1, 2]$ in half, getting $[1, \frac{3}{2}]$, and $[\frac{3}{2}, 2]$. The first of these intervals has infinitely many points of the sequence (the even-numbered terms), while the second has only one point. Thus we must choose $[1, \frac{3}{2}]$ to be our interval, and we choose a point in the sequence which is beyond $\frac{4}{3}$ and is in $[1, \frac{3}{2}]$. For example, $\frac{6}{5}$ will do. (So will $\frac{5}{4}$ or $\frac{7}{6}$, for instance.) Next, we cut $[1, \frac{3}{2}]$ in half, getting $[1, \frac{5}{4}]$ and $[\frac{5}{4}, \frac{3}{2}]$, and so on.

By this procedure, we get a subsequence $a_{i_1}, a_{i_2}, \ldots$ of the original sequence with the property that all the terms a_{i_j} with $j \geq n$ are contained in the nth subinterval chosen. This subinterval is of length $(b - a)/2^n$. Therefore we know that if $j, k > n$, then

$$|a_{i_j} - a_{i_k}| \leq \frac{b - a}{2^n}$$

Given any $\varepsilon > 0$, we can choose N so that $(b - a)/2^n < \varepsilon$. Then if $m, n > N$,

$$|a_{i_m} - a_{i_n}| \leq \frac{b - a}{2^N} < \varepsilon$$

and so the sequence $a_{i_1}, a_{i_2}, \ldots$ is a Cauchy sequence. By the completeness axiom, this sequence converges to some real number x. Because $[a, b]$ is closed, $x \in [a, b]$. We have therefore proved that $[a, b]$ is compact, as claimed. (Note, incidentally, that we have also finished the proof of the corollary to Theorem 3.3.)

We next generalize Theorem 4.2 to $\mathbf{R}^n$. For any $r > 0$, define the subset $A(r) \subset \mathbf{R}^n$ by

$$A(r) = \{(a_1, \ldots, a_n) : |a_j| \leq r \text{ for } j = 1, 2, \ldots, n\}$$

Proposition 4.3 *For any $r \geq 0$, the set $A(r)$ is compact.*

Proof. Let $\{\mathbf{w}_j\}$ be a sequence in $A(r)$. Then

$$\mathbf{w}_j = (a_{j1}, a_{j2}, \ldots, a_{jn}) \qquad j = 1, 2, \ldots$$

where $|a_{ji}| \leq r$. Consider the sequence $\{a_{j1}\}$ of real numbers. Since $|a_{j1}| \leq r$ for all j, each a_{j1} is contained in the closed interval $[-r, r]$. Thus, by Theorem 4.2 we can find a subsequence converging to a number a_1 in $[-r, r]$ This defines a subsequence of the sequence $\{\mathbf{w}_j\}$ with the property that the sequence of first components converges to a_1. For simplicity, we throw away the $\mathbf{w}_j$ not appearing in this subsequence and reindex so that

$$\lim_{j \to \infty} a_{j1} = a_1$$

Now consider the sequence of second components $\{a_{j2}\}$. A similar argument shows that we can find a subsequence of this sequence converging to a_2 in $[-r, r]$. This defines a subsequence of the sequence $\{\mathbf{w}_j\}$ whose second components converge to a_2. Note that the sequence of first components of this subsequence still converges to a_1, since it is a subsequence of a convergent sequence with limit a_1. (See Proposition 6.4, Chapter 3.)

Continuing in this way, we eventually obtain a subsequence of the original sequence $\{\mathbf{w}_j\}$ with the property that the kth components of the vectors in the subsequence converge to a number a_k in $[-r, r]$. Let $\mathbf{w} = (a_1, \ldots, a_n)$. Then the subsequence converges to $\mathbf{w}$, by Exercise 10 of Section 3.6. (The proof of this exercise is very much like the proof of Proposition 3.4, Chapter 3.) Since $\mathbf{w} \in A(r)$, Proposition 4.3 is proved.

We can now prove Theorem 4.1. Let C be a closed and bounded subset of $\mathbf{R}^n$. Then there is an $r > 0$ such that $C \subset B_r(\mathbf{0})$. It is easily seen that $B_r(\mathbf{0}) \subset A(r)$, so that C is a closed subset of a compact set and is therefore compact (by Proposition 1.5). This concludes the proof of Theorem 4.1.

Exercises

1. Which of the following are compact sets?
   (a) $A = \{x \in \mathbf{R} : |x - 3| \leq 7\}$

(b) $B = \{x \in \mathbf{R} : 0 < x \le 10\}$
(c) $C = \{(x, y) \in \mathbf{R}^2 : 0 < x^2 + y^2 \le 1\}$
(d) $D = \{(x, y) \in \mathbf{R}^2 : |x| + |y| \le 2\}$
(e) $E = \{(x, y) \in \mathbf{R}^2 : x^2 + y^2 = 1\}$
(f) $F = \{(x, y) \in \mathbf{R}^2 : xy = 1\}$
(g) $G = \{(x, y) \in \mathbf{R}^2 : 1 \le x^2 + y^2 \le 2\}$
(h) $H = \{(x, y, z) \in \mathbf{R}^3 : x^2 + 2y^2 + 3z^2 \le 2\}$
(i) $I = \{(x, y, z) \in \mathbf{R}^3 : x^2 + y^2 \le 2 \text{ and } z = xy\}$
(j) $J = \{(x, y, z) \in \mathbf{R}^3 : |x - 2| + |y - 1| \le 2\}$
(k) $K = \{(x, y, z) \in \mathbf{R}^3 : x^2 + y^2 - z = 0 \text{ and } x^2 + y^2 + z - 2 = 0\}$

*2. Show that if $\{a_n\}$ is an infinite sequence of real numbers in $[a, b]$, then $\limsup a_n \in [a, b]$ and $\limsup a_n$ is the limit of a subsequence of $\{a_n\}$. (This provides an alternate proof of Theorem 4.2; see Exercise 12 of Section 2 for the definition of $\limsup a_n$.)

*3. Prove Theorem 4.2 by proving that if $\{U_\lambda\}$ is any open covering of $[a, b]$, then there is a finite subcovering. (Then Exercise 11 of Section 1 applies. *Hint*: Let $S = \{x \in [a, b]$: the interval $[a, x]$ can be covered by finitely many $U_\lambda\}$. Let $c = \text{lub } S$; then $c \in U_{\lambda_0}$ for some λ_0. Assume $c < b$ and get a contradiction.)

5. Uniform Continuity

We now develop the notion of uniform continuity for functions between Euclidean spaces. This material will be needed later for the study of the integral.

Let X be a subset of $\mathbf{R}^n$, and $f: X \to \mathbf{R}^m$ a function. We say that f is *uniformly continuous on X* if for any $\varepsilon > 0$ there is a $\delta > 0$ such that

$$\| f(\mathbf{v}) - f(\mathbf{v}') \| < \varepsilon$$

for all $\mathbf{v}, \mathbf{v}'$ in X with $\|\mathbf{v} - \mathbf{v}'\| < \delta$.

Clearly any uniformly continuous function is continuous. The difference in this definition is that the number δ depends only on ε, not on the point under consideration. In the definition of continuity, we state that given ε and $\mathbf{v}$, there is a δ which may depend on ε and $\mathbf{v}$ such that $\| f(\mathbf{v}) - f(\mathbf{v}')\| < \varepsilon$ whenever $\|\mathbf{v} - \mathbf{v}'\| < \delta$. For example, the function $f: (0, 1) \to \mathbf{R}$ defined by $f(x) = 1/x$ is continuous but *not* uniformly continuous. To prove this we must show that there is an $\varepsilon > 0$ such that, no matter how small a δ we choose, we can find x, x' in $(0, 1)$ with $|x - x'| < \delta$ but $|f(x) - f(x')| \ge \varepsilon$. Set $\varepsilon = \frac{1}{2}$ and, for any $\delta > 0$, choose an integer n so that $1/n < \delta$. Then if $x = 1/n$ and $x' = 1/(n + 1)$,

$$|x - x'| = \frac{1}{n(n+1)} < \frac{1}{n} < \delta$$

but

$$|f(x) - f(x')| = 1 > \frac{1}{2}$$

The next result shows that the notions of continuity and uniform continuity coincide on compact sets.

Theorem 5.1 *Suppose that C is a compact subset of $\mathbf{R}^n$. Then any continuous function $f: C \to \mathbf{R}^m$ is uniformly continuous.*

Proof. We proceed by contradiction, and our first problem is how to state that f is not uniformly continuous. Again, we use the procedure given in Section 1.2. Write

$$f \text{ is uniformly continuous}$$

as $$(\forall \varepsilon > 0)(\exists \delta > 0)(\forall \mathbf{v}, \mathbf{w} \in C): \|\mathbf{v} - \mathbf{w}\| < \delta \Rightarrow \|f(\mathbf{v}) - f(\mathbf{w})\| < \varepsilon$$

Then one obtains the statement for "f is not uniformly continuous" by interchanging the $\forall$'s and $\exists$'s and denying the conclusion:

$$(\exists \varepsilon > 0)(\forall \delta > 0)(\exists \mathbf{v}, \mathbf{w} \in C): \|\mathbf{v} - \mathbf{w}\| < \delta \text{ and } \|f(\mathbf{v}) - f(\mathbf{w})\| \geq \varepsilon$$

or "there exists an $\varepsilon > 0$ such that for every $\delta > 0$ we can find $\mathbf{v}$, $\mathbf{w}$ in C with $\|\mathbf{v} - \mathbf{w}\| < \delta$ and $\|f(\mathbf{v}) - f(\mathbf{w})\| \geq \varepsilon$."

Having worked out what we are assuming, we can search for the contradiction. Choose the ε guaranteed by our last statement; then (setting $\delta = 1/n$) we can find points $\mathbf{v}_n$, $\mathbf{w}_n \in C$ such that $\|\mathbf{v}_n - \mathbf{w}_n\| < 1/n$ and $\|f(\mathbf{v}_n) - f(\mathbf{w}_n)\| \geq \varepsilon$. Because C is compact, some subsequence of $\{\mathbf{v}_n\}$ converges to a point in C; suppose that $\{\mathbf{v}_{n_k}\}$ converges to $\mathbf{v}_0 \in C$. Since f is continuous at $\mathbf{v}_0$, we can find $\delta > 0$ such that if $\|\mathbf{v} - \mathbf{v}_0\| < \delta$, $\|f(\mathbf{v}) - f(\mathbf{v}_0)\| < \varepsilon/2$.

Now choose k so large that

(a) $\|\mathbf{v}_0 - \mathbf{v}_{n_k}\| < \dfrac{\delta}{2}$

(b) $\dfrac{1}{n_k} < \dfrac{\delta}{2}$

(We can certainly satisfy these conditions, since $\mathbf{v}_{n_k}$ converges to $\mathbf{v}_0$.) Then since $\|\mathbf{v}_{n_k} - \mathbf{w}_{n_k}\| < 1/n_k$,

$$\|\mathbf{v}_0 - \mathbf{w}_{n_k}\| < \frac{\delta}{2} + \frac{1}{n_k} < \delta$$

Thus $$\|f(\mathbf{v}_0) - f(\mathbf{v}_{n_k})\| < \frac{\varepsilon}{2}$$

and $$\|f(\mathbf{v}_0) - f(\mathbf{w}_{n_k})\| < \frac{\varepsilon}{2}$$

Therefore $$\| f(\mathbf{v}_{n_k}) - f(\mathbf{w}_{n_k}) \| < \varepsilon$$

which contradicts our choice of $\mathbf{v}_{n_k}$, $\mathbf{w}_{n_k}$. The theorem follows.

Note. The form of the proof is one which is often useful when dealing with compact sets. To prove a theorem about compact sets, assume that it is false. Pick a sequence of points in the compact set making the theorem false. Extract a convergent subsequence. Then get a contradiction by looking at the limit of the subsequence.

Exercises

1. Prove directly from the definition that $f : \mathbf{R} \to \mathbf{R}$ defined by $f(x) = 7x + 4$ is uniformly continuous.
2. Prove directly from the definition that $f : \mathbf{R} \to \mathbf{R}$ defined by $f(x) = ax + b$ is uniformly continuous.
3. Prove that a bounded linear transformation $T : \mathbf{R}^n \to \mathbf{R}^m$ is uniformly continuous.
4. Prove directly from the definition that the function $f = [-a, a] \to \mathbf{R}$ defined by $f(x) = x^2$ is uniformly continuous.
5. Prove that the function $f : \mathbf{R} \to \mathbf{R}$ defined by $f(x) = x^2$ is not uniformly continuous.

6. Connected Sets

We now introduce two related concepts, connectedness and pathwise connectedness. These will be used to prove the intermediate value theorem.

A subset S in Euclidean space $\mathbf{R}^n$ is said to be *separated* if there are open subsets U_1 and U_2 of $\mathbf{R}^n$ such that $S \subset U_1 \cup U_2$, $U_1 \cap U_2 \cap S = \varnothing$ and neither $S \cap U_1$ or $S \cap U_2$ is empty. A set $S \subset \mathbf{R}^n$ is *connected* if it cannot be separated.

For example, the set $S = \{x \in R : x \neq 0\}$ is separated; set $U_1 = (-\infty, 0)$ and $U_2 = (0, \infty)$. On the other hand, we have the following.

Proposition 6.1 *Any closed interval* $[a, b] \subset \mathbf{R}$ *is connected..*

Proof. Suppose not. Then $[a, b] \subset U_1 \cup U_2$, where U_1 and U_2 are open subsets of $\mathbf{R}$ such that $U_1 \cap [a, b] \neq \varnothing$, $U_2 \cap [a, b] \neq \varnothing$ and $U_1 \cap U_2 \cap [a, b] = \varnothing$. Assume that $a \in U_1$ and define $c \in \mathbf{R}$ by

$$c = \text{glb}\,(U_2 \cap [a, b])$$

(If $a \in U_2$, the proof proceeds similarly.) Then $c \in [a, b]$ since $[a, b]$ is a closed set (by Proposition 3.3).

Suppose $c \in U_1$. Then, since U_1 is open, there is an $\varepsilon > 0$ such that $B_\varepsilon(c) = (c - \varepsilon, c + \varepsilon) \subset U_1$. Now, c is a lower bound for $U_2 \cap [a, b]$; so $x \geq c$ for all $x \in U_2$. However, since $(c - \varepsilon, c + \varepsilon) \subset U_1$ and $U_1 \cap U_2 = \varnothing$, we must have $x \geq c + \varepsilon$ for all $x \in U_2 \cap [a, b]$. This contradicts the fact that c is the *greatest* lower bound of $U_2 \cap [a, b]$. Thus c cannot be in U_1.

On the other hand, if $c \in U_2$, then $(c - \varepsilon, c + \varepsilon) \subset U_2$ for some $\varepsilon > 0$. Since $a \in U_1$, $c \neq a$ and we can choose ε small enough so that $(c - \varepsilon, c + \varepsilon) \subset U_2 \cap [a, b]$. In particular, the points between $c - \varepsilon$ and c are in $U_2 \cap [a, b]$. This contradicts the fact that c is a lower bound for $U_2 \cap [a, b]$.

It follows that $c \notin U_1 \cup U_2$ so that $[a, b] \not\subset U_1 \cup U_2$ which contradicts our original assumption that $[a, b]$ was separated.

We note that any interval in $\mathbf{R}$ can be shown to be connected, by a similar argument. This fact, together with the next result, characterizes the connected subsets of $\mathbf{R}$.

Proposition 6.2 *Let S be a connected subset of $\mathbf{R}$ and $a, b \in S$ with $a < b$. If $z \in \mathbf{R}$ with $a < z < b$, then $z \in S$.*

It follows that the connected subsets of $\mathbf{R}$ are exactly the intervals. (See Exercise 9.)

Proof. If $z \notin S$, define $U_1 = (-\infty, z)$ and $U_2 = (z, \infty)$. Then $U_1 \cap U_2 = \varnothing$, $S \subset U_1 \cup U_2$, $a \in U_1 \cap S$, $b \in U_2 \cap S$. Thus S is separated.

We next show that we can characterize separated sets in terms of closed sets.

Proposition 6.3 *A subset $S \subset \mathbf{R}^n$ is separated if and only if there are closed sets C_1 and C_2 in $\mathbf{R}^n$ with $S \subset C_1 \cup C_2$, $C_1 \cap C_2 \cap S = \varnothing$, $C_1 \cap S \neq \varnothing$, and $C_2 \cap S = \varnothing$.*

Proof. Suppose S separated by open sets U_1 and U_2. Define closed sets $C_1 = \mathbf{R}^n - U_1$, $C_2 = \mathbf{R}^n - U_2$. It follows easily that C_1 and C_2 satisfy the conditions of the proposition.

Conversely, if C_1 and C_2 are closed sets satisfying the conditions of the proposition, then the open sets $U_1 = \mathbf{R}^n - C_1$ and $U_2 = \mathbf{R}^n - C_2$ separate S. We leave the details as Exercise 6.

We now introduce the related notion of a pathwise connected set.

A subset $S \subset \mathbf{R}^n$ is said to be *pathwise connected* if, for any two points $\mathbf{v}$, $\mathbf{w} \in S$, there is a continuous function $\alpha : [0, 1] \to S$ such that $\alpha(0) = \mathbf{v}$ and $\alpha(1) = \mathbf{w}$. (The function α is called a *path* from $\mathbf{v}$ to $\mathbf{w}$.)

Proposition 6.4 *Any pathwise connected subset S of $\mathbf{R}^n$ is connected.*

Proof. We shall prove the contrapositive: if S is not connected, then S is not pathwise connected.

Suppose S is separated. Then, according to Proposition 6.3, we can find closed sets C_1 and C_2 in $\mathbf{R}^n$ such that $S \subset C_1 \cup C_2$, $S \cap C_1 \cap C_2 = \emptyset$, $S \cap C_1 \neq \emptyset$, and $S \cap C_2 = \emptyset$. Let $\mathbf{v}$ be a point of $S \cap C_1$ and $\mathbf{w}$ a point of $S \cap C_2$. If $\alpha : [0, 1] \to S$ is a path from $\mathbf{v}$ to $\mathbf{w}$, we define subsets $A_1, A_2 \subset [0, 1]$ by $A_1 = \alpha^{-1}(C_1)$, $A_2 = \alpha^{-1}(C_2)$. Then A_1 and A_2 are closed subsets of $\mathbf{R}$ contained in [0, 1] (see Proposition 7.4, Chapter 3 and Exercise 8 of Section 3.7), $0 \in A_1$ [since $f(0) = \mathbf{v} \in C_1$], and $1 \in A_2$ [since $f(1) = \mathbf{w} \in C_2$]. Furthermore, if t is any point in [0, 1], $f(t)$ is in C_1 or C_2. If $f(t) \in C_1$, then $t \in A_1$; if $f(t) \in C_2$, then $t \in A_2$. Thus $[0, 1] \subset A_1 \cup A_2$. Finally,

$$\begin{aligned} A_1 \cap A_2 &= \alpha^{-1}(C_1) \cap \alpha^{-1}(C_2) \\ &= \alpha^{-1}(C_1 \cap C_2) \\ &= \alpha^{-1}(\emptyset) \\ &= \emptyset \end{aligned}$$

Thus A_1 and A_2 separate [0, 1], which is not possible since [0, 1] is connected. It follows that there can be no path from $\mathbf{v}$ to $\mathbf{w}$ and S is not pathwise connected.

The next result relates pathwise connectivity and continuous functions.

Proposition 6.5 *Let S be a pathwise connected subset of $\mathbf{R}^n$ and $f : \mathbf{R}^n \to \mathbf{R}^m$ a continuous function. Then $f(S)$ is a pathwise connected subset of $\mathbf{R}^m$.*

Proof. Let $\mathbf{v}'$ and $\mathbf{w}'$ be any two points of $f(S)$. Then, there are points $\mathbf{v}$, $\mathbf{w} \in S$ such that $f(\mathbf{v}) = \mathbf{v}'$, $f(\mathbf{w}) = \mathbf{w}'$. Since S is pathwise connected, we can find a path $\alpha : [0, 1] \to S$ from $\mathbf{v}$ to $\mathbf{w}$. Then, the composite $f \circ \alpha : [0, 1] \to f(S)$ is a path from $\mathbf{v}'$ to $\mathbf{w}'$.

We can now prove the *intermediate value theorem.*

Theorem 6.6 *Let $f : [a, b] \to \mathbf{R}$ be a continuous function and let $z \in \mathbf{R}$ be any number between $f(a)$* and *$f(b)$. Then there is a $c \in [a, b]$ with $f(c) = z$.*

Proof. This follows immediately from Propositions 6.5 and 6.2. Since $[a, b]$ is pathwise connected, so is $f([a, b])$. Therefore $f([a, b])$ is connected, and Proposition 6.2 says that $c \in f([a, b])$.

Theorem 6.6 is another result which is commonly assumed in elementary calculus. It says, for instance, that if $f : [0, 1] \to \mathbf{R}$ is continuous, with $f(0) > 0$

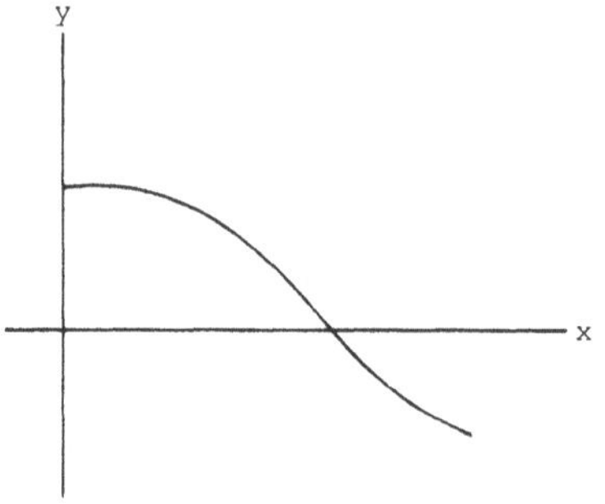

Figure 6.1

and $f(1) < 0$, then there exists a point $a \in [0, 1]$ with $f(a) = 0$. (See Figure 6.1.) Of course, the theorem gives no information about how to find a.

Exercises

1. Determine which of the following sets are connected.
   (a) $A = \{x \in \mathbf{R} : |x - 2| > 3\}$
   (b) $B = \{(x, y) \in \mathbf{R}^2 : 0 < x^2 + y^2 \leq 1\}$
   (c) $C = \{(x, y) \in \mathbf{R}^2 : y \geq x^2\}$
   (d) $D = \{(x, y) \in \mathbf{R}^2 : xy > 1\}$
   (e) $E = \{(x, y) \in \mathbf{R}^2 : y = x^2\}$
   (f) $F = \{(x, y) \in \mathbf{R}^2 : x^2 + y^2 = 1\}$
   (g) $G = \{(x, y) \in \mathbf{R}^2 : x^2 + y^2 = 1 \text{ and } |x| < 1\}$
   (h) $H = \{(x, y) \in \mathbf{R}^2 : xy = 1\}$
   (i) $I = \{(x, y) \in \mathbf{R}^2 : |x - 2| > 3\}$
   (j) $J = \{(x, y, z) \in \mathbf{R}^3 : |x - 1| + |y - 2| > 1\}$
   (k) $K = \{(x, y, z) \in \mathbf{R}^3 : z - x^2 - y^2 \geq 0\}$
2. (a) Give an example of connected sets S_1 and S_2 in $\mathbf{R}^2$ such that $S_1 \cup S_2$ is not connected.
   (b) Give an example of connected sets S_1 and S_2 in $\mathbf{R}^2$ such that $S_1 \cap S_2$ is not connected.
3. Let S_1 and S_2 be connected sets in $\mathbf{R}^n$ with $S_1 \cap S_2 \neq \varnothing$. Prove that $S_1 \cup S_2$ is connected.
4. Prove that any interval is connected.
5. A subset $C \subset \mathbf{R}^n$ is *convex* if whenever $\mathbf{v}, \mathbf{w} \in C$, the line segment from $\mathbf{v}$ to $\mathbf{w}$ is also in C. (Explicitly, $t\mathbf{v} + (1 - t)\mathbf{w} \in C$ for all $t \in [0, 1]$.) Prove that any convex set is pathwise connected.
6. Give a detailed proof of Proposition 6.3.
7. Let S be a connected subset of $\mathbf{R}^n$ and $f : S \to \mathbf{R}$ a continuous function. Let $\mathbf{v}_0$ and $\mathbf{v}_1$ be two points of S and $z \in \mathbf{R}$ with $f(\mathbf{v}_0) < z < f(\mathbf{v}_1)$. Prove that there is a $\mathbf{v} \in S$ such that $f(\mathbf{v}) = \mathbf{z}$.

8. Let $f:[0, 1]\to[0, 1]$ be continuous. Show that there exists $x\in[0, 1]$ with $f(x)=x$. [*Hint*: Consider $g(x)=f(x)-x$.] This result is the *Brouwer fixed point theorem* for [0, 1].
9. Let $S\subset \mathbf{R}$ be nonempty, and suppose that whenever $x\in S$, $y\in S$, and $x<z<y$, then $z\in S$. Prove that S is an interval. {*Hint*: If S is bounded above and below, let $a=\text{glb } S$ and $b=\text{lub } S$, and prove that $(a, b)\subset S\subset[a, b]$. If S is bounded above but not below, let $b=\text{lub } S$ and prove that $S=(-\infty, b)$ or $[-\infty, b]$. Do the other cases similarly.} Note: $\mathbf{R}$ itself is an interval.

*10. Prove that a connected open subset of $\mathbf{R}^n$ is pathwise connected.

*11. Let $S\subset\mathbf{R}^2$ be the set defined by

$$S=\{(x, y): y=\sin\frac{1}{x},\ 0<x\le 1\}\cup\{(0, y):\ -1\le y\le 1\}$$

Prove that S is connected but not pathwise connected.

12. In this exercise, we sketch a different proof of Theorem 6.6. Let $f:[a, b]\to\mathbf{R}$ be continuous, and assume that $f(a)<z<f(b)$. Let $S=\{x\in[a, b]: f(x)\le z\}$.
(a) Show that S has a least upper bound c.
(b) Assume that $f(c)\ne z$ and derive a contradiction.
(c) Now give a proof for the case where $f(a)>z>f(b)$.
13. Let $F:\mathbf{R}\to\mathbf{R}$ be differentiable. The function f' need not be continuous (see Exercise 3 of Section 4.6). Show that f' has the intermediate value property in any case: if $a<b$ and α is between $f'(a)$ and $f'(b)$, then there is a number c with $a<c<b$ such that $f'(c)=\alpha$. [Hint: Either α is between $f'(a)$ and $a_1[f'(b)-f'(a)]/(b-a)$, or α is between a_1 and b. In the first case, let $g(x)=[f(x)-f(a)]/(x-a)$, $g(a)=f'(a)$. Show that g is continuous and use Theorem 6.6 and the mean value theorem, Exercise 9 of Section 4.1.]

8

Maxima and Minima

In many applications of mathematics, one wishes to maximize (or minimize) some function. In economics, for example, it is usual to try to maximize profit or minimize cost. Finding the extreme values of a function of one variable is one of the applications of elementary calculus. In this chapter, we develop criteria for finding the points at which a function of several variables takes on its maximum or minimum.

Just as in the case of functions of one variable, two cases need to be considered. To illustrate, let $f: [-1, 1] \to \mathbf{R}$ be defined by $f(x) = x^2$. The minimum value of f is at $x = 0$, where $f'(x) = 2x$ is zero. On the other hand, f has a maximum at $x = -1$, but $f'(x) \neq 0$ at these points. Thus we can use the vanishing of the derivative to find maxima and minima on the "interior" of $[-1, 1]$, but not at the "boundary" points $\{-1, 1\}$ of $[-1, 1]$. In this case, of course, there are only two boundary points, while for a function of several variables there are infinitely many. As a result, we need to develop a special procedure for dealing with maxima and minima at boundary points. This procedure, called the method of *Lagrange multipliers*, is one of the major topics of this chapter. The other is the analysis of maxima and minima at interior points; as we shall see, this analysis for $\mathbf{R}^n$ is similar to the case of $\mathbf{R}^1$.

1. Maxima and Minima at Interior Points

Let S be a subset of $\mathbf{R}^n$ and $f: S \to \mathbf{R}$ a differentiable function. If S is closed and bounded, we know from Theorems 7.3.3 and 7.4.1 that S has a maximum and a minimum somewhere in S. These theorems, however, give us no information about where the maximum or minimum is. In this section, we discuss a method of locating the maximum and minimum values.

We shall solve only a part of the problem here, and we need some definitions to describe what we shall do. We say that a point $\mathbf{v} \in S$ is an *interior* point of S if some open ball of positive radius containing $\mathbf{v}$ is contained in S; a point $\mathbf{v} \in S$ is a *boundary point* if each open ball containing $\mathbf{v}$ meets both S and its complement. (See Exercises 12 to 20 of Section 3.5.) Here are some examples.

1. Let S be the closed unit disk in $\mathbf{R}^2$: $S = \{\mathbf{v} = (x_1, x_2) : x_1^2 + x_2^2 \leq 1\}$. (See Figure 1.1.) A point $\mathbf{v}_0 \in S$ with $\|\mathbf{v}_0\| < 1$ is an interior point, since the open ball of radius $1 - \|\mathbf{v}_0\|$ about $\mathbf{v}_0$ is in S. On the other hand, a point $\mathbf{v}_1 \in S$ with $\|\mathbf{v}_1\| = 1$ is a boundary point; any ball about $\mathbf{v}_1$ contains points $\mathbf{w}$ with $\|\mathbf{w}\| > 1$, and therefore no ball about $\mathbf{v}_1$ is in S.
2. Let S be the closed rectangle in $\mathbf{R}^2$ shown in Figure 1.2:

$$S = \{\mathbf{v} = (x_1, x_2) : |x_1| \leq 1 \text{ and } |x_2| \leq 1\}$$

Then any point $\mathbf{v}_0 = (x_1, x_2)$ with $|x_1| < 1$ and $|x_2| < 1$ is in the interior of S, since, if $|x_1| \leq 1 - \varepsilon$ and $|x_2| \leq 1 - \varepsilon$, then $B_\varepsilon(\mathbf{v}_0) \subset S$. (See Exercise 2.) On the other hand, if $\mathbf{v}_1 = (x_1, x_2)$ with, say, $x_2 = -1$, then every ball about $\mathbf{v}_1$ contains vectors $\mathbf{w} = (y_1, y_2)$ with $|y_2| > 1$. Therefore any such point $\mathbf{v}_1$ is a boundary point.
3. Let S be the closed unit ball in $\mathbf{R}^n$: $S = \{\mathbf{v} \in \mathbf{R}^n : \|\mathbf{v}\| \leq 1\}$. Then, just as in Example 1, the interior points are the points $\mathbf{v}$ with $\|\mathbf{v}\| < 1$ and the boundary points are those with $\|\mathbf{v}\| = 1$. (See Exercise 3.)

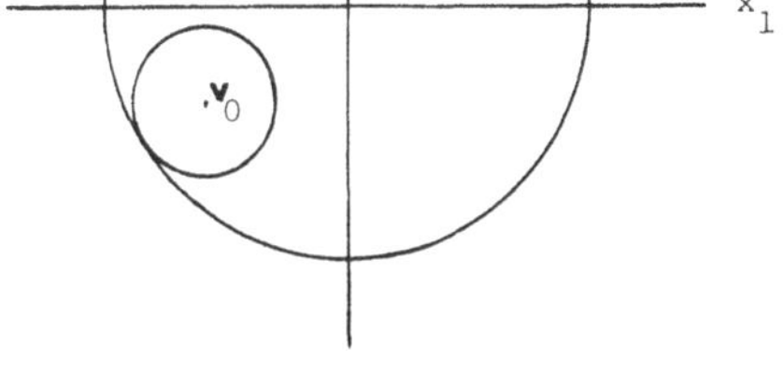

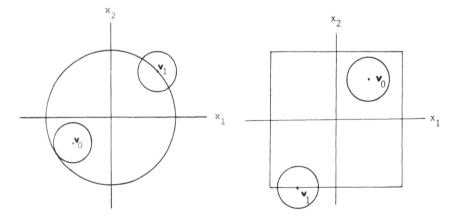

Figure 1.1 An interior point and a boundary point of a disk in $\mathbf{R}^2$.

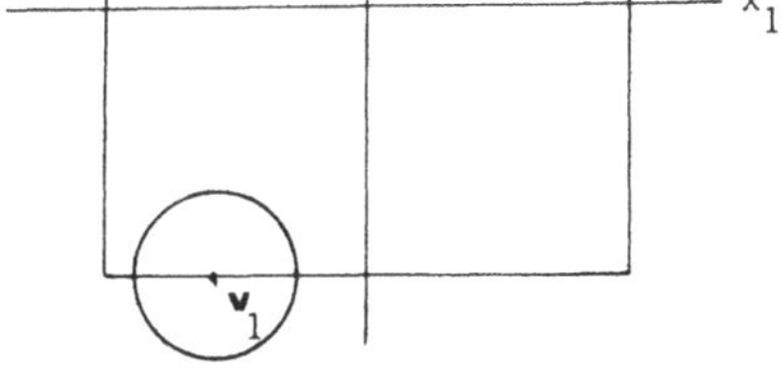

Figure 1.2 An interior point and a boundary point of a rectangle in $\mathbf{R}^2$.

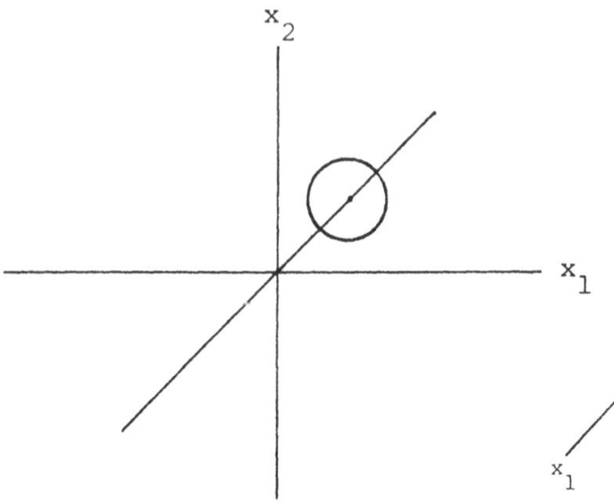

Figure 1.3 A line in $\mathbf{R}^2$ has no interior points.

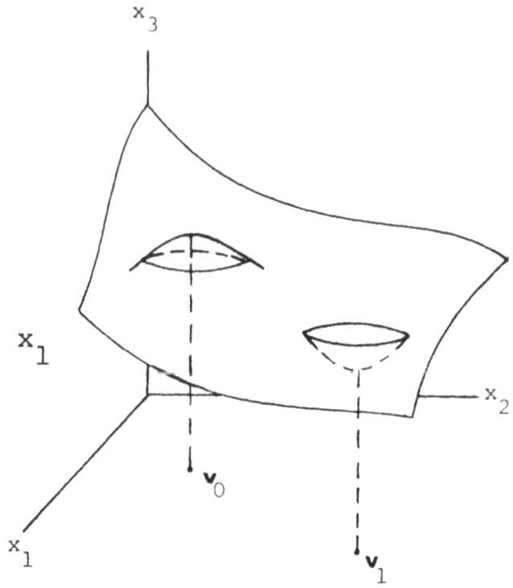

Figure 1.4 The graph of a function with a local maximum at $\mathbf{v}_0$ and a local minimum at $\mathbf{v}_1$.

4. If S is open, then every point $\mathbf{v}$ of S is in an open ball contained in S; hence, every point of S is an interior point.
5. If $S = \{\mathbf{v} \in \mathbf{R}^2 : \mathbf{v} = (x, x)\}$, then no point of S is in an open ball contained in S (see Figure 1.3), and so every point of S is a boundary point.

We shall be concerned here with maxima and minima at interior points; we study the corresponding question for boundary points in Sections 4, 5, and 6.

As in the case of functions of one variable, we say that an interior point $\mathbf{v}_0 \in S$ is a *local maximum for* f if $f(\mathbf{v}) \leq f(\mathbf{v}_0)$ for all $\mathbf{v}$ in some ball about $\mathbf{v}_0$; similarly, $\mathbf{v}_1$ is a *local minimum for* f if $f(\mathbf{v}) \geq f(\mathbf{v}_1)$ for all $\mathbf{v}$ in some ball about $\mathbf{v}_1$. (See Figure 1.4.) We call $\mathbf{v}_0$ an *extremum for* f if it is either a local maximum or a local minimum.

Just as in the case of functions of one variable, we say that the interior point $\mathbf{v}_0$ is a *critical point* of f if $f'(\mathbf{v}_0) = D_{\mathbf{v}_0} f = 0$ (the zero transformation). One reason for this definition is the following theorem.

Theorem 1.1 *If* $\mathbf{v}_0$ *is an interior local maximum or local minimum point of* $f: S \to \mathbf{R}$ *and if* f *is differentiable near* $\mathbf{v}_0$, *then* $\mathbf{v}_0$ *is a critical point of* f.

Proof. Let $\mathbf{v}_0 = (a_1, \ldots, a_n)$ be an interior point of S which is a local maximum for f and set $\mathbf{v} = (x_1, a_2, \ldots, a_n)$. Then $f(\mathbf{v}_0) \geq f(\mathbf{v})$ for x_1 near a_1; that is, f, regarded as a function of its first variable only, has a local maximum at $\mathbf{v}_0$. A standard calculus result states that

$$\frac{\partial f}{\partial x_1}(\mathbf{v}_0) = 0$$

(See Exercise 7, Section 4.1.) The same reasoning, of course, applies to the other components; thus

$$\frac{\partial f}{\partial x_1}(\mathbf{v}_0) = \cdots = \frac{\partial f}{\partial x_n}(\mathbf{v}_0) = 0$$

and it follows from Proposition 4.3.1 that $f'(\mathbf{v}_0) = D_{\mathbf{v}_0} f = 0$. Thus $\mathbf{v}_0$ is a critical point of f. The same reasoning works when $\mathbf{v}_0$ is a local minimum.

For some purposes, Theorem 1.1 suffices to find the maximum and minimum of a function f. We often know in practice that S is a compact set, that $f(\mathbf{v}) = 0$ if $\mathbf{v}$ is a boundary point for S, and that $f > 0$ somewhere in S. (If, for instance, S is determined by physical constraints, these conditions are often met. See Exercises 6 through 10 for other examples.) Now we know that the maximum value of f is attained at an interior point of S. As the maximum is a local maximum, we know that it is attained at a critical point of S. We need only evaluate f at every critical point and see where it is biggest.

We do not always have such information at hand, however, and it is useful to have a method of distinguishing among maxima, minima, and critical points which are neither. Undoubtedly the reader has seen examples of critical points which are not extrema. For instance, the function $f(x) = x^3$ has a critical point at $x = 0$ but neither a maximum or a minimum. With more variables, more complicated phenomena can occur. For example, the function $f(x, y) = x^2 - y^2$ has a critical point at $(0, 0)$ which is neither a maximum or minimum, since f has a maximum along the y axis and a minimum along the x axis. (This surface is called a *saddle*, for obvious reasons; see Figure 1.5.) In the next sections, we shall develop methods for describing the behavior of a function near a critical point.

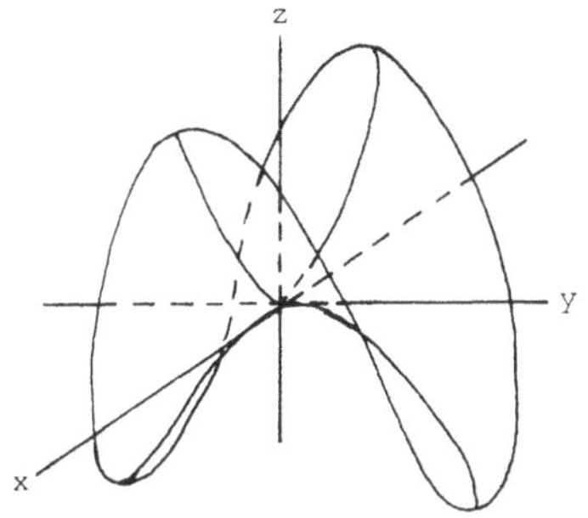

Figure 1.5 A saddle surface.

Exercises

1. Determine the interior points and the boundary points of the following sets.
   (a) $[a, b) \subset \mathbf{R}^1$
   (b) The set of rational numbers in $\mathbf{R}^1$
   (c) $\{(x, y) \in \mathbf{R}^2 : xy = 1\}$
   (d) $\{(x, y) \in \mathbf{R}^2 : xy < 1\}$
   (e) $\{(x, y) \in \mathbf{R}^2 : x > 0 \text{ and } y = \sin(1/x)\}$
   (f) $\{(x, y, z) \in \mathbf{R}^3 : x + y + z = 0\}$
   (g) $\{(x, y, z) \in \mathbf{R}^3 : x + y + z < 0\}$
   (h) $\{(x, y, z) \in \mathbf{R}^3 : xy \leq 1\}$
2. Let S be the cube in $\mathbf{R}^n$ defined by

$$S = \{(x_1, \ldots, x_n) : |x_i| \leq 1, 1 \leq i \leq n\}$$

   Prove that the interior of S is the set

$$\text{Int } S = \{(x_1, \ldots, x_n) : |x_i| < 1, 1 \leq i \leq n\}$$

   and that all other points of S are boundary points. (See Example 2 of this section.)
3. Prove that the interior of the closed ball $\bar{B}_r(\mathbf{v}_0)$ in $\mathbf{R}^n$ is the open ball $B_r(\mathbf{v}_0)$ and that all other points of $\bar{B}_r(\mathbf{v}_0)$ are boundary points.
4. Find all critical points of the following functions.
   (a) $f: \mathbf{R} \to \mathbf{R}, f(x) = 2x^3 - 9x^2 + 12x$
   (b) $f: \mathbf{R} \to \mathbf{R}, f(x) = e^{2x}$
   (c) $f: \mathbf{R} \to \mathbf{R}, f(x) = \sin x$
   (d) $f: \mathbf{R}^2 \to \mathbf{R}, f(x, y) = 2x + 3y - 6$
   (e) $f: \mathbf{R}^2 \to \mathbf{R}, f(x, y) = 3 - x^2 - y^2$
   (f) $f: \mathbf{R}^2 \to \mathbf{R}, f(x, y) = xy$
   (g) $f: \mathbf{R}^2 \to \mathbf{R}^2, f(x, y) = e^{x(y-1)}$
   (h) $f: \mathbf{R}^2 \to \mathbf{R}, f(x, y) = 2x^3y - 10x^2y + 12xy$
   (i) $f: \mathbf{R}^2 \to \mathbf{R}, f(x, y) = \cos(x + y)$
5. For each of the following functions, (0, 0) is a critical point. Determine whether it is a local maximum, a local minimum, or neither.
   (a) $f(x, y) = x^4 + y^4$
   (b) $f(x, y) = x^4 - y^4$
   (c) $f(x, y) = x^4 + y^3$
   (d) $f(x, y) = -x^4 - y^4$
6. Find the shortest distance from the point (0, 1) to the graph of the parabola $y = x^2$.
7. Find the shortest distance from the point (0, 5, 1) to the graph of the circular paraboloid $z = 2x^2 + 2y^2$.
8. Find the point on a given line such that the sum of the distances from two given fixed points is a minimum. (See Figure 1.6.) The solution to this exercise is related to the optical law of reflection. According to *Fermat's principle*, a light ray travels along a path that minimizes time.

*9. (*The method of least squares*)
   (a) Suppose that $(x_1, y_1), \ldots, (x_n, y_n)$ are n points in the plane. We wish to find the line L, given by the equation $y = ax + b$ which minimizes the sum of the

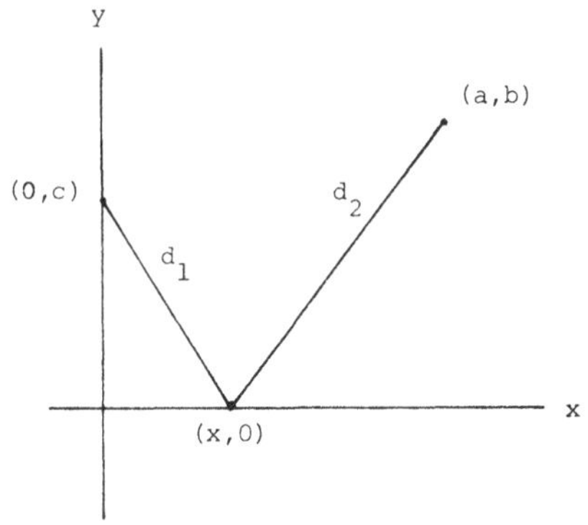

Figure 1.6 Minimize $d_1 + d_2$.

squares of the distances between the points and the line, measured along the y axis. That is, we want to minimize

$$F(a, b) = \sum_{j=1}^{n} (ax_j + b - y_j)^2$$

Show that the minimum of F is attained at

$$a = \left(\bar{x}\bar{y} - \frac{1}{n}\sum_{j=1}^{n} x_j y_j\right)\left(\bar{x}^2 - \frac{1}{n}\sum_{j=1}^{n} x_j^2\right)^{-1}$$

$$b = \bar{y} - a\bar{x}$$

where $\bar{x}$, $\bar{y}$ are the *means* of the x_j, y_j, respectively:

$$\bar{x} = \frac{1}{n}\sum_{j=1}^{n} x_j$$

$$\bar{y} = \frac{1}{n}\sum_{j=1}^{n} y_j$$

The line $y = ax + b$, with a and b chosen as above, is generally regarded in statistics as the line which comes closest to fitting the data points $(x_1, y_1), \ldots,$ (x_n, y_n).

(b) Consider the corresponding problem in $\mathbf{R}^3$, with points

$$(x_1, y_1, z_1), \ldots, (x_n, y_n, z_n)$$

and the plane $P : z = ax + by + c$, with

$$F(a, b, c) = \sum_{j=1}^{n} (z_j - ax_j - by_j - c)^2$$

10. Let λ_1, λ_2 be positive real numbers with $\lambda_1 + \lambda_2 = 1$. Let $S = \{(x_1, x_2) \in \mathbf{R}^2 : x_1 > 0, x_2 > 0\}$ and define $f: S \to \mathbf{R}$ by

$$f(x_1, x_2) = \tfrac{1}{2} x_1^{-\lambda_1} x_2^{-\lambda_2} (x_1 + x_2)$$

(a) Find all critical points for f.
(b) Show that they are all minima. (*Hint*: Set $z = x_1/x_2$.)
(c) Use (b) to prove that

$$x_1^{\lambda_1} x_2^{\lambda_2} \leq \tfrac{1}{2}(x_1 + x_2)$$

11. Let $\lambda_1, \ldots, \lambda_n$ be positive real numbers with $\lambda_1 + \cdots + \lambda_n = 1$. Let $S = \{(x_1, \ldots, x_n) \in \mathbf{R}^n : x_i > 0, 1 \leq i \leq n\}$. Prove that

$$x_1^{\lambda_1} x_2^{\lambda_2} \cdots x_n^{\lambda_n} < \frac{1}{n}(x_1 + \cdots + x_n)$$

for any $(x_1, \ldots, x_n) \in S$. (See Exercise 10.)
12. Find the minimum distance from (6, 3, 4) to the paraboloid $x_3 = x_1^2 + x_2^2$.

2. Quadratic Forms

Because this section may seem pointless without some explanation of purpose, we begin with some motivational material which will not be used in the rest of the section.

Suppose that $f: \mathbf{R} \to \mathbf{R}$ is a differentiable function and that a is a critical point of f. Assuming that f has a continuous second derivative, we can expand f near a by Taylor's theorem (Theorem 2.1, Chapter 6):

$$f(x) = f(a) + (x - a) f'(a) + \tfrac{1}{2}(x - a)^2 f''(c) \tag{2.1}$$

c between a and x. Since a is a critical point, $f'(a) = 0$; thus we may rewrite (2.1) as

$$f(x) - f(a) = (x - a)^2 f''(c) \tag{2.2}$$

Now suppose that $f''(a) > 0$. Then $f''(c) > 0$ if c is sufficiently near a (since f'' is continuous), and so $f(x) - f(a) > 0$ if $x \neq a$ is sufficiently near a [by (2.2)]. Thus a is a local minimum if $f''(a) > 0$. Similarly, a is a local maximum if $f''(a) < 0$. This result is usually referred to as the *second derivative test* in elementary calculus. [If $f''(a) = 0$, then the test fails; a can be a local maximum, a local minimum, or neither. See Exercise 10.]

We wish to use a similar procedure for critical points in $\mathbf{R}^n$. Now, however, we need to use Taylor's theorem for functions of several variables (Theorem 4.1, Chapter 6) and the expression for the second derivative is more complicated. To analyze it, we first discuss quadratic forms.

A *quadratic form* in n variables $x_1, \ldots, x_n$ is an expression of the form

$$Q(x_1, \ldots, x_n) = \sum_{i=1}^{n} \sum_{j=1}^{n} a_{ij} x_i x_j$$

For instance, if $n = 2$, then $Q(x_1, x_2) = a_{11}x_1^2 + (a_{12} + a_{21})x_1x_2 + a_{22}x_2^2$. If $i \neq j$, there are two terms involving x_ix_j, $a_{ij}x_ix_j$ and $a_{ji}x_ix_j$, *and we always assume that* $a_{ij} = a_{ji}$. (This assumption does not restrict the form Q at all, of course. It simply makes expressions more symmetric.)

The quadratic form Q is called *positive definite* if $Q(x_1, \ldots, x_n) > 0$ unless $x_1 = \cdots = x_n = 0$, *negative definite* if $Q(x_1, \ldots, x_n) < 0$ unless $x_1 = \cdots = x_n = 0$, *indefinite* if $Q(x_1, \ldots, x_n)$ takes on positive and negative values.

Examples

1. The quadratic form $Q_1(x_1, x_2) = x_1^2 + x_2^2$ is positive definite.
2. The quadratic form $Q_2(x_1, x_2) = -x_1^2 - 2x_2^2$ is negative definite.
3. The quadratic form $Q_3(x_1, x_2) = x_1^2 - 3x_2^2$ is indefinite.
4. An $n \times n$ matrix $A = (a_{ij})$ is symmetric if $a_{ij} = a_{ji}$ for all i, j. If A is any symmetric $n \times n$ matrix, we denote by Q_A the quadratic form

$$Q_A(x_1, \ldots, x_n) = \sum_{i=1}^{n} \sum_{j=1}^{n} a_{ij} x_i x_j$$

In this context, we say that the matrix A is positive definite when Q_A is positive definite and similarly for the other kinds of definiteness. For instance, if

$$A = \begin{bmatrix} -1 & 0 \\ 0 & -2 \end{bmatrix}$$

then Q_A is the same Q as in Example 2 above, and we say that A is negative definite.

Note that if $T: \mathbf{R}^n \to \mathbf{R}^n$ is the linear transformation defined by A and $\mathbf{v} = (x_1, \ldots, x_n)$, then

(2.3) $$Q_A(\mathbf{v}) = \langle T(\mathbf{v}), \mathbf{v} \rangle$$

We should also remark that there are quadratic forms which are not positive definite, negative definite, or indefinite. The quadratic form

$$Q_4(x_2, x_2) = x_1^2 + 2x_1x_2 + x_2^2 = (x_1 + x_2)^2$$

for instance, is always ≥ 0, but it is not positive definite, since $Q_4(1, -1) = 0$. We say that the quadratic form Q is *positive semidefinite* if $Q(\mathbf{v}) \geq 0$ for all $\mathbf{v}$; thus Q_4 is positive semidefinite, but not positive definite. Similarly, Q is called *negative semidefinite* if $Q(\mathbf{v}) \leq 0$ for all $\mathbf{v}$. We shall not be concerned with such quadratic forms in the rest of this section.

Deciding whether a symmetric matrix A is positive definite or not can be difficult. In the case of 2×2 matrices, however, there is a fairly simple criterion.

Theorem 2.1 *The symmetric* 2×2 *matrix* $A = (a_{ij})$ *is*
(a) *Positive definite if* $a_{11} > 0$ *and* $a_{11}a_{22} - a_{12}^2 > 0$.
(b) *Negative definite if* $a_{11} < 0$ *and* $a_{11}a_{22} - a_{12}^2 > 0$.
(c) *Indefinite if* $a_{11}a_{22} - a_{12}^2 < 0$.

Proof. We begin by writing out Q_A, assuming that $a_{11} \neq 0$:

$$\begin{aligned}
(2.4) \quad Q_A(x_1, x_2) &= a_{11}x_1^2 + 2a_{12}x_1x_2 + a_{22}x_2^2 \\
&= a_{11}\left(x_1^2 + \frac{2a_{12}}{a_{11}}x_1x_2\right) + a_{22}x_2^2 \\
&= a_{11}\left[x_1^2 + \frac{2a_{12}}{a_{11}}x_1x_2 + \left(\frac{a_{12}}{a_{11}}\right)^2 x_2^2\right] + \left(a_{22} - \frac{a_{12}^2}{a_{11}}\right)x_2^2 \\
&= a_{11}\left[\left(x_1 + \frac{a_{12}}{a_{11}}x_2\right)^2 + \frac{1}{a_{11}^2}(a_{11}a_{22} - a_{12}^2)x_2^2\right]
\end{aligned}$$

Suppose now that $a_{11} > 0$ and $a_{11}a_{22} - a_{12}^2 > 0$. Then (2.4) is a sum of positive multiples of squares, and therefore $Q_A(x_1, x_2) \geq 0$ for all (x_1, x_2). If $Q_A(x_1, x_2) = 0$, then both terms of (2.4) must be 0; in particular, $(1/a_{11})(a_{11}a_{22} - a_{12})x_2^2 = 0$. It follows that $x_2 = 0$. Going back to (2.4), we see that $a_{11}x_1^2 = 0$ so that $x_1 = 0$. It follows that $Q_A(x_1, x_2) > 0$ if either x_1 or x_2 is nonzero, and (a) is proved. Part (b) is proved in the same way.

For (c), there are a number of cases. If, for instance, $a_{11} > 0$, then (2.4) shows that $Q_A(1, 0) > 0$ and $Q_A(-a_{12}/a_{11}, 1) < 0$. The case where $a_{11} < 0$ is similar. If $a_{11} = 0$, then we cannot use (2.4), and we must go back to the original expression for Q_A:

$$\begin{aligned}
Q_A(x_1, x_2) &= 2a_{12}x_1x_2 + a_{22}x_2^2 \\
&= x_2(2a_{12}x_1 + a_{22}x_2)
\end{aligned}$$

In this expression, $a_{12} \neq 0$, since our assumption is that $-a_{12}^2 < 0$. It is easy to check (Exercise 12) that we can choose x_1 and x_2 to make $Q_A(x_1, x_2) > 0$ or < 0. Thus Q_A is indefinite in this case, too, and the theorem is proved.

The quantity $a_{11}a_{22} - a_{12}^2$ is called the *determinant* of the symmetric 2×2 matrix A and denoted Det A. We shall define the determinant of an arbitrary $n \times n$ matrix in Appendix A; for a general 2×2 matrix $A = (a_{ij})$, the formula is

$$\text{Det } A = a_{11}a_{22} - a_{12}a_{21}$$

(This reduces to $a_{11}a_{22} - a_{12}^2$ if $a_{12} = a_{21}$.) For a 3×3 matrix A,

$$\begin{aligned}\text{Det } A &= a_{11}(a_{22}a_{33} - a_{23}a_{32}) - a_{12}(a_{21}a_{33} - a_{23}a_{31}) \\ &\quad + a_{13}(a_{21}a_{32} - a_{22}a_{31})\end{aligned}$$

The extension of Theorem 2.1 to $n \times n$ matrices involves determinants; we state it here without proof.

Theorem 2.2 *Let Q be a quadratic form on $\mathbf{R}^n$ with matrix $A = (a_{ij})$. Let A_j, $1 \le j \le n$, be the $j \times j$ matrix formed by the first j rows and columns of A. Thus $A_1 = (a_{11})$,*

$$A_2 = \begin{bmatrix} a_{11} & a_{12} \\ a_{21} & a_{22} \end{bmatrix}$$

and $A_n = A$. Then

(a) *Q is positive definite if and only if* Det $A_j > 0$ *for* $1 \le j \le n$.
(b) *Q is negative definite if and only if* $(-1)^j$ Det $A_j > 0$ *for* $1 \le j \le n$.
(c) *If* Det $A_j \ne 0$ *for* $1 \le j \le n$ *and neither* (a) *or* (b) *holds, then Q is indefinite.*

For example, consider the 3×3 matrix

$$A = \begin{bmatrix} 3 & 2 & 5 \\ 2 & 3 & 4 \\ 5 & 4 & 9 \end{bmatrix}$$

Then $A_1 = (3)$ and $A_2 = \begin{bmatrix} 3 & 2 \\ 2 & 3 \end{bmatrix}$, while $A_3 = A$. We have

$$\text{Det } A_1 = 3 \qquad \text{Det } A_2 = 5 \qquad \text{Det } A_3 = 2$$

Since all determinants are positive, A is positive definite. In fact, some algebra shows that

$$Q_A(x, y, z) = 2(x + y + 2z)^2 + (x + z)^2 + y^2$$

In this form, it is easy to check that Q is positive definite.

When $n = 1$, $A = (a)$ is given by a number, and the quadratic form is $Q_A(x) = ax^2$. It is positive definite when $a > 0$, negative definite when $a < 0$. (Note that these results agree with Theorem 2.2.) When $a = 0$, the form is only semidefinite, and Theorem 2.2 does not apply.

We shall need another result about quadratic forms in the next section; it says, for instance, that if A and B are symmetric matrices, A is positive definite, and B is "close" to A, then B is positive definite. We shall merely state the result here, leaving the proof to Section 7. First, however, we need to define what "close" means for matrices. This is not difficult. Let $A = (a_{ij})$. We may regard A as an element of $\mathbf{R}^{n^2}$:

$$A = (a_{11}, \ldots, a_{1n}, a_{21}, \ldots, a_{2n}, \ldots, a_{nn})$$

Thus we define $\| A \|$ by*

$$\| A \| = \left(\sum_{i=1}^{n} \sum_{j=1}^{n} a_{ij}^2 \right)^{1/2}$$

Now we can state the result.

Proposition 2.3 *Let A be a symmetric $n \times n$ matrix.*

(a) *If A is positive definite, then there is an $\varepsilon > 0$ such that any symmetric $n \times n$ matrix B with $\| A - B \| < \varepsilon$ is positive definite.*
(b) *If A is negative definite, then there is an $\varepsilon > 0$ such that any symmetric $n \times n$ matrix B with $\| A - B \| < \varepsilon$ is negative definite.*
(c) *If A is indefinite and $\mathbf{v}_1, \mathbf{v}_2$ are unit vectors with $Q_A(\mathbf{v}_1) > 0$ and $Q_A(\mathbf{v}_2) < 0$, then there is an $\varepsilon > 0$ such that if B is any symmetric $n \times n$ matrix with $\| A - B \| < \varepsilon$, then $Q_B(\mathbf{v}_1) > 0$ and $Q_B(\mathbf{v}_2) < 0$. In particular, B is indefinite.*

Exercises

1. Find the quadratic forms associated to these matrices:

(a) $\begin{bmatrix} 1 & 2 \\ 2 & 1 \end{bmatrix}$ (b) $\begin{bmatrix} 3 & -1 \\ -1 & -2 \end{bmatrix}$

(c) $\begin{bmatrix} 3 & 2 \\ 2 & 4 \end{bmatrix}$ (d) $\begin{bmatrix} -2 & 1 \\ 1 & -3 \end{bmatrix}$

(e) $\begin{bmatrix} 3 & 2 \\ 2 & 7 \end{bmatrix}$ (f) $\begin{bmatrix} 4 & 3 \\ 3 & 2 \end{bmatrix}$

2. Which of the above forms are positive definite, which are negative definite, and which indefinite?
3. Find the quadratic forms associated to these matrices:

(a) $\begin{bmatrix} 1 & 0 & 0 \\ 0 & 1 & 0 \\ 0 & 0 & 1 \end{bmatrix}$ (b) $\begin{bmatrix} 3 & 1 & 2 \\ 1 & 5 & 3 \\ 2 & 3 & 7 \end{bmatrix}$

*This definition of $\| A \|$ is not the same as that given in Exercise 7 of Section 7.2. We use the current definition only in this section and Section 7.

(c) $\begin{bmatrix} -4 & 0 & 1 \\ 0 & -3 & 2 \\ 1 & 2 & -5 \end{bmatrix}$ (d) $\begin{bmatrix} -3 & 1 & 0 \\ 1 & -6 & 1 \\ 0 & 1 & 7 \end{bmatrix}$

(e) $\begin{bmatrix} 4 & 2 & 0 & 0 \\ 2 & 5 & 3 & 1 \\ 0 & 3 & 5 & 0 \\ 0 & 1 & 0 & 3 \end{bmatrix}$ (f) $\begin{bmatrix} 1 & -1 & 0 & 0 \\ -1 & 2 & 0 & 0 \\ 0 & 0 & 3 & 1 \\ 0 & 0 & 1 & 4 \end{bmatrix}$

4. Determine which of the matrices in Exercise 3 are positive definite, which are negative definite, and which are indefinite.
5. Find the matrix A corresponding to each of the quadratic forms Q_A given below:
   (a) $Q_A(x_1, x_2) = x_1^2 + 6x_1x_2 - 3x_2^2$
   (b) $Q_A(x_1, x_2) = 4x_1x_2 + x_2^2$
   (c) $Q_A(x_1, x_2) = x_1^2 + x_1x_2 + x_2^2$
   (d) $Q_A(x_1, x_2) = 4x_1x_2 - 7x_1^2 - 3x_2^2$
   (e) $Q_A(x_1, x_2) = 6(x_1^2 + x_2^2) - 5x_1x_2$
   (f) $Q_A(x_1, x_2) = x_1x_2$
6. Determine which of the above quadratic forms are positive definite, which are negative definite, and which are indefinite.
7. Find the matrix A corresponding to each of the following forms Q_A:
   (a) $Q_A(x_1, x_2, x_3) = x_1^2 - 2x_2^2 + 4x_3^2$
   (b) $Q_A(x_1, x_2, x_3) = 3x_1^2 - 2x_1x_2 + x_2^2 + 4x_2x_3 + 8x_3^2$
   (c) $Q_A(x_1, x_2, x_3) = 2x_1x_2 + 4x_1x_3 + 6x_2x_3$
   (d) $Q_A(x_1, x_2, x_3) = x_1x_2 + x_2x_3 - 4(x_1^2 + x_2^2 + x_3^2)$
   (e) $Q_A(x_1, x_2, x_3, x_4) = x_1^2 + x_2^2 - x_3^2 + x_4^2 - 6x_1x_2 + x_2x_3$
8. Determine which of the above forms are positive definite, which are negative definite, and which are indefinite.
9. Give examples of symmetric 3×3 matrices which are
   (a) Positive definite
   (b) Positive semidefinite but not positive definite
   (c) Indefinite
   (d) Negative definite
   (e) Negative semidefinite but not negative definite
10. Construct examples of functions $f: \mathbf{R} \to \mathbf{R}$ with $f'(0) = f''(0) = 0$ such that 0 is
    (a) A local maximum
    (b) A local mimimum
    (c) Neither
11. Let Q be a quadratic form on $\mathbf{R}^n$. Prove that $Q(\lambda \mathbf{v}) = \lambda^2 Q(\mathbf{v})$ for any $\mathbf{v} \in \mathbf{R}^n$, $\lambda \in \mathbf{R}$.
12. Complete the proof of Theorem 2.1. (*Hint*: Choose $x_2 > 0$; then let x_1 be alternately >0 and <0, with $|x_1|$ large.)
*13. Prove Theorem 2.2 for $n = 3$.

3. Criteria for Local Maxima and Minima

Let S be a subset of $\mathbf{R}^n$, let $f: S \to \mathbf{R}$ be a function with a continuous second derivative, and let $\mathbf{v}_0$ be an interior point of S which is a critical point of f. We would like to have a method of determining whether $\mathbf{v}_0$ is a local maximum, a local minimum, or neither. Just as in the case of functions on $\mathbf{R}^1$, we look at the second derivative for a criterion.

Recall (from Section 6.1) that the Hessian of f at a point $\mathbf{v}$, $H(f)(\mathbf{v})$, is a matrix formed from the second partial derivatives of f. If the partial derivatives are continuous, as we are assuming, then $H(f)(\mathbf{v})$ is a symmetric matrix. (See Proposition 1.1, Chapter 6.)

Theorem 3.1 *Let U be an open set in $\mathbf{R}^n$ and $f: U \to \mathbf{R}$ a function having continuous second-order partial derivatives. Let $\mathbf{v}_0$ be a critical point of f and let $H(f)(\mathbf{v}_0)$ be the Hessian of f at $\mathbf{v}_0$.*

(a) *If $H(f)(\mathbf{v}_0)$ is positive definite, then $\mathbf{v}_0$ is a local minimum.*
(b) *If $H(f)(\mathbf{v}_0)$ is negative definite, then $\mathbf{v}_0$ is a local maximum.*
(c) *If $H(f)(\mathbf{v}_0)$ is indefinite, then $\mathbf{v}_0$ is neither a local maximum nor a local minimum.*

Proof. The proof of Theorem 3.1 is like that for functions on $\mathbf{R}$, sketched at the beginning of Section 2. At a few points, however, we need Propositions 2.3 and 2.4.

We begin with Taylor's theorem (Theorem 6.3.1), with $m = 1$. We expand f about the critical point $\mathbf{v}_0$. Since $(\partial f/\partial x_i)(\mathbf{v}_0) = 0$, $1 \le i \le n$, we find that

$$(3.1) \qquad f(\mathbf{v}) - f(\mathbf{v}_0) = \frac{1}{2} \sum_{i=1}^{n} \sum_{j=1}^{n} (x_i - a_i)(x_j - a_j) \frac{\partial^2 f}{\partial x_i \partial x_j}(\mathbf{v}_1)$$

where $\mathbf{v}_1$ is on the line segment from $\mathbf{v}$ to $\mathbf{v}_0$. The right side of (3.1) is, except for the factor $\frac{1}{2}$, just the quadratic form associated with the Hessian at $\mathbf{v}_1$ (evaluated at $\mathbf{v} - \mathbf{v}_0$). Let us call this quadratic form $Q(\mathbf{v}_1)$; then we may rewrite (3.1) as

$$(3.2) \qquad f(\mathbf{v}) - f(\mathbf{v}_0) = \tfrac{1}{2} Q(\mathbf{v}_1)(\mathbf{v} - \mathbf{v}_0)$$

(a) Suppose $H(f)(\mathbf{v}_0)$ is positive definite. Then, by Proposition 2.3, there is an $\varepsilon > 0$ such that if $\| A - H(f)(\mathbf{v}_0)\| < \varepsilon$, then A is positive definite. Since the second derivative of f is continuous, we may pick $\delta > 0$ such that if $\|\mathbf{v} - \mathbf{v}_0\| < \delta$, then $\mathbf{v} \in U$ and $\| H(f)(\mathbf{v}_1) - H(f)(\mathbf{v}_0)\| < \varepsilon$. Since $\mathbf{v}_1$ is on the line segment from $\mathbf{v}$ to $\mathbf{v}_0$, we have

$$\|\mathbf{v} - \mathbf{v}_0\| < \delta \Rightarrow \|\mathbf{v}_1 - \mathbf{v}_0\| < \delta$$

Therefore $H(f)(\mathbf{v}_1)$ is positive and $Q(\mathbf{v}_1)(\mathbf{v} - \mathbf{v}_0) > 0$ when $0 < \|\mathbf{v} - \mathbf{v}_0\| < \delta$. Now (3.2) tells us that $f(\mathbf{v}) - f(\mathbf{v}_0) > 0$ when $0 < \|\mathbf{v} - \mathbf{v}_0\| < \delta$, so that $\mathbf{v}_0$ is a local minimum for f.

(b) This is essentially the same as (a), with appropriate sign changes.

(c) We know that $Q(\mathbf{v}_0)$ is indefinite; therefore we can find vectors $\mathbf{w}_1$ and $\mathbf{w}_2$ with $Q(\mathbf{v}_0)(\mathbf{w}_1) > 0$ and $Q(\mathbf{v}_0)(\mathbf{w}_2) < 0$. We shall show, essentially, that f increases in the direction of $\mathbf{w}_1$ and decreases in the direction of $\mathbf{w}_2$. Thus $\mathbf{v}_0$ will not be an extremum.

Using Proposition 2.3, we may choose $\varepsilon > 0$ such that if A is any symmetric matrix with

$$\|A - H(f)(\mathbf{v}_0)\| < \varepsilon$$

then $Q_A(\mathbf{w}_1) > 0$ and $Q_A(\mathbf{w}_2) < 0$. Choose $\delta > 0$ so that, if $\|\mathbf{v} - \mathbf{v}_0\| < \delta$, then $\mathbf{v} \in U$ and

$$\|H(f)(\mathbf{v}) - H(f)(\mathbf{v}_0)\| < \varepsilon$$

We now let $\mathbf{u}_1$ be a vector in the direction $\mathbf{w}_1$ from $\mathbf{v}_0$, but within δ of $\mathbf{v}_0$; that is, we let

$$\mathbf{u}_1 = \mathbf{v}_0 + \lambda \mathbf{w}_1$$

where $\lambda > 0$ and

$$\|\mathbf{u}_1 - \mathbf{v}_0\| = \|\lambda \mathbf{w}_1\| < \delta$$

(See Figure 3.1.) Then, according to formula (3.2),

$$f(\mathbf{u}_1) - f(\mathbf{v}_0) = \frac{1}{2} Q(\mathbf{v}_1)(\mathbf{u}_1 - \mathbf{v}_0)$$

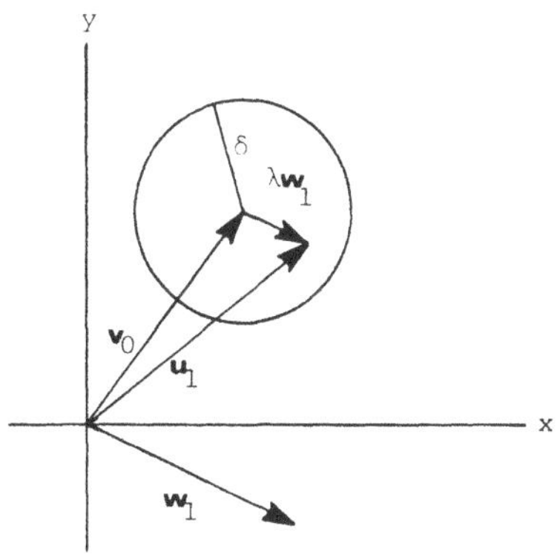

Figure 3.1

where $\mathbf{v}_1$ is on the line segment from $\mathbf{v}_0$ to $\mathbf{u}_1$. Thus,

$$f(\mathbf{u}_1) - f(\mathbf{v}_0) = \frac{1}{2}Q(\mathbf{v}_1)(\lambda \mathbf{w}_1) = \frac{\lambda^2}{2}Q(\mathbf{v}_1)(\mathbf{w}_1) > 0$$

since

$$\|\mathbf{v}_1 - \mathbf{v}_0\| < \|\mathbf{u}_1 - \mathbf{v}_0\| < \delta$$

Hence $\mathbf{v}_0$ is not a local maximum.

On the other hand, if $\mu > 0$ is chosen such that the vector $\mathbf{u}_2 = \mathbf{v}_0 + \mu \mathbf{w}_2$ satisfies

$$\|\mathbf{u}_2 - \mathbf{v}_0\| = \|\mu \mathbf{w}_2\| < \delta$$

then

$$f(\mathbf{u}_2) - f(\mathbf{v}_0) = \frac{1}{2}Q(\mathbf{v}_2)(\mathbf{u}_2 - \mathbf{v}_0)$$

where $\mathbf{v}_2$ is on the line segment from $\mathbf{v}_0$ to $\mathbf{u}_2$. Therefore,

$$f(\mathbf{u}_2) - f(\mathbf{v}_0) = \frac{1}{2}Q(\mathbf{v}_2)(\mu \mathbf{w}_2) = \frac{\mu^2}{2}Q(\mathbf{v}_2)(\mathbf{w}_2) < 0$$

since

$$\|\mathbf{v}_2 - \mathbf{v}_0\| < \|\mathbf{u}_2 - \mathbf{v}_0\| < \delta$$

It follows that $\mathbf{v}_0$ is not a local minimum either, and part (c) of Theorem 3.2 is proved.

Let us see some examples using Theorem 3.1.

1. Let $f\colon \mathbf{R}^2 \to \mathbf{R}$ be defined by

$$f(x, y) = x^3 - 6xy + 3y^2 - 24x + 4$$

Since

$$\frac{\partial f}{\partial x} = 3x^2 - 6y - 24 \qquad \frac{\partial f}{\partial y} = -6x + 6y$$

$f'(x, y) = 0$ when $x = y$ and $3x^2 - 6y - 24 = 0$. The critical points are therefore (4,4) and $(-2, -2)$.

Next, we compute the second partial derivatives of f:

$$\frac{\partial^2 f}{\partial x^2} = 6x \qquad \frac{\partial^2 f}{\partial x \partial y} = \frac{\partial^2 f}{\partial y \partial x} = -6 \qquad \frac{\partial^2 f}{\partial y^2} = 6$$

Thus

$$H(f)(x, y) = \begin{bmatrix} 6x & -6 \\ -6 & 6 \end{bmatrix}$$

In particular,

$$H(f)(4, 4) = \begin{bmatrix} 24 & -6 \\ -6 & 6 \end{bmatrix} \qquad H(f)(-2, -2) = \begin{bmatrix} -12 & -6 \\ -6 & 6 \end{bmatrix}$$

Theorem 2.1 tells us that $H(f)(4, 4)$ is positive definite and $H(f)(-2, -2)$ is indefinite; hence $(4, 4)$ is a local minimum and $(-2, -2)$ is not a local extremum.

2. Next, consider the function $g : \mathbf{R}^2 \to \mathbf{R}$:

$$g(x, y) = x^2 - 8xy + 2y^2 - 6x - 4y + 9$$

Since

$$\frac{\partial g}{\partial x} = 2x - 8y - 6 \qquad \frac{\partial g}{\partial y} = -8x + 4y + 4$$

the critical points of g occur where

$$2x - 8y - 6 = 0 \qquad -8x + 4y + 4 = 0$$

or at $(\frac{1}{7}, \frac{-5}{7})$. Furthermore,

$$\frac{\partial^2 g}{\partial x^2} = 2 \qquad \frac{\partial^2 g}{\partial x \partial y} = \frac{\partial^2 g}{\partial y \partial x} = -8 \qquad \frac{\partial^2 f}{\partial y^2} = 4$$

so

$$H(g)(x, y) = \begin{bmatrix} 2 & -8 \\ -8 & 4 \end{bmatrix} \tag{3.3}$$

In particular, $H(g)(\frac{1}{7}, \frac{-5}{7})$ is given by (3.3). Applying Theorem 2.1, we see that $H(g)(\frac{1}{7}, \frac{-5}{7})$ is indefinite. Thus $(\frac{1}{7}, \frac{-5}{7})$ is not an extremum of g.

The point $(\frac{1}{7}, \frac{-5}{7})$ is often called a *saddle point* of g and $(-2, 2)$ is called a saddle point of f. (See Figure 1.5 for a similar surface.) In dimensions higher than 2, there is a considerable variety of types of saddle points possible. (Unfortunately, it is difficult to sketch them.)

Theorem 3.1 does not describe the nature of all critical points, since not all quadratic forms are positive definite, negative definite, or indefinite. A few simple examples show why. The functions

$$f_1(x_1, x_2) = x_1^2 + x_2^4$$
$$f_2(x_1, x_2) = x_1^2 - x_2^4$$
$$f_3(x_1, x_2) = x_1^2 + x_2^3$$

all have critical points at $(0, 0)$, and in each case the Hessian is

$$\begin{bmatrix} 1 & 0 \\ 0 & 0 \end{bmatrix}$$

However, f_1 has a local minimum at the origin, f_2 has a saddle point there, and the origin is neither a saddle point nor a local minimum or maximum for f_3. When the test fails, one is usually forced to test nearby points to determine the nature of the critical points. (There are more mathematically sophisticated procedures, but they are cumbersome.) In practice, however, Theorem 3.1 takes care of most cases.

Exercises

1. Find and classify the critical points of the following.
   (a) $f(x, y) = x^2 - 4x + 2y^2 + 7$
   (b) $f(x, y) = 3xy - x^2 - y^2$
   (c) $f(x, y) = e^{-x^2-y^2}$
   (d) $f(x, y) = x^2 - 2xy + \frac{1}{3}y^3 - 3y + 8$
   (e) $f(x, y) = xy + 4/x + 2/y$
   (f) $f(x, y) = (\cosh x)(\sin y)$
   (g) $f(x, y) = 12x^3 - 36xy - 2y^3 + 15y^2 - 36x^2 + 72y + 5$
   (h) $f(x, y, z) = x^2 + y^2 + z^2 + xz + yz + 2x - 2y - 4z + 10$
2. Determine the critical points and the local maxima and minima of the function
$$f(x, y) = x^3 - 3axy + y^3$$
   for all values of a.
3. Show that for all the functions of Section 1, Exercise 5, the test of Theorem 3.1 fails at the critical point (0, 0).
4. Show that $f(x, y) = x^3 - xy^2$ has a critical point at (0, 0) which is not an extremum. Sketch the graph of f near (0, 0).
5. Let $g: \mathbf{R}^2 \to \mathbf{R}$ be given by $g(x, y) = (x - y^2)(x - 2y^2)$.
   (a) Sketch the regions in the plane where g is positive, where g is negative, and where g is zero.
   (b) Show that (0, 0) is a critical point for g and that, restricted to any line through (0, 0), g has a local minimum at (0, 0).
   (c) Show that (0, 0) is not a local minimum for g.
6. Find the dimensions of the rectangular box of maximum volume whose surface area is 150 inches.
7. A rectangular parallelepiped has three of its faces on the coordinate planes. If one vertex is to be on the plane $3x + 2y + z = 4$, find the dimensions for maximum volume.
8. A somewhat fancy rectangular box with a volume of 1000 cubic inches has sides and bottom made out of tin and a top made out of platinum. If sheet tin costs \$0.50 a square inch and sheet platinum \$500 a square inch, what dimensions minimize the cost?
9. Find three positive numbers x, y, and z whose sum is 24 which maximize xy^2z^3.
10. Find three positive numbers x, y, and z whose sum is 33 which minimize $x^2 + 2y^2 + 3z^2$.
11. Let $f: \mathbf{R}^2 \to \mathbf{R}$ be a function whose Laplacian Δf is everywhere negative. (See Exercise 8 of Section 6.1.) Show that f has no local minimum.

*12. Bob and Larry are engaged in a game of chance. The rules of the game specify that Bob picks a number x, then Larry picks a number y and pays Bob $f(x, y)$

dollars, where $f(x, y) = y^2 - x^2 + 6y + 8x - 2$. Describe the best strategy for each player. What happens?

*13. Let $(x_1, y_1), \ldots, (x_n, y_n)$ be points in the plane. Find the equation of the parabola $y = ax^2 + bx + c$ such that the quantity

$$F(a, b, c) = \sum_{j=1}^{n} (y_j - ax_j^2 - bx_j^2 - bx_j - c)^2$$

is minimized. (This parabola is the best fit to the data in the sense of least squares; compare Exercise 9 of Section 1.)

4. Constrained Maxima and Minima: I

In our discussion of maxima and minima in the previous sections, we had to defer any investigation of points on the boundary of a set which might be extreme points. In this section, we see how to deal with these points.

Let g: $\mathbf{R}^k \to \mathbf{R}^m$ be a continuously differentiable function and define $S \subset \mathbf{R}^k$ by

$$S = \{\mathbf{v} \in \mathbf{R}^k : g(\mathbf{v}) = \mathbf{0}\}$$

Since $S = g^{-1}(\{\mathbf{0}\})$, S is closed. (See Proposition 7.4, Chapter 3.) Let U be an open set containing S, and let f: $U \to \mathbf{R}$ be a continuously differentiable function on U. The question we shall study here is the following: how can we find the maximum and minimum values (if any) of f restricted to S?

To see the relationship of this question to the question of extreme points on the boundary, consider an example mentioned in Section 1: we want to maximize a function on the compact set $S = \{\mathbf{v} \in \mathbf{R}^n : \|\mathbf{v}\| \le 1\}$. S is not a set of the sort described above. We know, however, how to find all extreme points on the interior of S, the set $\{\mathbf{v} \in \mathbf{R}^n : \|\mathbf{v}\| < 1\}$. That leaves only the set

$$\partial S = \{\mathbf{v} \in \mathbf{R}^n : \|\mathbf{v}\| = 1\}$$

(customarily, the boundary of a set S is denoted by ∂S), and ∂S is a set of the appropriate kind. [We can rewrite the criterion for membership in ∂S as $\langle \mathbf{v}, \mathbf{v} \rangle - 1 = 0$, and $g(\mathbf{v}) = \langle \mathbf{v}, \mathbf{v} \rangle - 1$ is continuously differentiable.]

Not all boundaries can be expressed in the appropriate form; for instance, the boundary of the unit square in $\mathbf{R}^2$ cannot be. In many cases, however, the methods we shall now develop for finding extreme points can be applied anyway. We shall discuss this matter later.

To explain how the method works, we consider a special case. First of all, we assume that g: $\mathbf{R}^3 \to \mathbf{R}^1$. (Later, we shall generalize this somewhat.) Secondly, we assume that g is of the special form

$$g(x_1, x_2, x_3) = h(x_1, x_2) - x_3 \tag{4.1}$$

where h is a continuously differentiable function. While this assumption may seem quite drastic, we shall see later that it really is not.

Under these assumptions, we need to maximize $f(x_1, x_2, x_3)$ subject to the constraint

$$x_3 = h(x_1, x_2)$$

That is, we effectively need to maximize the function

$$F(x_1, x_2) = f(x_1, x_2, h(x_1, x_2))$$

We can do this with the help of Theorem 1.1 and the chain rule. Let $H : \mathbf{R}^2 \to \mathbf{R}^3$ be defined by

$$H(x_1, x_2) = (x_1, x_2, h(x_1, x_2))$$

Then $F = f \circ H$, and therefore

$$F'(x_1, x_2) = f'(H(x_1, x_2)) \circ H'(x_1, x_2)$$

Since f' has the matrix

$$\begin{bmatrix} \dfrac{\partial f}{\partial x_1} & \dfrac{\partial f}{\partial x_2} & \dfrac{\partial f}{\partial x_3} \end{bmatrix}$$

and H' the matrix

$$\begin{bmatrix} 1 & 0 \\ 0 & 1 \\ \dfrac{\partial h}{\partial x_1} & \dfrac{\partial h}{\partial x_2} \end{bmatrix}$$

we see that

$$F'(\mathbf{v}) = \begin{bmatrix} \dfrac{\partial f}{\partial x_1} + \dfrac{\partial f}{\partial x_3}\dfrac{\partial h}{\partial x_1} & \dfrac{\partial f}{\partial x_2} + \dfrac{\partial f}{\partial x_3}\dfrac{\partial h}{\partial x_2} \end{bmatrix}$$

where derivatives of h are evaluated at $\mathbf{v}$, and derivatives of f are evaluated at $H(\mathbf{v})$. Hence at a critical point of F,

$$\text{(4.2)} \qquad \frac{\partial f}{\partial x_j}(H(\mathbf{v})) + \frac{\partial f}{\partial x_3}(H(\mathbf{v}))\frac{\partial h}{\partial x_j}(\mathbf{v}) = 0 \qquad \text{for } j = 1, 2$$

We can rewrite these equations more symmetrically by recalling the function g. Using (4.1), we have

$$\frac{\partial g}{\partial x_1} = \frac{\partial h}{\partial x_1} \qquad \frac{\partial g}{\partial x_2} = \frac{\partial h}{\partial x_2} \qquad \frac{\partial g}{\partial x_3} = -1$$

Thus (4.2) becomes

$$\frac{\partial f}{\partial x_j}\frac{\partial g}{\partial x_3} = \frac{\partial f}{\partial x_3}\frac{\partial g}{\partial x_j} \qquad \text{for } j = 1, 2 \tag{4.3}$$

We add on one trivial equation,

$$\frac{\partial f}{\partial x_3}\frac{\partial g}{\partial x_3} = \frac{\partial f}{\partial x_3}\frac{\partial g}{\partial x_3}$$

Now (4.3), with this last equation, becomes

$$\frac{\partial g}{\partial x_3}\operatorname{grad} f = \frac{\partial f}{\partial x_3}\operatorname{grad} g \tag{4.4}$$

(See Section 5.6 for the definition of the gradient.) We may summarize this result by saying that *if* $\mathbf{v}$ *is an extreme point of* f *on* S, *then* $(\operatorname{grad} f)(\mathbf{v})$ *and* $(\operatorname{grad} g)(\mathbf{v})$ *are proportional.*

Let us pause for an example. We shall maximize the function $x + y + z$ on the subset of $\mathbf{R}^3$ given by $z = 1 - 7x^2 - 3y^2$. In this case

$$f(x, y, z) = x + y + z \qquad g(x, y, z) = 1 - 7x^2 - 3y^2 - z$$

and

$$\operatorname{grad} f = (1, 1, 1) \qquad \operatorname{grad} g = (-14x, -6y, -1)$$

If grad f and grad g are proportional, then their ratio must be -1, since that is the ratio of their third components. Thus we must have

$$(-14x, -6y, -1) = -(1, 1, 1)$$

or

$$14x = 1 \qquad 6y = 1$$

It follows that $x = \frac{1}{14}$ and $y = \frac{1}{6}$. From the equation defining the constraint, we have

$$z = 1 - \frac{7}{14^2} - \frac{3}{6^2} = \frac{20}{21}$$

Therefore, the point $\mathbf{v}_0 = (\frac{1}{14}, \frac{1}{6}, \frac{20}{21})$ is a "critical point" for f given the constraint imposed by g; it could be a maximum, a minimum, or neither. In fact, $\mathbf{v}_0$ is the point where f is largest (given the constraint imposed by g), as one can see by evaluating f at nearby points.

We assumed in our previous work that g was of the form $g(x_1, x_2, x_3) = h(x_1, x_2) - x_3$; put differently, we have assumed that our constraint expressed x_3 as a function of x_1 and x_2. If we had assumed instead that x_2 was a function of x_1 and x_3, for instance, we would arrive at the same result. The reason is that our conclusion (grad f and grad g are proportional) does not treat any one coordinate differently from any other. (The skeptical reader may go through the calculation; see Exercise 11.)

This remark, that the criterion for a constrained maximum holds if any coordinate is a function of the rest, may not seem to make the conclusion hold much more generally, but it does. As we shall see in the next chapter, a result called the *implicit function theorem* states that if $g : \mathbf{R}^3 \to \mathbf{R}$ is continuously differentiable and if $g(\mathbf{v}_0) = 0$, grad $g(\mathbf{v}_0) \neq 0$, then the constraint $g(\mathbf{v}) = 0$ may be regarded as expressing one of the coordinates of $\mathbf{v}$ as a function of the rest (at least when $\mathbf{v}$ is near $\mathbf{v}_0$). This theorem, therefore, says that our reasoning applies wherever $(\text{grad } g)(\mathbf{v}_0) \neq 0$.

Assuming for the moment the implicit function theorem, we may sum up our discussion as follows:

Theorem 4.1 *Let $g : \mathbf{R}^3 \to \mathbf{R}$ be a continuously differentiable function, and let $S = \{\mathbf{v} \in \mathbf{R}^3 : g(\mathbf{v}) = 0\}$. Let U be an open set in $\mathbf{R}^3$ containing S, and let $f\colon U \to \mathbf{R}$ be continuously differentiable. Suppose that* grad g *is nonzero everywhere on S. Then, if $\mathbf{v}$ is a point of S where f has a maximum or minimum,* $(\text{grad } f)(\mathbf{v})$ *and* $(\text{grad } g)(\mathbf{v})$ *are proportional.*

We discussed the situation above where we wanted a maximum; of course, the discussion for a minimum is similar. There is a test like Theorem 3.1 to distinguish maxima, minima, and critical points which are neither; it is fairly cumbersome, and it is almost never used. Instead, one determines the nature of the critical point by testing nearby points or by using other information.

Here is an example of Theorem 4.1: we maximize $2x + 3y + 6z$ subject to the constraint $x^2 + y^2 + z^2 = 1$. In this case

$$f(x, y, z) = 2x + 3y + 6z$$

$$\text{grad } f = (2, 3, 6)$$

and

$$g(x, y, z) = x^2 + y^2 + z^2 - 1$$

$$\text{grad } g = (2x, 2y, 2z)$$

Notice that grad $g = 0$ only at $(0, 0, 0)$, and $g(0, 0, 0) \neq 0$. Thus the conditions of Theorem 4.1 are met. We conclude that at a maximum or minimum of f (under the constraint), $(2, 3, 6)$ and $(2x, 2y, 2z)$ are proportional. That is, $(x, y, z) = \lambda(2, 3, 6)$, where λ is some real number. We determine λ by using the constraint: since $g(x, y, z) = 0$, we have

$$\lambda^2(2^2 + 3^2 + 6^2) = 1$$

or $\lambda^2 = \frac{1}{49}$, $\lambda = \pm\frac{1}{7}$. We need, therefore, to consider the points

$$\mathbf{v}_1 = \left(\frac{2}{7}, \frac{3}{7}, \frac{6}{7}\right) \qquad \mathbf{v}_2 = \left(\frac{-2}{7}, \frac{-3}{7}, \frac{-6}{7}\right)$$

As is easily checked, $f(\mathbf{v}_1) = 7$, $f(\mathbf{v}_2) = -7$. We know that f has a maximum and a minimum on the set of points satisfying $g(\mathbf{v}) = 0$, by Theorem 3.3, Chapter 7, since this set is closed and bounded and therefore compact (by Theorem 4.1, Chapter 7). Since $\mathbf{v}_1$ and $\mathbf{v}_2$ are the only possible extreme points, $\mathbf{v}_1$ is a maximum and $\mathbf{v}_2$ is a minimum.

We have assumed throughout this discussion that f and g were functions from $\mathbf{R}^3$ to $\mathbf{R}^1$. Exactly the same reasoning applies, however, for functions from $\mathbf{R}^{n+1}$ to $\mathbf{R}^1$. (Exercise 12 gives a sketch of the steps.) Furthermore, the implicit function theorem still applies in this general situation. Thus we may state an analogue of Theorem 4.1 for $\mathbf{R}^{n+1}$.

Theorem 4.2 *Let $g: \mathbf{R}^{n+1} \to \mathbf{R}$ be a continuously differentiable function, and let $S = \{\mathbf{v} \in \mathbf{R}^{n+1} : g(\mathbf{v}) = 0\}$. Let U be an open set in $\mathbf{R}^{n+1}$ containing S, and let $f: U \to \mathbf{R}$ be continuously differentiable. Suppose that* grad g *is nonzero everywhere on S. Then if $\mathbf{v}$ is a point of S where f has a maximum or minimum,* $(\text{grad } f)\mathbf{v}$ *and* $(\text{grad } g)\mathbf{v}$ *are proportional.*

We shall give a geometric interpretation of this result in the next section.

We note finally that g need not be defined on all of $\mathbf{R}^{n+1}$; the theorem also applies if g is defined on an open subset of $\mathbf{R}^{n+1}$.

Exercises

1. Maximize $f(x, y, z) = 6z - x^2 - y^2$ subject to the condition $x + y + z = 3$. (There are two ways to do this problem: by using Theorem 4.1 or by solving for z and reducing to a one-variable problem. Try it both ways.)
2. Find the points at which the function $f(x, y) = xy$ takes its maximum and minimum values subject to the condition $x^2 + y^2 = 4$.
3. Find the point on the unit sphere $x^2 + y^2 + z^2 = 1$ which is furthest from the point $(1, 2, -2)$.
4. Do Exercise 6 of the previous section by the methods of this section.
5. Do Exercise 7 of the previous section by the methods of this section.
6. Do Exercise 8 of the previous section by the methods of this section.
7. Do Exercise 9 of the previous section by the methods of this section.
8. Do Exercise 10 of the previous section by the methods of this section.
9. A rectangular box has three of its faces on the coordinate planes. Find the dimensions of the box of largest volume if the vertex of the box in the first octant lies on the ellipsoid

$$\frac{x^2}{4} + \frac{y^2}{9} + \frac{z^2}{25} = 1$$

10. A rectangular box has three of its faces on the coordinate planes. Find the dimensions of the box of largest volume if the vertex of the box in the first octant lies on the surface

$$x^3 + y^3 + z^3 = 1$$

11. Give the proof of Theorem 4.1 for the case where g has the form

$$g(x_1, x_2, x_3) = h(x_2, x_3) - x_1$$

*12. Prove Theorem 4.2. *Hint*: Assume first that

$$g(x_1, \ldots, x_{n+1}) = h(x_1, \ldots, x_n) - x_{n+1}$$

Thus the problem is to maximize

$$f(x_1, \ldots, x_n, h(x_1, \ldots, x_n))$$

Define $$H(x_1, \ldots, x_n) = (x_1, \ldots, x_n, h(x_1, \ldots, x_n))$$

Now find the critical points of $F = f \circ H$. These give equations like (4.2). Use the expression for g to get equations like (4.3). Add on the equation

$$\frac{\partial f}{\partial x_{n+1}} \frac{\partial g}{\partial x_{n+1}} = \frac{\partial f}{\partial x_{n+1}} \frac{\partial g}{\partial x_{n+1}}$$

and conclude that an equation like (4.4) holds. Now apply the implicit function theorem, as in the proof of Theorem 4.1, to complete the proof; see Example 3 in Section 9.5 for a discussion of the implicit function theorem.

5. The Method of Lagrange Multipliers

There is a standard way of arranging the calculations which arise when Theorem 4.1 or Theorem 4.2 is used, the method of Lagrange multipliers.

Let U be an open subset of $\mathbf{R}^{n+1}$, let $f, g\colon U \to \mathbf{R}$ be continuously differentiable, and let $S = g^{-1}(0)$, as in Theorem 4.2. At a critical point $\mathbf{v}$ of f on S, $(\operatorname{grad} f)(\mathbf{v}) = \lambda (\operatorname{grad} g)(\mathbf{v})$, where λ is some constant. Thus

(5.1) $$\frac{\partial f}{\partial x_j}(\mathbf{v}) = \lambda \frac{\partial g}{\partial x_j}(\mathbf{v}) \qquad 1 \le j \le n+1$$

Moreover,

(5.2) $$g(\mathbf{v}) = 0$$

since $\mathbf{v} \in S$.

Now consider the function F, where

$$F(x_1, \ldots, x_{n+1}, \lambda) = f(x_1, \ldots, x_{n+1}) - \lambda g(x_1, \ldots, x_{n+1})$$

Then $$\frac{\partial F}{\partial x_j} = \frac{\partial f}{\partial x_j} - \lambda \frac{\partial g}{\partial x_j} \qquad 1 \le j \le n+1$$

and $$\frac{\partial F}{\partial \lambda} = -g$$

Thus we may summarize (5.1) and (5.2) as

$$\frac{\partial F}{\partial x_1} = \frac{\partial F}{\partial x_2} = \cdots = \frac{\partial F}{\partial x_{n+1}} = \frac{\partial F}{\partial \lambda} = 0 \tag{5.3}$$

That is, we solve the problem of finding extrema of f subject to the constraint g by setting $F = f - \lambda g$ and setting all derivatives of F equal to 0. This precedure is called the *method of Lagrange multipliers*.

Here is an example. We determine the volume of the largest rectangular solid with faces parallel to the coordinate planes inscribed in the ellipsoid $x^2 + 4y^2 + 9z^2 = 1$. If the rectangle touches the ellipse at (x, y, z) with $x > 0$, $y > 0$, and $z > 0$, then its length, width, and height are $2x$, $2y$, and $2z$, respectively; thus we need to maximize $8xyz$ subject to the constraint $x^2 + 4y^2 + 9z^2 = 1$. It is easy to check that Theorem 4.1 applies.

Let $F(x, y, z, \lambda) = 8xyz - \lambda(x^2 + 4y^2 + 9z^2 - 1)$;

then
$$8yz - 2\lambda x = 0$$
$$8xz - 8\lambda y = 0$$
$$8xy - 18\lambda z = 0$$
and
$$x^2 + 4y^2 + 9z^2 - 1 = 0$$

Solving these equations may seem difficult, but it is not so hard if we exploit their symmetry. Multiply the first equation by $x/2$, the second by $y/2$, and the third by $z/2$; we then have

$$4xyz = \lambda x^2 = 4\lambda y^2 = 9\lambda z^2$$

Hence $x^2 = 4y^2 = 9z^2$, and the last equation shows that they are all $\frac{1}{3}$. Thus $x = 1/\sqrt{3}$, $y = 1/2\sqrt{3}$, $z = 1/3\sqrt{3}$, and the volume is

$$8xyz = \frac{4}{9\sqrt{3}}$$

(We know that there is a maximum, by Theorems 7.3.3 and 7.4.1.)

Notice that we never found λ. There is no need to, of course, since λ does not appear in the eventual answer. When possible, one uses λ to find symmetries in the equations, as above.

For a second example, we maximize $x^2 - xy + 2y^2$ subject to the constraint $x^2 + y^2 = 1$. Thus

$$f(x, y) = x^2 - xy + 2y^2$$
$$g(x, y) = x^2 + y^2 - 1$$

Now $(\text{grad } g)(x, y) = (2x, 2y)$, which is nonzero whenever $g(x, y) = 0$. Thus Theorem 4.1 applies.

Let $F(x, y, \lambda) = x^2 - xy + 2y^2 - \lambda(x^2 + y^2 - 1)$.

Then
$$\frac{\partial F}{\partial x} = 2x - y - 2\lambda x = 0$$
$$\frac{\partial F}{\partial y} = -x + 4y - 2\lambda y = 0$$
and
$$\frac{\partial F}{\partial \lambda} = -(x^2 + y^2 - 1) = 0$$

at a critical point of f subject to the constraint. From the first two equations, we have

$$2\lambda xy = 2xy - y^2$$
$$2\lambda xy = 4xy - x^2$$

Thus $2xy - y^2 = 4xy - x^2$, or equivalently,

$$x^2 - 2xy - y^2 = 0$$

Solving this equation (with the help of the quadratic formula), we obtain

$$x = (1 \pm \sqrt{2})y$$

In addition, we know that $x^2 + y^2 = 1$, so we have

$$[(1 \pm \sqrt{2})^2 + 1]y^2 = 1$$

or
$$y^2 = \frac{1}{4 \pm 2\sqrt{2}}$$

Hence
$$x^2 = (1 \pm \sqrt{2})^2 y^2 = \frac{3 \pm \sqrt{2}}{4 \pm 2\sqrt{2}}$$

and
$$xy = (1 \pm \sqrt{2})y^2 = \frac{1 \pm \sqrt{2}}{4 \pm 2\sqrt{2}}$$

It follows that

$$f(x, y) = \frac{3 \pm \sqrt{2} - (1 \pm \sqrt{2}) + 2}{4 \pm 2\sqrt{2}} = \frac{2}{2 \pm \sqrt{2}}$$

at the critical points. Since f must attain its maximum and minimum on S (again by Theorems 7.3.3 and 7.4.1), we know that f has a maximum on the set, and clearly it is $2/(2 - \sqrt{2}) = 2 + \sqrt{2}$; it is reached when $y^2 = 1/(4 - 2\sqrt{2}) = (2 + \sqrt{2})/4$. Similarly, the minimum value of f is $2/(2 + \sqrt{2}) = 2 - \sqrt{2}$.

The reader may wonder what Theorem 4.1 (or 4.2) means geometrically. Here is one interpretation. First of all, consider the relationship between grad $g(\mathbf{v})$ and S, the set where $g(\mathbf{v}) = 0$. The directional derivative of g at $\mathbf{v}$ in the direction $\mathbf{w}$ is equal to $(D_{\mathbf{v}} g)(\mathbf{w})$ (by Proposition 5.1, Chapter 5). However, it is easily seen that

$$(D_{\mathbf{v}} g)(\mathbf{w}) = \langle (\text{grad } g)\mathbf{v}, \mathbf{w} \rangle$$

(See Proposition 6.2, Chapter 5.) The function g is identically 0 on S; therefore the directional derivatives in directions along S should be 0. That is, (grad g) $(\mathbf{v})$ should be orthogonal to S at $\mathbf{v}$, or, equivalently, (grad g) $(\mathbf{v})$ *points in the direction of the normal to S at* $\mathbf{v}$. (See Figure 5.1.)

Suppose now that $\mathbf{v}$ is a local maximum (or minimum) for f on S. Then the same sort of argument as in Section 1 shows that the directional derivatives of f in directions along f should be 0. Therefore (grad f) $(\mathbf{v})$, too, must point in the direction of the normal to S at $\mathbf{v}$. This means that (grad f) $(\mathbf{v})$ is a multiple of (grad g) $(\mathbf{v})$, as Theorem 4.1 states.

Exercises

1–10. Do each of the exercises of the previous section by the method of Lagrange multipliers.

11. Find the minimum value of the function

$$f(x, y) = x^2 + 8xy - 5y^2$$

subject to the constraint $x^2 + y^2 = 25$.

12. Find the maximum value of the function

$$f(x, y, z) = x^2 + xy + y^2 + xz + z^2$$

subject to the constraint $x^2 + y^2 + z^2 = 4$.

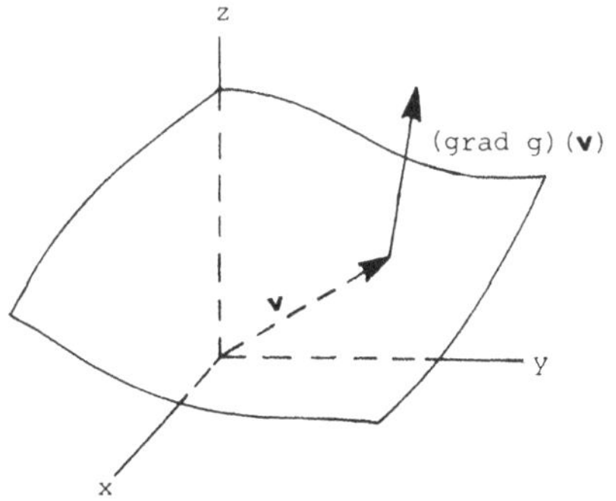

Figure 5.1 The gradient as a normal.

13. Maximize the function $g(x, y, z) = xyz$ subject to the constraint $x + y + z = 3$, $x > 0$, $y > 0$, $z > 0$.
14. A rectangular box has three of its faces on the coordinate planes. Find the dimension of the box of largest volume if the vertex of the box in the first octant lies on the ellipsoid

$$\frac{x^2}{a^2} + \frac{y^2}{b^2} + \frac{z^2}{c^2} = 9$$

15. A cylindrical can is to contain 1 liter of liquid (1 liter = 1000 cm^3). Find the dimensions which minimize the surface area if
    (a) The can has a top.
    (b) The can has no top.
16. A box has a volume of 4 liters. Find the dimensions of the box which minimize surface area if
    (a) The box has a top.
    (b) The box has no top.
    (c) The box has no top and is twice as long as it is wide.
17. A box with volume 3 liters is to be made with tin sides, a copper bottom, and no top. If sheet tin costs \$0.50 a square inch and sheet copper costs \$2.00 a square inch, find the dimensions of the least expensive box.
18. French wine sells at wholesale for \$5 a bottle, German wine for \$4 a bottle, and California wine for \$3 a bottle. Assume that the profit on x bottles of French wine is $2x - x^2/120$, on y bottles of German wine $3y - y^2/60$, and on z bottles of California wine is z. How much of each wine should a merchant buy to maximize his profit if he has \$600?
19. The load that a rectangular beam of given length can support is proportional to the width and the square of the thickness. Suppose that a cylindrical log with a circular cross section of radius r is to be pared down until it has a rectangular cross section. What dimensions make it strongest? (See Figure 5.2.)
20. Maximize $(x_1 x_2 \cdots x_n)^2$ subject to $\sum_{j=1}^{n} x_j^2 = 1$, and show that the result implies that the geometric mean of n positive numbers is no greater than their arithmetic mean.
21. If x, y, z, u, v, w are positive and $x/u + v/v + z/w = 1$, show that $27xyz \leq uvw$.

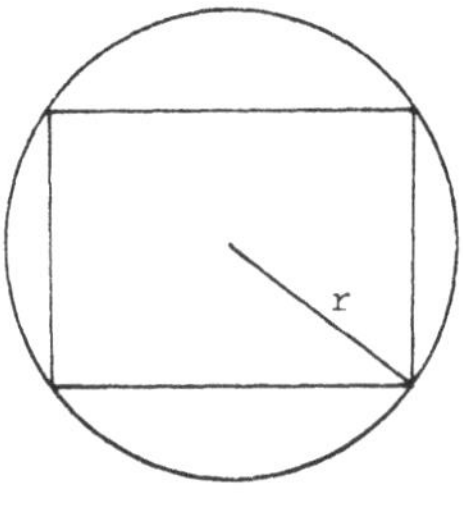

Figure 5.2 Cutting a beam from a log.

22. Let p and q be positive.
(a) Find the minimum of the function

$$\frac{x^p}{p} + \frac{y^q}{q}$$

subject to the constraint $xy = 1$.
*(b) Use (a) to prove that, if $1/p + 1/q = 1$ and $x > 0$, $y > 0$, then

$$xy \leq \frac{x^p}{p} + \frac{y^q}{q}$$

6. Constrained Maxima and Minima: II

In the previous section, we investigated extrema of functions f subject to the condition $g(\mathbf{v}) = 0$, where $g : \mathbf{R}^{n+1} \to \mathbf{R}^1$; that is, subject to one constraint. In this section, we extend this discussion to constraints $G : \mathbf{R}^{n+k} \to \mathbf{R}^k$. For simplicity, we shall deal with the case $G : \mathbf{R}^3 \to \mathbf{R}^2$, but the reasoning holds more generally.

We assume, then, that $G : \mathbf{R}^3 \to \mathbf{R}^2$ is a continuously differentiable function, and we let $S = \{\mathbf{v} \in \mathbf{R}^3 : G(\mathbf{v}) = \mathbf{0}\}$. We wish to maximize (or minimize) the continuously differentiable function f on S. Ordinarily S will be a curve in $\mathbf{R}^3$. For instance, if

$$G(x_1, x_2, x_3) = (x_2 - x_1^2, x_3 - x_1^3)$$

then

$$S = \{(x_1, x_1^2, x_1^3) : x_1 \in \mathbf{R}\}$$

Let us write $G(\mathbf{v}) = (g_1(\mathbf{v}), g_2(\mathbf{v}))$; then S is the set of points satisfying $g_1(\mathbf{v}) = g_2(\mathbf{v}) = 0$. We can use the geometric explanation at the end of the previous section to see what the procedure should be. At a typical point $\mathbf{v} \in S$, there is a plane of vectors normal to S. By the same reasoning as in Section 5, $(\text{grad } g_1)(\mathbf{v})$ should be normal to S (g_1 is identically 0 on S, and therefore the directional derivative of g_1 in the direction tangent to S should be 0). Similarly, $(\text{grad } g_2)(\mathbf{v})$ should be normal to S. If $(\text{grad } g_1)(\mathbf{v})$ and $(\text{grad } g_2)(\mathbf{v})$ are linearly independent, therefore, they should span the plane normal to S. At an extremum of f on S, the directional derivative of f along the tangent to S will be 0. Therefore $(\text{grad } f)(\mathbf{v})$ will be orthogonal to S, and therefore $(\text{grad } f)(\mathbf{v})$ should be a linear combination of $(\text{grad } g_1)(\mathbf{v})$ and $(\text{grad } g_2)(\mathbf{v})$. The next theorem makes this result precise. We state it for arbitrary n and k.

Theorem 6.1 *Let $G : \mathbf{R}^{n+k} \to \mathbf{R}^k$ be continuously differentiable, and let g_1, $\ldots$, g_k be the coordinate functions of G. Assume that* $(\text{grad } g_1)(\mathbf{v})$, $\ldots$, (grad g_k) ($\mathbf{v}$) *are linearly independent whenever* $G(\mathbf{v}) = \mathbf{0}$. *Let* $S = \{\mathbf{v} : G(\mathbf{v}) = \mathbf{0}\}$, *and let* $f\colon U \to \mathbf{R}$ *be a continuously differentiable function defined on the open set U containing S. If* $\mathbf{v}_0$ *is a point of S where f (restricted to S) has a maximum or minimum, then* (grad f) ($\mathbf{v}_0$) *is a linear combination of* (grad g_1) ($\mathbf{v}_0$), . . . , (grad g_k) ($\mathbf{v}_0$).

We defer the proof of Theorem 6.1 briefly, and first illustrate how to apply it. Again, the most efficient method is to use Lagrange multipliers. Suppose, for instance, that $G : \mathbf{R}^3 \to \mathbf{R}^2$. At an extreme point $\mathbf{v}_0$, we know that there are real numbers λ_1 and λ_2 such that

$$(\text{grad } f)(\mathbf{v}_0) = \lambda_1 (\text{grad } g_1)(\mathbf{v}_0) + \lambda_2 (\text{grad } g_2)(\mathbf{v}_0) \tag{6.1}$$

Furthermore,

$$g_1(\mathbf{v}_0) = g_2(\mathbf{v}_0) = 0 \tag{6.2}$$

Set

$$F(\mathbf{v}, \lambda_1, \lambda_2) = f(\mathbf{v}) - \lambda_1 g_1(\mathbf{v}) - \lambda_2 g_2(\mathbf{v})$$

Then

$$\frac{\partial F}{\partial x_j} = \frac{\partial f}{\partial x_j} - \lambda_1 \frac{\partial g_1}{\partial x_j} - \lambda_2 \frac{\partial g_2}{\partial x_j} \qquad j = 1, 2, 3$$

and

$$\frac{\partial F}{\partial \lambda_1} = -g_1(\mathbf{v}) \qquad \frac{\partial F}{\partial \lambda_2} = -g_2(\mathbf{v})$$

Hence (6.1) and (6.2) become

$$\frac{\partial F}{\partial x_1} = \frac{\partial F}{\partial x_2} = \frac{\partial F}{\partial x_3} = \frac{\partial F}{\partial \lambda_1} = \frac{\partial F}{\partial \lambda_2} = 0 \tag{6.3}$$

This gives the conditions for an extreme point. In practice, solving the resulting equations is tedious at least.

Here is a relatively simple example of a problem with two constraints. We wish to find the highest point on the curve given by the intersection of $x^2 + y^2 + z^2 = 1$ with $4z = 3xy$. That is, we wish to maximize

$$f(x, y, z) = z$$

with the constraints

$$g_1(x, y, z) = x^2 + y^2 + z^2 - 1 = 0$$

and

$$g_2(x, y, z) = 3xy - 4z = 0$$

It is not hard to check that the conditions of Theorem 6.1 are met. We set

$$F(x, y, z, \lambda_1, \lambda_2) = z - \lambda_1 (x^2 + y^2 + z^2 - 1) - \lambda_2 (3xy - 4z)$$

The condition for an extreme point is that all derivatives of F are 0. Thus

$$-2\lambda_1 x - 3\lambda_2 y = 0$$

$$-2\lambda_1 y - 3\lambda_2 x = 0$$

$$1 - 2\lambda_1 z - 4\lambda_2 = 0$$

$$-(x^2 + y^2 + z^2 - 1) = 0$$

and

$$-(3xy - 4z) = 0$$

The first two equations say that

$$\frac{2\lambda_1}{3\lambda_2} = \frac{-y}{x} = -\frac{x}{y}$$

Thus

$$x^2 = y^2 \qquad y = \pm x$$

The last equation then asserts that

$$z = \pm\frac{3x^2}{4}$$

and the fourth equation gives

$$2x^2 + \frac{9x^4}{16} - 1 = 0$$

or

$$9x^4 + 32x^2 - 16 = 0$$

This is a quadratic equation in x^2. Solving, we find that

$$x^2 = -4 \quad \text{or} \quad \frac{4}{9}$$

The first root does not give a solution for x; the second gives

$$x = \pm\frac{2}{3}$$

Then, since $y = \pm x$ and $z = 3xy/4$, we find that the critical points are

$$\left(\frac{2}{3}, \frac{2}{3}, \frac{1}{3}\right)\left(\frac{2}{3}, \frac{-2}{3}, \frac{-1}{3}\right)\left(\frac{-2}{3}, \frac{-2}{3}, \frac{1}{3}\right)\left(\frac{-2}{3}, \frac{2}{3}, \frac{-1}{3}\right)$$

Since the set S satisfying the constraints is a closed subset of the unit sphere, S is compact (by Theorem 4.1, Chapter 7) and f has a maximum on S (by Theorem 3.3, Chapter 7). It is now clear that the maximum is $\frac{1}{3}$, attained at $(\frac{2}{3}, \frac{2}{3}, \frac{1}{3})$ and $(\frac{-2}{3}, \frac{-2}{3}, \frac{1}{3})$.

We mentioned at the beginning of the previous section that not all boundaries of sets S are given by constraints like $G(\mathbf{v}) = \mathbf{0}$, where G is continuously differentiable. For instance, if S is the unit cube in $\mathbf{R}^3$,

$$S = \{(x, y, z) \in \mathbf{R}^3 : 0 \leq x, y, z \leq 1\}$$

then ∂S consists of all $(x, y, z) \in S$ such that at least one of x, y, z is 0 or 1. ∂S cannot be described in the way needed to apply Theorem 4.1 or 6.1. (For one thing, ∂S has corners, and is therefore not smooth enough.) However, S is made up of pieces, e.g., the square

$$\{(x, y, z) \in \mathbf{R}^3 : x = 0, 0 < y, z < 1\}$$

or the line

$$\{(x, y, z) \in \mathbf{R}^3 : x = y = 1, 0 \leq z \leq 1\}$$

for each of which Theorem 4.2 or Theorem 6.1 is applicable. Thus we can apply our theorems by cutting ∂S into enough smaller pieces. In practice, most problems involving constrained maxima and minima can be solved in this way.

We conclude this section by proving Theorem 6.1 in the special case where $G : \mathbf{R}^3 \to \mathbf{R}^2$; the proof we give generalizes, however. We assume first that $G = (g_1, g_2)$, where

$$g_1(x_1, x_2, x_3) = h_1(x_1) - x_2$$

and

$$g_2(x_1, x_2, x_3) = h_2(x_1) - x_3$$

Then S is the set of points of the form

$$(x, h_1(x_1), h_2(x_1))$$

Define $H : \mathbf{R}^1 \to \mathbf{R}^3$ by

$$H(x) = (x, h_1(x), h_2(x))$$

Then, as in the proof of Theorem 4.1, finding the extreme points of f subject to the conditions $G(\mathbf{v}) = \mathbf{0}$ is equivalent to finding the extreme points of $f \circ H$.

We apply the chain rule: at a critical point of $f \circ H$,

$$\text{(6.4)} \qquad \frac{d}{dx}(f \circ H) = \frac{\partial f}{\partial x_1} + \frac{\partial f}{\partial x_2}\frac{\partial h_1}{\partial x_1} + \frac{\partial f}{\partial x_3}\frac{d h_2}{d x_1} = 0$$

We add two more equations,

$$\text{(6.5)} \qquad \begin{aligned} \frac{\partial f}{\partial x_2} - \frac{\partial f}{\partial x_2} + 0 &= 0 \\ \frac{\partial f}{\partial x_3} + 0 - \frac{\partial f}{\partial x_3} &= 0 \end{aligned}$$

Since

$$\frac{\partial g_1}{\partial x_1} = \frac{d h_1}{d x_1} \qquad \frac{\partial g_1}{\partial x_2} = -1 \qquad \frac{\partial g_1}{\partial x_3} = 0$$

and

$$\frac{\partial g_2}{\partial x_1} = \frac{d h_2}{d x_1} \qquad \frac{\partial g_2}{\partial x_2} = 0 \qquad \frac{\partial g_2}{\partial x_3} = -1$$

we can rewrite (6.4) and (6.5) together as

$$(\text{grad } f)(\mathbf{v}) + \frac{\partial f}{\partial x_2}(\text{grad } g_1)(\mathbf{v}) + \frac{\partial f}{\partial x_3}(\text{grad } g_2)(\mathbf{v}) = 0$$

if $\mathbf{v}$ is a critical point of f. That is, at a critical point $\mathbf{v}$, $(\text{grad } f)(\mathbf{v})$ is a linear combination of $(\text{grad } g_1)(\mathbf{v})$ and $(\text{grad } g_2)(\mathbf{v})$.

The theorem will follow in general if we can show that the constraint $G(\mathbf{v}) = \mathbf{0}$ can be regarded as defining two of the coordinates of $\mathbf{v}$ as functions of the third. We shall see in the next chapter that the implicit function theorem permits this. Thus the proof is complete except for this one step, which we take care of in the next chapter.

Exercises

1. Find the maximum and minimum of $f(x, y, z) = z$ subject to the constraints $x^2 + y^2 + z^2 = 1$ and $x + y + z = -1$.
2. Find the local extrema of $f(x, y, z) = xyz$ subject to the constraints $x + y + z = 3$ and $x - 2y + 2z = 1$.
3. Find the minimum value of $xy + yz$ given that $x^2 + y^2 = 2$ and $x^2 + z^2 = 2$.
4. Find the point on the line $x = y = 2z$ closest to $(3, 4, 5)$.
5. Find the point on the line $x = y = z$ closest to the line $x = 0$, $y + z = 4$.
6. Find the point on the curve in $\mathbf{R}^3$ given by

$$x^2 + y^2 = 1 \qquad \text{and} \qquad z^2 - x^2 = 1$$

nearest the origin.
7. Find the distance between the closest points on the curves $y = x^2$ and $x^2 - y^2 = \frac{27}{16}$. (*Hint*: Regard the square of the distance as a function of four variables.)
8. What point on the intersection of the unit sphere with $x + y + z = 1$ is closest to $(2, 3, \frac{1}{3})$?
9. Maximize $x_1y_1 + x_2y_2$, given that $x_1^2 + x_2^2 = y_1^2 + y_2^2 = 1$. What does the answer mean?

*10. Give the proof of Theorem 6.1 in the general case, assuming the implicit function theorem, Theorem 9.5.1. *Hint*: The implicit function theorem says, in effect, that we may assume that the constraints are given by functions of the form

$$\begin{aligned} g_1(\mathbf{v}) &= h_1(x_1, \ldots, x_n) - x_{n+1} \\ g_2(\mathbf{v}) &= h_2(x_1, \ldots, x_n) - x_{n+2} \qquad (6.6) \\ &\cdots\cdots\cdots \\ g_k(\mathbf{v}) &= h_k(x_1, \ldots, x_n) - x_{n+k} \end{aligned}$$

Define

$$H(x_1, \ldots, x_n) = (x_1, \ldots, x_n, h_1(x_1, \ldots, x_n), \ldots, h_k(x_1, \ldots, x_n))$$

Show that we need to maximize (or minimize)

$$F(x_1, \ldots, x_n) = (f \circ H)(x_1, \ldots, x_n)$$

Find the condition for grad F to be 0. Then show, as in the proof of Theorems 4.1 and 4.2, and the special case of Theorem 6.1, that if can be rewritten as

$$\operatorname{grad} f + \frac{\partial f}{\partial x_{n+1}} \operatorname{grad} g_1 + \cdots + \frac{\partial f}{\partial x_{n+k}} \operatorname{grad} g_k = 0$$

Show that this proves the theorem for the constraints in the form (6.6). (As mentioned above, the implicit function theorem then gives the result in general.)

*11. Try to maximize $f(x, y, z) = xyz$ subject to the constraints $xy = 1$, $x^2 + z^2 = 1$. What goes wrong and why?

7. The Proof of Proposition 2.3

The proof of Proposition 2.3 depends on two lemmas. The first gives a criterion for a matrix to be positive (and negative) definite, and the second gives an inequality concerning $\|A\|$.

Lemma 7.1 (a) *The symmetric matrix A is positive definite if there is an $\varepsilon > 0$ such that $Q_A(\mathbf{v}) > 0$ for all $\mathbf{v}$ with $0 < \|\mathbf{v}\| < \varepsilon$.*

(b) *The symmetric matrix A is positive definite if $Q_A(\mathbf{v}) > 0$ for all $\mathbf{v}$ with $\|\mathbf{v}\| = 1$.*

Similar statements hold for negative definite and indefinite A. The proofs are easy; we shall do (a), leaving (b) and the other statements as exercises.

Proof of (a). Let $\mathbf{v} \in \mathbf{R}^n$ be a nonzero vector. Then, if

$$\mathbf{w} = \frac{\varepsilon \mathbf{v}}{2\|\mathbf{v}\|}$$

$\|\mathbf{w}\| < \varepsilon$ so that $Q_A(\mathbf{w}) > 0$. However,

$$\begin{aligned} Q_A(\mathbf{v}) &= \langle T_A \mathbf{v}, \mathbf{v} \rangle \\ &= \left\langle T_A\left(\frac{2\|\mathbf{v}\|}{\varepsilon}\mathbf{w}\right), \frac{2\|\mathbf{v}\|}{\varepsilon}\mathbf{w} \right\rangle \\ &= \frac{4\|\mathbf{v}\|^2}{\varepsilon^2}\langle T_A \mathbf{w}, \mathbf{w} \rangle \\ &= \frac{4\|\mathbf{v}\|^2}{\varepsilon^2} Q_A(\mathbf{w}) > 0 \end{aligned}$$

where $T_A : \mathbf{R}^n \to \mathbf{R}^n$ is the linear transformation defined by A. Thus A is positive definite.

Lemma 7.2 *If $T = T_A : \mathbf{R}^n \to \mathbf{R}^n$ is the linear transformation with matrix $A = (a_{ij})$, then*

$$\| T_A(\mathbf{v}) \| \le \| A \|$$

for any unit vector $\mathbf{v} \in \mathbf{R}^n$.

Proof. Let $\mathbf{v} = \sum_{i=1}^n b_i \mathbf{e}_i$ be a unit vector in $\mathbf{R}^n$, where $\mathbf{e}_1, \ldots, \mathbf{e}_n$ is the standard basis for $\mathbf{R}^n$. Then

$$\begin{aligned} \| T\mathbf{v} \|^2 &= \langle T\mathbf{v}, T\mathbf{v} \rangle \\ &= \sum_{i=1}^n \sum_{j=1}^n b_i b_j \langle T\mathbf{e}_i, T\mathbf{e}_j \rangle \\ &\le \sum_{i=1}^n \sum_{j=1}^n b_i b_j \| T\mathbf{e}_i \| \cdot \| T\mathbf{e}_j \| \end{aligned}$$

by the Schwarz inequality (Theorem 2.1, Chapter 5). This last expression can be rewritten as

$$\begin{aligned} \left(\sum_{j=1}^n b_j \| T\mathbf{e}_j \| \right)^2 &= \left\langle \mathbf{v}, \sum_{j=1}^n \| T\mathbf{e}_j \| \mathbf{e}_j \right\rangle^2 \\ &\le \| \mathbf{v} \|^2 \left(\sum_{j=1}^n \| T\mathbf{e}_j \|^2 \right)^2 \\ &= \left(\sum_{j=1}^n \| T\mathbf{e}_j \|^2 \right)^2 \end{aligned}$$

again by the Schwarz inequality (recall that $\mathbf{v}$ is a unit vector). Thus

$$\| T\mathbf{v} \| \le \sum_{j=1}^n \| T\mathbf{e}_j \|^2$$

Since $T\mathbf{e}_j = \sum_{i=1}^n a_{ij} \mathbf{e}_i$,

$$\| T\mathbf{e}_j \|^2 = \sum_{i=1}^n (a_{ij})^2$$

and
$$\sum_{j=1}^n \| T\mathbf{e}_j \|^2 = \sum_{i=1}^n \sum_{j=1}^n (a_{ij})^2$$

Thus $\| T\mathbf{v} \|^2 \le \| T \|^2$, which proves the lemma.

We now use these lemmas to prove Proposition 2.3.

Statements (a) and (b) have essentially identical proofs; we prove only (a). Let $C = \{\mathbf{v} \in \mathbf{R}^n : \|\mathbf{v}\| = 1\}$. Then C is compact, by Theorem 4.1 of Chapter 7, and the function $Q_A : \mathbf{R}^n \to \mathbf{R}$ is continuous; by Theorem 3.3 of Chapter 7, Q_A has a minimum value ε on C. We know that $\varepsilon > 0$, since A is positive definite.

Now suppose that B is a symmetric matrix with $\| B - A \| < \varepsilon$, and let $\mathbf{v}$ be any unit vector. Then

$$\begin{aligned} |Q_B(\mathbf{v}) - Q_A(\mathbf{v})| &= |\langle B\mathbf{v}, \mathbf{v}\rangle - \langle A\mathbf{v}, \mathbf{v}\rangle| \\ &= |\langle (B - A)\mathbf{v}, \mathbf{v}\rangle| \\ &\leq \|\mathbf{v}\| \, \|(B - A)\mathbf{v}\| \qquad \text{(by the Schwarz inequality)} \\ &= \|(B - A)\mathbf{v}\| \\ &\leq \| B - A \| \qquad \text{(by Lemma 7.2)} \\ &< \varepsilon \end{aligned}$$

Thus $Q_B(\mathbf{v}) - Q_A(\mathbf{v}) > -\varepsilon$, and $Q_B(\mathbf{v}) > Q_A(\mathbf{v}) - \varepsilon > 0$ for all $\mathbf{v} \in C$. By Lemma 7.1, B is positive definite.

To prove (c), pick ε such that $Q_A(\mathbf{v}_1) \geq \varepsilon$, $Q_A(\mathbf{v}_2) \leq -\varepsilon$. The same calculation as in the preceding proof shows that

$$|Q_A(\mathbf{v}_1) - Q_B(\mathbf{v}_1)| < \varepsilon \qquad |Q_A(\mathbf{v}_2) - Q_B(\mathbf{v}_2)| < \varepsilon$$

which means that

$$Q_B(\mathbf{v}_1) - Q_B(\mathbf{v}_1) > -\varepsilon \qquad Q_B(\mathbf{v}_2) - Q_A(\mathbf{v}_2) < \varepsilon$$

Therefore

$$Q_B(\mathbf{v}_1) > 0 \qquad Q_B(\mathbf{v}_2) < 0$$

as claimed.

Note. For semidefinite transformations (positive or negative), the corresponding result is false. For instance,

$$A = \begin{bmatrix} 1 & 0 \\ 0 & 0 \end{bmatrix}$$

is positive semidefinite, but if ε is positive, then

$$A_\varepsilon = \begin{bmatrix} 1 & 0 \\ 0 & \varepsilon \end{bmatrix}$$

is positive definite, and if ε is negative, then A_ε is indefinite. The most that can be proved is that if A is nonzero and positive semidefinite, then no sufficiently close B is negative definite. (See Exercise 3.)

Exercises

1. Complete the proof of Lemma 7.1.
2. State and prove the analogue of Lemma 7.1 for indefinite symmetric matrices.
3. Prove that if A is nonzero and positive semidefinite, then there is an $\varepsilon > 0$ such that whenever B is negative definite, $\| A - B \| \geq \varepsilon$.
4. Use Exercise 3 to show that if $H(f)(\mathbf{v})$ is positive semidefinite and $H(f)(\mathbf{v}) \neq 0$, then $\mathbf{v}$ is not a local maximum for f. State and prove a similar result about nonzero negative semidefinite transformations.

*5. In Exercise 9 of Section 7.2, we defined a different norm on linear transformations $T : V \to W$, where V and W are normed vector spaces:

$$\| T \|_0 = \operatorname*{lub}_{\|\mathbf{v}\| = 1} \| T\mathbf{v} \|$$

(The subscript 0 is to distinguish this norm from the one used in the rest of this section.) If A is an $n \times n$ matrix, it defines a linear transformation $T_A : \mathbf{R}^n \to \mathbf{R}^n$ (one uses the standard basis); define

$$\| A \|_0 = \| T_A \|_0$$

(where $\mathbf{R}^n$ has the standard norm). Prove that Lemma 7.2 and Proposition 2.3 still hold if we use the $\| \quad \|_0$ norm on the $n \times n$ matrices.

9

The Inverse and Implicit Function Theorems

We have seen (Chapter 4) that the derivative of a function at a point provides a good linear approximation to the function near the point. This fact suggests that properties of the derivative at a point should be reflected in properties of the function near the point. The inverse function theorem and implicit function theorem are both instances of this phenomenon. The inverse function theorem states that, if the derivative of a function $f: \mathbf{R}^n \to \mathbf{R}^n$ is invertible at $\mathbf{v}_0$, then so is f, at least near $\mathbf{v}_0$. The implicit function theorem gives conditions (on the Jacobian of a function) under which a collection of equations can be solved for some of the variables in terms of the rest. We note that the implicit function theorem is exactly what is needed to complete the discussion of constrained maxima and minima of the previous chapter.

Both these theorems are quite important in mathematics, and the reader should, at the least, learn what they say. The proofs, however, are quite difficult, and the reader may wish to skip them on first reading.

1. Inverse Functions

Before we can state an inverse function theorem, we need to know what an inverse function is. Let U and V be open subsets of $\mathbf{R}^n$, and let $f: U \to V$ be a function. We say that a function $g: V \to U$ is an *inverse* for f if $(g \circ f)(\mathbf{v}) = \mathbf{v}$ for all $\mathbf{v} \in U$ and $(f \circ g)(\mathbf{w}) = \mathbf{w}$ for all $\mathbf{w} \in V$. Clearly g is an inverse for f if and only if f is an inverse for g. It is also immediate that $f: U \to V$ has an inverse if and only if f is bijective (that is, both $1 - 1$ and onto); see Exercise 1.

For example, the function $f\colon (0, 1) \to (0, 1)$ given by $f(x) = x^2$ has an inverse $g(y) = \sqrt{y}$ since

$$(g \circ f)(x) = g(x^2) = \sqrt{x^2} = x$$

and

$$(f \circ g)(y) = f(\sqrt{y}) = (\sqrt{y})^2 = y$$

However, the function $h: (-1, 1) \to (0, 1)$ defined by $h(x) = x^2$ does not have an inverse since h is not injective: if k is an inverse, then $k(\frac{1}{4}) = k(h(\frac{1}{2})) = \frac{1}{2}$ and $k(\frac{1}{4}) = k(h(-\frac{1}{2})) = -\frac{1}{2}$. Thus we have two values for $k(\frac{1}{4})$, which is impossible.

We begin with an easy result.

Proposition 1.1 *Let $g_1, g_2\colon V \to U$ be inverses for the function $f\colon U \to V$. Then $g_1 = g_2$.*

Note If $f\colon U \to V$ has an inverse, the inverse is usually denoted by $f^{-1}\colon V \to U$.

Proof of Proposition 1.1. We use the definition of the inverse: For any $\mathbf{v} \in V$, we have

$$\begin{aligned} g_1(\mathbf{v}) &= (g_1 \circ (f \circ g_2))(\mathbf{v}) \\ &= ((g_1 \circ f) \circ g_2)(\mathbf{v}) \\ &= g_2(\mathbf{v}) \end{aligned}$$

Consider now the linear transformation $T\colon \mathbf{R} \to \mathbf{R}$ defined by $T(x) = ax$. It is immediate that T has an inverse if and only if $a \neq 0$, in which case T^{-1} is given by $T^{-1}(y) = (1/a)y$. Thus, we see that if $T: \mathbf{R} \to \mathbf{R}$ is a linear transformation which has an inverse, then this inverse is a linear transformation. The next result generalizes this observation.

Proposition 1.2 *Let $T: \mathbf{R}^n \to \mathbf{R}^n$ be a linear transformation which has an inverse function $T^{-1}: \mathbf{R}^n \to \mathbf{R}^n$. Then this inverse function is a linear transformation.*

Proof. Let $\mathbf{w}_1, \mathbf{w}_2 \in \mathbf{R}^n$ be vectors with $T^{-1}(\mathbf{w}_1) = \mathbf{v}_1$ and $T^{-1}(\mathbf{w}_2) = \mathbf{v}_2$. We must show that

$$T^{-1}(a\mathbf{w}_1 + b\mathbf{w}_2) = a\mathbf{v}_1 + b\mathbf{v}_2$$

for any $a, b \in \mathbf{R}$. To do this, we first note that

$$T(\mathbf{v}_1) = T(T^{-1}(\mathbf{w}_1)) = \mathbf{w}_1$$

$$T(\mathbf{v}_2) = T(T^{-1}(\mathbf{w}_2)) = \mathbf{w}_2$$

by the definition of inverse function. Therefore,

$$\begin{aligned} T^{-1}(a\mathbf{w}_1 + b\mathbf{w}_2) &= T^{-1}(aT(\mathbf{v}_1) + bT(\mathbf{v}_2)) \\ &= T^{-1}(T(a\mathbf{v}_1 + b\mathbf{v}_2)) \text{ (since } T \text{ is a linear transformation)} \\ &= a\mathbf{v}_1 + b\mathbf{v}_2 \end{aligned}$$

since T^{-1} is the inverse of T. Thus T^{-1} is a linear transformation.

A linear transformation which has an inverse is said to be *invertible* and is called an *isomorphism*. If $I_n : \mathbf{R}^n \to \mathbf{R}^n$ is the identity transformation, then it follows from Proposition 1.2 and the definition of inverse function that *the linear transformation $T : \mathbf{R}^n \to \mathbf{R}^n$ is invertible if and only if there is a linear transformation $S : \mathbf{R}^n \to \mathbf{R}^n$ such that*

$$T \circ S = S \circ T = I_n$$

We know from Section 2.6 that a linear transformation is determined by its matrix. We now give a condition for a linear transformation to be invertible in terms of its matrix.

Let I_n denote the $n \times n$ identity matrix,*

$$I_n = \begin{bmatrix} 1 & 0 & 0 & \cdots & 0 & 0 \\ 0 & 1 & 0 & \cdots & 0 & 0 \\ \vdots & & & & & \vdots \\ 0 & 0 & 0 & \cdots 0 & & 1 \end{bmatrix}$$

An $n \times n$ matrix A is *invertible* if there is an $n \times n$ matrix B such that

$$AB = BA = I_n$$

(See Exercises 7 to 9 in Section 2.8 and 11 to 14 in Section 2.9.) The matrix B is called the *inverse* of the matrix A and usually denoted by A^{-1}.

Proposition 1.3 *Let A be an invertible $n \times n$ matrix and let B and C be $n \times n$ matrices which are inverses for A. Then $B = C$.*

The proof of this proposition is similar to the proof of Proposition 1.1. We leave it as an exercise for the reader.

We can now state the condition on the invertibility of a linear transformation in terms of its matrix.

Proposition 1.4 *Let $T : \mathbf{R}^n \to \mathbf{R}^n$ be a linear transformation with $n \times n$ matrix A. Then T is invertible if and only if A is invertible.*

*We use the same symbol for the identity matrix and the identity linear transformation. This should cause no confusion.

Proof. Suppose first of all that $S : \mathbf{R}^n \to \mathbf{R}^n$ is an inverse for T, so that $S \circ T = T \circ S = I_n$ and let B be the matrix of S. Then according to Proposition 9.1 of Chapter 2, the matrix of $S \circ T$ is BA and the matrix of $T \circ S$ is AB. Since the matrix of the identity linear transformation is the identity matrix, it follows that $BA = AB = I_n$ and A is invertible.

Conversely, suppose A is an invertible matrix and let B be an $n \times n$ matrix with $AB = BA = I_n$. Let $S : \mathbf{R}^n \to \mathbf{R}^n$ be the linear transformation whose matrix is B. Then, just as above, it follows that $S \circ T = T \circ S = I_n$ so that T is an invertible linear transformation.

Let us look at two examples of Proposition 1.4.

1. Let $T: \mathbf{R}^2 \to \mathbf{R}^2$ be the linear transformation defined by $T(x, y) = (x + y, y)$. The matrix of T is the 2×2 matrix A,

$$A = \begin{bmatrix} 1 & 0 \\ 1 & 1 \end{bmatrix}$$

It is easily seen that A is invertible with A^{-1} given by

$$A^{-1} = \begin{bmatrix} 1 & 0 \\ -1 & 1 \end{bmatrix}$$

so that T is invertible with $T^{-1}: \mathbf{R}^2 \to \mathbf{R}^2$ defined by $T^{-1}(x, y) = (x - y, y)$.

2. Let $T : \mathbf{R}^2 \to \mathbf{R}^2$ be defined by $T(x, y) = (2x + y, x + y)$. Then the matrix of T is the 2×2 matrix

$$\begin{bmatrix} 2 & 1 \\ 1 & 1 \end{bmatrix}$$

If A is to be invertible, we need to find a 2×2 matrix

$$B = \begin{bmatrix} a & b \\ c & d \end{bmatrix}$$

with $AB = BA = I_2$. In particular, we need

$$\begin{bmatrix} 2 & 1 \\ 1 & 1 \end{bmatrix}\begin{bmatrix} a & b \\ c & d \end{bmatrix} = \begin{bmatrix} 2a + c & 2b + d \\ a + c & b + d \end{bmatrix} = \begin{bmatrix} 1 & 0 \\ 0 & 1 \end{bmatrix}$$

or, equivalently,

$$2a + c = 1$$

$$2b + d = 0$$

$$a + c = 0$$

$$b + d = 1$$

The second and third equations tell us that $d = -2b$ and $c = -a$. Feeding this into the first and fourth equation, we have

$$2a - a = 1 \qquad b - 2b = 1$$

or $a = 1$ and $b = -1$. Thus, $d = 2$ and $c = -1$, so that

$$B = \begin{bmatrix} 1 & -1 \\ -1 & 2 \end{bmatrix}$$

(One can check that both $AB = I_2$ and $BA = I_2$.) Therefore T is invertible with $T^{-1}: \mathbf{R}^2 \to \mathbf{R}^2$ given by

$$T^{-1}(x, y) = (x - y, \; -x + 2y)$$

The reader may wonder why we considered the inverse for linear transformations $T: \mathbf{R}^n \to \mathbf{R}^p$ only when $n = p$. It is a standard result in linear algebra that if the linear transformation $T: \mathbf{R}^n \to \mathbf{R}^p$ has an inverse, then $n = p$. In fact, this result is true more generally; see the remarks at the end of the next section.

In the next section, we relate the invertibility of a differentiable function f to the invertibility of the linear transformation $D_{\mathbf{v}} f$.

Exercises

1. Prove that a function $f: U \to V$ has an inverse if and only if f is bijective. (That is, f is $1-1$ and onto.)
2. Which of the following functions have inverses? Determine the inverse function when it exists.
   (a) $f: (2, 6) \to (10, 42)$, $f(x) = x^2 + 6$.
   (b) $f: (-3, -1) \to (7, 15)$, $f(x) = x^2 + 6$.
   (c) $f: (-3, 1) \to (7, 15)$, $f(x) = x^2 + 6$.
   (d) $f: (0, 3) \to (0, 81)$, $f(x) = x^4$.
   (e) $f: (-7, 10) \to (-344, 999)$, $f(x) = x^3 - 1$.
   (f) $f: \mathbf{R} \to \mathbf{R}$, $f(x) = x^m$, m odd.
   (g) $f: (0, \pi) \to (-1, 1)$, $f(x) = \cos x$.
   (h) $f: (-\pi/2, \pi/2) \to \mathbf{R}$, $f(x) = \tan x$.
   (i) $f: (0, \pi) \to (0, 1]$, $f(x) = \sin x$.
3. Determine the inverse linear transformation and its matrix for each of the following linear transformations.
   (a) $T: \mathbf{R} \to \mathbf{R}$, $T(x) = 3x$.
   (b) $T: \mathbf{R}^2 \to \mathbf{R}^2$, $T(x, y) = (y, x)$.
   (c) $T: \mathbf{R}^2 \to \mathbf{R}^2$, $T(x, y) = (x + 2y, y)$.
   (d) $T: \mathbf{R}^2 \to \mathbf{R}^2$, $T(x, y) = (x + ay, y)$, $a \in \mathbf{R}$.
   (e) $T: \mathbf{R}^2 \to \mathbf{R}^2$, $T(x, y) = (x + y, x)$.
   (f) $T: \mathbf{R}^2 \to \mathbf{R}^2$, $T(x, y) = (2x + y, x + y)$.
   (g) $T: \mathbf{R}^2 \to \mathbf{R}^2$, $T(x, y) = (ax + cy, bx + dy)$, $a, b, c, d \in \mathbf{R}$, $\Delta = ad - bc = 1$.

(h) $T: \mathbf{R}^2 \to \mathbf{R}^2$, $T(x, y) = (ax + cy, bx + dy)$, $a, b, c, d \in \mathbf{R}$, $\Delta = ad - bc \neq 0$.
(i) $T: \mathbf{R}^3 \to \mathbf{R}^3$, $T(x, y, z) = (z, x, y)$.
(j) $T: \mathbf{R}^3 \to \mathbf{R}^3$, $T(x, y, z) = (x, ax + y, bx + cy + z)$.

4. Prove Proposition 1.3.
5. Let $T: \mathbf{R}^2 \to \mathbf{R}^2$ be a linear transformation with matrix

$$\begin{bmatrix} a & b \\ c & d \end{bmatrix}$$

and suppose $(b, d) = \lambda(a, c)$, $\lambda \in \mathbf{R}$. Prove that T is not invertible. [*Hint*: Show that $T(\lambda, -1) = \mathbf{0}$ so that T is not one-to-one.]
6. Let $T: \mathbf{R}^n \to \mathbf{R}^n$ be an invertible linear transformation with matrix A. Prove that the columns of A are linearly independent (as vectors in $\mathbf{R}^n$).

2. The Inverse Function Theorem

In this section, we state the inverse function theorem for a function $f: U \to V$, where U and V are open subsets of $\mathbf{R}^n$. Since the result is stronger when $n = 1$, we give a separate statement for this case. The proof of the theorem for $n = 1$ is given in the next section; the proof for arbitrary n is given in Appendix B.

We begin with a necessary condition for a differentiable function to have a differentiable inverse.

Proposition 2.1 *Let U and V be open subsets of $\mathbf{R}^n$ and $g: V \to U$ an inverse for the function f; $U \to V$. If f is differentiable at $\mathbf{v} \in U$ and g is differentiable at $\mathbf{w} = f(\mathbf{v}) \in V$, then both $D_{\mathbf{v}} f$ and $D_{\mathbf{w}} g$ are invertible, and*

$$D_{\mathbf{w}} g = (D_{\mathbf{v}} f)^{-1}$$

Proof. Let $I: U \to U$ be the identity mapping, $I(\mathbf{v}) = \mathbf{v}$. Then $D_{\mathbf{v}} I = I_n$, the identity linear transformation in $\mathbf{R}^n$. On the other hand, we have

$$D_{\mathbf{v}} I = D_{\mathbf{v}}(g \circ f) = (D_{\mathbf{w}} g) \circ (D_{\mathbf{v}} f)$$

from the chain rule. In the same way, we can show that $(D_{\mathbf{v}} f) \circ (D_{\mathbf{w}} g) = I_n$ and the proposition follows.

We now state the basic result on inverse functions for the case $n = 1$.

Theorem 2.2 *Let $f: (a, b) \to \mathbf{R}$ be a function with a continuous derivative. Suppose $f'(x) \neq 0$ for all $x \in (a, b)$. Then*

(a) *The function f is injective on (a, b).*
(b) *The image of f is an open interval (c, d).*
(c) *The inverse $f^{-1}: (c, d) \to (a, b)$ to f is differentiable with derivative*

$$(f^{-1})'(y) = \frac{1}{f'(x)}$$

where $y = f(x)$.

The proof will be given in the next section.

In view of Proposition 2.1, Theorem 2.2 is the best result that could be expected; if f is to have a differentiable inverse, then $f'(x)$ must be nonzero every point in (a, b)

We now consider a few examples from the point of view of Theorem 2.2.

1. Let $f: (0, 1) \to \mathbf{R}$ be given by $f(x) = x^2$. Then f has a differentiable inverse g given by $g(y) = \sqrt{y}$, $y \in (0, 1)$. This is consistent with Theorem 2.2 since $f'(x) = 2x$ is nonzero on $(0, 1)$.
2. Let f: $(-1, 1) \to \mathbf{R}$ be given by given by $f(x) = x^2$. As we saw at the beginning of the last section, f is not invertible. This is again consistent with Theorem 2.2 since $f'(x) = 2x = 0$ when $x = 0$.
3. Let f: $\mathbf{R} \to \mathbf{R}$ be given by $f(x) = x^3$. Then f has an inverse g: $\mathbf{R} \to \mathbf{R}$, $g(y) = y^{1/3}$. However,

$$g'(y) = \frac{1}{3y^{2/3}}$$

   for $y \neq 0$ and $g'(y)$ does not exist at $y = 0$. This is not surprising since $f'(x) = 3x^2 = 0$ when $x = 0$.
4. Theorem 2.2 can be used to define interesting functions. For example, let f: $(-\pi/2, \pi/2) \to \mathbf{R}$ be given by $f(x) = \sin x$. Then $f'(x) = \cos x \neq 0$ on $(-\pi/2, \pi/2)$, so we know from Theorem 2.2 that f has a differentiable inverse $g(y) = \arcsin y$ defined for $y \in (-1, 1)$ [since $\sin(\pi/2) = 1$ and $\sin(-\pi/2) = -1$]. Furthermore, we know that

$$g'(y) = \frac{1}{f'(x)} = \frac{1}{\cos x}$$

   where $\sin x = y$. However,

$$\begin{aligned}\cos x &= (1 - \sin^2 x)^{1/2}\\ &= (1 - y^2)^{1/2}\end{aligned}$$

   so we have

$$\frac{d}{dy}\arcsin y = (1 - y^2)^{-1/2}$$

   This is the usual formula from elementary calculus.

The analogue of Theorem 2.2 for $\mathbf{R}^n$ $(n > 1)$ might seem to be that if f is

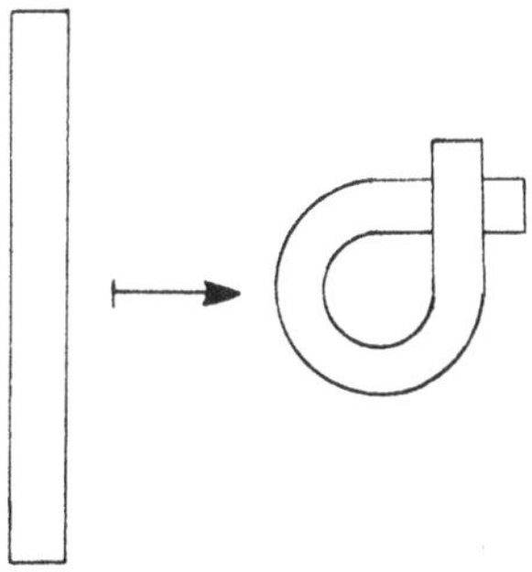

Figure 2.1 A noninvertible mapping.

differentiable on a connected open set U and $f'(\mathbf{v})$ is invertible for all $\mathbf{v} \in U$, then f has an inverse. This statement is false, however. Figure 2.1 should illustrate what can go wrong. (An explicit example is given in Exercise 2.) The problem is that in higher dimensions, functions have too many directions to move around in. The reader should convince himself that no such problem exists in $\mathbf{R}^1$. The difference is that a point moving in $\mathbf{R}^1$ cannot return to its starting point without stopping somewhere, while in $\mathbf{R}^2$ there is the possibility of going around in circles.

For higher dimensions, then, we need to lower our sights. We do this by asking only for a *local inverse*. That is, if $\mathbf{v}_0 \in U$, then we ask that the restriction of f to some small open set contained in U be invertible. In the case of Figure 2.1, for instance, the function f does indeed have such "local inverses."

The content of the general inverse function theorem is that local inverses exist when f' is invertible.

Theorem 2.3 (Inverse Function Theorem) *Let U_1 be an open set in $\mathbf{R}^n$, and let $f: U_1 \to \mathbf{R}^n$ be a function with a continuous derivative. Suppose that $\mathbf{v}_0$ is a point in U_1 such that $D_{\mathbf{v}_0} f$ is invertible. Then there is an open subset U of U_1 containing $\mathbf{v}_0$ such that*

(a) *The function f is injective on U.*
(b) *The set $V = f(U)$ is open.*
(c) *The inverse $f^{-1}: V \to U$ to f is differentiable with derivative*

$$D_{\mathbf{w}} f^{-1} = (D_{\mathbf{v}} f)^{-1}$$

where $\mathbf{w} = f(\mathbf{v})$.

We give the proof of this result in Appendix B.

We close this section with a few remarks about this theorem.

1. Note that the inverse function theorem applies only to *some* open set U about $\mathbf{v}_0$. Since U is not specified, the theorem is hard to apply to specific examples. It is most useful as a theoretical tool, in cases where it is simply important that *some* inverse exist.
2. We have discussed inverses of functions f mapping $\mathbf{R}^n$ (or an open set in $\mathbf{R}^n$) to $\mathbf{R}^n$. The reader may have wondered about inverses of a function $f: U \to \mathbf{R}^m$ (where U is an open set in $\mathbf{R}^n$) when $m \neq n$. A partial answer is that if f^{-1} is defined on an open set in $\mathbf{R}^m$ and both f and f^{-1} are differentiable, then $m = n$. The proof is a simple application of the chain rule, and is left as Exercise 1.

In fact, the hypothesis that f and f^{-1} are differentiable is convenient but unnecessary. It can be shown that if $f: U \to \mathbf{R}^m$ (where $U \subseteq \mathbf{R}^n$ is open) is continuous and 1-1, and if $f(U)$ is an open set in $\mathbf{R}^m$, then $m = n$. This result is fundamental in the subject of dimension theory; its proof is difficult and will not be given here.

It may not seem too surprising that there is no continuous 1-1 function mapping an open interval in $\mathbf{R}^1$ onto an open disk in $\mathbf{R}^2$, for instance. However, if one removes the requirement that f be 1-1, then such maps exist. Exercise 7 gives an example. (The sets involved there are closed, for convenience, but similar examples exist for open sets.)

Exercises

1. Suppose that $f: U \to \mathbf{R}^m$ and $g: U' \to \mathbf{R}^n$ (where $U \subseteq \mathbf{R}^n$ and $U' \subseteq \mathbf{R}^m$) are differentiable and $f \circ g$ and $g \circ f$ are both the identity. Show that $n = m$. (You may assume the fact that, if $T: \mathbf{R}^n \to \mathbf{R}^m$ is an invertible linear transformation, then $n = m$.)
2. Show that the function $f(r, \theta) = (r\cos\theta, r\sin\theta)$, defined for $r > 0$ and $\theta \in \mathbf{R}$, has a local inverse near every point in its image, but no global inverse.
3. Show that each of the following functions has a differentiable inverse and obtain a formula for the derivative of the inverse function. (See Example 4 above.)
   (a) $f: (0, \pi) \to \mathbf{R}$, $f(x) = \cos x$.
   (b) $f: (\frac{-\pi}{2}, \frac{\pi}{2}) \to \mathbf{R}$, $f(x) = \tan x$.
   (c) $f: (0, \frac{\pi}{2}) \to \mathbf{R}$, $f(x) = \sec x$.
   (d) $f: \mathbf{R} \to \mathbf{R}$, $f(x) = e^x$.
   (e) $f: (-\frac{1}{2}, \frac{1}{2}) \to \mathbf{R}$, $f(x) = x^2 - x$.
   (f) $f: (0, \infty) \to \mathbf{R}$, $f(x) = \cosh x = (e^x + e^{-x})/2$.
   (g) $f: \mathbf{R} \to \mathbf{R}$, $f(x) = \sinh x = (e^x - e^{-x})/2$.
4. The functions below have differentiable inverses when restricted to some interval including the point given. Find the largest such open interval.
   (a) $f(x) = x^2 - 7x + 6$, $x_0 = 1$.
   (b) $f(x) = x^3 - 5x^2 + 3x + 3$, $x_0 = 0$.

(c) $f(x) = \sin x$, $x_0 = 3$.
(d) $f(x) = \sin x$, $x_0 = -1$.
(e) $f(x) = \tan x$, $x_0 = 0$.
(f) $f(x) = e^x$, $x_0 = 2$.
(g) $f(x) = 16 + e^{x^2}$, $x_0 = -1$.
(h) $f(x) = \sin x + \cos x$, $x_0 = 0$.

5. Compute $(f^{-1})'(y_0)$, for each of the functions in Exercise 4. Here $y_0 = f(x_0)$.
6. Let $T: \mathbf{R}^m \to \mathbf{R}^n$ be a linear transformation, and let $\mathbf{w}_0$ a fixed vector in $\mathbf{R}^m$. Define $f: \mathbf{R}^m \to \mathbf{R}^n$ by $f(\mathbf{v}) = T(\mathbf{v}) + \mathbf{w}_0$, and prove that f is invertible if and only if T is.

*7. (*The Polya space-filling curve*) Let S be a nonisosceles right triangle (including the inside) in $\mathbf{R}^2$. We shall define a continuous function P mapping the interval $[0, 1]$ onto S.

Any point in the interval may be written as a binary "decimal" (where everything is in the base 2 instead of 10). For instance, $\frac{1}{3} = 0.01010101\cdots$, $\frac{1}{2} = 0.10000\cdots = 0.01111\cdots$, and so on. Given a number $x = .a_1a_2a_3\cdots$, we define $P(x)$ as follows: the altitude to the hypotenuse of S divides S into two triangles, similar to S and of different sizes. If $a_1 = 1$, we work with the bigger of the two, if $a_1 = 0$, with the smaller. Next, look at a_2 and repeat the process, dividing the appropriate triangle in the same way. Keep repeating the process. (Figure 2.2 shows an early stage in the construction.) The triangles get smaller and smaller, and their intersection will be a single point. This point is $P(x)$.

(a) Prove that the intersection of the triangles is a single point. (*Hint*: Use Exercise 3 of Section 5.1.)
(b) As noted earlier, $\frac{1}{2}$ has two different binary expansions. Show that the two expansions lead to the same value for $P(\frac{1}{2})$. [If $x = a/2^n$, where a and n are positive integers and $a/2^n < 1$, then x has two binary expansions, but $P(x)$ is well defined for essentially the same reason that $P(\frac{1}{2})$ is.]
(c) Show that P is continuous. (*Hint*: If x_0 and x_1 are sufficiently close, they have binary expansions which agree to a large number of places.)
(d) Show that P is onto S.
(e) (For the especially brave.) Show that P is nowhere differentiable if the smaller acute angle of the triangle is $> 30°$ (and $< 45°$, of course).

*8. Let U and V be open subsets of $\mathbf{R}^n$. We say U and V are *diffeomorphic* if there is a differentiable function f taking U onto V with differentiable inverse. We call the mapping f a *diffeomorphism*.

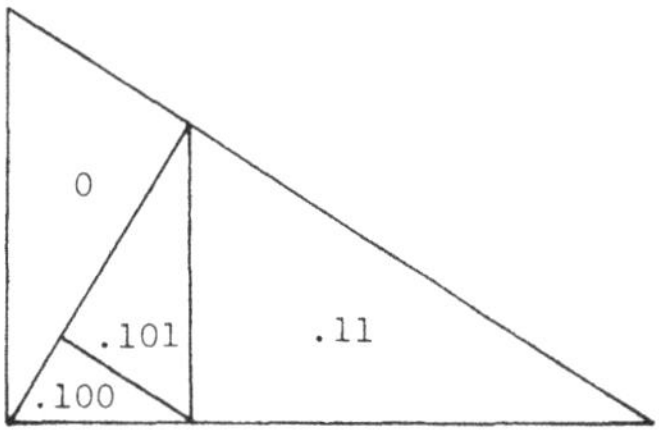

Figure 2.2 The Polya space-filling curve.

(a) Show that any two nonempty open balls in $\mathbf{R}^n$ are diffeomorphic.
(b) Suppose U and V are as above and that $f: U \to V$ is differentiable and surjective with $D_{\mathbf{v}} f$ invertible for all $\mathbf{v} \in U$. Is f a diffeomorphism?
(c) If we assume in addition that f is injective, can we conclude f is a diffeomorphism?

3. The Proof of Theorem 2.2

This section is devoted to proving the inverse function theorem for functions from $\mathbf{R}^1$ to $\mathbf{R}^1$. As we shall see, part (a) is fairly easy, but (b) and (c) will take more work. This work will, however, be useful preparation for the proof of Theorem 2.3 given in Appendix B.

Let x_1 and x_2 be two distinct points in (a, b). According to the mean value theorem (Exercise 9, Section 4.1), there is a point ξ between x_1 and x_2 such that

$$f(x_1) - f(x_2) = f'(\xi)(x_1 - x_2)$$

Since $f'(\xi) \neq 0$, $f(x_1) - f(x_2) \neq 0$, which means that $f(x_1) \neq f(x_2)$. Thus f is injective and (a) is proved.

To prove part (b), we use the intermediate value theorem (Theorem 6.6, Chapter 7). Let x_1 and x_2 be points in (a, b), as above, and let $c = f(x_1)$ and $d = f(x_2)$. Suppose that y is any number between x_1 and x_2. Then Theorem 6.6 of Chapter 7 states that for some x between x_1 and x_2, $y = f(x)$. That is, the set of points $I = \{f(x) : x \in (a, b)\}$ has the property that if c and d are any two points in I, then every point between c and d is in I. We saw in Section 7.6 that I is therefore an interval. (See Exercise 9 of Section 7.6.)

To finish the proof of (b), we need to show that I is an open interval. If not, then I is of the form $[c, d)$, $[c, d)$, $(c, d]$, $(-\infty, d]$, or $[c, -\infty)$. In every one of these cases, I has either a largest or smallest element. Thus f takes on either a maximum or a minimum at some point $x_0 \in (a, b)$. We know (Exercise 7, Section 4.1) that $f'(x_0) = 0$. This contradicts the hypothesis about f; therefore I is open, and (b) is proved.

For part (c), choose a point y_0 in (c, d), and pick $x_0 \in (a, b)$ with $f(x_0) = y_0$. Let a', b' be points satisfying $a < a' < x < b' < b$, and let $J = \{f(x) : x \in [a', b']\}$. Then the same proof as in part (b) shows that J is an interval. However, J must be a compact set, by Proposition 3.2 of Chapter 7. Therefore $J = [c', d']$ for some numbers c', d' with $c < c' < d' < d$. Moreover, $y \neq c'$ and $y \neq d'$. Again, the reason is like one used in part (b). If, say, $y = c'$, then x_0 is a local minimum for f, and so $f'(x_0) = 0$. This contradicts the hypothesis for f. Similarly, $y \neq d'$.

We know from Theorem 3.3 of Chapter 7 that $|f'(x)|$ has a minimum value on the interval $[a', b']$; call this minimum m. By hypothesis, $m \neq 0$. Given $y \in [c', d']$, we let $y = f(x)$, where $x \in [a', b']$. For some number ξ between x_0 and x, we have (by the mean value theorem)

$$|y - y_0| = |f(x) - f(x_0)| = |x - x_0|\,|f'(\xi)| \geq m|x - x_0|$$

or [since $x = f^{-1}(y)$, $x_0 = f^{-1}(y_0)$]

$$|f^{-1}(y) - f^{-1}(y_0)| \leqq \frac{1}{m}|y - y_0|$$

We can now show that f^{-1} is continuous at y_0. Given $\varepsilon > 0$, we choose $\delta > 0$ such that $\delta < m\varepsilon$ and $(y_0 - \delta, y_0 + \delta) \subset [c', d']$. (This last condition can be met because $y_0 \neq c'$ and $y_0 \neq d'$.) Then if $|y - y_0| < \delta$, we know that $y \in [c', d']$, so that

$$|f^{-1}(y) - f^{-1}(y_0)| \leqq \frac{1}{m}|y - y_0| < \frac{1}{m}(m\varepsilon) = \varepsilon$$

and f^{-1} is continuous at y_0. Since y_0 is arbitrary, f^{-1} is continuous.

Now let x_0 and y_0 be as above. We will show that

$$\lim_{y \to y_0} \left| \frac{f^{-1}(y) - f^{-1}(y_0)}{y - y_0} - \frac{1}{f'(x_0)} \right| = 0$$

This will prove that

$$(f^{-1})'(y_0) = \frac{1}{f'(x_0)}$$

which is statement (c). Given $\varepsilon > 0$, we may choose $\delta_2 > 0$ such that

$$|f(x) - f(x_0) - (x - x_0) f'(x_0)| < m\varepsilon |f'(x_0)|\,|x - x_0|$$

when $0 < |x - x_0| < \delta_2$. [This is true because $f'(x_0)$ is the derivative of f at x_0.] Now choose $\delta > 0$ such that $\delta < m\varepsilon$ and $(y_0 - \delta, y_0 + \delta) \subset [c', d']$. If $0 < |y - y_0| < \delta$, then, as we saw above,

$$|x - x_0| = |f^{-1}(y) - f^{-1}(y_0)| \leqq \frac{1}{m}|y - y_0| < \frac{1}{m}(m\varepsilon) = \varepsilon$$

and so

$$\begin{aligned}
\left| f^{-1}(y) - f^{-1}(y_0) - \frac{1}{f'(x_0)}(y - y_0) \right| &= \left| x - x_0 - \frac{y - y_0}{f'(x_0)} \right| \\
&= \frac{1}{|f'(x_0)|} |f'(x_0)(x - x_0) \\
&\quad - [f(x) - f(x_0)]| \\
&\leq m\varepsilon |x - x_0| \leqq m\varepsilon \left(\frac{1}{m}|y - y_0| \right) \\
&= \varepsilon |y - y_0|
\end{aligned}$$

which is what we had to prove. This completes the proof of statement (c) and of Theorem 2.2.

Exercises

*1. Show that if U is an open interval in $\mathbf{R}$ and $f: U \to \mathbf{R}$ is a continuous function with the property that $f(a) < f(b)$ if $a < b$ (and $a, b \in U$), then f has a continuous inverse on U.

2. Use Exercise 1 to show that if n is odd, then $f(x) = x^n$ has a continuous inverse on all of $\mathbf{R}$.

4. The Implicit Function Theorem: I

It is often necessary to deal with relations between variables which do not define functions. Consider, for instance, the set of points $(x, y) \in \mathbf{R}^2$ satisfying

$$x^2 + y^2 - 1 = 0 \tag{4.1}$$

These points lie on the unit circle about the origin (see Figure 4.1). Notice that (4.1) does not define y as a function of x, since there are two values of y satisfying (4.1) for every x with $-1 < x < 1$. Similarly, x is not a function of y.

We can use (4.1) to define y as a function of x (for $-1 < x < 1$) by the simple expedient of choosing one of the values of y for each x. If, as is reasonable, we want the function to be continuous, then we have two ways of defining the function, one corresponding to the upper half of the circle and one corresponding to the lower half: $y = (1 - x^2)^{1/2}$ or $y = -(1 - x^2)^{1/2}$ $(-1 < x < 1)$. There is no particular reason for picking one of these or the other. If, however, we start out with a point on the curve $[(-\frac{3}{5}, -\frac{4}{5})$, say], then there is a natural choice: we pick the function whose graph contains the given point [in our case, we let $y = g(x) = -(1 - x^2)^{1/2}$].

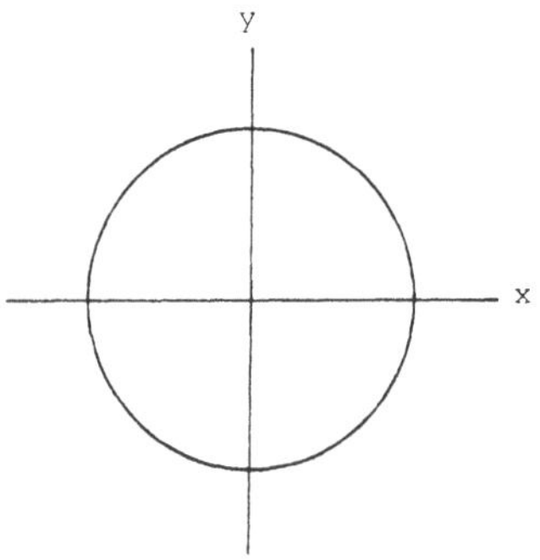

Figure 4.1 The unit circle.

The points (1, 0) and $(-1, 0)$ are special: we do not have a good way of defining y as a function of x on any open interval containing one of these points. At (1, 0), for instance, we have two problems:

(a) For $x > 1$, there is no value of y such that (x, y) satisfies (4.1).
(b) For $x < 1$, we have a choice of functions (as above), and no way to decide which to take.

Now we consider a slightly more general situation. Let $f: \mathbf{R}^2 \to \mathbf{R}$ be a continuously differentiable function; consider the set of points (x, y) such that

$$f(x, y) = 0 \tag{4.2}$$

We would like to use (4.2) to define y explicitly as a function of x. When is that possible?

The first point to notice is that generally we do not have precisely one y satisfying (4.2) for each x. Thus we cannot simply use (4.2) to define the function. What we really hope to have is a continuous function $g(x)$ such that

$$f(x, g(x)) = 0 \tag{4.3}$$

Then we can say that (4.2) gives y implicitly as a function of x, the function being g.

As the example showed, we often have a choice of functions g. To restrict our choice, we assume that we know a point (x_0, y_0) satisfying (4.2), and we require this point to be in the graph of g:

$$g(x_0) = y_0$$

We have not yet specified the domain of g, but since $g(x_0)$ is defined, g should be defined at least on some open interval containing x_0.

Does g exist? The example shows that we need some further condition. [We cannot find g for (4.1) if $(x_0, y_0) = (1, 0)$, for instance.] The condition is fairly simple: a certain partial derivative must not vanish.

Theorem 4.1 *Let $f: \mathbf{R}^2 \to \mathbf{R}^1$ be continuously differentiable, let $f(x_0, y_0) = 0$, and suppose that $(\partial f/\partial y)(x_0, y_0) \neq 0$. Then there is an open interval U about y_0 with the following property: there exists a continuous function $g: U \to \mathbf{R}^1$ such that $g(x_0) = y_0$ and*

$$f(x, g(x)) = 0$$

These properties determine g uniquely. Moreover, g is differentiable, and

$$g'(x) = -\left(\frac{\partial f}{\partial x}\right)\left(\frac{\partial f}{\partial y}\right)^{-1} \tag{4.4}$$

where the partial derivatives are evaluated at $(x, g(x))$.

Theorem 4.1 is a special case of the implicit function theorem, which we shall soon state more generally. Its proof, like that of the general implicit function theorem, is based on the inverse function theorem. We shall give the proof of Theorem 4.1 separately, however, because it is somewhat more explicit and easier to follow.

First, however, we note something about formula (4.4). The reader may have learned about differentiation of implicit functions in elementary calculus. For instance, suppose that y is defined implicitly as a function of x by

$$f(x, y) = x^2 + xy + y^2 - 7 = 0$$

The usual rule for computing dy/dx is to differentiate both sides of this equation according to the usual rules, remembering that y is a function of x:

$$2x + y + x\frac{dy}{dx} + 2y\frac{dy}{dx} = 0$$

or

$$(2x + y) + (x + 2y)\frac{dy}{dx} = 0 \tag{4.5}$$

Now it is easy to compute dy/dx by solving:

$$\frac{dy}{dx} = -\frac{2x + y}{x + 2y}$$

This procedure is equivalent to (4.4). Notice that in (4.5), the terms involving dy/dx always come from differentiating some function of y. In fact, the coefficient of dy/dx in (4.5) is just $\partial f/\partial y$. Similarly, the terms not involving dy/dx are just $\partial f/\partial x$. Thus (4.5) amounts to

$$\frac{\partial f}{\partial x} + \frac{\partial f}{\partial y}\frac{dy}{dx} = 0$$

and this equation gives (4.4).

Proof of Theorem 4.1. Define $F : \mathbf{R}^2 \to \mathbf{R}^2$ by

$$F(x, y) = (x, f(x, y))$$

Then F' has matrix

$$\begin{bmatrix} 1 & 0 \\ \dfrac{\partial f}{\partial x} & \dfrac{\partial f}{\partial y} \end{bmatrix}$$

which is invertible at (x_0, y_0) since $(\partial f/\partial y)(x_0, y_0) \neq 0$. (Verify this!) The inverse function theorem says that there is a ball B of radius $r > 0$ about $(x_0, 0) = F(x_0, y_0)$ in $\mathbf{R}^2$ and a function $G : B \to \mathbf{R}^2$ which is a differentiable inverse for F. Let

$$G(x, y) = (g_1(x, y), g_2(x, y))$$

Then

$$(x, y) = F(G(x, y)) = F(g_1(x, y), g_2(x, y)) = (g_1(x, y), f(g_1(x, y), g_2(x, y))) \tag{4.6}$$

Equating coordinates, we find that

$$x = g_1(x, y)$$

and

$$y = f(g_1(x, y), g_2(x, y)) = f(x, g_2(x, y))$$

Let $U = \{x : |x - x_0| < r\}$, so that $(x, 0) \in B$ if $x \in U$. Set

$$g(x) = g_2(x, 0)$$

Then g is differentiable, and

$$0 = f(x, g_2(x, 0)) = f(x, g(x))$$

Furthermore, since

$$G(x_0, 0) = (x_0, y_0)$$

we have

$$g(x_0) = g_2(x_0, 0) = y_0$$

Therefore g is the required function.

To finish the proof, we compute $g'(x)$. We know that

$$\begin{bmatrix} 1 & 0 \\ 0 & 1 \end{bmatrix} = J_{F \circ G}(x, y) = F'(G(x, y))G'(x, y)$$

$$= \begin{bmatrix} 1 & 0 \\ \dfrac{\partial f}{\partial x}(G(x, y)) & \dfrac{\partial f}{\partial y}(G(x, y)) \end{bmatrix} \begin{bmatrix} \dfrac{\partial g_1}{\partial x} & \dfrac{\partial g_1}{\partial y} \\ \dfrac{\partial g_2}{\partial x} & \dfrac{\partial g_2}{\partial y} \end{bmatrix}$$

by the chain rule. Since $g_1(x, y) = x$, we have

$$\frac{\partial g_1}{\partial x} = 1 \qquad \frac{\partial g_1}{\partial y} = 0$$

Now compute the lower left-hand term of the product:

$$0 = \frac{\partial f}{\partial x}(G(x, y)) + \frac{\partial f}{\partial y}(G(x, y))\frac{\partial g_2}{\partial x}$$

Let $y = 0$, so that $g_2(x, 0) = g(x)$; we get

$$0 = \frac{\partial f}{\partial x}(x, g(x)) + \frac{\partial f}{\partial y}(x, g(x))\frac{dg}{dx}$$

or

$$\frac{dg}{dx} = -\frac{\partial f/\partial x}{\partial f/\partial y}$$

[where the derivatives are evaluated at $(x, g(x))$]. This completes the proof.

Notice that in the example at the beginning of the section, there are two points, $(1, 0)$ and $(-1, 0)$, near which y is not defined implicitly as a function of x; at those points, $\partial f/\partial y = 0$. However, $\partial f/\partial x \neq 0$ there, so that near those points x is defined implicitly as a function of y.

Notice also that the function f need not be defined on all of $\mathbf{R}^2$; the theorem holds even if f is defined only on an open set of $\mathbf{R}^2$.

Exercises

1. Each of the following equations define y as a function of x on an interval including the point given. Compute dy/dx at that point.
   (a) $x^2 + xy + y^3 - 11 = 0$; $(1, 2)$
   (b) $\sin x + \cos y - 1 = 0$; $(\frac{\pi}{2}, \frac{\pi}{2})$
   (c) $x^2 + y^3 - 12 = 0$; $(-2, 2)$
   (d) $e^x + \tan y - 1 = 0$; $(0, 0)$
   (e) $xy - x^2 + y^6 + 1 = 0$; $(2, 1)$
2. Reguard x as a function of y and compute dx/dy for each of the equations of Exercise 1 at the points indicated.
3. Let $f(x, y) = x^4 + y^4 - 1$. At what points (x, y) on the set $S = \{(x, y) : f(x, y) = 0\}$ does the condition $f(x, y) = 0$ fail to define y implicitly as a function of x? Where is x not implicitly a function of y?
4. Let $f(x, y) = xy - \cos y + x^2 + 1$. At what points (x, y) on the set $\{(x, y) : f(x, y) = 0\}$ does the condition $f(x, y) = 0$ fail to define either of x or y as a function of the other?
5. (a) For what points on the lemniscate $(x^2 + y^2)^2 - 2a^2(x^2 - y^2) = 0$ $(a > 0)$ is it impossible to express either x or y as a function of the other?
   (b) Find the points on the lemniscate where y can be regarded as a function of x and compute dy/dx at these points.

*6. Let $f: \mathbf{R}^2 \to \mathbf{R}$ have a continuous derivative, and let $S = \{(x, y) : f(x, y) = 0\}$. Suppose that there exists $\varepsilon > 0$, $N > 0$ such that $(\partial f/\partial y)_{(x, y)} \geq \varepsilon$ and $(\partial f/\partial x)_{(x, y)} \leq N$ for all $(x, y) \in S$ with $a < x < b$, and let $f(x_0, y_0) = 0$, $x_0 \in (a, b)$. Show that there is a differentiable function $g : (a, b) \to \mathbf{R}$ such that $g(x_0) = y_0$ and $f(x, g(x)) = 0$. (The point of the problem is that it specifies an interval on which the implicit function is defined.)

5. The Implicit Function Theorem: II

The general implicit function theorem deals with functions from $\mathbf{R}^{n+k}$ to $\mathbf{R}^n$. To see what it should say, consider a linear transformation $T: \mathbf{R}^{n+k} \to \mathbf{R}^n$. We regard $\mathbf{R}^{n+k}$ as $\mathbf{R}^k \times \mathbf{R}^n$, think of T as a function of two variables $\mathbf{x} \in \mathbf{R}^k$, $\mathbf{y} \in \mathbf{R}^n$, and write $T(\mathbf{x}, \mathbf{y}) = T_1(\mathbf{x}) + T_2(\mathbf{y})$, where $T_1: \mathbf{R}^k \to \mathbf{R}^n$ and $T_2: \mathbf{R}^n \to \mathbf{R}^n$ are linear transformations. For instance, if $T: \mathbf{R}^3 \to \mathbf{R}^2$ is defined by the matrix

$$\begin{bmatrix} 1 & 3 & 5 \\ 2 & 4 & 6 \end{bmatrix}$$

then we regard a point (x_1, x_2, x_3) in $\mathbf{R}^3$ as $(x_1, (x_2, x_3))$, where $\mathbf{x} = x_1 \in \mathbf{R}$ and $\mathbf{y} = (x_2, x_3) \in \mathbf{R}^2$; T_1 is given by the matrix

$$\begin{bmatrix} 1 \\ 2 \end{bmatrix}$$

and T_2 by the matrix

$$\begin{bmatrix} 3 & 5 \\ 4 & 6 \end{bmatrix}$$

We would like to know when the equation $T(\mathbf{x}, \mathbf{y}) = \mathbf{0}$ defines $\mathbf{y}$ implicitly as a function of $\mathbf{x}$. To begin, we consider the example above. We know that

$$\begin{aligned} T(x_1, x_2, x_3) &= T_1(x_1) + T_2(x_2, x_3) \\ &= \begin{bmatrix} 1 \\ 2 \end{bmatrix} x_1 + \begin{bmatrix} 3 & 5 \\ 4 & 6 \end{bmatrix} \begin{bmatrix} x_2 \\ x_3 \end{bmatrix} \end{aligned}$$

Given x_1, we want to find x_2 and x_3 so that

$$T(x_1, x_2, x_3) = \mathbf{0}$$

that is, so that

$$\begin{bmatrix} 3 & 5 \\ 4 & 6 \end{bmatrix} \begin{bmatrix} x_2 \\ x_3 \end{bmatrix} = -\begin{bmatrix} 1 \\ 2 \end{bmatrix} x_1 = \begin{bmatrix} -x_1 \\ -2x_1 \end{bmatrix} \tag{5.1}$$

The matrix $\begin{bmatrix} 3 & 5 \\ 4 & 6 \end{bmatrix}$ is invertible; in fact, its inverse is

$$\begin{bmatrix} -3 & \frac{5}{2} \\ 2 & -\frac{3}{2} \end{bmatrix}$$

as is easily checked. Multiplying both sides of (5.1) by this matrix, we obtain

$$\begin{bmatrix} x_2 \\ x_3 \end{bmatrix} = \begin{bmatrix} -3 & \frac{5}{2} \\ 2 & -\frac{3}{2} \end{bmatrix} \begin{bmatrix} -x_1 \\ -2x_1 \end{bmatrix} = \begin{bmatrix} -2x_1 \\ x_1 \end{bmatrix} \tag{5.2}$$

This gives $\mathbf{y} = (x_2, x_3)$ as a function of $x_1 = \mathbf{x}$; this is the one function defined implicitly by $T(\mathbf{x}, \mathbf{y}) = \mathbf{0}$. The critical step in deriving (5.2) was inverting the linear transformation T_2.

Now we return to the general case, where $T : \mathbf{R}^{n+k} \to \mathbf{R}^n$. Suppose, as was the case in the example, that T_2 is invertible. Then if $\mathbf{x}$ is any vector in $\mathbf{R}^k$, $T_1(\mathbf{x}) \in \mathbf{R}^n$. We want to find $\mathbf{y} \in \mathbf{R}^n$ such that

$$\mathbf{0} = T(\mathbf{x}, \mathbf{y}) = T_1(\mathbf{x}) + T_2(\mathbf{y}) \tag{5.3}$$

that is,

$$T_2(\mathbf{y}) = -T_1(\mathbf{x})$$

or

$$y = T_2^{-1}(-T_1(\mathbf{x})) = -T_2^{-1} T_1(\mathbf{x}) \tag{5.4}$$

Equation (5.4) defines $\mathbf{y}$ uniquely, and thus makes $\mathbf{y}$ a function of $\mathbf{x}$. Working backwards, we see that this choice of $\mathbf{y}$ lets us satisfy (5.3). That is, (5.3) defines $\mathbf{y}$ implicitly as a function of $\mathbf{x}$, and this function is given explicitly by (5.4).

Remark. It is a standard result from linear algebra that an $n \times n$ matrix is invertible if and only if its columns are independent (as vectors in $\mathbf{R}^n$). (See Exercises 5 and 6 of Section 1 for a proof of half of this statement.) Using this fact, we can summarize the discussion above as follows. Let $T : \mathbf{R}^{n+k} \to \mathbf{R}^n$ be a linear transformation considered as a function $T(\mathbf{x}, \mathbf{y})$, $\mathbf{x} \in \mathbf{R}^k$, $\mathbf{y} \in \mathbf{R}^n$. Then, the equation $T(\mathbf{x}, \mathbf{y}) = \mathbf{0}$ can be solved for $\mathbf{y}$ as a function of $\mathbf{x}$ if the last n columns of the matrix of T are independent.

Now let $f: \mathbf{R}^{n+k} \to \mathbf{R}^n$ be a function with continuous partial derivatives near a point $\mathbf{v}_0$ and suppose that $f(\mathbf{v}_0) = \mathbf{0}$. Since f can be approximated near $\mathbf{v}_0$ by the linear transformation $f'(\mathbf{v}_0)$, we might hope that the above analysis applies to f whenever it applies to $f'(\mathbf{v}_0)$. The statement that it does (at least near $\mathbf{v}_0$) is the implicit function theorem.

Theorem 5.1 (The Implicit Function Theorem) *Let U be an open subset of $\mathbf{R}^{n+k}$ and $f : \mathbf{R}^{n+k} \to \mathbf{R}^n$ a function with continuous partial derivatives on U. [We write $f(\mathbf{x}, \mathbf{y})$, where $\mathbf{x} \in \mathbf{R}^k$, $\mathbf{y} \in \mathbf{R}^n$.] Suppose that $f(\mathbf{x}_0, \mathbf{y}_0) = \mathbf{0}$ and that the last n columns of the Jacobian of f at $(\mathbf{x}_0, \mathbf{y}_0)$ are independent. Then there is an open set U_1 in $\mathbf{R}^k$ containing $\mathbf{x}_0$ and a unique function $h : U_1 \to \mathbf{R}^n$ with $h(\mathbf{x}_0) = \mathbf{y}_0$ and $f(\mathbf{x}, h(\mathbf{x})) = \mathbf{0}$ for all $\mathbf{x}$ in U_1. Furthermore, the function h is differentiable. In fact, if one writes the matrix for $f'(\mathbf{x}_0, \mathbf{y}_0)$ as (A, B), where A is the matrix defined by the first k columns of the Jacobian of f at $(\mathbf{x}_0, \mathbf{y}_0)$ and B is the matrix defined by the last n columns, then the Jacobian matrix for h at $\mathbf{x}_0$ is $-B^{-1} A$.*

While we have stated the theorem in general, we shall be concerned primarily with the case where $n = 1$. Here are some examples.

1. Suppose that $f: \mathbf{R}^3 \to \mathbf{R}$ is given by

$$f(x_1, x_2, x_3) = x_1^2 + x_2^2 + x_3^2 - 1$$

We let $\mathbf{x} = (x_1, x_2)$ and $\mathbf{y} = x_3$. The point $(\frac{1}{3}, \frac{2}{3}, -\frac{2}{3}) = (\mathbf{x}_0, \mathbf{y}_0)$ satisfies

$$f(\mathbf{x}_0, \mathbf{y}_0) = 0$$

Moreover, $\partial f/\partial x_3 = 2x_3 \neq 0$ at $(\mathbf{x}_0, \mathbf{y}_0)$. (In this example, $n = 1$, and the condition on the Jacobian matrix is that the last entry should be nonzero.) Thus the conditions of Theorem 5.1 are met, and we can conclude that the equation $f(x_1, x_2, x_3) = 0$ defines x_3 implicitly as a function of x_1 and x_2 near the point $(\frac{1}{3}, \frac{2}{3}, -\frac{2}{3})$. In this case, of course, we can easily write the function explicitly:

$$x_3 = -(1 - x_1^2 - x_2^2)^{1/2}$$

2. More generally, let $f: \mathbf{R}^3 \to \mathbf{R}$ be a continuously differentiable function, and suppose that

$$f(a_1, a_2, a_3) = 0 \qquad \frac{\partial f}{\partial x_3}(a_1, a_2, a_3) \neq 0$$

Then the conditions of Theorem 5.1 are met at $(\mathbf{x}_0, \mathbf{y}_0)$, where $\mathbf{x}_0 = (a_1, a_2)$ and $\mathbf{y}_0 = a_3$. We conclude that $f(x_1, x_2, x_3) = 0$ defines x_3 as an implicit function of x_1 and x_2 near the point (a_1, a_2, a_3).
3. Similarly, if $f: \mathbf{R}^{k+1} \to \mathbf{R}$ is continuously differentiable and $f(\mathbf{x}_0, \mathbf{y}_0) = 0$, $(\partial f/\partial x_{k+1})(\mathbf{x}_0, y_0) \neq 0$, then Theorem 5.1 says that the equation $f(\mathbf{x}, \mathbf{y}) = 0$ implicitly defines $\mathbf{y}$ as a function of $\mathbf{x}$ near $(\mathbf{x}_0, \mathbf{y}_0)$.

In the above examples, we always expressed the last variable as a function of the others. There is, of course, nothing special about the last variable. If, for instance, we had $(\partial f/\partial x_1)(a_1, a_2, a_3) \neq 0$ in Example 2, we could then use $f(x_1, x_2, x_3) = 0$ to express x_1 implicitly as a function of x_2 and x_3.

This fact is often quite useful. For example, let $f: \mathbf{R}^k \to \mathbf{R}^1$ be a continuously differentiable function with the property that if $f(\mathbf{v}) = 0$, then $(\operatorname{grad} f)(\mathbf{v}) \neq \mathbf{0}$. Let $S = \{\mathbf{v} : f(\mathbf{v}) = 0\}$. If $\mathbf{v}_0 \in S$, then one of the partial derivatives of f at $\mathbf{v}_0$ is nonzero (since $\operatorname{grad} f \neq \mathbf{0}$ on S), and it follows that near $\mathbf{v}_0$, the set S can be regarded as defining one coordinate as a function of the rest. Thus S can be regarded as made up of pieces of the graphs of functions from $\mathbf{R}^k$ to $\mathbf{R}^1$. This is exactly what is needed to complete the proof of Theorem 4.1 of Chapter 8.

Similarly, we have used the general implicit function theorem in Theorem 6.1 of Chapter 8. We had a function $G: \mathbf{R}^{n+k} \to \mathbf{R}^n$; we assumed that at each point $\mathbf{y}_0 \in S = \{\mathbf{v} : f(\mathbf{v}) = 0\}$, the gradients of the coordinate functions were independent. That assumption means that there exist k linearly independent columns of G' at each point of S, and thus we can express some k coordinates of $\mathbf{v} \in S$ as a function of the other n, at least near $\mathbf{v}_0$. (This last statement amounts to a theorem in linear algebra; while it is not hard, proving it here would take us too far afield. For a proof, see L. Corwin and R. Szczarba, *Calculus in Vector Spaces*, Marcel Dekker, New York, 1979, Theorem 6.4.3.)

We shall not prove the implicit function theorem here (though we sketch the proof in Exercise 5). The proof is similar to the proof of Theorem 3.1, but the linear algebra involved is more complicated.

Exercises

1. Which of the following equations implicitly define z as a function of x and y near the given point? When the function does, compute $\partial z/\partial x$ and $\partial z/\partial y$ at the point.
   (a) $x^2 + y^2 + z^2 = 49$; $(6, -3, -2)$
   (b) $xy + yz + xz - xyz - 4 = 0$; $(2, 2, -3)$
   (c) $xye^z + z\cos(x^2 + y^2) = 0$; $(0, 0, 0)$
   (d) $xy^6 - yz^3 + z - xy - 11 = 0$; $(1, 2, 3)$
   (e) $2xy + e^{xz} - z\ln y - 1 = 0$; $(0, 1, 1)$
   (f) $x + y + z + \cos xyz - 1 = 0$; $(0, 0, 0)$
2. Which of the following equations define z and w as functions of x and y near the given point (x, y, z, w)? When the function does, compute $\partial z/\partial x$, $\partial z/\partial y$, $\partial w/\partial x$, and $\partial^2 z/\partial x\partial y$ at the point.
   (a) $z^2 + w^2 = x^2 + y$, $z + w = x^2 - y$; $(2, 1, 1, 2)$
   (b) $x - y^2 + z^2 + 2w^2 = 6$, $x^2 + y^2 = z^2 + w^2$; $(1, 2, -1, 2)$
   (c) $x^2 + y^2 - z + w = 3$, $xy + zw + 1 = 0$; $(1, 0, -1, 1)$
   (d) $xe^w + yz - z^2 = 0$, $y = \cos w + x^2 - z^2 = 1$; $(2, 1, 2, 0)$
3. The point $(0, 2, 4)$ is on the graph of
$$x^2 + y^2 + z^2 = 20$$
$$x - xy + z = 4$$
   Can you use these equations to define y and z as functions of x near $(0, 2, 4)$? How about x and z as functions of y, or x and y as functions of z?

*4. Let U be an open subset of $\mathbf{R}^n$ and $f\colon U \to \mathbf{R}^m$ a function whose component functions have continuous partial derivatives. We say that f is an *immersion* if $D_{\mathbf{v}} f$ is injective for all $\mathbf{v}$ in U and a *submersion* if $D_{\mathbf{v}} f$ is surjective for all $\mathbf{v}$ in U.
   (a) Suppose that $f\colon U \to \mathbf{R}^m$ is an immersion. Prove that for each $\mathbf{v}$ in U, we can find an open set V of U containing $\mathbf{v}$, an open set W of $\mathbf{R}^m$ containing $f(\mathbf{v})$, and a diffeomorphism h of W onto an open subset of $\mathbf{R}^m$ such that
$$hf(x_1, \ldots, x_n) = (x_1, \ldots, x_n, 0, \ldots, 0)$$
   (b) Suppose that $f\colon U \to \mathbf{R}^m$ is a submersion. Prove that for each $\mathbf{v}$ in U, we can find an open subset V of U containing $\mathbf{v}$, an open subset W of $\mathbf{R}^m$ containing $f(\mathbf{v})$, and a diffeomorphism h of W onto an open subset of $\mathbf{R}^m$ such that

$$hf(x, \ldots, x_n) = (x_{i_1}, \ldots, x_{i_m})$$

for some $i_1 < \cdots < i_m$.

5. Supply the details to the following outline of a proof of Theorem 5.1. Let $F: \mathbf{R}^{n+k} \to \mathbf{R}^{n+k}$ be defined by $F(\mathbf{x}, \mathbf{y}) = (\mathbf{x}, f(\mathbf{x}, \mathbf{y}))$. The Jacobian of F is

$$\begin{bmatrix} I_k & 0 \\ \multicolumn{2}{c}{J_f} \end{bmatrix}$$

where I_k is the $k \times k$ identity matrix, 0 is a $k \times n$ matrix of zeros, and J_f is the Jacobian matrix of f. The condition on $D_{(\mathbf{x}_0, \mathbf{y}_0)} f$ makes $D_{(\mathbf{x}_0, \mathbf{y}_0)} F$ invertible. Now apply the inverse function theorem to get an inverse G for F in some ball about $(\mathbf{x}_0, \mathbf{0}) = F(\mathbf{x}_0, \mathbf{y}_0)$. Set $G(\mathbf{x}, \mathbf{y}) = (g_1(\mathbf{x}, \mathbf{y}), g_2(\mathbf{x}, \mathbf{y}))$, where $g_1(\mathbf{x}, \mathbf{y}) \in \mathbf{R}^k$ and $g_2(\mathbf{x}, \mathbf{y}) \in \mathbf{R}^n$. Show that $g_1(\mathbf{x}, \mathbf{y}) = \mathbf{x}$ and $f(\mathbf{x}, g_2(\mathbf{x}, \mathbf{y})) = \mathbf{y}$. Set $h(\mathbf{x}) = g_2(\mathbf{x}, 0)$ and show that h satisfies the conditions of the theorem. To compute the derivative of h at $\mathbf{x}_0$, define $H(\mathbf{x}) = (\mathbf{x}, h(\mathbf{x}))$. Then the matrix of $D_{\mathbf{x}} H$ is

$$\begin{bmatrix} I_k \\ J_h \end{bmatrix}$$

where $J_h(\mathbf{x})$ is the Jacobian matrix of h. Use the chain rule and the fact that $f(H(\mathbf{x})) = \mathbf{0}$ to compute $J_h(\mathbf{x}_0)$.

10

Integration

Integration, as covered in a first-year calculus course, divides into two topics: the theory of integration (the definition, general properties, and the like) and techniques of integration (such as substitution and integration by parts). In this chapter, we begin the task of describing the integral for functions of n variables; in the next chapter, we shall see how to reduce integration problems in $\mathbf{R}^n$ to problems in $\mathbf{R}^1$.

The theory of the integral in $\mathbf{R}^n$ follows the same lines as for $\mathbf{R}^1$, and we therefore begin with a review of integration of functions of one variable. We then define integrals in $\mathbf{R}^n$, prove some basic properties of integrals, and demonstrate that continuous functions are integrable. The sections defining the integral are important, since the integral often arises in practical applications as a limit of the sorts of sums found in the definition.

1. Integration of Functions of One Variable

In this section, we discuss the theory of integration of functions of one variable. The theory is built on two basic principles. One is that there is an obvious definition for the integral of a step function; the other is that if $f \geq g$, then the integral of f is greater than or equal to the integral of g.

We define a *partition* P of a closed interval $[a, b]$ to be a sequence of real numbers $\{t_0, \ldots, t_p\}$ with

$$a = t_0 < t_1 < \cdots < t_p = b$$

For example, $\{0, \frac{1}{2}, \frac{3}{4}, 2\}$ is a partition of the interval $[0, 2]$.

Let f: $[a, b] \to \mathbf{R}$ be a bounded function and $P = \{t_0, t_1, \ldots, t_p\}$ a partition of the interval $[a, b]$. Define numbers M_i and m_i, $1 \leq i \leq p$, by

$$M_i = \text{lub}\,\{f(x) : t_{i-1} \leq x < t_i\}$$

$$m_i = \text{glb}\,\{f(x) : t_{i-1} \leq x < t_i\}$$

The *upper sum of f on $[a, b]$ relative to the partition P is defined by*

$$U_a^b(f, P) = \sum_{i=1}^{p} M_i(t_i - t_{i-1})$$

The *lower sum of f on $[a, b]$ relative to the partition P is defined by*

$$L_a^b(f, P) = \sum_{i=1}^{p} m_i(t_i - t_{i-1})$$

Unless there is a need to make explicit the interval of definition, we shall denote $U_a^b(f, P)$ by $U(f, P)$ and $L_a^b(f, P)$ by $L(f, P)$. Notice that $L(f, P) \leq U(f, P)$, since $m_i \leq M_i$ for all i.

To understand what these sums mean, assume that f is a nonnegative function, and define $h : [a, b] \to \mathbf{R}$ by

$$h(x) = \begin{cases} M_i & \text{if } t_{i-1} \leq x < t_i \\ M_p & \text{if } x = b \end{cases}$$

Then h is a *step function* (see Figure 1.1), and its integral, which represents the area under its graph, can be thought of as a sum of areas of rectangles; in fact, the area under the graph of h is precisely $U(f, P)$. Similarly, if we define $g : [a, b] \to \mathbf{R}$ by

$$g(x) = \begin{cases} m_i & \text{if } t_{i-1} \leq x < t_i \\ m_p & \text{if } x = b \end{cases}$$

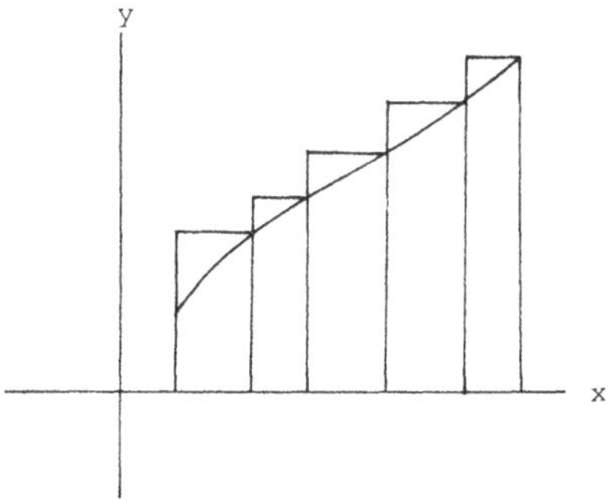

Figure 1.1 A step function $\geq f$ approximating f.

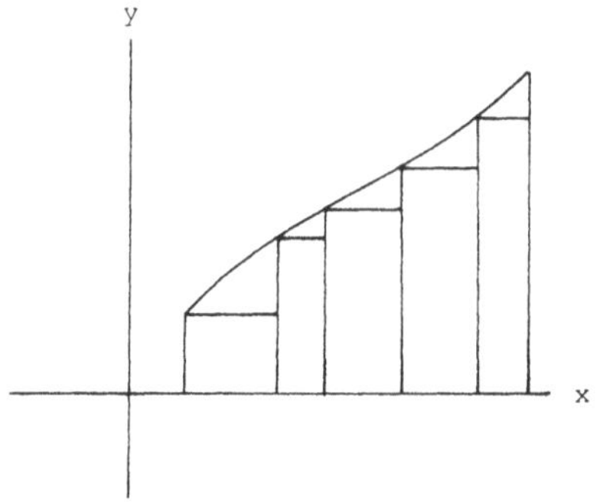

Figure 1.2 A step function $\leq f$ approximating f.

then g is a step function (see Figure 1.2), and the area under the graph of g is $L(f, P)$. Since $g \leq f \leq h$, the integral of f from a to b should lie between $L(f, P)$ and $U(f, P)$, the integrals of g and h, respectively. The idea behind defining the integral of f is to use a sequence of partitions P_k which make $L(f, P_k) - U(f, P_k)$ arbitrarily small, and then to define the integral as the common limit of $L(f, P_k)$ and $U(f, P_k)$.

Here are some examples of upper and lower sums.

1. Let $f(x) = x^2 - 2$ on the interval $[-1, 2]$ and let P be the partition $\{-1, 0, \frac{1}{2}, \frac{3}{2}, 2\}$. (See Figure 1.3.) We have

$$M_1 = -1 \qquad M_2 = -7/4 \qquad M_3 = 1/4 \qquad M_4 = 2$$

$$m_1 = -2 \qquad m_2 = -2 \qquad m_3 = -7/4 \qquad m_4 = 1/4$$

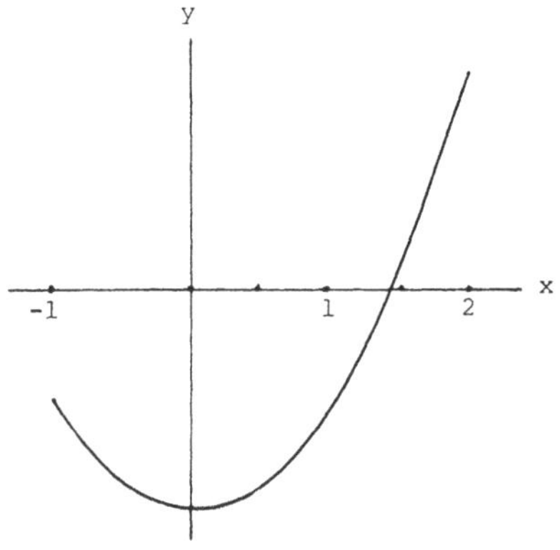

Figure 1.3 The parabola $y = x^2 - 2$.

Then

$$U(f, P) = (-1) + (-\tfrac{7}{4})(\tfrac{1}{2}) + (\tfrac{1}{4})1 + 2(\tfrac{1}{2}) = -\tfrac{3}{8}$$

$$L(f, P) = (-3)1 + (-2)(\tfrac{1}{2}) + (-\tfrac{7}{4})1 + (\tfrac{1}{4})(\tfrac{1}{2}) = -\tfrac{37}{8}$$

2. Let $f(x) = x$ on $[0, 1]$ and $P_m = \{0, 1/m, 2/m, \ldots, (m-1)/m, 1\}$. Then $M_i = i/m$ and $m_i = (i-1)/m$, so that

$$U(f, P_m) = \sum_{i=1}^{m} \left(\frac{i}{m}\right)\frac{1}{m} = \frac{1}{m^2}\sum_{i=1}^{m} i = \frac{1}{m^2}\left[\frac{m(m+1)}{2}\right]$$

(see Proposition 3.1, Chapter 1)

$$= \frac{m+1}{2m}$$

and

$$L(f, P_m) = \sum_{i=1}^{m} \left(\frac{i-1}{m}\right)\frac{1}{m} = \frac{1}{m^2}\sum_{i=1}^{m} (i-1) = \frac{m-1}{2m}$$

Note that both $U(f, P_m)$ and $L(f, P_m)$ approach $\frac{1}{2}$ as m approaches infinity.

If P and P' are two partitions of the same interval, we say that P' is a *refinement* of P whenever P' is obtained from P by introducing some additional points. For example, $P' = \{0, \frac{1}{2}, \frac{2}{3}, 1, \frac{4}{3}, \frac{3}{2}, 2\}$ is a refinement of $P = \{0, \frac{1}{2}, 1, \frac{3}{2}, 2\}$.

If P_1 and P_2 are any two partitions of the same interval, we say that P is a *common refinement* of P_1 and P_2 if P is a refinement of both P_1 and P_2. Given P_1 and P_2, we can obtain a common refinement P by taking the union of the points of P_1 and P_2. For example, if

$$P_1 = \{0, \tfrac{1}{3}, 1, \tfrac{4}{3}, 2\} \qquad P_2 = \{0, \tfrac{1}{4}, 1, \tfrac{7}{4}, 2\}$$

then

$$P = \{0, \tfrac{1}{4}, \tfrac{1}{3}, 1, \tfrac{4}{3}, \tfrac{7}{4}, 2\}$$

is a common refinement of P_1 and P_2.

We now prove two useful results about upper and lower sums.

Lemma 1.1 *Let $f\colon [a, b] \to \mathbf{R}$ be a bounded function and let P, P' be two partitions of $[a, b]$. Then if P' is a refinement of P, we have*

$$U(f, P') \le U(f, P)$$

$$L(f, P') \ge L(f, P)$$

Proof. We shall consider the case in which P' is obtained from P by adding a single point; the general case follows by induction (see Exercise 9).

Let $P = \{t_0, \ldots, t_m\}$ and $P' = \{t_0, \ldots, t_{i-1}, s, t_i, \ldots, t_m\}$, where

$$t_{i-1} < s < t_i$$

Most of the terms of $U(f, P)$ and $U(f, P')$ are the same. Subtracting, we see that

$$U(f, P) - U(f, P') = M_i(t_i - t_{i-1}) - M_i'(s - t_{i-1}) - M_i''(t_i - s)$$

where

$$M_i = \text{lub}\,\{ f(x) : t_{i-1} \leq x < t_i\}$$

$$M_i' = \text{lub}\,\{ f(x) : t_{i-1} \leq x < s\}$$

$$M_i'' = \text{lub}\,\{ f(x) : s \leq x < t_i\}$$

(See Figure 1.4.) Then $M_i \geq M_i'$ and $M_i \geq M_i''$ by Exercise 7 of Section 7.2, so that

$$U(f, P) - U(f, P') \geq M_i(t_i - t_{i-1}) - M_i(s - t_{i-1}) - M_i(t_i - s) = 0$$

A similar argument holds for the lower sums. (See Exercise 10.)

Lemma 1.2 *Let* $f : [a, b] \to \mathbf{R}$ *be a bounded function and* P_1, P_2 *two partitions of* $[a, b]$. *Then*

$$L(f, P_1) \leq U(f, P_2)$$

Proof. Let P be a common refinement of P_1 and P_2. Then

$$L(f, P_1) \leq L(f, P)$$

$$U(f, P) \leq U(f, P_2)$$

by Lemma 1.1. Since $L(f, P) \leq U(f, P)$, we have

$$L(f, P_1) \leq L(f, P) \leq U(f, P_2)$$

as claimed.

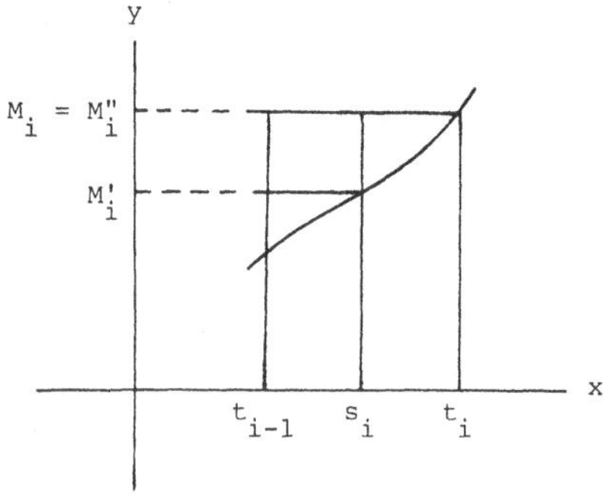

Figure 1.4 The upper sum of a refinement.

Define two sets of real numbers $U(f)$ and $L(f)$ by

$$U(f) = \{U(f, P) : P \text{ any partition of } [a, b]\}$$

$$L(f) = \{L(f, P) : P \text{ any partition of } [a, b]\}$$

If P_0 is any fixed partition of $[a, b]$, then for any partition P of $[a, b]$,

$$L(f, P) \leq U(f, P_0)$$

$$U(f, P) \geq L(f, P_0)$$

by Lemma 1.2. Thus $L(f)$ is bounded above, so that the least upper bound of $L(f)$ exists by Theorem 2.2 of Chapter 7. Similarly, glb $U(f)$ exists.

A bounded function f: $[a, b] \to \mathbf{R}$ is said to be *integrable* if

$$\text{glb } U(f) = \text{lub } L(f)$$

This common value is called the *integral* or *definite integral* of f on the interval $[a, b]$ and is denoted by

$$\int_a^b f(x)\,dx$$

Here are some examples.

3. Let f: $[a, b] \to \mathbf{R}$ be the constant function $f(x) = c$. Then $U(f, P) = L(f, P) = c(b - a)$ for any partition P, so that glb $U(f) =$ lub $U(f) = c(b - a)$. Therefore, f is integrable, with

$$\int_a^b f(x)\,dx = c(b - a)$$

4. Let f : $[0, 1] \to \mathbf{R}$ be defined by $f(x) = x$. We proved in Example 2 earlier in the section that for any positive integer m, there is a partition P_m with

$$U(f, P_m) = \frac{m+1}{2m} \qquad L(f, P_m) = \frac{m-1}{2m}$$

We can use these partitions to prove that

$$\text{glb } U(f) = \text{lub } L(f) = \frac{1}{2} = \int_0^1 x\,dx$$

We prove only that lub $L(f) = \frac{1}{2}$; the proof of the fact that glb $U(f) = \frac{1}{2}$ is similar.

First of all, we claim that $\frac{1}{2}$ is an upper bound for $L(f)$. If not, then we can find a partition P of $[0, 1]$ such that $L(f, P) = r > \frac{1}{2}$. However, we can now find an integer m with $U(f, P_m) < r = L(f, P)$, which contradicts Lemma 1.2. Explicitly, we choose

$$m > \frac{1}{2r-1}$$

Then $$U(f, P_m) = \frac{m+1}{2m} = \frac{1}{2} + \frac{1}{2m} < \frac{1}{2} + \frac{2r-1}{2} = r$$

Next, we claim that $\frac{1}{2}$ is the least upper bound of $L(f)$. If not, then lub $L(f) = r < \frac{1}{2}$. Choose an integer m such that

$$m > \frac{1}{1-2r}$$

Then $$L(f, P_m) = \frac{m-1}{2m} = \frac{1}{2} - \frac{1}{2m} > \frac{1}{2} - \frac{1-2r}{2} = r$$

which contradicts the fact that r is an upper bound for $L(f)$. Thus lub $L(f) = \frac{1}{2}$.

The argument given in this example suggests the following result.

Proposition 1.3 *Let $f: [a, b] \to \mathbf{R}$ be a bounded function and $P_1, P_2, \ldots, P_k, \ldots$ a sequence of partitions of $[a, b]$ such that*

$$\lim_{k\to\infty} U(f, P_k) = \lim_{k\to\infty} L(f, P_k) = I$$

Then f is integrable and

$$\int_a^b f(x)\,dx = I$$

We leave the proof as Exercise 8.

Here is one more example.

5. Suppose $b > 0$ and $f: [0, b] \to \mathbf{R}$ is given by $f(x) = x^2$. For each positive integer k, let P_k be the partition of $[0, b]$ into k equal subintervals:

$$P_k = \left\{0, \frac{b}{k}, \frac{2b}{k}, \ldots, \frac{(k-1)b}{k}, b\right\}$$

Since f is increasing on $[0, b]$, it follows that

$$M_j = \text{lub}\left\{f(x) : \frac{(j-1)b}{k} \le x < \frac{jb}{k}\right\} = \left(\frac{jb}{k}\right)^2$$

$$m_j = \text{glb}\left\{f(x) : \frac{(j-1)b}{k} \le x < \frac{jb}{k}\right\} = \left[\frac{(j-1)b}{k}\right]^2$$

Therefore $$U(f, P_k) = \sum_{j=1}^{k} \frac{j^2 b^2}{k^2} \cdot \frac{b}{k} = \frac{b^3}{k^3} \sum_{j=1}^{k} j^2$$

and $$L(f, P_k) = \sum_{j=1}^{k} \frac{(j-1)^2 b^2}{k^2} \cdot \frac{b}{k} = \frac{b^3}{k^3} \sum_{j=1}^{k} (j-1)^2$$

Using Exercise 1 of Section 1.3, we have

$$U(f, P_k) = \frac{b^3}{k^3} \frac{k(k+1)(2k+1)}{6} = \frac{b^3}{6}\left(1 + \frac{1}{k}\right)\left(2 + \frac{1}{k}\right)$$

and $$L(f, P_k) = \frac{b^3}{k^3} \frac{(k-1)k(2k-1)}{6} = \frac{b^3}{6}\left(1 - \frac{1}{k}\right)\left(2 - \frac{1}{k}\right)$$

Thus $$\lim_{k \to \infty} U(f, P_k) = \lim_{k \to \infty} L(f, P_k) = \frac{b^3}{3}$$

so that

$$\int_0^b x^2\,dx = \frac{b^3}{3}$$

by Proposition 1.3.

Exercises

1. Prove directly from the definition that

$$\int_a^b x\,dx = \frac{1}{2}(b^2 - a^2)$$

(See Proposition 3.1, Chapter 1.)

2. Prove directly from the definition that

$$\int_0^1 x^3\,dx = \frac{1}{4}$$

(See Exercise 3 of Section 1.3.)

3. Prove directly from the definition that

$$\int_0^b x^3\,dx = \frac{b^4}{4}$$

4. (a) Show (using the definition of the derivative) that for $a > 0$,

$$\lim_{n \to 0} \frac{a^h - 1}{h} = \ln a$$

(b) Use (a) and the definition of the derivative to compute $\int_1^c a^x\,dx$ (for $a > 0$, $a \neq 1$).

5. Use the result of Exercise 4 (a) to compute $\int_1^a (1/x)\,dx$, $a > 1$. (*Hint*: It pays to take the points of the partition in a geometric progression.)
6. Prove that $\lim_{n\to\infty} \sum_{j=1}^{n} n/(n^2 + j^2) = \pi/4$. [*Hint*: Consider $\int_0^1 dx/(1 + x^2)$.]
7. (a) Prove that $\lim_{n\to\infty} \sum_{j=1}^{n} 1/(n + j) = \ln 2$.

   *(b) Prove that $\lim_{n\to\infty} \sum_{j=1}^{n} (-1)^{j-1}/j = \ln 2$. (*Hint*: Group all terms of the form $1/2^k l$, l odd.)
8. Prove Proposition 1.3.
9. Use induction to complete the proof of the assertion of Lemma 1.1 involving upper sums.
10. Prove the assertion of Lemma 1.1 involving lower sums.

*11. Let $f: [a, b] \to \mathbf{R}$ be a function, and define f^+, f^- by

$$f^+(x) = \begin{cases} f(x) & \text{if } f(x) > 0 \\ 0 & \text{otherwise} \end{cases}$$

$$f^-(x) = \begin{cases} -f(x) & \text{if } f(x) < 0 \\ 0 & \text{otherwise} \end{cases}$$

(a) Show that f^+ and f^- are nonnegative and that $f = f^+ - f^-$

(b) Show that f is integrable if and only if f^+ and f^- are.

*12. (a) Let f and g be nonnegative bounded integrable functions on $[c, d]$. Prove that fg is integrable. (*Hint*: Suppose that on an interval $[x_j, x_{j+1})$, $\sup\{f(x)\} - \inf\{f(x)\} = a$ and $\sup\{g(x)\} - \inf\{g(x)\} = b$. Show that

$$\sup\{f(x)g(x)\} - \inf\{f(x)g(x)\} \leq K(a + b)$$

where K is a bound for f and g.)

(b) Use the result of Exercise 11 to show that if F and G are bounded integrable functions, then FG is integrable.

*13. Let $f: [a, b] \to \mathbf{R}$ be integrable and suppose there is a number $m > 0$ such that $|f(x)| \geq m$ for all $x \in [a, b]$. Prove that the function $g(x) = 1/f(x)$ is integrable.

14. Show that the function $f: [0, 1] \to \mathbf{R}$ defined by

$$f(x) = \begin{cases} 0 & \text{if } x \text{ is irrational} \\ 1 & \text{if } x \text{ is rational} \end{cases}$$

is not integrable. (*Hint*: Every interval contains rational and irrational numbers.)

*15. Show that the function $g: [0, 1] \to \mathbf{R}$ defined by

$$g(x) = \begin{cases} 0 & \text{if } x \text{ is irrational} \\ \dfrac{1}{q} & \text{if } x \text{ is the rational number } \dfrac{p}{q} \text{ in lowest terms} \end{cases}$$

is integrable. [*Hint*: Given $\varepsilon > 0$, $f(x) > \varepsilon/2$ at only finitely many points. Choose a partition to isolate these points in intervals of total length $< \varepsilon/2$.]

2. Properties of the Integral

The definition of the integral which we have given is somewhat cumbersome to use. The following proposition gives a useful characterization of integrability.

Proposition 2.1 *A bounded function $f : [a, b] \to \mathbf{R}$ is integrable if and only if for any $\varepsilon > 0$, there is a partition P of $[a, b]$ such that $U(f, P) - L(f, P) < \varepsilon$.*

Proof. Suppose first of all that f is integrable and set

$$I = \int_a^b f(x)\,dx = \text{glb } U(f) = \text{lub } L(f)$$

By properties of the least upper bound and greatest lower bound, we can find partitions P_1 and P_2 so that

$$U(f, P_1) - I < \frac{\varepsilon}{2}$$

and

$$I - L(f, P_2) < \frac{\varepsilon}{2}$$

Let P be a common refinement of P_1 and P_2. Then

$$U(f, P) - I < \frac{\varepsilon}{2}$$

and

$$I - L(f, P) < \frac{\varepsilon}{2}$$

by Lemma 1.1. Thus

$$U(f, P) - L(f, P) = [U(f, P) - I] + [I - L(f, P)] < \varepsilon$$

Conversely, if for any $\varepsilon > 0$, we can find a partition P of $[a, b]$ with $U(f, P) - L(f, P) < \varepsilon$, it follows that

$$\text{glb } U(f) - \text{lub } L(f) \leqslant U(f, P) - L(f, P) < \varepsilon$$

Since ε is arbitrary, it follows that glb $U(f) =$ lub $L(f)$, so that f is integrable.

The next result gives some of the elementary properties of the integral.

Proposition 2.2 *Let f, g: $[a, b] \to \mathbf{R}$ be integrable functions and p, q real numbers.*

(a) *Then*

$$\int_a^b [p f(x) + q g(x)]\, dx = p \int_a^b f(x)\, dx + q \int_a^b g(x)\, dx$$

(b) *If* $f(x) \geq g(x)$ *for all* $x \in [a, b]$, *then*

$$\int_a^b f(x)\, dx \geq \int_a^b g(x)\, dx$$

(c) *If* $m \leq f(x) \leq M$ *for all* $x \in [a, b]$, *then*

$$m(b-a) \leq \int_a^b f(x)\, dx \leq M(b-a)$$

(d) *If* $c \in [a, b]$, *then*

$$\int_a^c f(x)\, dx + \int_c^b f(x)\, dx = \int_a^b f(x)\, dx$$

Proof. As we shall prove a more general theorem in Section 5, we shall not give a complete proof here. We sketch the proof of (a) in the case $p = q = 1$, leaving the rest as exercises.

Note first of all that if $[t_{i-1}, t_i]$ is any subinterval of P, then, on this subinterval,

$$\text{lub}\,\{f(x) + g(x)\} \leq \text{lub}\,\{f(x)\} + \text{lub}\,\{g(x)\}$$

and

$$\text{glb}\,\{f(x) + g(x)\} \leq \text{glb}\,\{f(x)\} + \text{glb}\,\{g(x)\}$$

(See Proposition 3.4, Chapter 7.) Hence, if P is any partition of $[a, b]$,

$$\begin{aligned} L(P, f) + L(P, g) &\leq L(P, f+g) \\ &\leq U(P, f+g) \leq U(P, f) + U(P, g) \end{aligned} \tag{2.1}$$

Now let $\int_a^b f(x)\, dx = I$ and $\int_a^b g(x)\, dx = J$. Given any $\varepsilon > 0$, we may choose partitions P_1, P_2 so that

$$I - L(P_1, f) < \frac{\varepsilon}{4} \qquad U(P_1, f) - I < \frac{\varepsilon}{4}$$

$$J - L(P_2, g) < \frac{\varepsilon}{4} \qquad U(P_2, g) - J < \frac{\varepsilon}{4}$$

Let P be a common refinement of P_1 and P_2. Then

$$\begin{aligned} &I - L(P, f) < \frac{\varepsilon}{4} \qquad U(P, f) - I < \frac{\varepsilon}{4} \\ &J - L(P, g) < \frac{\varepsilon}{4} \qquad U(P, g) - J < \frac{\varepsilon}{4} \end{aligned} \tag{2.2}$$

We see from (2.1) that

$$(2.3)\qquad \begin{aligned} (I+J) - L(P, f+g) &< \frac{\varepsilon}{2} \\ U(P, f+g) - (I+J) &< \frac{\varepsilon}{2} \end{aligned}$$

In particular,

$$U(P, f+g) - L(P, f+g) < \varepsilon$$

Thus $f+g$ is integrable by Proposition 2.1. Moreover, (2.3) says that

$$I + J - \text{lub}\,\{L(P, f+g) : P \text{ a partition of } [a, b]\} \leq \frac{\varepsilon}{2}$$

$$\text{glb}\,\{U(P, f+g) : P \text{ a partition of } [a, b]\} - (I+J) \leq \frac{\varepsilon}{2}$$

Hence
$$\left| I + J - \int_a^b (f+g)(x)\,dx \right| \leq \frac{\varepsilon}{2} < \varepsilon$$

and since ε is arbitrary,

$$\int_a^b (f+g)(x)\,dx = I + J = \int_a^b f(x)\,dx + \int_a^b g(x)\,dx$$

as claimed.

The basic result for the integral of a function of one variable is the following.

Theorem 2.3 (The Fundamental Theorem of Calculus) *Let $f : [a, b] \to \mathbf{R}$ be a continuous function and define $F : [a, b] \to \mathbf{R}$ by*

$$F(t) = \int_a^t f(x)\,dx$$

Then F is differentiable and $F'(t) = f(t)$.

We have not yet proved that continuous functions are integrable on $[a, b]$. We shall assume this fact here; a proof will be given in Section 6. (Also see Exercise 13.)

Proof. This is an application of parts (c) and (d) of Proposition 2.2. We need to compute

$$\lim_{h \to 0} \frac{F(t+h) - F(t)}{h}$$

Assume for the moment that $h > 0$. Then, from (d) of Proposition 2.2,

$$F(t+h) - F(t) = \int_t^{t+h} f(x)\,dx$$

Given $\varepsilon > 0$, we may pick $\delta > 0$ such that if $|x - t| < \delta$, then

$$|f(x) - f(t)| < \varepsilon$$

or

$$f(t) - \varepsilon < f(x) < f(t) + \varepsilon$$

(Here, of course, we use the continuity of f.) Therefore

$$h[f(t) - \varepsilon] < \int_t^{t+h} f(x)\,dx < h[f(t) + \varepsilon]$$

or

(2.4)
$$f(t) - \varepsilon < \frac{F(t+h) - F(t)}{h} < f(t) + \varepsilon$$

when $0 < h < \delta$.

When $h < 0$, we have

$$F(t+h) - F(t) = -\int_{t+h}^{t} f(x)\,dx$$

and similar reasoning shows that (2.4) still holds for $-\delta < h < 0$. Therefore if $0 < |h| < \delta$,

(2.5)
$$\left| \frac{F(t+h) - F(t)}{h} - f(t) \right| < \varepsilon$$

Since ε is arbitrary, (2.5) implies that

$$\lim_{h \to 0} \frac{F(t+h) - F(t)}{h} = f(t)$$

which proves the theorem.

Theorem 2.3 is useful in evaluating integrals. If $f : [a, b] \to \mathbf{R}$ is continuous, a function $G : [a, b] \to \mathbf{R}$ is said to be an *antiderivative* (or a *primitive*) for f if $G'(x) = f(x)$ for $x \in [a, b]$. For example, $G(x) = x^3/3$ is an antiderivative for $f(x) = x^2$. If $G : [a, b] \to \mathbf{R}$ is an antiderivative for f and $F : [a, b] \to \mathbf{R}$ is defined as in Theorem 2.3, then $(G - F)' = G' - F' = 0$, so F and G differ by a constant c:

$$G(t) = F(t) + c$$

{The reason is that if $t \in [a, b]$, then $(G - F)(t) - (G - F)(a) = 0$; see

Exercise 10, Section 4.1.} In particular,

$$G(a) = F(a) + c = c$$

since $F(a) = 0$. Thus $G(t) - G(a) = F(t)$ for all $t \in [a, b]$. It follows that

$$\int_a^b f(x)\,dx = F(b) = G(b) - G(a)$$

where G is *any* antiderivative for f. We often write

$$G(x)\Big|_a^b = G(b) - G(a)$$

For example,

$$\int_a^b x^2\,dx = \frac{x^3}{3}\Big|_a^b = \frac{b^3}{3} - \frac{a^3}{3}$$

Finally, if $a > b$, we define $\int_a^b f(x)\,dx$ to be $-\int_b^a f(x)\,dx$.

Exercises

(In all exercises except the last, assume that continuous functions are integrable.)

1. Let $f: [a, b] \to \mathbf{R}$ be a continuous function and $g: [c, d] \to [a, b]$ a differentiable function. Prove that

$$\int_{g(c)}^{g(d)} f(x)\,dx = \int_c^d f(g(x))g'(x)\,dx$$

This formula is the basis for the technique of integration usually called *substitution*. (*Hint*: Use the chain rule and the discussion following Theorem 2.3.)

2. Use the formula of Exercise 1 to evaluate the following.

(a) $\int_1^2 2x(x^2 + 1)^{1/2}\,dx$ [Let $g(x) = x^2 + 1$ and $f(u) = u^{1/2}$.]
(b) $\int_0^{\sqrt{\pi/2}} x \sin x^2\,dx$
(c) $\int_0^{\pi/2} \sin x\, e^{\cos x}\,dx$
(d) $\int_0^1 e^{x+e^x}\,dx$
(e) $\int_0^1 \frac{x\,dx}{x^4 + 1}$

3. Prove the "integration by parts" formula: If f' and g' are continuous, then

$$\int_a^b f(x)g'(x)\,dx = f(x)g(x)\Big|_a^b - \int_a^b f'(x)g(x)\,dx$$

4. Evaluate the following:

(a) $\int_0^{\pi/2} x \sin x\,dx$
(b) $\int_0^1 xe^x\,dx$
(c) $\int_0^1 \arctan x\,dx$
(d) $\int_0^{\pi/3} x^2 \cos x\,dx$
(e) $\int_0^{5\pi/6} e^x \sin x\,dx$ (*Hint*: Integrate by parts twice and combine terms.)

5. Prove Proposition 2.2.
6. Let $f: [a, b] \to \mathbf{R}$ be a continuous function. Prove that there is a number $c \in [a, b]$ such that

$$\int_a^b f(x)\,dx = f(c)(b-a)$$

This is the *first mean value theorem for integrals.* [*Hint*: Combine the corollary to Theorem 3.3 of Chapter 7 with part (c) of Proposition 2.2 and apply the intermediate value theorem (Theorem 6.6, Chapter 7).]
7. Let $f: [a, b] \to \mathbf{R}$ be a continuous function with $f(x) \geq 0$ for all $x \in [a, b]$ and $f(x_0) > 0$ for some $x_0 \in [a, b]$. Prove that

$$\int_a^b f(x)\,dx > 0$$

(*Hint*: Use Exercise 5, Section 3.2.)
8. Complete the proof of Theorem 2.3 (by dealing with the case $h < 0$).
9. Show that f is integrable on $[a, b]$ and if $a < c < b$, then f is integrable on $[a, c]$ and on $[c, b]$.
*10. Let h be continuous and nonnegative, and let f be continuous. Show that there is a number $c \in [a, b]$ such that

$$\int_a^b f(x)h(x)\,dx = f(c)\int_a^b h(x)\,dx$$

[This result is similar to the one in Exercise 6; in fact, Exercise 6 is the special case $h(x) = 1$.] *Hint*: If m is the minimum value and M the maximum value of h on $[a, b]$, then $m f(x) \leq h(x) \leq M f(x)$. Now use the hint for Exercise 6.
*11. Let $f: [a, b]$ be monotone and continuously differentiable, and let $g: [a, b] \to \mathbf{R}$ be continuous. Prove that there is a number $c \in [a, b]$ such that

$$\int_a^b f(x)g(x)\,dx = f(a)\int_a^c g(x)\,dx + f(b)\int_c^b g(x)\,dx$$

This is called the *second mean value theorem for integrals.* (*Hint*: Let $G(t) = \int_a^t g(x)\,dx$. We may assume that f is monotone increasing, so that $f' \geq 0$. Show that

$$\int_a^b f(x)g(x)\,dx = f(b)G(b) - \int_a^b f'(x)G(x)\,dx$$

Apply the result of Exercise 10 and simplify.)
12. What is wrong with the following reasoning? Let f be a function such that $f(2x) = \frac{1}{2}f(x)$. Then

$$\int_0^2 f(x)\,dx = \int_0^1 2f(2t)\,dt \qquad (x = 2t)$$

$$= \int_0^1 f(t)\,dt$$

from the equation for f. Thus

$$\int_1^2 f(x)\,dx = \int_0^2 f(x)\,dx - \int_0^1 f(t)\,dt = 0$$

Now let $f(x) = 1/x$:

$$\int_1^2 \frac{dx}{x} = 0$$

or $\ln 2 = 0$. Thus $2 = 1$.

*13. Prove that if f is continuous on $[a, b]$, then f is integrable on $[a, b]$. (*Hint*: f is uniformly continuous, by Theorem 5.1 of Chapter 7. Given $\varepsilon > 0$, partition $[a, b]$ so finely that the lub and glb of f on each subinterval differ by less than $\varepsilon/(b - a)$. Now apply Proposition 2.1.)

3. The Integral of a Function of Two Variables

The definition of the integral of a function of several variables is a straightforward extension of the one variable case. However, as the number of variables increases, so does the complexity of the notation. Because of this, we give a detailed treatment of the integral of a function of two variables here. In the next section, we give a sketch of the definition of the integral for functions of more than two variables.

Let $\mathbf{a} = (a_1, a_2)$, $\mathbf{b} = (b_1, b_2)$ be points in $\mathbf{R}^2$ with $a_1 < b_1$, $a_2 < b_2$ and let $R(\mathbf{a}, \mathbf{b})$ be the *rectangle*

$$R(\mathbf{a}, \mathbf{b}) = \{(x, y) : a_1 \le x \le b_1,\ a_2 \le y \le b_2\}^*$$

If $P_1 = \{t_0, \ldots, t_m\}$ is a partition of $[a_1, b_1]$ and $P_2 = \{s_0, \ldots, s_k\}$ is a partition of $[a_2, b_2]$, then the sets R_{ij} given by

$$R_{ij} = \{(x, y) : t_{i-1} \le x \le t_i \text{ and } s_{j-1} \le y \le s_j\}$$

define a decomposition of $R(\mathbf{a}, \mathbf{b})$ into subrectangles. We call this decomposition a *partition* of $R(\mathbf{a}, \mathbf{b})$ and denote it by $P_1 \times P_2$ (see Figure 3.1).

Note that the decomposition pictured in Figure 3.2 is *not* a partition according to our definition. Our partitions of rectangles must arise from partitions of the "edges" as described above.

*For convenience here, we use (x, y) for the coordinates in $\mathbf{R}^2$.

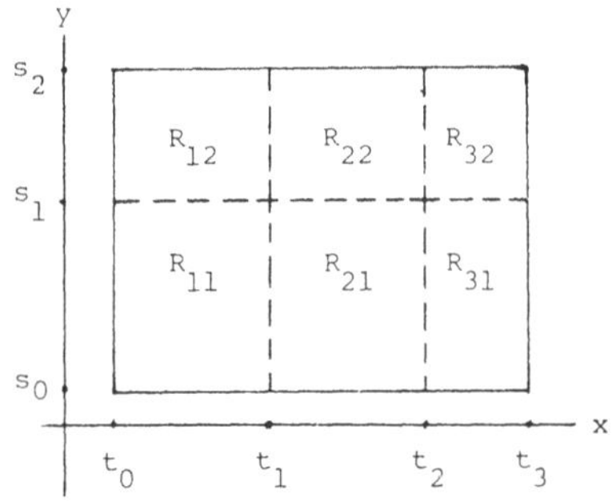

Figure 3.1 A partition of a rectangle in $\mathbf{R}^2$.

Suppose now that $f: R(\mathbf{a}, \mathbf{b}) \to \mathbf{R}$ is a bounded function and $P_1 \times P_2$ a partition of $R(\mathbf{a}, \mathbf{b})$. We define numbers M_{ij}, m_{ij}, $1 \leq i \leq p$, $1 \leq j \leq k$, by

$$M_{ij} = \text{lub}\,\{f(\mathbf{v}) : \mathbf{v} \in R_{ij}\}$$

$$m_{ij} = \text{glb}\,\{f(\mathbf{v}) : \mathbf{v} \in R_{ij}\}$$

The *upper sum* of f on $R(\mathbf{a}, \mathbf{b})$ relative to the partition $P_1 \times P_2$ is defined by

$$U(f, P_1 \times P_2) = \sum_{i=1}^{p} \sum_{j=1}^{k} M_{ij}\mu(R_{ij})$$

where $\mu(R_{ij})$ is the "area" of the rectangle R_{ij},

$$\mu(R_{ij}) = (t_i - t_{i-1})(s_j - s_{j-1})$$

Similarly, the *lower sum* of f on $R(\mathbf{a}, \mathbf{b})$ relative to the partition $P_1 \times P_2$ is defined by

$$L(f, P_1 \times P_2) = \sum_{i=1}^{p} \sum_{j=1}^{k} m_{ij}\mu(R_{ij})$$

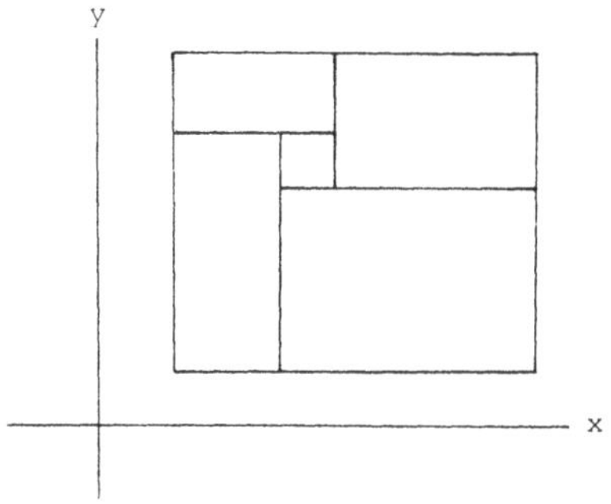

Figure 3.2 A decomposition which is not a partition.

Clearly $L(f, P_1 \times P_2) \leq U(f, P_1 \times P_2)$.

Suppose now that $P_1 \times P_2$ and $P_1' \times P_2'$ are two partitions of $R(\mathbf{a}, \mathbf{b})$. We say that $P_1' \times P_2'$ is a *refinement* of $P_1 \times P_2$ if P_1' is a refinement of P_1 and P_2' is a refinement of P_2. Equivalently, $P_1' \times P_2'$ is a refinement of $P_1 \times P_2$ if each of the rectangles of $P_1' \times P_2'$ is contained in a rectangle of $P_1 \times P_2$. For example, if

$$P = \{0, 1, 2\} \qquad P' = \{0, \tfrac{1}{2}, 1, 2\}$$

then $P \times P'$ is a refinement of $P \times P$.

If $P_1 \times P_2$ and $P_1' \times P_2'$ are partitions of $R(\mathbf{a}, \mathbf{b})$, then $P_1'' \times P_2''$ will be a common refinement of $P_1 \times P_2$ and $P_1' \times P_2'$ if P_1'' is a common refinement of P_1 and P_1' and P_2'' is a common refinement of P_2 and P_2'. One way to produce a common refinement of $P_1 \times P_2$ and $P_1' \times P_2'$ is to regard each partition as a transparency with lines on it; one common refinement is obtained by putting one transparency on the other. (See Figure 3.3.)

We can now state the analogues of Lemmas 1.1 and 1.2.

Lemma 3.1 *Let* $f\colon R(\mathbf{a}, \mathbf{b}) \to \mathbf{R}$ *be a bounded function and let* $P_1 \times P_2$, $P_1' \times P_2'$ *be partitions of* $R(\mathbf{a}, \mathbf{b})$. *Then*

(i) $L(f, P_1 \times P_2) \leq U(f, P_1' \times P_2')$

(ii) If $P_1' \times P_2'$ is a refinement of $P_1 \times P_2$, then

$$U(f, P_1' \times P_2') \leq U(f, P_1 \times P_2)$$

$$L(f, P_1' \times P_2') \geq L(f, P_1 \times P_2)$$

The proof of this result proceeds just as did the proofs of Lemmas 1.1 and 1.2. We leave it as an exercise for the reader.

Define two sets $U(f)$ and $L(f)$ by

$$U(f) = \{U(f, P_1 \times P_2) : P_1 \times P_2 \text{ a partition of } R(\mathbf{a}, \mathbf{b})\}$$

$$L(f) = \{L(f, P_1 \times P_2) : P_1 \times P_2 \text{ a partition of } R(\mathbf{a}, \mathbf{b})\}$$

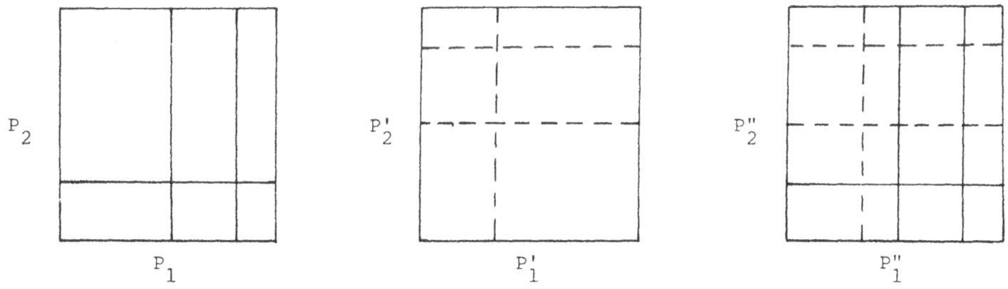

Figure 3.3 A common refinement.

Then, just as in Section 1, $U(f)$ is bounded from below and $L(f)$ is bounded from above, so that glb $U(f)$ and lub $L(f)$ both exist. A bounded function $f\colon R(\mathbf{a}, \mathbf{b}) \to \mathbf{R}$ is said to be *integrable* if

$$\text{glb } U(f) = \text{lub } L(f)$$

This common value is called the *integral* of f on $R(\mathbf{a}, \mathbf{b})$ and denoted

$$\int_{R(\mathbf{a}, \mathbf{b})} f$$

We shall see in Section 6 that, for example, all continuous functions are integrable.

Here are a few examples.

1. Let $f\colon R(\mathbf{a}, \mathbf{b}) \to \mathbf{R}$ be a constant function: $f(\mathbf{v}) = c$. Then $L(f, P) = U(f, P) = c(b_1 - a_1)(b_2 - a_2)$ for all partitions P, and it follows that

$$\int_{R(\mathbf{a}, \mathbf{b})} f = c(b_1 - a_1)(b_2 - a_2)$$

2. Let $S = R(\mathbf{a}, \mathbf{b})$, where $\mathbf{a} = (0, 0)$ and $\mathbf{b} = (1, 1)$ and let $f\colon S \to \mathbf{R}$ be defined by $f(x, y) = xy$. Let P_m be the partition of $[0, 1]$ defined in Example 2 at the beginning of Section 1, $P_m = \{0, 1/m, 2/m, \ldots, (m - 1)/m, 1\}$, and consider the partition $P_m \times P_m$ of S. Then

$$R_{ij} = \left\{(x, y) \in \mathbf{R}^2 : \frac{i - 1}{m} \le x \le \frac{i}{m}, \frac{j - 1}{m} \le y \le \frac{j}{m}\right\}$$

and $\mu(R_{ij}) = 1/m^2$ for all i, j. Furthermore, it is easy to see that

$$M_{ij} = \frac{i}{m}\frac{j}{m} = \frac{ij}{m^2}$$

$$m_{ij} = \frac{i - 1}{m}\frac{j - 1}{m} = \frac{(i - 1)(j - 1)}{m^2}$$

Thus

$$\begin{aligned} U(f, P_m \times P_m) &= \sum_{i=1}^{m} \sum_{j=1}^{m} \frac{ij}{m^2} \cdot \frac{1}{m^2} \\ &= \frac{1}{m^4} \sum_{i=1}^{m} i \sum_{j=1}^{m} j \\ &= \frac{1}{m^4} \frac{m(m + 1)}{2} \frac{m(m + 1)}{2} \\ &= \frac{1}{4}\left(1 + \frac{1}{m}\right)^2 \end{aligned}$$

and
$$L(f, P_m \times P_m) = \sum_{i-1}^{m} \sum_{j-1}^{m} \frac{(i-1)(j-1)}{m^2} \cdot \frac{1}{m^2}$$
$$= \frac{1}{m^4} \sum_{i=1}^{m} (i-1) \sum_{j=1}^{m} (j-1)$$
$$= \frac{1}{m^4} \frac{(m-1)m}{2} \frac{(m-1)m}{2}$$
$$= \frac{1}{4}\left(1 - \frac{1}{m}\right)^2$$

It follows that

$$\lim_{m \to \infty} U(f, P_m \times P_m) = \lim_{m \to \infty} L(f, P_m \times P_m) = \frac{1}{4}$$

and, just as in Example 4 of Section 1, we see that

$$\int_S xy = \frac{1}{4}$$

3. Suppose $S = R(\mathbf{a}, \mathbf{b})$, where $\mathbf{a} = (0, 0)$ and $\mathbf{b} = (b, c)$, and let $f: S \to \mathbf{R}$ be defined by $f(x, y) = x^2y$. Let P_k be the partition of $[0, b]$ defined in Example 5 of Section 1:

$$P_k = \left\{0, \frac{b}{k}, \frac{2b}{k}, \ldots, \frac{(k-1)b}{k}, b\right\}$$

and let P_k' be the analogous partition of $[0, c]$:

$$P_k' = \left\{0, \frac{c}{k}, \frac{2c}{k}, \ldots, \frac{(k-1)}{k}, c\right\}$$

Then, if $P_k'' = P_k \times P_k'$,

$$R_{ij} = \left\{(x, y) : \frac{(i-1)b}{k} \le x \le \frac{ib}{k}, \frac{(j-1)c}{k} \le y \le \frac{jc}{k}\right\}$$

and $\mu(R_{ij}) = bc/k^2$. Furthermore,

$$M_{ij} = \left(\frac{ib}{k}\right)^2 \frac{jc}{k}$$

and
$$m_{ij} = \left[\frac{(i-1)b}{k}\right]^2 \frac{(j-1)c}{k}$$

Thus
$$U(f, P_k'') = \sum_{i=1}^{k} \sum_{j=1}^{k} \frac{i^2 b^2}{k^2} \frac{jc}{k} \frac{bc}{k^2}$$
$$= \frac{b^3 c^2}{k^5} \sum_{i=1}^{k} i^2 \sum_{j=1}^{k} j$$
$$= \frac{b^3 c^2}{k^5} \frac{k(k+1)(2k+1)}{6} \frac{k(k+1)}{2}$$
$$= \frac{b^3 c^2}{12}\left(1 + \frac{1}{k}\right)^2 \left(2 + \frac{1}{k}\right)$$

Similarly,
$$L(f, P_k'') = \frac{b^3 c^2}{12}\left(1 - \frac{1}{k}\right)^2 \left(2 - \frac{1}{k}\right)$$

and it follows that
$$\lim_{k\to\infty} U(f, P_k'') = \lim_{k\to\infty} L(f, P_k'') = \frac{b^3 c^2}{6}$$

Now, just as in Example 4 of Section 1, we have
$$\int_S x^2 y = \frac{b^3 c^2}{6}$$

One problem with our definition of the integral is that it lets us integrate functions only over rectangles. We may need to integrate over more general sets—for instance, triangles or disks. For integrals over such sets, we proceed as follows. Let S be a closed and bounded subset of $\mathbf{R}^2$, and let $f : S \to \mathbf{R}$ be a bounded function. Since S is bounded, we can find a rectangle S_1 containing S. Define $f_1 : S_1 \to \mathbf{R}$ by
$$f_1(\mathbf{v}) = \begin{cases} f(\mathbf{v}) & \text{if } \mathbf{v} \in S \\ 0 & \text{if } \mathbf{v} \notin S \end{cases}$$

We say that $f : S \to \mathbf{R}$ is *integrable* if f_1 is and we define the *integral* of f over S by
$$\int_S f = \int_{S_1} f_1$$

We need to show that this definition does not depend on the particular rectangle S_1 which we choose containing S. This is not hard to see. For example, if S_2 is a second rectangle containing S and $f_2 : S_2 \to \mathbf{R}$ is defined as is f_1, then the intersection $S_0 = S_1 \cap S_2$ is a rectangle containing

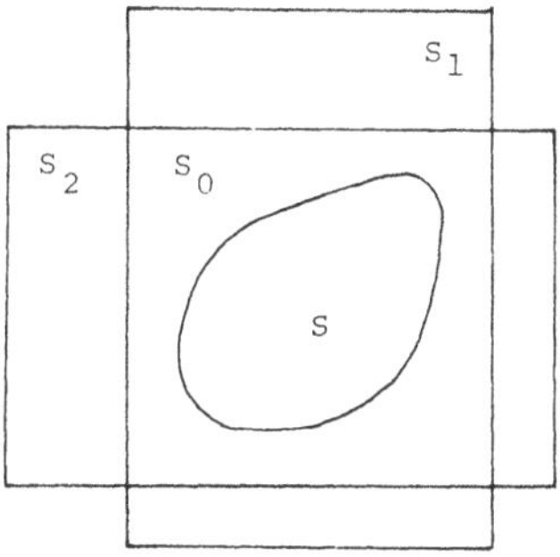

Figure 3.4 Rectangles containing S.

S. Furthermore, both S_1 and S_2 differ from S_0 by a union of rectangles on which the functions f_1 and f_2 vanish. (See Figure 3.4.) Thus, if $f_0 : S_0 \to \mathbf{R}$ is defined in the same way as f_1 and f_2, it follows readily that f_i is integrable if and only if f_0 is integrable, $i = 1, 2$, and

$$\int_{S_1} f_1 = \int_{S_0} f_0 = \int_{S_2} f_2$$

(See Exercises 7 and 8.)

We conclude this section by remarking that the result analogous to Proposition 1.2 holds in $\mathbf{R}^2$ as well. We leave its statement and proof as Exercise 2.

Exercises

1. Let $R(\mathbf{a}, \mathbf{b})$ be a rectangle in $\mathbf{R}^2$, and let $f: R(\mathbf{a}, \mathbf{b}) \to \mathbf{R}$ be the constant function $f(\mathbf{v}) = c$, $\mathbf{v} \in R(\mathbf{a}, \mathbf{b})$. Prove that f is integrable and that

$$\int_{R(\mathbf{a},\mathbf{b})} f = c\mu(R(\mathbf{a}, \mathbf{b}))$$

2. State and prove the analogue of Proposition 2.1 for functions of two variables.
3. Let $f: \mathbf{R}^2 \to \mathbf{R}$ be defined by $f(x, y) = x$. Use Exercise 2 to prove that

$$\int_S f = \frac{1}{2}$$

where S is the rectangle defined by $(0, 0)$ and $(1, 1)$. (See Example 2 of Section 1.)
4. Let $S = R(\mathbf{a}, \mathbf{b})$ the rectangle in $\mathbf{R}^2$ defined by the points $\mathbf{a} = (a_1, a_2)$, $\mathbf{b} = (b_1, b_2)$ and let $f: S \to \mathbf{R}$ be the function $f(x, y) = x$. Prove directly from the definition that

$$\int_S f = \frac{1}{2}(b_2 - a_2)(b_1^2 - a_1^2)$$

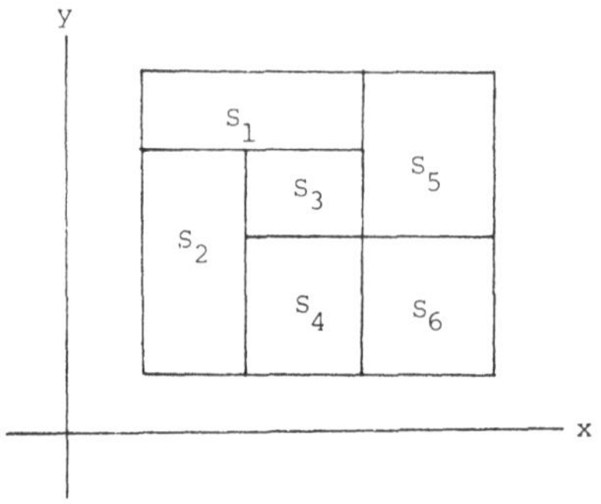

Figure 3.5 A decomposition of a rectangle.

5. Determine, directly from the definition, the integral of the function f over the rectangle S of Exercise 3, where
   (a) $f(x, y) = 2x + y$
   (b) $f(x, y) = xy$
   (c) $f(x, y) = x^2 + y^2$
6. Prove Lemma 3.1.
*7. Let $S = R(\mathbf{a}, \mathbf{b})$ be the rectangle in $\mathbf{R}^2$ defined by the points $\mathbf{a} = (a_1, a_2)$, $\mathbf{b} = (b_1, b_2)$. The *interior* of S is the set

$$\text{int } S = \{(x, y) : a_1 < x < b_1,\ a_2 < y < b_2\}$$

Suppose $S = \cup_{j=1}^{m} S_j$, where each S_j is a rectangle and $(\text{int } S_i) \cap (\text{int } S_j) = \varnothing$ for $i \neq j$. (See Figure 3.5 for an example.) Let $f\colon S \to \mathbf{R}$ be an integrable function. Prove that $f\colon S_j \to \mathbf{R}$ is integrable, $1 \leq j \leq m$, and

$$\int_S f = \sum_{j=1}^{m} \int_{S_j} f$$

8. Suppose $S \subset \mathbf{R}^2$ and $f\colon S \to \mathbf{R}$ a function. Let S_1 and S_2 be rectangles containing S, $S_0 = S_1 \cap S_2$, and define $f_i\colon S_i \to \mathbf{R}$, $i = 0, 1, 2$, by

$$f_i(\mathbf{v}) = \begin{cases} f(\mathbf{v}) & \text{for } \mathbf{v} \in S \\ 0 & \text{for } \mathbf{v} \notin S \end{cases}$$

Prove that each of the functions f_1 and f_2 is integrable if and only if f_0 is integrable and

$$\int_{S_1} f_1 = \int_{S_0} f_0 = \int_{S_2} f_2$$

9. Let $f(x, y) = x$. Compute $\int_S f$, where S is the triangle bounded by the x axis and the lines $x = 1$ and $x = y$. (See Figure 3.6.)

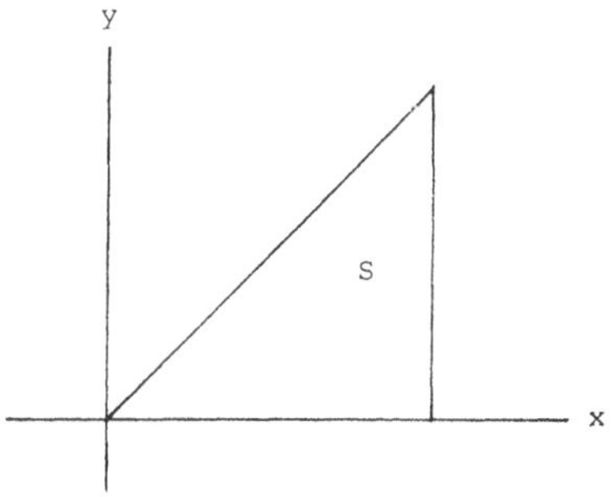

Figure 3.6 A triangular region.

4. The Integral of a Function of *n* Variables

The procedure for defining the integral of a function of n variables will hardly be a surprise after the previous sections. The main problem is the notation.

Let $\mathbf{a} = (a_i)$ and $\mathbf{b} = (b_i)$ be points in $\mathbf{R}^n$ with $a_i < b_i$, $1 \leq i \leq n$, and define the *rectangle* $R(\mathbf{a}, \mathbf{b})$ by

$$R(\mathbf{a}, \mathbf{b}) = \{\mathbf{v} = (x_i) \in \mathbf{R}^n : a_i \leq x_i \leq b_i,\ i = 1, 2, \ldots, n\}$$

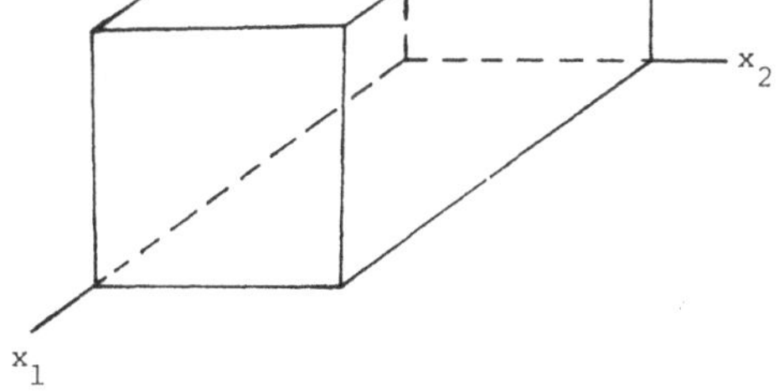

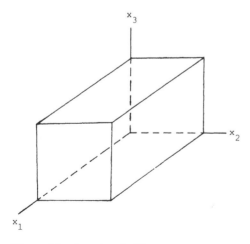

Figure 4.1 A rectangle $\mathbf{R}^3$.

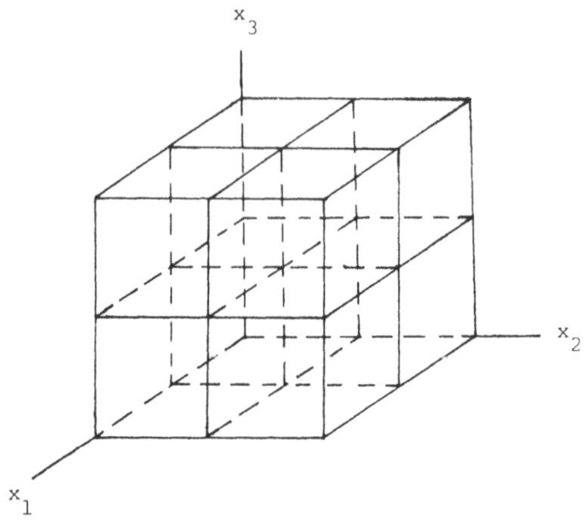

Figure 4.2 A partition of a rectangle in $\mathbf{R}^3$.

The "volume" $\mu(R(\mathbf{a}, \mathbf{b}))$ of the rectangle $R(\mathbf{a}, \mathbf{b})$ is defined to be

$$\mu(R(\mathbf{a}, \mathbf{b})) = (b_1 - a_1)(b_2 - a_2) \cdots (b_n - a_n)$$

For example, if $r = (0, 0, 0)$ and $b = (2, 1, 1)$, the rectangle $R(\mathbf{a}, \mathbf{b})$ is pictured in Figure 4.1. Its volume is 2.

If P_j is a partition of the interval $[a_j, b_j]$, $j = 1, 2, \ldots, n$, we define a *partition* $P = P_1 \times P_2 \times \cdots \times P_n$ of $R(\mathbf{a}, \mathbf{b})$ into subrectangles just as in the case $n = 2$. For example, if $n = 3$ and P_1, P_2, P_3 are each the partition of the interval $[0, 1]$ into two equal pieces, $0 < \frac{1}{2} < 1$, we obtain the partition of the unit cube in $\mathbf{R}^3$ into eight congruent subcubes. (See Figure 4.2.)

Suppose that $f\colon R(\mathbf{a}, \mathbf{b}) \to \mathbf{R}$ is a bounded function and $P = P_1 \times P_2 \times \cdots \times P_n$ a partition of $R(\mathbf{a}, \mathbf{b})$. Let $R_1, \ldots, R_q$ be all of the subrectangles of $R(\mathbf{a}, \mathbf{b})$ occurring in the partition P and define m_i, M_i, $1 \leq i \leq q$, by

$$m_i = \text{glb}\,\{f(\mathbf{v}) : \mathbf{v} \in R_i\}$$

$$M_i = \text{lub}\,\{f(\mathbf{v}) : \mathbf{v} \in R_i\}$$

Then, just as in the previous section, we define $L(f, P)$ and $U(f, P)$ by

$$L(f, P) = \sum_{i=1}^{q} m_i \mu(R_i)$$

$$U(f, P) = \sum_{i=1}^{q} M_i \mu(R_i)$$

The analogue of Lemma 3.1 is as follows.

Lemma 4.1 *Let $f: R(\mathbf{a}, \mathbf{b}) \to \mathbf{R}$ be a bounded function and let $P = P_1 \times \cdots \times P_n$, $P' = P_1' \times \cdots \times P_n'$ be partitions of $R(\mathbf{a}, \mathbf{b})$. Then*

(i) $L(f, P) \leq U(f, P')$.
(ii) If P' is a refinement of P, then

$$U(f, P') \leq U(f, P)$$
$$L(f, P') \geq L(f, P)$$

The proof of this result proceeds just as did the proofs of Lemmas 1.1 and 1.2. We leave it as an exercise for the reader.

Finally, we define subsets of $\mathbf{R}$ by

$$U(f) = \{U(f, P) : P = P_1 \times P_2 \times \cdots \times P_n \text{ is a partition of } R(\mathbf{a}, \mathbf{b})\}$$
$$L(f) = \{L(f, P) : P = P_1 \times P_2 \times \cdots \times P_n \text{ is a partition of } R(\mathbf{a}, \mathbf{b})\}$$

As before, these sets are bounded. We say that f is *integrable* if

$$\text{glb } U(f) = \text{lub } L(f)$$

The *integral* of f is defined to be this common value and denoted by

$$\int_{R(\mathbf{a}, \mathbf{b})} f$$

The analogue of Proposition 2.1 is true for functions of n variables. We state it here, leaving the proof as an exercise.

Proposition 4.2 *Let S be a rectangle in $\mathbf{R}^n$ and $f: S \to \mathbf{R}$ a bounded function. Then f is integrable if and only if for every $\varepsilon > 0$, there is a partition P of S such that*

$$U(f, P) - L(f, P) < \varepsilon$$

Here are some examples.

1. Let $f: R[\mathbf{a}, \mathbf{b}] \to \mathbf{R}$ be the constant function $f(\mathbf{v}) = c$. Then

$$L(f, P) = U(f, P) = c\mu(R(\mathbf{a}, \mathbf{b}))$$

for any partition P. Thus

$$\int_{R(\mathbf{a}, \mathbf{b})} f = c\mu(R(\mathbf{a}, \mathbf{b}))$$

2. Suppose S is the rectangle in $\mathbf{R}^3$ defined by the vectors $\mathbf{a} = (0, 0, 0)$ and $\mathbf{b} = (\mathrm{b}, \mathrm{c}, \mathrm{d})$, and $f: S \to \mathbf{R}$ is defined by $f(x, y, z) = xyz$. Let k be a positive integer and $\tilde{P}_k$ the partition of S into rectangles R_{ijl}, $1 \leq i, j, l \leq k$, where

R_{ijl} consists of all (x, y, z) with

$$\frac{(i-1)b}{k} \le x \le \frac{ib}{k} \qquad \frac{(j-1)c}{k} \le y \le \frac{jc}{k} \qquad \frac{(l-1)l}{k} \le z \le \frac{ld}{k}$$

One checks easily that

$$\mu(R_{ijl}) = \frac{bcd}{k^3}$$

$$M_{ijl} = \text{lub}\,\{f(\mathbf{v}) : \mathbf{v} \in R_{ijl}\}$$
$$= \frac{ib}{k}\frac{jc}{k}\frac{ld}{k}$$
$$= \frac{bcd}{k^3} ijl$$

and
$$m_{ijl} = \text{glb}\,\{f(\mathbf{v}) : \mathbf{v} \in R_{ijl}\}$$
$$= \frac{bcd}{k^3}(i-1)(j-1)(l-1)$$

Therefore,

$$U(f, \tilde{P}_k) = \sum_{i=1}^{k} \sum_{j=1}^{k} \sum_{l=1}^{k} \frac{bcd}{k^3} ijl \frac{bcd}{k^3}$$
$$= \frac{b^2c^2d^2}{k^6} \sum_{i=1}^{k} i \sum_{j=1}^{k} j \sum_{l=1}^{k} l$$
$$= \frac{b^2c^2d^2}{k^6} \frac{k^3(k+1)^3}{8}$$
$$= \frac{b^2c^2d^2}{8}\left(1 + \frac{1}{k}\right)^3$$

and, similarly,

$$L(f, \tilde{P}_k) = \frac{b^2c^2d^2}{8}\left(1 - \frac{1}{k}\right)^3$$

It follows that

$$\lim_{k\to\infty} U(f, \tilde{P}_k) = \lim_{k\to\infty} L(f, \tilde{P}_k) = \frac{b^2c^2d^2}{8}$$

and, just as in Example 4 of Section 1,

$$\int_S xyz = \frac{b^2c^2d^2}{8}$$

In $\mathbf{R}^n$, too, one may wish to integrate over regions other than $\mathbf{R}^n$. The procedure is like that for $\mathbf{R}^2$. If S is a closed, bounded subset of $\mathbf{R}^n$, we let S_1 be a rectangle containing S. Define $f_1: S_1 \to \mathbf{R}$ by

$$f_1(\mathbf{v}) = \begin{cases} f(\mathbf{v}) & \text{if } \mathbf{v} \in S \\ 0 & \text{if } \mathbf{v} \notin S \end{cases}$$

We then say that $f: S \to \mathbf{R}$ is integrable if f_1 is integrable on S_1, and we define

$$\int_S f = \int_{S_1} f$$

It is not hard to show, just as in the case $n = 2$, that this definition does not depend on our choice of S_1. (See Exercise 8.)

Exercises

1. Let S be a rectangle in $\mathbf{R}^n$ and $f: S \to \mathbf{R}$ the constant function $f(\mathbf{v}) = c$, $\mathbf{v} \in S$. Prove that f is integrable and that

$$\int_S f = c\mu(S)$$

2. State and prove the analogue of Proposition 2.1 for functions of n variables.
3. Let $f: \mathbf{R}^n \to \mathbf{R}$ be defined by $f(x_1, \ldots, x_n) = x_1$. Use Exercise 2 to prove that

$$\int_S f = \frac{1}{2}$$

where S is the rectangle defined by $(0, 0, \ldots, 0)$ and $(1, 1, \ldots, 1)$. (See Example 2 of Section 1.)
4. Let S be the rectangle in $\mathbf{R}^n$ defined by the points $\mathbf{a} = (a_i)$, $\mathbf{b} = (b_i)$ and let $f: S \to \mathbf{R}$ be the function $f(x_1, \ldots, x_n) = x_1$. Prove directly from the definition that

$$\int_S f = \frac{1}{2}(b_n - a_n) \cdots (b_2 - a_2)(b_1^2 - a_1^2)$$

5. Determine, directly from the definition, the integral of f over the rectangle S in $\mathbf{R}^3$ defined by $(0, 0, 0)$ and $(1, 1, 1)$ when
   (a) $f(x, y, z) = 2x + y + z$
   (b) $f(x, y, z) = xyz$
   (c) $f(x, y, z) = x^2 + y^2 + z^2$
6. Prove Lemma 4.1.
*7. State and prove the analogue of Exercise 7 of Section 3 for $\mathbf{R}^n$.
8. State and prove the analogue of Exercise 8 of Section 3 for $\mathbf{R}^n$.
*9. Let $f: \mathbf{R}^3 \to \mathbf{R}$ be given by $f(x, y, z) = x$ and compute $\int_S f$, where S is the region in the first octant bounded by the coordinate planes and the plane $x + y + z = 1$. (See Figure 4.3.)

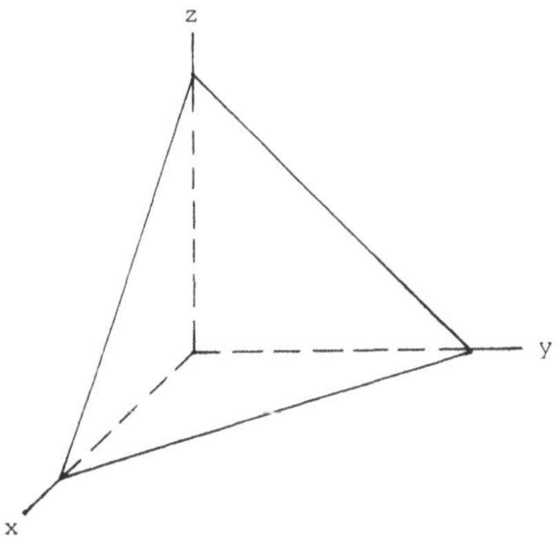

Figure 4.3

10. Let S be a rectangle in $\mathbf{R}^n$, let $f: S \to \mathbf{R}$ be integrable, and define f^+ and f^- by

$$f^+(\mathbf{v}) = \max\{f(\mathbf{v}), 0\}$$
$$f^-(\mathbf{v}) = \max\{-f(\mathbf{v}), 0\}$$

Show that f^+ and f^- are integrable, that $|f(\mathbf{v})| = f^+(\mathbf{v}) + f^-(\mathbf{v})$, and that $f(\mathbf{v}) = f^+(\mathbf{v}) - f^-(\mathbf{v})$. Use this to prove that if f is integrable, then $|f|$ is integrable.

*11. Show that if f and g are bounded integrable functions on the rectangle S, then fg is also integrable on S. (See Exercises 11 and 12 of Section 1.)

*12. We have defined integrals only for bounded functions on rectangles. In this exercise, we define the integral of certain unbounded functions.

Let S be a rectangle in $\mathbf{R}^n$, and let $f: S \to \mathbf{R}$ be a nonnegative function. For each positive integer n, define $f_n; S \to \mathbf{R}$ by

$$f_n(\mathbf{v}) = \min\{f(\mathbf{v}), n\}$$

(a) Show that for every $\mathbf{v} \in S$, $\lim_{n\to\infty} f_n(\mathbf{v}) = f(\mathbf{v})$.

(b) Suppose that for every n, the function f_n is integrable on S, and suppose that the set $\{\int_S f_n : n = 1, 2, \ldots\}$ is bounded. Show that

$$\lim_{n\to\infty} \int_S f_n$$

exists.

We define

$$\int_S f = \lim_{n\to\infty} \int_S f_n$$

when the limit exists. If f is an arbitrary function (not necessarily nonnegative), then we define

$$\int_S f = \int_S f^+ - \int_S f^-$$

(see Exercise 10) if $\int_S f^+$ and $\int_S f^-$ both exist. (Note that both f^+ and f^- are nonnegative.)

(c) Show that this definition coincides with the usual one if f is bounded.

(d) Show that the analogue of Proposition 2.1 still holds when we extend the definition as above. (See Proposition 5.1.)

Note. The integral defined in this exercise is often called an *improper integral* when f is unbounded.

(e) Show that $\int_0^1 f(x)\,dx$ is defined for $f(x) = x^k$ if and only if $k > -1$. [Define $f(0) = 0$ for $k \leq 0$.]

(f) Prove the analogue of Exercise 7, Section 3, for these integrals.

(g) Show that $|f|$ is integrable if f is.

*13. Another possible definition of integrals for unbounded functions is the following: if f is as in the preceding problem, define

$$f^{(n)}(\mathbf{v}) = \begin{cases} f(\mathbf{v}) & \text{if } |f(\mathbf{v})| \leq n \\ n & \text{if } f(\mathbf{v}) > n \\ -n & \text{if } f(\mathbf{v}) < -n \end{cases}$$

and let $\int_S' f = \lim_{n\to\infty} \int_S f^{(n)}$ if all the functions $f^{(n)}$ are integrable and the limit exists. (We write $\int_S' f$ to emphasize that this is *not* a definition we shall use.)

(a) Show that if $\int_S f$ exists (possibly as an improper integral, in the sense described in the previous problem), then $\int_S' f$ exists and equals $\int_S f$.

(b) Show that if $S = [-1, 1] \subset \mathbf{R}$ and $f(x) = 1/x$ [$f(0) = 0$], then $\int_S' f$ exists, but $\int_S f$ does not.

(c) Parts (a) and (b) may make it appear that the new definition is better than the old one because it is more general. Show, however, that the analogue of Exercise 7, Section 3, fails for $\int'$. [If $S \subset \mathbf{R}^1$, this means that part (a) of Proposition 2.2 fails to hold.]

*14. Let $f\colon \mathbf{R}^n \to \mathbf{R}$ be a function. We can sometimes define a different sort of improper integral of f,

$$\int_{\mathbf{R}^n} f$$

The difference is that in the previous case (Exercise 11), f became unbounded, while here the rectangle becomes unbounded.

Suppose first that f is nonnegative. Let $\{S_m\}$ be a sequence of rectangles such that $S_1 \subset S_2 \subset S_3 \subset \cdots$ and $\cup_{m=1}^{\infty} S_m = \mathbf{R}^n$. [For instance, S_m might be $R(-\mathbf{v}_m, \mathbf{v}_m)$, where $\mathbf{v}_m = (m, m, \ldots, m)$.] Define

$$\int_{\mathbf{R}^n} f = \lim_{m\to\infty} \int_{S_m} f$$

if all the integrals $\int_{S_m} f$ are defined and the limit exists.

(a) Show that $\int_{\mathbf{R}^n} f$ is defined if and only if all the integrals $\int_{S_m} f$ are defined and the sequence $\int_{S_m} f$ is bounded.
(b) Show that if $\int_{\mathbf{R}^n} f$ is defined, then $\int_S f$ is defined for any rectangle $S \subset \mathbf{R}^n$, and the value of $\int_{\mathbf{R}^n} f$ does not depend on the particular sequence S_m chosen.
(c) Now let f be arbitrary; define

$$\int_{\mathbf{R}^n} f = \int_{\mathbf{R}^n} f^+ - \int_{\mathbf{R}^n} f^-$$

if both integrals on the right are defined. Prove results analogous to those in parts (d) and (f) of Exercise 11. (See Exercise 10.)
(d) Show that $\int_{\mathbf{R}} 1/(1+x^2)\,dx$ is defined, but that $\int_{\mathbf{R}} x/(1+x^2)\,dx$ is not.
(e) Show that if f is integrable on $\mathbf{R}^n$, so is $|f|$.

Note. We can define $\int_S f$ similarly when f is an infinite rectangle other than $\mathbf{R}^n$; simply define f to be 0 off S, and then set

$$\int_S f = \int_{\mathbf{R}^n} f$$

*15. On $\mathbf{R} = \mathbf{R}^1$, one sometimes defines improper integrals slightly differently. We shall discuss integrals of the form $\int_a^\infty f(x)\,dx$; there are similar results for $\int_{-\infty}^a f(x)\,dx$. What is special about $\mathbf{R}$, incidentally, is that there are only two ways for a rectangle in $\mathbf{R}$ to become unbounded.

Let $f\colon [a, \infty] \to \mathbf{R}$ be a function such that $\int_a^t f(x)\,dx$ exists for all $t \geq a$. We define

$$\int_a^\infty f(x)\,dx = \lim_{t\to\infty} \int_a^t f(x)\,dx$$

if the limit exists. [More explicitly, $\int_a^\infty f(x)dx = K$ if and only if for every $\varepsilon > 0$ there is a number T such that

$$\left| K - \int_a^t f(x)\,dx \right| < \varepsilon$$

if $t > T$.]
(a) Show that Proposition 2.2 holds for this integral.
(b) Find a function f such that $\int_a^\infty f(x)\,dx$ is defined (for some a), but $\int_a^\infty |f(x)|\,dx$ is not. (This shows that this new definition differs from the one in the previous problem.)
(c) For what values of n is $\int_1^\infty x^n\,dx$ defined?
(d) If f is defined on all of $\mathbf{R}$, then one defines $\int_{\mathbf{R}} f(x)\,dx = \int_{-\infty}^\infty f(x)\,dx$ to be $\int_{-\infty}^a f(x)\,dx + \int_a^\infty f(x)\,dx$, if both integrals exist. Show that the choice of the number a does not affect the definition.
(e) Show that

$$\int_{-\infty}^\infty \frac{x\,dx}{1+x^2}$$

is not defined, although

$$\lim_{t\to\infty} \int_{-t}^{t} \frac{x\,dx}{1+x^2}$$

exists. (This shows that it makes a difference in defining $\int_{-\infty}^{\infty} f(x)\,dx$ whether or not the endpoints of the region of integration become infinite independently.)

5. Properties of the Integral

We now develop some of the elementary properties of the integral of a function of n variables.

Proposition 5.1 *Let S be a bounded subset of $\mathbf{R}^n$, c a real number, and $f, g: S \to \mathbf{R}$ integrable functions. Then the functions $f+g$, $cf: S \to \mathbf{R}$ are integrable, with*

$$\int_S (f+g) = \int_S f + \int_S g$$

$$\int_S cf = c\int_S f$$

Proof. We prove the part of the proposition relating to the function $f+g$, leaving the proof of the assertion involving the function cf as Exercise 1. Since the integral of a function over a set that is not a rectangle is defined to be the integral of a related function over a rectangle, it follows that we can assume S is a rectangle.

Let ε be any positive number and let P be a partition of S into subrectangles $S_1, \ldots, S_m$ such that

$$U(f, P) - L(f, P) < \frac{\varepsilon}{2}$$

$$U(g, P) - L(g, P) < \frac{\varepsilon}{2}$$

This is possible because both f and g are integrable. Define numbers M_i, M_i', M_i'', $1 \le i \le m$, by

$$M_i = \text{lub}\,\{f(\mathbf{v}) + g(\mathbf{v}) : \mathbf{v} \in S_i\}$$

$$M_i' = \text{lub}\,\{f(\mathbf{v}) : \mathbf{v} \in S_i\}$$

$$M_i'' = \text{lub}\,\{g(\mathbf{v}) : \mathbf{v} \in S_i\}$$

so that

$$U(f+g, P) = \sum_{i=i}^{m} M_i \mu(S_i)$$

$$U(f, P) = \sum_{i=1}^{m} M_i' \mu(S_i)$$

$$U(g, P) = \sum_{i=1}^{m} M_i'' \mu(S_i)$$

Since $M_i \leq M_i' + M_i''$ (see Proposition 3.4, Chapter 7),

$$U(f+g, P) \leq U(f, P) + U(g, P) \tag{5.1}$$

Similarly, it follows that

$$L(f+g, P) \geq (f, P) + L(g, P) \tag{5.2}$$

Thus
$$\begin{aligned} U(f+g, P) - L(f+g, P) &\leq [U(f, P) + U(g, P)] \\ &\quad - [L(f, P) + L(g, P)] \\ &\leq [U(f, P) - L(f, P)] \\ &\quad + [U(g, P) - L(g, P)] \\ &< \varepsilon \end{aligned}$$

and $f+g$ is then integrable by Proposition 4.2.

In addition, the inequalities (5.1) and (5.2) imply that

$$\begin{aligned} \int_S (f+g) = \text{glb } U(f+g) &\leq \text{glb } U(f) + \text{glb } U(g) \\ &= \int_S f + \int_S g \end{aligned}$$

and
$$\begin{aligned} \int_S (f+g) = \text{lub } L(f+g) &\geq \text{lub } L(f) + \text{lub } L(g) \\ &= \int_S f + \int_S g \end{aligned}$$

Therefore
$$\int_S (f+g) = \int_S f + \int_S g$$

as desired.

A straightforward induction argument establishes the following.

Corollary *Let S be a subset of $\mathbf{R}^n$, $f_j : S \to \mathbf{R}$ integrable functions, and a_j real numbers, $j = 1, \ldots, k$. Then $a_1 f_1 + \cdots + a_k f_k$ is integrable with*

$$\int_S \sum_{j=1}^{k} a_j f_j = \sum_{j=1}^{k} a_j \int_S f_j$$

We now prove that the integral of a nonnegative function is nonnegative.

Proposition 5.2 *Let S be a subset of $\mathbf{R}^n$ and $f : S \to \mathbf{R}$ a nonnegative integrable function on S. Then*

$$\int_S f \geq 0$$

Proof. Again, we can assume S is a rectangle. Let P be the trivial partition of S, the one consisting of exactly one subrectangle, and let

$$m = \text{glb}\,\{f(\mathbf{v}) \mid \mathbf{v} \in S\}$$

Then $L(f, P) = m\mu(S) \geq 0$ since $m \geq 0$. As a consequence,

$$\int_S f = \text{lub}\; L(f) \geq L(f, P) \geq 0$$

and the result is proved.

Corollary 1 Let S be a subset of $\mathbf{R}^n$ and $f, g : S \to \mathbf{R}$ integrable functions with $f(x) \geq g(x)$ for all x in S. Then

$$\int_S f \geq \int_S g$$

Proof. Apply Proposition 5.2 to the function $f - g$ and use Proposition 5.1.

Corollary 2 *Let S be a rectangle in $\mathbf{R}^n$ and f: $S \to \mathbf{R}$ an integrable function. Suppose m and M are numbers with*

$$m \leq f(\mathbf{v}) \leq M$$

for all $\mathbf{v} \in S$. Then

$$m\mu(S) \leq \int_S f \leq M\mu(S)$$

The proof is immediate.

Let $f : S \to \mathbf{R}$ be a nonnegative integrable function which is *not* identically zero. One might expect that, in this case, the integral of f over S would be positive. This is not generally true. (See Exercise 4.) However, one can prove the following.

Proposition 5.3 *Let S be a rectangle in $\mathbf{R}^n$ and $f\colon S \to \mathbf{R}$ a nonnegative continuous integrable function which is not identically zero. Then*

$$\int_S f > 0$$

Proof. In the interest of simplifying notation, we prove this proposition for the case $n = 2$. The general proof is similar.

Let the rectangle be $R = R(\mathbf{a}, \mathbf{b})$, where $\mathbf{a} = (a_1, a_2)$ and $\mathbf{b} = (b_1, b_2)$. Suppose that $f(\mathbf{v}_0) = c > 0$; for simplicity, assume that $\mathbf{v}_0 = (x_0, y_0)$ is not on the boundary of the rectangle (so that $a_1 < x_0 < b_1$, $a_2 < y_0 < b_2$). Choose $\delta > 0$ so that $|f(\mathbf{v}) - f(\mathbf{v}_0)| < c/2$ when $\mathbf{v} \in B = B_\delta(\mathbf{v}_0)$. It follows that

$$\begin{aligned} f(\mathbf{v}) = |f(\mathbf{v})| &= |f(\mathbf{v}_0) + f(\mathbf{v}) - f(\mathbf{v}_0)| \\ &\geq |f(\mathbf{v}_0)| - |f(\mathbf{v}) - f(\mathbf{v}_0)| \\ &\geq c - \frac{c}{2} = \frac{c}{2} \end{aligned}$$

for $\mathbf{v} \in B \cap R$. Furthermore, we may assume (by possibly making δ smaller) that $B \subset R$.

Now let $R_0 = \{\mathbf{v} = (x, y) : 0 \leq x - x_0 \leq \delta/2,\ 0 \leq y - y_0 \leq \delta/2\}$. Then R_0 is a rectangle, and $R_0 \subset B$. Let $g\colon S \to \mathbf{R}$ be defined by

$$g(\mathbf{v}) = \begin{cases} \dfrac{c}{2} & \text{if } \mathbf{v} \in R_0 \\ 0 & \text{if } \mathbf{v} \notin R_0 \end{cases}$$

Then $f(\mathbf{v}) \geq g(\mathbf{v})$. Moreover, if we take a partition P of R so that R_0 is one of the rectangles in P, then $U(P, g) = L(P, g) = c\delta^2/8$. By Proposition 3.2, g is integrable. Clearly

$$\int_S g = \frac{c\delta^2}{8} > 0$$

Hence $\int_S f \geq \int_S g > 0$, as claimed.

Proposition 5.4 *Let S be a subset in $\mathbf{R}^n$ and $f\colon S \to \mathbf{R}$ an integrable function. Then the function $|f|$ is integrable with*

$$\left|\int_S f\right| \leq \int_S |f|$$

Both assertions of this proposition follow by comparing the upper and lower sums of f and $|f|$. We leave the details to the reader as Exercise 5. (See also Exercise 8.)

Exercises

1. Prove the second half of Proposition 5.1.
2. Prove the corollary to Proposition 5.1.
3. Prove the second corollary to Proposition 5.2.
4. Let S be the rectangle in $\mathbf{R}^2$ defined by the points $(0, 0)$ and $(1, 1)$. Define $f: S \to \mathbf{R}$ by

$$f(x, y) = \begin{cases} 1 & \text{if } y = 0 \\ 0 & \text{if } y \neq 0 \end{cases}$$

Prove that f is integrable with

$$\int_S f = 0$$

5. Prove Proposition 5.4.
6. Let S be a compact pathwise connected subset of $\mathbf{R}^n$, and let $f: S \to \mathbf{R}$ a continuous integrable function. Prove that there is a point $\mathbf{v}_0 \in S$ such that

$$\int_S f = \mu(S)\, f(\mathbf{v}_0)$$

This is the *first mean value theorem for integrals*. (*Hint*: Combine Corollary 2 to Proposition 5.2 with the corollary to Theorem 3.3 of Chapter 7 and apply Exercise 9 of Section 7.6.)

7. Let S be a compact subset of $\mathbf{R}^n$; let $f, h: S \to \mathbf{R}$ be continuous integrable functions with $h \geq 0$. Show that there is a point $\mathbf{v}_0 \in S$ with

$$\int_S f(\mathbf{v})\, h(\mathbf{v}) = f(\mathbf{v}_0) \int_S h(\mathbf{v})$$

(See Exercise 11 of Section 2. You may assume that fh is integrable.)

8. Let S be a rectangle in $\mathbf{R}^n$, and let $f: S \to \mathbf{R}^n$ be a function. Is it true that if $|f|$ is integrable, then f is?
9. Let S be a rectangle in $\mathbf{R}^n$, and $f: S \to \mathbf{C}$ be a complex-valued function. We say that f is integrable if $\mathrm{Re}(f)$ and $\mathrm{Im}(f)$ are integrable, and define

$$\int_S f = \int_S \mathrm{Re}(f) + i \int_S \mathrm{Im}(f)$$

(a) Prove the analogue of Proposition 5.1 for such functions f. (Note: c is allowed to be complex.)

*(b) Show that if f is integrable, then so is $|f|$, and $|\int_S f| \leq \int_S |f|$.

6. Integrable Functions

The basic result of this section is the assertion that a continuous function on a rectangle is integrable. For the integrals of a function over a set which is not a rectangle, however, we need to know that certain discontinuous functions are integrable. We shall describe a result of this sort, using the notion of *content zero*. (The proof will be given in the next section.)

Theorem 6.1 *Let S be rectangle in $\mathbf{R}^n$, and let $f\colon S \to \mathbf{R}$ be a continuous function. Then f is integrable.*

Before proving this result, we introduce two useful notions.

Let $A \subset \mathbf{R}^n$ be a bounded set. The *diameter of A*, diam A, is defined by

$$\operatorname{diam} A = \operatorname{lub} \{ \|\mathbf{v} - \mathbf{w}\| : \mathbf{v}, \mathbf{w} \in A \}$$

For example, if $A = B_r(\mathbf{v}_0)$ is the ball of radius r about $\mathbf{v}_0$, then diam $A = 2r$. (Think of diametrically opposed points.) Every bounded set A has a diameter, as the set

$$\{ \|\mathbf{v} - \mathbf{w}\| : \mathbf{v}, \mathbf{w} \in A \}$$

is bounded and has a least upper bound by the completeness of the real numbers. (See Theorem 2.2, Chapter 7.)

Let P be a partition of the rectangle S. We define the *mesh of P*, mesh P, to be the largest of the diameters of the subrectangles making up P. It follows that if $\mathbf{v}$ and $\mathbf{w}$ are two points of S in the same subrectangle of P, then

$$\|\mathbf{v} - \mathbf{w}\| \le \operatorname{mesh} P$$

Proof of Theorem 6.1. We prove that, for any $\varepsilon > 0$, there is a partition P of S so that

$$U(f, P) - L(f, P) < \varepsilon$$

Theorem 6.1 will then follow from Proposition 4.2.

Suppose $\varepsilon > 0$. Since S is compact and f is continuous, it follows that f is uniformly continuous. (See Theorem 5.1, Chapter 7.) Thus, there is a $\delta > 0$ such that

$$|f(\mathbf{v}) - f(\mathbf{w})| < \frac{\varepsilon}{\mu(S)}$$

whenever $\|\mathbf{v} - \mathbf{w}\| < \delta$.

Let P be a partition of S with mesh $P < \delta$. We denote the subrectangles of S by $S_1, \ldots, S_k$ and define numbers M_i and m_i, $1 \le i \le k$, by

$$M_i = \text{lub } \{f(\mathbf{v}) : \mathbf{v} \in S_i\}$$

$$m_i = \text{glb } \{f(\mathbf{v}) : \mathbf{v} \in S_i\}$$

Then
$$U(f, P) = \sum_{i=1}^{k} M_i \mu(S_i)$$

and
$$L(f, P) = \sum_{i=1}^{k} m_i \mu(S_i)$$

Now f is continuous and each S_i is compact; so according to Theorem 3.3 of Chapter 7, we can find points $\mathbf{v}_i, \mathbf{w}_i \in S_i$ with $M_i = f(\mathbf{v}_i)$ and $m_i = f(\mathbf{w}_i)$, $1 \le i \le k$. Furthermore, since mesh $P < \delta$, we have $\|\mathbf{v}_i - \mathbf{w}_i\| < \delta$, $1 \le i \le k$. It follows that

$$M_i - m_i = f(\mathbf{v}_i) - f(\mathbf{w}_i) < \frac{\varepsilon}{\mu(S)}$$

Thus
$$\begin{aligned} U(f, P) - L(f, P) &= \sum_{i=1}^{k} (M_i - m_i)\mu(S_i) \\ &< \sum_{i=1}^{k} \frac{\varepsilon}{\mu(S)} \mu(S_i) \\ &= \frac{\varepsilon}{\mu(S)} \sum_{i=1}^{k} \mu(S_i) = \varepsilon \end{aligned}$$

This completes the proof of Theorem 6.1.

The requirement in Theorem 6.1 that f be continuous can be weakened; it is only necessary that f be continuous at "most" points of S. To make this precise, we need the notion of content zero.

A set $B \subset \mathbf{R}^n$ is said to have *content zero* (or *volume zero*) if, for any $\varepsilon > 0$, we can find a finite collection of rectangles in $\mathbf{R}^n$ containing B in their union and having total volume less that ε. For example, any line segment in $\mathbf{R}^2$ has content zero. (See Figure 6.1.) In the same way, an arc of a circle in $\mathbf{R}^2$ has content zero. (See Figure 6.2.) More generally, "lower dimensional sets"—curves in $\mathbf{R}^2$, surfaces in $\mathbf{R}^3$, and the like, have zero content. (*Warning*: Any theorem of this sort needs to be carefully phrased, since there are many pathological counterexamples. See Exercise 7, Section 9.2.)

The strengthened form of Theorem 6.1 is the following.

Theorem 6.2 *Let S be a rectangle in $\mathbf{R}^n$ and $f\colon S \to \mathbf{R}$ a bounded function. Suppose that $B \subset S$ is a set of content zero and that f is continuous on $S - B$. Then f is integrable.*

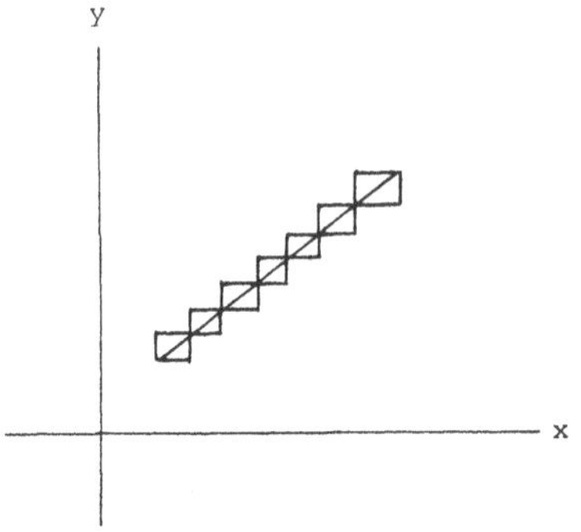

Figure 6.1 A line segment has content zero.

We shall prove this theorem in the next section. The idea behind the proof is, as usual, to use Proposition 4.2. We enclose B in rectangles of very small volume, so that upper and lower sums for f on B cannot differ by much; on the rest of S, f is continuous and we can reason as in Theorem 6.1.

As we remarked above, Theorem 6.2 can be used to prove the integrability of continuous functions on sets other than rectangles. For example, let S be the subset in $\mathbf{R}^2$ defined by the curves $y = 1 - x^2$ and $y = x^2$ and let S_1 be a rectangle containing S. (See Figure 6.3.) Let $f\colon S \to \mathbf{R}$ be any continuous function and define $f_1\colon S_1 \to \mathbf{R}$ by

$$f_1(\mathbf{v}) = \begin{cases} f(\mathbf{v}) & \text{if } \mathbf{v} \in S \\ 0 & \text{if } \mathbf{v} \notin S \end{cases}$$

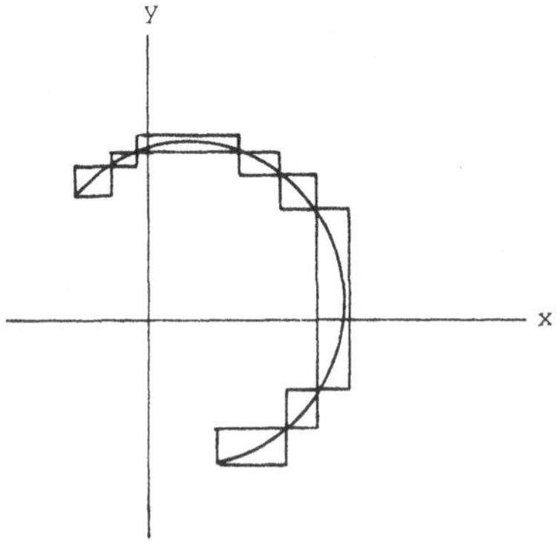

Figure 6.2 A circle has content zero.

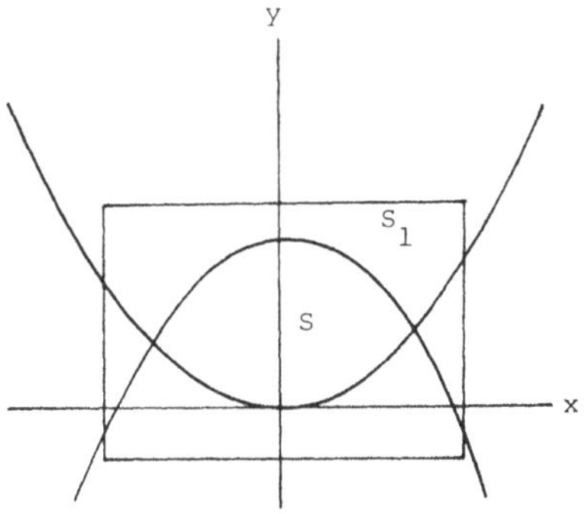

Figure 6.3 A rectangle containing S.

Then f_1 is continuous except for the points on the two curves $y = 1 - x^2$ and $y = x^2$. Since these curves have content zero, Theorem 6.2 applies to say that f_1 is integrable. Thus, by definition, $f\colon S \to \mathbf{R}$ is integrable.

Exercises

1. Prove that, if $A \subset B$, then diam $A \leq$ diam B.
2. Prove diam $B_r(\mathbf{v}_0) = 2r$.
3. Let P and P' be partitions of a rectangle S in $\mathbf{R}^n$. Prove that if P' is a refinement of P, then mesh $P' \leq$ mesh P.
4. Prove that a line segment in $\mathbf{R}^2$ has content zero.
5. Prove that the finite union of sets of content zero has content zero.
6. Let $f\colon [a, b] \to \mathbf{R}$ be an integrable function, and let $G(f)$ be the graph of f,

$$G(f) = \{(x, y) \in \mathbf{R}^2 : y = f(x)\}$$

Show that $G(f)$ has zero content in $\mathbf{R}^2$. [*Hint*: Given ε, let P be a partition of $[a, b]$ such that $U(f, P) - L(f, P) < \varepsilon$. Use the graphs of the step functions for $U(f, P)$ and $L(f, P)$ to enclose $G(f)$ in a union of rectangles of area less than ε.]

*7. Let S be a rectangle in $\mathbf{R}^n$ and $f\colon S \to \mathbf{R}$ an integrable function. Prove that the graph of f,

$$G(f) = \{(\mathbf{x}, y) \in \mathbf{R}^n \times \mathbf{R}^1 : y = f(\mathbf{x})\}$$

has zero content in $\mathbf{R}^{n+1} = \mathbf{R}^n \times \mathbf{R}^1$.

8. Let $\{\mathbf{v}_n\}$ be a convergent sequence in $\mathbf{R}^n$. Show that the set $\{\mathbf{v}_1, \mathbf{v}_2, \ldots, \mathbf{v}_n, \ldots\}$ has content zero.

*9. Let $f\colon [a, b] \to \mathbf{R}$ be a bounded increasing function. [That is, $f(x_1) \geq f(x_2)$ whenever $x_1 \geq x_2$.] Show that f is integrable.

7. The Proof of Theorem 6.2

The proof uses Proposition 4.2; we wish to show that for every $\varepsilon > 0$ there is a partition P of S such that

$$U(f, P) - L(f, P) < \varepsilon$$

For the moment, let P be any partition, with subrectangles $S_1, \ldots, S_m$, and let $S_1, \ldots, S_k$ be the subrectangles of P which intersect B. Then $B \subset \cup_{i=1}^{k} S_k$; and, if

$$M_i = \text{lub } \{f(\mathbf{v}) : \mathbf{v} \in S_i\}$$
$$m_i = \text{glb } \{f(\mathbf{v}) : \mathbf{v} \in S_i\}$$

we have

(7.1)

$$U(f, P) - L(f, P) = \sum_{i=1}^{k} (M_i - m_i)\mu(S_i) + \sum_{i=k+1}^{m} (M_i - m_i)\mu(S_i)$$

We shall show that for an appropriate choice of P, we can make the first sum small because the total volume of $\cup_{i=1}^{k} S_i$ can be made small, while the second sum can be made small (as in Theorem 6.1) because f is continuous on the remaining rectangles.

To begin with, let

$$M = \text{lub } \{f(\mathbf{v}) : \mathbf{v} \in S\}$$
$$m = \text{glb } \{f(\mathbf{v}) : \mathbf{v} \in S\}$$

and choose the rectangles $S_1, \ldots, S_k$ so that $B \subset S_1 \cup \cdots \cup S_k$ and

$$\sum_{i=1}^{k} \mu(S_i) < \frac{\varepsilon}{2(M - m)} \tag{7.2}$$

Let $S_{k+1}, \ldots, S_m$ be additional subrectangles so that $S_1, \ldots, S_k, S_{k+1}, \ldots, S_m$ define a partition P of S. We can, by slightly enlarging the rectangles $S_1, \ldots, S_k$ if necessary, assume that $B \subset \text{Int } S_1 \cup \cdots \cup \text{Int } S_k$ so that f is continuous on the compact set $S' = S_{k+1} \cup \cdots \cup S_{p_m}$. (See Figure 7.1.)

Now, according to Theorem 5.1, Chapter 7, f is uniformly continuous on S'; so we can choose $\delta > 0$ such that

$$|f(\mathbf{v}) - f(\mathbf{w})| < \frac{\varepsilon}{2\mu(S)}$$

for $\mathbf{v}, \mathbf{w} \in S'$ $\|\mathbf{v} - \mathbf{w}\| < \delta$. We can assume, subdividing if necessary, that diam $S_i < \delta$ for $i = k + 1, \ldots, m$. (This subdivision may subdivide the

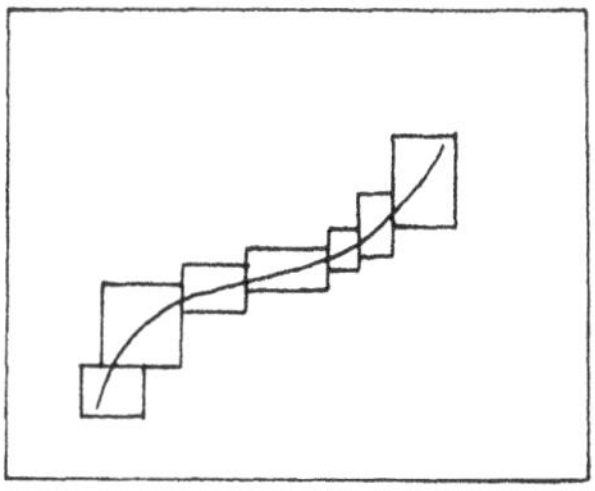

Figure 7.1

rectangles $S_1, \ldots, S_k$, too, but for notational convenience we assume that $S_1, \ldots, S_k$ remain unchanged.) Then

$$M_i - m_i < \frac{\varepsilon}{2\mu(S)} \tag{7.3}$$

for $i = k + 1, \ldots, m$. Using (7.2) and (7.3), we have

$$\begin{aligned} U(f, P) - L(f, P) &= \sum_{i=1}^{k} (M_i - m_i)\mu(S_i) + \sum_{i=k+1}^{m} (M_i - m_i)\mu(S_i) \\ &< \sum_{i=1}^{k} (M - m)\mu(S_i) + \sum_{i=k+1}^{m} \frac{\varepsilon}{2\mu(S)}\mu(S_i) \\ &= (M - m) \sum_{i=1}^{k} \mu(S_i) + \frac{\varepsilon}{2\mu(S)} \sum_{i=k+1}^{m} \mu(S_i) \\ &< (M - m)\frac{\varepsilon}{2(M - m)} + \frac{\varepsilon}{2\mu(S)}\mu(S) \\ &= \varepsilon \end{aligned}$$

This completes the proof of Theorem 6.2.

Theorem 6.2 can also be extended. Define a set $B \subset S$ to be of *measure zero* if for every $\varepsilon > 0$ we can find a set of rectangles $R_1, R_2, \ldots$, possibly overlapping, such that

$$B \subset \bigcup_{i=1}^{\infty} R_i \qquad \sum_{i=1}^{\infty} \mu(R_i) \leq \varepsilon$$

[The second statement means that $\sum_{i=1}^{n} \mu(R_i) \leq \varepsilon$ for every n.]

Theorem 7.1 *The bounded function $f: S \to \mathbf{R}$ is integrable if and only if the set of points on which f is discontinuous is of measure* 0.

The interested reader may wish to prove Theorem 7.1 by working through Exercises 3 through 7.

Exercises

1. Prove that a compact set with measure zero has content zero.
2. Show that the set $S = \{(x, y) \in \mathbf{R}^2 : x = 0\}$ has measure zero but does not have content zero.

*3. Let S be a compact rectangle in $\mathbf{R}^n$, and let $f: S \to \mathbf{R}$ be bounded. We define the *oscillation* of f at $\mathbf{v}_0$, $\omega(f, \mathbf{v}_0)$, by

$$\omega(f, \mathbf{v}_0) = \lim_{\delta > 0} (\text{lub}\,\{f(\mathbf{v}) : \|\mathbf{v} - \mathbf{v}_0\| < \delta\} - \text{glb}\,\{f(\mathbf{v}) : \|\mathbf{v} - \mathbf{v}_0\| < \delta\})$$

Prove that the above limit always exists and that f is continuous at $\mathbf{v}_0$ if and only if $\omega(f, \mathbf{v}_0) = 0$. (See Exercises 13 and 14 of Section 7.2.)

*4. Let f and S be as in Exercise 3, and suppose that f is integrable. For $\varepsilon > 0$, set

$$K_\varepsilon = \{\mathbf{v}_0 \in S : \omega(f, \mathbf{v}_0) \geq \varepsilon\}$$

Show that K_ε has content 0. (*Hint*: If not, then $\exists \delta > 0$: any finite union of rectangles containing K_ε has volume $\geq \delta$. Show that any upper and lower sums for f differ by at least $\delta\varepsilon$.)

*5. Let f and S be as in Exercise 3, and suppose that $K \subset S$ is a compact set such that

$$\omega(f, \mathbf{v}_0) < \varepsilon$$

for all $\mathbf{v}_0 \in K$. Show that $\exists \delta > 0$ such that if $\mathbf{v}_0, \mathbf{v}_1 \in K$ and $\|\mathbf{v}_0 - \mathbf{v}_1\| < \delta$, then $|f(\mathbf{v}_0) - f(\mathbf{v}_1)| < \varepsilon$. (Note the similarity of this result to uniform continuity. That should give a hint of how to do this problem.)

*6. Let f and S be as in Exercise 3, and let K_ε be defined as in Exercise 4. Show that if K_ε has content 0 for all $\varepsilon > 0$, then f is integrable on S. (*Hint*: The proof is like that for Theorem 6.2, but the result of Exercise 5 replaces Theorem 5.1, Chapter 7.) Note that this is the converse of Exercise 4.

*7. Prove Theorem 7.1. (*Hint*: The set of points on which f is discontinuous is $\cup_{n=1}^{\infty} K_{1/n}$. It is also useful to remember that $\sum_{n=1}^{\infty} \varepsilon/2^n = \varepsilon$.)

11

Iterated Integrals and the Fubini Theorem

In the previous chapter we introduced the notion of the integral for a function of n variables. We now take up the problem of computing this integral. The basic result is the Fubini theorem, which effectively reduces the problem of integrating a function of n variables to the evaluation of n ordinary integrals of a function of one variable. We give two forms of the Fubini theorem and show how to apply it to some problems.

We also discuss two other topics in this chapter. One is the differentiation of a function defined by an integral; we show that, under certain circumstances, differentiating an integral is the same as integrating a derivative. The other, the formula for changing variables in an integral, is an n-dimensional analogue of integration by substitution.

1. The Fubini Theorem

In Section 10.6 we proved that a large class of functions, the continuous functions, are integrable. However, we still have no convenient way of evaluating the integral. We rectify this situation now by reducing the problem of integrating a function of n variables to that of performing n integrations of a function of a single variable. We begin with functions of two variables.

Let $S = R(\mathbf{a}, \mathbf{b})$ be the rectangle with $\mathbf{a} = (a_1, a_2)$ and $\mathbf{b} = (b_1, b_2)$, as in Figure 1.1, and let $f: S \to \mathbf{R}$ be a continuous function. If we fix a point y in $[a_2, b_2]$, we can think of $f(x, y)$ as a function from $[a_1, b_1]$ into $\mathbf{R}$. In fact, this function is continuous and we can integrate it. The integral depends on y;

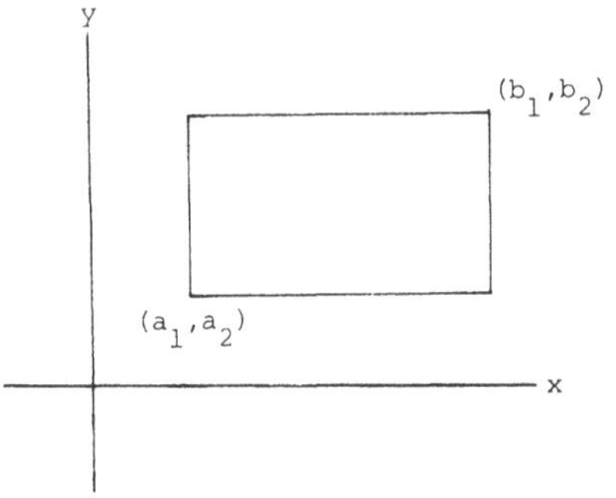

Figure 1.1 A rectangle in $\mathbf{R}^2$.

thus it gives a function $g : [a_2, b_2] \to \mathbf{R}$,

$$g(y) = \int_{a_1}^{b_1} f(x, y)\, dx$$

For example, if $f(x, y) = 2xy - y^2$,

$$\begin{aligned} g(y) &= \int_{a_1}^{b_1} (2xy - y^2)\, dx \\ &= (x^2 y - xy^2)\Big|_{x=a_1}^{x=b_1} \\ &= b_1^2 y - b_1 y^2 - a_1^2 y + a_1 y^2 \\ &= (b_1^2 - a_1^2) y - (b_1 - a_1) y^2 \end{aligned}$$

Lemma 1.1 *Let f be a continuous function or a rectangle $S = R(\mathbf{a}, \mathbf{b})$, $\mathbf{a} = (a_1, a_2)$, $\mathbf{b} = (b_1, b_2)$. Define $g : [a_2, b_2] \to \mathbf{R}$ by*

$$g(y) = \int_{a_1}^{b_1} f(x, y)\, dx$$

Then g is continuous.

Proof. Let ε be an arbitrary positive number. Since f is uniformly continuous on S (Theorem 5.1, Chapter 7), we can find a $\delta > 0$ so that $|f(x, y) - f(x', y')| < \varepsilon/(b_1 - a_1)$ whenever $\|(x, y) - (x', y')\| < \delta$. Then, if $|y - y'| < \delta$, we have

$$\begin{aligned}|g(y) - g(y')| &= \left|\int_{a_1}^{b_1} (f(x, y) - f(x, y'))\,dx\right| \\ &\leq \int_{a_1}^{b_1} |f(x, y) - f(x, y')|\,dx \\ &\text{(by Proposition 5.4, Chapter 10)} \\ &< \int_{a_1}^{b_1} \frac{\varepsilon}{b_1 - a_1}\,dx = \varepsilon\end{aligned}$$

since $\|(x, y) - (x, y')\| = |y - y'| < \delta$. It follows that g is continuous.

Now, since $g : [a_2, b_2] \to \mathbf{R}$ is continuous, it is integrable. The integral of g,

$$\int_{a_2}^{b_2} g(y)\,dy = \int_{a_2}^{b_2} \left[\int_{a_1}^{b_1} f(x, y)\,dx\right] dy$$

is called an *iterated integral.*

In the example above,

$$\begin{aligned}\int_{a_2}^{b_2} \left[\int_{a_1}^{b_1} (2xy - y^2)\,dx\right] dy &= \int_{a_2}^{b_2} [(b_1^2 - a_1^2)y - (b_1 - a_1)y^2]\,dy \\ &= (b_1^2 - a_1^2)\frac{y^2}{2} - (b_1 - a_1)\frac{y^3}{3}\Bigg|_{a_2}^{b_2} \\ &= \frac{1}{2}(b_1^2 - a_1^2)(b_2^2 - a_2^2) \\ &\quad - \frac{1}{3}(b_1 - a_1)(b_2^3 - a_2^3)\end{aligned}$$

Of course, we could also have integrated f with respect to y first and then with respect to x. In that case, the calculation goes as follows:

$$\begin{aligned}\int_{a_2}^{b_2} (2xy - y^2)\,dy &= \left(xy^2 - \frac{y^3}{3}\right)\Bigg|_{y=a_2}^{y=b_2} \\ &= b_2^2 x - \frac{b_2^3}{3} - a_2^2 x + \frac{a_2^3}{3} \\ &= (b_2^2 - a_2^2)x - \left(\frac{b_2^3}{3} - \frac{a_2^3}{3}\right)\end{aligned}$$

$$\begin{aligned}\int_{a_1}^{b_1}\left[\int_{a_2}^{b_2}(2xy-y^2)\,dy\right]dx &= \int_{a_1}^{b_1}\left[(b_2^2-a_2^2)x-\left(\frac{b_2^3}{3}-\frac{a_2^3}{3}\right)\right]dx\\ &= \left[(b_2^2-a_2^2)\frac{x^2}{2}-\left(\frac{b_2^3}{3}-\frac{a_2^3}{3}\right)x\right]\Bigg|_{a_1}^{b_1}\\ &= \frac{1}{2}(b_2^2-a_2^2)(b_1^2-a_1^2)\\ &\quad -\frac{1}{3}(b_2^3-a_2^3)(b_1-a_1)\end{aligned}$$

the same answer as above. The next result shows that this is not a coincidence.

Theorem 1.2 (The Fubini Theorem) *Let f be a continuous function on the rectangle $S = R(\mathbf{a}, \mathbf{b})$ in $\mathbf{R}^n$, where $\mathbf{a} = (a_1, \ldots, a_n)$ and $\mathbf{b} = (b_1, \ldots, b_n)$. Let $S' = R(\mathbf{a}', \mathbf{b}')$ be the rectangle in $\mathbf{R}^{n-1}$, where $\mathbf{a}' = (a_1, \ldots, a_{n-1})$ and $\mathbf{b}' = (b_1, \ldots, b_{n-1})$, and let $g : S' \to \mathbf{R}$ be defined by*

$$g(x_1, \ldots, x_{n-1}) = \int_{a_n}^{b_n} f(x_1, \ldots, x_n)\,dx_n$$

Then g is continuous and

$$\int_S f = \int_{S'} g = \int_{S'}\left[\int_{a_n}^{b_n} f(x_1, \ldots, x_n)\,dx_n\right] \tag{1.1}$$

We shall prove this theorem in Section 4. First, we examine what it means in some special cases.

1. Assume that $n = 2$. Then the rectangle $S = R(\mathbf{a}, \mathbf{b})$ is the one in Figure 1.1, and S' is the interval $[a_1, b_1]$. The theorem says that

$$\int_S f = \int_{a_1}^{b_1}\left[\int_{a_2}^{b_2} f(x_1, x_2)\,dx_2\right]dx_1$$

or, if we use x and y as the variables,

$$\int_S f = \int_{a_1}^{b_1}\left[\int_{a_2}^{b_2} f(x, y)\,dy\right]dx \tag{1.2}$$

Usually we write (1.2) with fewer parentheses:

$$\int_S f = \int_{a_1}^{b_1}\int_{a_2}^{b_2} f(x, y)\,dy\,dx$$

While Theorem 1.2 gives only this conclusion, the proof makes it clear that we can get the same result by integrating first with respect to x:

$$\int_S f = \int_{a_2}^{b_2} \int_{a_1}^{b_1} f(x, y)\, dx\, dy$$

In particular, we see that the two iterated integrals of f are equal (assuming that f is continuous). We saw one example of this earlier. It is useful to state this as a separate result:

Corollary to Theorem 1.2 *Let f be a continuous function on the rectangle $S = R(\mathbf{a}, \mathbf{b})$ in R^2, where $\mathbf{a} = (a_1, a_2)$ and $\mathbf{b} = (b_1, b_2)$. Then*

$$\int_S f = \int_{a_1}^{b_1} \int_{a_2}^{b_2} f(x, y)\, dy\, dx = \int_{a_2}^{b_2} \int_{a_1}^{b_1} f(x, y)\, dx\, dy$$

2. Assume now that $n = 3$. Then $S = R(\mathbf{a}, \mathbf{b})$ is a rectangular solid, as in Figure 1.2 and S' is the rectangle given by (a_1, a_2) and (b_1, b_2). Using x, y, and z as the variables, we get

$$\int_S f = \int_{S'} \left[\int_{a_3}^{b_3} f(x, y, z)\, dz \right]$$

However, using the discussion above, we can write the integral over S' as an iterated integral:

$$\int_S f = \int_{a_1}^{b_1} \left\{ \int_{a_2}^{b_2} \left[\int_{a_3}^{b_3} f(x, y, z)\, dz \right] dy \right\} dx$$

or, without parentheses,

$$\int_S f = \int_{a_1}^{b_1} \int_{a_2}^{b_2} \int_{a_3}^{b_3} f(x, y, z)\, dz\, dy\, dx$$

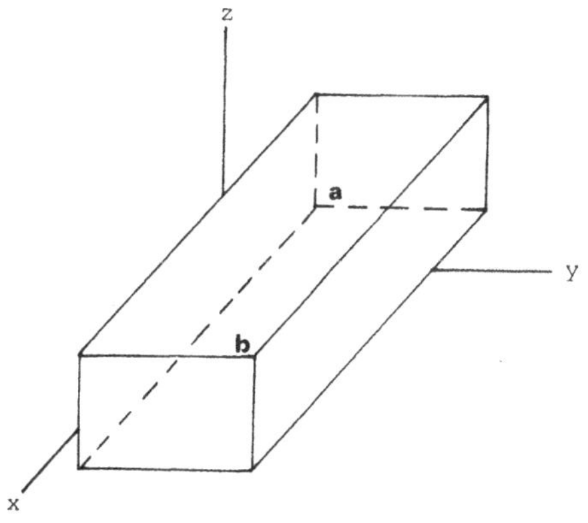

Figure 1.2 A rectangle in $\mathbf{R}^3$.

The order $dz\,dy\,dx$ means that one integrates first with respect to the z variable (holding x and y fixed), then with respect to y (with x fixed), and finally with respect to x. For instance, if $\mathbf{a} = (0, 0, 0)$, $\mathbf{b} = (1, 2, 3)$, and $f(x, y, z) = xy^2z^3$, then

$$\begin{aligned}\int_S f &= \int_0^1 \int_0^2 \int_0^3 xy^2z^3\,dz\,dy\,dx \\ &= \int_0^1 \int_0^2 \left(xy\frac{2z^4}{4}\right)\Bigg|_{z=0}^{z=3} dy\,dx \\ &= \int_0^1 \int_0^2 \frac{81}{4}xy^2\,dy\,dx \\ &= \int_0^1 \left(\frac{27}{4}xy^3\right)\Bigg|_{y=0}^{y=2} dx \\ &= \int_0^1 54x\,dx = 27x^2\Big|_0^1 = 27\end{aligned}$$

Again, we can take the iterated integrals in other orders; for instance,

$$\begin{aligned}\int_S f &= \int_{a_2}^{b_2} \int_{a_1}^{b_1} \int_{a_3}^{b_3} f(x, y, z)\,dz\,dx\,dy \\ &= \int_{a_3}^{b_3} \int_{a_1}^{b_1} \int_{a_2}^{b_2} f(x, y, z)\,dy\,dx\,dz\end{aligned}$$

That is, a result like the corollary to Theorem 1.2 holds for $n = 3$.

We could continue in the same way, giving examples for $n = 4$, $n = 5$, and so on. However, the idea is the same and the cases $n = 2$ and $n = 3$ are the most important in practice. The same principle holds in higher dimensions, too: as long as f is continuous, all the iterated integrals are equal. *That is, the order in which we integrate the variables does not affect the result.*

Our statement of Theorem 1.2 is by no means the strongest possible; for instance, it is not necessary to assume that f is continuous. (We shall give a stronger result in the next section.) However, it is not sufficient merely to assume that f is integrable. For instance, let S be the rectangle in $\mathbf{R}^2$ defined by (0, 0) and (1, 1), and define $f\colon S \to \mathbf{R}$ by

$$f(x, y) = \begin{cases} 1 & \text{if either } x \text{ or } y \text{ is irrational} \\ 1 - \dfrac{1}{q} & \text{if } x \text{ and } y \text{ are rational and } x = \dfrac{p}{q} \text{ in lowest terms} \end{cases}$$

Then f is integrable (see Exercise 8), but $\int_0^1 f(x, y)\,dy$ does not exist when

x is rational. Hence we cannot even define the iterated integral $\int_0^1 \int_0^1 f(x, y)\, dy\, dx$.

Exercises

1. Evaluate the following.

(a) $\displaystyle\int_3^5 \int_1^2 (y - 2x)\, dx\, dy$ (b) $\displaystyle\int_1^2 \int_3^5 (y - 2x)\, dy\, dx$

(c) $\displaystyle\int_0^2 \int_0^3 xy^2\, dx\, dy$ (d) $\displaystyle\int_0^3 \int_0^2 xy^2\, dy\, dx$

(e) $\displaystyle\int_0^2 \int_{-1}^1 e^{2x+y}\, dx\, dy$ (f) $\displaystyle\int_0^\pi \int_0^{\pi/2} \cos(x + y)\, dy\, dx$

(g) $\displaystyle\int_0^1 \int_0^1 \frac{x}{1 + xy}\, dx\, dy$ (h) $\displaystyle\int_0^1 \int_0^1 \frac{x}{1 + xy}\, dy\, dx$

2. Evaluate the following.

(a) $\displaystyle\int_{-2}^4 \int_4^8 \int_0^2 xyz\, dx\, dy\, dz$ (b) $\displaystyle\int_{-2}^6 \int_{-4}^4 \int_0^1 (xy + z^2)\, dx\, dz\, dy$

(c) $\displaystyle\int_0^2 \int_0^{\pi/3} \int_0^{\pi/2} \cos(x + z)\, dz\, dx\, dy$ (d) $\displaystyle\int_1^3 \int_0^2 \int_{-1}^1 yze^{xz}\, dx\, dz\, dy$

(e) $\displaystyle\int_0^{\pi/3} \int_1^2 \int_{-\pi/2}^0 y \cos x \cos z\, dz\, dy\, dx$ (f) $\displaystyle\int_0^1 \int_{-3}^0 \int_0^2 \frac{yz^2}{1 + x^2}\, dy\, dz\, dx$

3. Use Theorem 1.2 to find $\int_{R(\mathbf{a}, \mathbf{b})} f$, where
(a) $f(x, y) = y - 2x$, $\mathbf{a} = (1, 3)$, $\mathbf{b} = (2, 5)$
(b) $f(x, y) = y - 2x$, $\mathbf{a} = (-2, 4)$, $\mathbf{b} = (1, 5)$
(c) $f(x, y) = xy^2$, $\mathbf{a} = (0, 0)$, $\mathbf{b} = (2, 2)$
(d) $f(x, y) = xy^2$, $\mathbf{a} = (-1, 1)$, $\mathbf{b} = (2, 2)$
(e) $f(x, y) = x(1 + xy)^{-1}$, $\mathbf{a} = (0, 0)$, $\mathbf{b} = (1, 1)$
(f) $f(x, y) = y \sin xy$, $\mathbf{a} = (0, 0)$, $\mathbf{b} = (\pi/4, 1)$
(g) $f(x, y, z) = xyz$, $\mathbf{a} = (0, 4, -2)$, $\mathbf{b} = (2, 8, 4)$
(h) $f(x, y, z) = xy + z^2$, $\mathbf{a} = (0, -2, -4)$, $\mathbf{b} = (1, 6, 4)$
(i) $f(x, y, z) = x \sin(xy + z)$, $\mathbf{a} = (0, 1, 0)$, $\mathbf{b} = (\pi/6, 2, \pi/2)$
(j) $f(x, y, z) = yz^2(1 + x^2)^{-1}$, $\mathbf{a} = (0, -3, 0)$, $\mathbf{b} = (1, 0, 2)$
(k) $f(x_1, x_2, x_3, x_4) = x_1x_3 - x_2x_4$, $\mathbf{a} = (0, 0, 0, 0)$, $\mathbf{b} = (1, 2, 2, 1)$
(l) $f(x_1, x_2, x_3, x_4) = 2x_1x_3^2 - x_3x_4 \sin x_2$, $\mathbf{a} = (0, 0, 0, 0)$, $\mathbf{b} = (1, \pi/3, 2, 4)$

*4. Prove the analogue of Lemma 1.1 for functions of n variables.

5. Let $f(x, y) = f_1(x) f_2(y)$, where f_1 and f_2 are continuous. Verify directly that

$$\int_{a_2}^{b_2} \left[\int_{a_1}^{b_1} f(x, y)\, dx \right] dy = \int_{a_1}^{b_1} \left[\int_{a_2}^{b_2} f(x, y)\, dy \right] dx$$

*6. Prove that, in the situation of Exercise 5,

$$\int_{a_2}^{b_2}\left[\int_{a_1}^{b_1} f(x, y)\,dx\right]dy = \int_S f$$

where S is the rectangle defined by $(a_1, a_2), (b_1, b_2)$. (That is, prove Theorem 1.2 for this case.)

7. Let $f: [-1, 1] \to \mathbf{R}$ be continuous, and define g by

$$g(x) = \int_0^1 f(xy)\,dy \qquad -1 \le x \le 1$$

Show that g is continuous.

*8. (a) Show that the function f, defined on the unit square in $\mathbf{R}^2$ by

$$f(x, y) = \begin{cases} 1 & \text{if } x \text{ or } y \text{ is irrational} \\ 1 - \dfrac{1}{q} & \text{if } x \text{ and } y \text{ are both rational and } x = \dfrac{p}{q} \text{ in lowest terms} \end{cases}$$

is integrable. *Hint*: Use Exercise 6 or 7 of Section 10.7. Show that

$$\omega(f; (x_0, y_0)) = \begin{cases} 0 & \text{if } x_0 \text{ is irrational} \\ \dfrac{1}{q} & \text{if } x_0 = \dfrac{p}{q} \text{ in lowest terms} \end{cases}$$

(b) Show that $\int_0^1 \int_0^1 f(x, y)\,dx\,dy$ exists and equals $\int_S f$.

(c) Show that $\int_0^1 f(x, y)\,dy$ is not defined if x is rational.

2. Integrals over Nonrectangular Regions

We saw in the previous section how to integrate continuous functions over rectangles. There are many cases, however, in which we wish to integrate over regions which are not rectangular—over circular disks, for instance. In Chapter 10, we defined such integrals in terms of integrals over rectangles: if we want to find $\int_S f$, where S is a bounded set, we simply put S in a rectangle S_1, extend f by defining

$$f_1(\mathbf{v}) = \begin{cases} f(\mathbf{v}) & \mathbf{v} \in S \\ 0 & \mathbf{v} \notin S \end{cases}$$

and find $\int_{S_1} f_1$. Actually computing $\int_{S_1} f_1$, though, involves us in two difficulties, one theoretical and one more practical. We take care of the theoretical one with a theorem; the practical one will occupy us for this section and the next.

Consider the following problem: find the integral of the function $f(x, y) = xy$ over the triangular region S bounded by the x axis, y axis, and

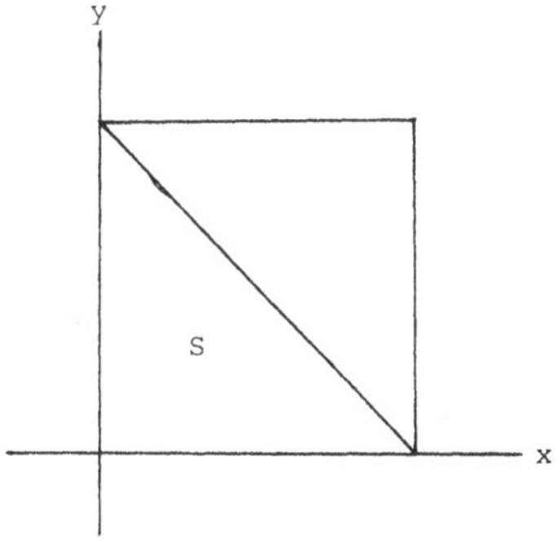

Figure 2.1

the line $x + y = 1$. (See Figure 2.1.) We may take S_1 to be the rectangle defined by (0, 0) and (1, 1); if we define f_1 as above, then, by definition,

$$\int_S f = \int_{S_1} f_1$$

and we now need to integrate f over a rectangle. However, Theorem 1.2 does not apply, because the function f_1 is not continuous.

This is the theoretical difficulty we spoke of earlier. To get around it, we give an extension of Theorem 1.2.

Theorem 2.1 *Let f be an integrable function on the rectangle $S = R(\mathbf{a}, \mathbf{b})$ in $\mathbf{R}^n$, where $\mathbf{a} = (a_1, \ldots, a_n)$ and $\mathbf{b} = (b_1, \ldots, b_n)$. Let $S' = R(\mathbf{a}', \mathbf{b}')$ be the rectangle in $\mathbf{R}^{n-1}$ with $\mathbf{a}' = (a_1, \ldots, a_{n-1})$ and $\mathbf{b}' = (b_1, \ldots, b_{n-1})$. Suppose that*

$$g(x_1, \ldots, x_{n-1}) = \int_{a_n}^{b_n} f(x_1, \ldots, x_n)\, dx_n$$

exists for all $(x_1, \ldots, x_{n-1}) \in S'$; that is, suppose that f can always be integrated in the nth coordinate. Suppose further that g is integrable. Then

$$\int_S f = \int_{S'} g = \int_{S'} \left[\int_{a_n}^{b_n} f(x_1, \ldots, x_n)\, dx_n \right] \tag{2.1}$$

Remarks

1. As with Theorem 1.2, we have stated the theorem in a way which singles out the last variable. The same result holds if we integrate out any other variable; that is, any iterated integral is equal to $\int_S f$ (assuming that f is

integrable and the iterated integral makes sense.) We have not stated the theorem in this more general form simply because the notation becomes unwieldy.

2. Theorem 2.1 can be improved somewhat; for instance, the hypothesis that g is integrable is unnecessary, since it can be proved from the other hypotheses. (An even better result is given in Exercise 3 of Section 4.) However, Theorem 2.1 is good enough for most cases arising in practice. We shall prove Theorem 2.1 in Section 4.

We now return to the example above, namely

$$f_1(x, y) = \begin{cases} xy & \text{if } (x, y) \in S \\ 0 & \text{if } (s, y) \notin S \end{cases}$$

where S is the triangle of Figure 2.1. To apply Theorem 2.1, we first need to know that, for fixed x, the integral

$$\int_0^1 f_1(x, y)\, dy$$

exists. However, for fixed x, $f_1(x, y)$ is a continuous function of y except at $y = 1 - x$, so this integral exists by Theorem 6.2, Chapter 10. In fact,

$$\int_0^1 f_1(x, y)\, dy = \int_0^{1-x} xy\, dy \qquad [\text{since } f_1(x, y) = 0 \text{ for } y > 1 - x]$$

$$= \frac{xy^2}{2}\Bigg|_{y=0}^{y=1-x}$$

$$= \frac{1}{2}(x - 2x^2 + x^3)$$

This function is clearly integrable; so

$$\int_S f = \int_{S_1} f_1 = \int_0^1 \int_0^1 f(x, y)\, dy\, dx$$

$$= \int_0^1 \frac{1}{2}(x - 2x^2 + x^3)\, dx$$

$$= \left(\frac{x^2}{4} - \frac{x^3}{3} + \frac{x^4}{8}\right)\Bigg|_0^1 = \frac{1}{24}$$

and the problem is solved.

In practice, one rarely bothers to extend f to a rectangle as we did above; instead, one writes down the iterated integral and computes directly. It is here that the practical problem arises. The limits of integration in the iterated integral may depend on some of the variables which have not been integrated out, and it may require some care to set up the integral correctly.

Here is how the above problem would generally be done: we want to write the integral in the form

$$\int\int f(x, y)\,dy\,dx$$

where we have not yet determined the limits of integration. Since the first integral is with respect to y, we fix x. Now we notice that for fixed $x \in [0, 1]$, the piece of the line through $(x, 0)$ and parallel to the y axis which lies in S corresponds to the interval $0 \le y \le 1 - x$. (See Figure 2.2.) Thus y varies between 0 and $1 - x$ in the first integral. The x coordinates of points in S varies from 0 to 1, and therefore the second integral (over x) should be from 0 to 1. In short,

$$\int_S f = \int_0^1 \int_0^{1-x} xy\,dy\,dx$$

Now the computation is much as before:

$$\int_0^1 \left(\int_0^{1-x} xy\,dy \right) dx = \int_0^1 \left(\frac{1}{2} xy^2 \Big|_{y=0}^{y=1-x} \right) dx$$

$$= \int_0^1 \frac{1}{2}(x - 2x^2 + x^3)\,dx = \frac{1}{24}$$

as above. We could also do this integral by integrating first with respect to x. In this case, we first fix y. We now need to integrate $f(x, y)$ as a function of x. The integral is over the values of x which makes (x, y) lie in S—that is, for

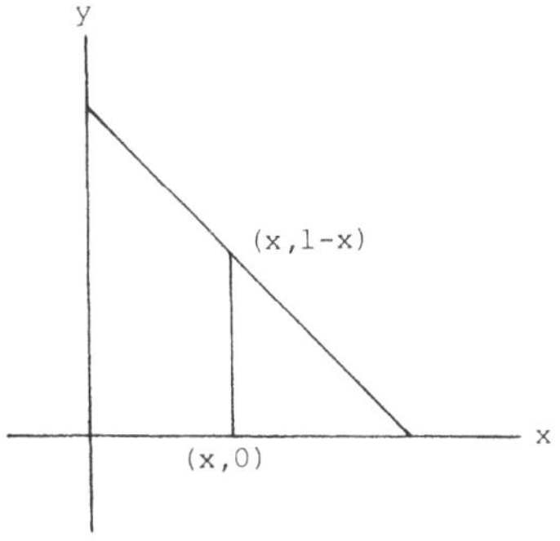

Figure 2.2 Integrating first with respect to y.

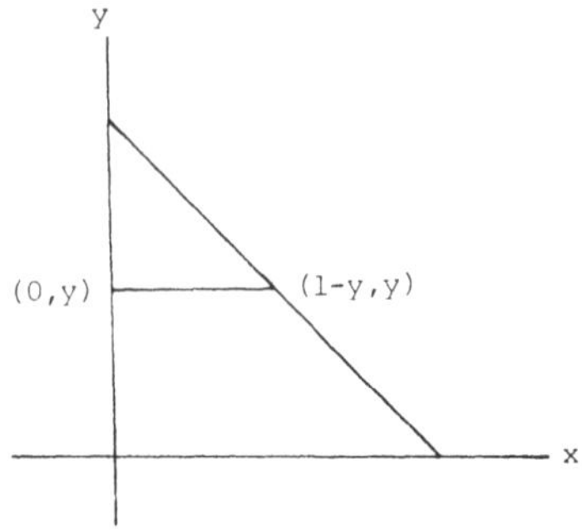

Figure 2.3 Integrating first with respect to x.

$0 \leq x \leq 1 - y$. (See Figure 2.3.) The possible y coordinates in S vary from 0 to 1. Therefore

$$\int_S f = \int_0^1 \int_0^{1-y} xy\,dx\,dy$$

and the computation just as above.

The major problem, then, in these computations is getting the limits of integration correct. (Of course, there is also the problem of actually performing the integrations, but that problem is treated in elementary calculus.) We shall see a number of further examples, but the method is always the same. Given an integral

$$\int\int_S f(x, y)\,dx\,dy$$

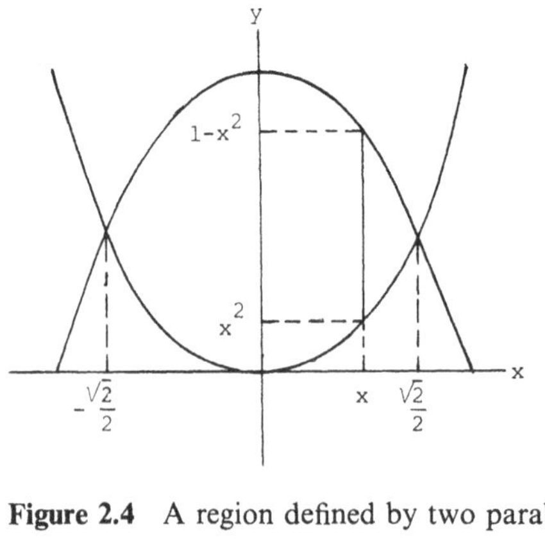

Figure 2.4 A region defined by two parabolas.

for example, where x is integrated first, we fix y and determine the limits on the integral in x by seeing which points (x, y) are in S. Then we see which values of y appear in S. For integrals in $\mathbf{R}^3$, the method is similar, but we shall postpone this discussion until the next section.

We close this section with another example. More examples are given in the next section.

Suppose S is the subset of $\mathbf{R}^2$ defined by the two curves $y = 1 - x^2$ and $y = x^2$ (see Figure 2.4), and $f(x, y) = x + y$. The curves intersect where $1 - x^2 = x^2$, or $x = \pm\sqrt{2}/2$. (The y coordinate is $\frac{1}{2}$, but we shall not need this fact.) For fixed x in $[-\sqrt{2}/2, \sqrt{2}/2]$, y ranges from x^2 to $1 - x^2$; so

$$\begin{aligned} h(x) &= \int_{x^2}^{1-x^2} (x + y)\,dy \\ &= xy + \frac{y^2}{2}\bigg|_{y=x^2}^{y=1-x^2} \\ &= x(1 - x^2) + \frac{1}{2}(1 - x^2)^2 - x^3 - \frac{1}{2}x^4 \\ &= \frac{1}{2} + x - x^2 - 2x^3 \end{aligned}$$

It follows that

$$\begin{aligned} \int_{-\sqrt{2}/2}^{\sqrt{2}/2} \left[\int_{x^2}^{1-x^2} (x + y)\,dy\right] dx &= \int_{-\sqrt{2}/2}^{\sqrt{2}/2} \left(\frac{1}{2} + x - x^2 - 2x^3\right) dx \\ &= \frac{\sqrt{2}}{2} - \frac{\sqrt{2}}{6} = \frac{\sqrt{2}}{3} \end{aligned}$$

We can do this integral as an iterated integral in the other order, but it is considerably more complicated. The reason is that for fixed y, x varies between $-\sqrt{y}$ and $\sqrt{y}$ if $0 \le y \le \frac{1}{2}$ and between $-\sqrt{1 - y}$ and $\sqrt{1 - y}$ if $\frac{1}{2} \le y \le 1$. (See Figure 2.4 again; if $y = 1 - x^2$, then $x^2 = 1 - y$, or $x = \pm\sqrt{1 - y}$.) Thus the iterated integral needs to be broken into two pieces:

$$\int_S f = \int_0^{\frac{1}{2}} \int_{-\sqrt{y}}^{\sqrt{y}} (x + y)\,dx\,dy + \int_{\frac{1}{2}}^{1} \int_{-\sqrt{1-y}}^{\sqrt{1-y}} (x + y)\,dx\,dy$$

Of course, evaluating the left-hand side gives the same answer as above. (See Exercise 4.) It often pays in problems like this to experiment with setting up both iterated integrals, in order to see which will be less troublesome to compute.

Exercises

1. Evaluate the following.

(a) $\int_1^2 \int_1^{x^2} (x^2 + y^2)\,dy\,dx$ (b) $\int_{-1}^1 \int_{-x^2}^{x^2} (x^2 - y)\,dy\,dx$

(c) $\int_{-1}^1 \int_{x^2}^{2-x^2} x^2 y^2\,dy\,dx$ (d) $\int_0^1 \int_0^{2-2y} (x^2 + y^2)\,dx\,dy$

(e) $\int_{-1}^1 \int_0^{\sqrt{1-y^2}} xy^2\,dx\,dy$ (f) $\int_{-1}^1 \int_{x^2}^1 \sqrt{1-y}\,dy\,dx$

(g) $\int_0^1 \int_0^{1-y} \int_0^{1-x-y} y\,dz\,dx\,dy$ (h) $\int_0^1 \int_0^{1-x} \int_0^{1-y^2-x} xz\,dz\,dy\,dx$

(i) $\int_{-1}^1 \int_0^{1-z^2} \int_{-\sqrt{y}}^{\sqrt{y}} xyz^2\,dx\,dy\,dz$ (j) $\int_0^1 \int_0^x \int_{3-y}^3 (x+y)^2 z\,dz\,dy\,dx$

2. Let S_1 be the region in the plane bounded by the x axis, the y axis, and the line $x + y = 1$, and let S_2 be the region bounded by the curves $y = 1 - x^2$ and $y = x^2$. Compute:

(a) $\int_{S_1} (x^2 - xy)$ (b) $\int_{S_2} (x^2 + y^3 + 2xy)$

(c) $\int_{S_1} (x - 2x^2)\,e^{xy}$ (d) $\int_{S_2} x \sin y$

(e) $\int_{S_2} (1 - x^2)(1 + y^2)$ (f) $\int_{S_1} \frac{y(1-x)}{1+y^2}$

3. Compute $\int_S xy - y^2$, where
   (a) S is the region bounded by the x axis, the y axis, and the line $x + 2y = 4$.
   (b) S is the region bounded by the x axis, the line $y = x$, and the line $x = 2$.
   (c) S is the region bounded by the y axis, the line $y = x$, and the line $y = 2$.
   (d) S is the region bounded by the lines $y = x$, $y = -2x$, and $y = 3$.
4. Complete the calculation of $\int_S f$ in the last example of the section, integrating first with respect to x.

3. More Examples

We give here some further examples of iterated integrals of functions of two and three variables.

1. Let S be the region in $\mathbf{R}^2$ defined by the curves $y = x^2$ and $x = y^2$ (see Figure 3.1) and let $f\colon S \to \mathbf{R}$ be the function defined by $f(x, y) = 3x + 2y$. The curves intersect at $(0, 0)$ and $(1, 1)$. For fixed x (with

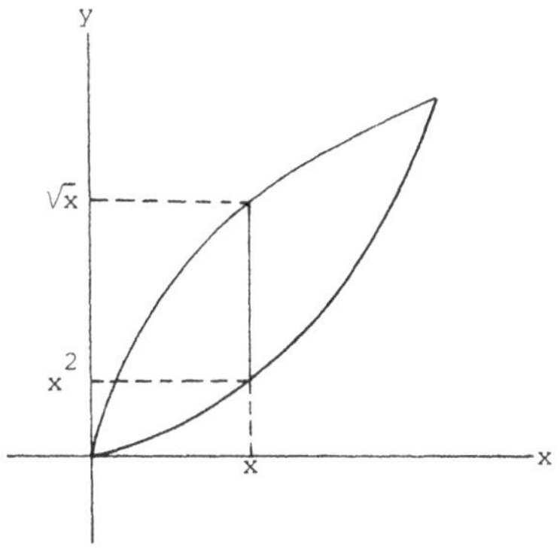

Figure 3.1 A region defined by two parabolas.

$0 \leq x \leq 1$), the points (x, y) which are in S are those with $x^2 \leq y \leq \sqrt{x}$, as the figure shows. Thus

$$\int_S f = \int_0^1 \int_{x^2}^{\sqrt{x}} (3x + 2y)\, dy\, dx$$

We compute the inner integral:

$$\int_{x^2}^{\sqrt{x}} (3x + 2y)\, dy = (3xy + y^2)\Big|_{y=x^2}^{y=\sqrt{x}}$$

$$= 3x^{3/2} + x - 3x^3 - x^4$$

It follows that

$$\int_S f = \int_0^1 (3x^{3/2} + x - 3x^3 - x^4)\, dx$$

$$= \frac{6}{5}x^{5/2} + \frac{x^2}{2} - \frac{3}{4}x^4 - \frac{1}{5}x^5\Big|_0^1 = \frac{3}{4}$$

2. According to the remark at the end of Section 1, the order of integration in the iterated integral of a continuous function is immaterial. However, it can happen that one order is preferable over the other. We saw an example of this at the end of the previous section. Here is another.

   Let S be the region in the first quadrant in $\mathbf{R}^2$ defined by $2y = x^2$ and $x = 2$, and $f\colon S \to \mathbf{R}$ be the function defined by

$$f(x, y) = \frac{2x}{(1 + x^2 + y^2)^2}$$

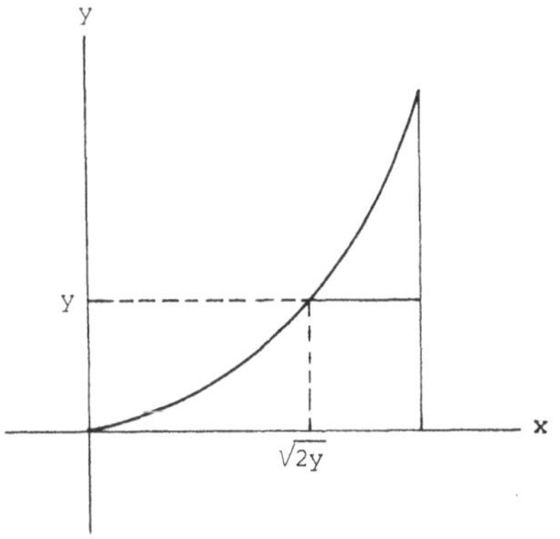

Figure 3.2

(See Figure 3.2.) In this case (as in Example 1),

$$\int_S f = \int_0^2 \int_0^{x^2/2} \frac{2x}{(1+x^2+y^2)^2}\,dy\,dx$$

or

$$\int_S f = \int_0^2 \int_{\sqrt{2y}}^{2} \frac{2x}{(1+x^2+y^2)^2}\,dy\,dx$$

The first of these iterated integrals involves integrating the function $2x/(1 + x^2 + y^2)^2$ with respect to y. This is extremely complicated. The second iterated integral is much easier:

$$\begin{aligned}\int_0^2 \int_{\sqrt{2y}}^{2} \frac{2x}{(1+x^2+y^2)^2}\,dx\,dy &= \int_0^2 \left(\frac{-1}{1+x^2+y^2}\bigg|_{x=\sqrt{2y}}^{x=2} \right) dy \\ &= \int_0^2 \left(\frac{-1}{5+y^2} + \frac{1}{1+2y+y^2} \right) dy \\ &= \left(-\frac{1}{\sqrt{5}} \arctan \frac{y}{\sqrt{5}} - \frac{1}{1+y} \right)\bigg|_0^2 \\ &= -\frac{1}{\sqrt{5}} \arctan \frac{2}{\sqrt{5}} - \frac{1}{3} + 1\end{aligned}$$

3. If S is a region in $\mathbf{R}^2$ and $f: S \to \mathbf{R}$ is a nonnegative integrable function, we can think of the upper and lower sums of f as giving approximations for the volume of the region under the graph of f and over S. It follows that the integral of f over S can be interpreted as giving the exact value for this volume.

For example, let $f(x, y) = 2 - x - y$ and S be the region in the first quadrant bounded by the curve $x^2 + y = 1$. (See Figure 3.3.) The volume of the region under the graph of f over S is

$$\begin{aligned}\int_S f &= \int_0^1 \int_0^{1-x^2} (2 - x - y)\,dy\,dx \\ &= \int_0^1 \left\{ \left[(2-x)y - \frac{y^2}{2} \right] \Bigg|_{y=0}^{y=1-x^2} \right\} dx \\ &= \int_0^1 \left(\frac{3}{2} - x - x^2 + x^3 - \frac{1}{2}x^4 \right) dx \\ &= \frac{3}{2} - \frac{1}{2} - \frac{1}{3} + \frac{1}{4} - \frac{1}{10} = \frac{49}{60}\end{aligned}$$

4. The integration of functions of three variables is a bit more complicated. Let S be the region in $\mathbf{R}^3$ defined by

$$S = \{(x, y, z) : x + y + z = 1, x \geq 0, y \geq 0, z \geq 0\}$$

and $f : S \to \mathbf{R}$ the function defined by $f(x, y, z) = y - z$. Then, for fixed y and z, x ranges along the line parallel to the x axis from the point $(0, y, z)$ to the point $(1 - y - z, y, z)$ which is on the plane $x + y + z = 1$. (See Figure 3.4.) Once the integration with respect to x has been performed, if we fix z, y will range from 0 to $1 - z$. [One way of looking at it is this: after the first integral in x, we have a function of y and z defined for those pairs (y, z) in

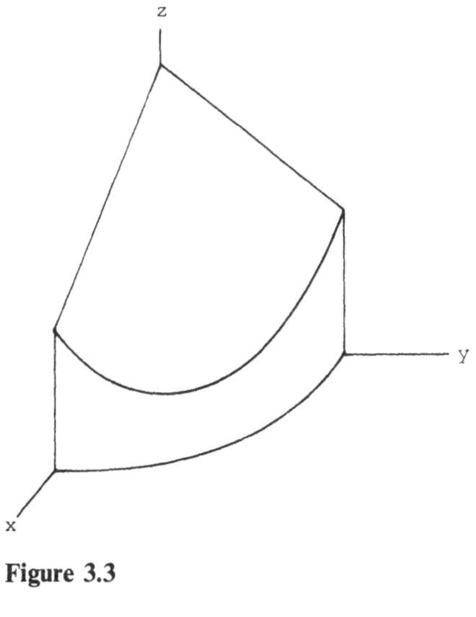

Figure 3.3

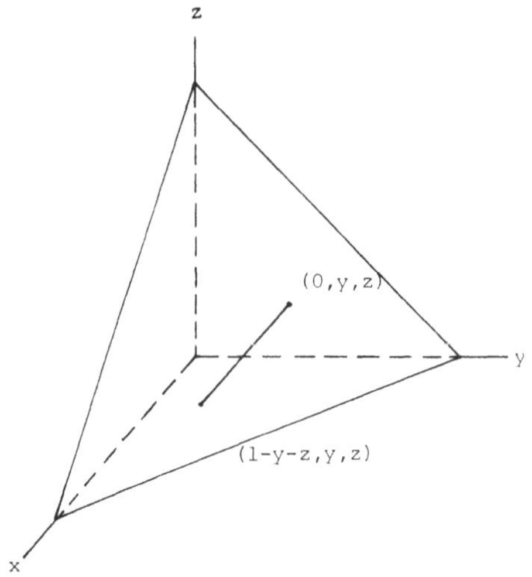

Figure 3.4

the projection S' of S on the $y-z$ plane. For fixed z, the points $y \in S'$ range from 0 to $1-z$.] Finally, z ranges from 0 to 1. Thus,

$$\begin{aligned}
\int_S f &= \int_0^1 \int_0^{1-z} \int_0^{1-y-z} (y-z)\,dx\,dy\,dz \\
&= \int_0^1 \int_0^{1-z} \left[(y-z)x \Big|_{x=0}^{x=1-y-z} \right] dy\,dz \\
&= \int_0^1 \int_0^{1-z} (y - z - y^2 + z^2)\,dy\,dz \\
&= \int_0^1 \left[\left(\frac{1}{2}y^2 - zy - \frac{1}{3}y^3 + z^2 y \right) \Big|_{y=0}^{y=1-z} \right] dz \\
&= \int_0^1 \left(\frac{1}{6} - z + \frac{3}{2}z^2 - \frac{2}{3}z^3 \right) dz \\
&= \frac{1}{6} - \frac{1}{2} + \frac{1}{2} - \frac{1}{6} = 0
\end{aligned}$$

In setting up these limits, it is useful to remember that the limits of integration can depend on the variables which have not yet been integrated out, but not on the variables which have been integrated out or the variable being integrated. For instance,

$$\int_0^1 \int_0^{1-z} \int_0^{1-y^2-z^2} f(x, y, z)\,dx\,dy\,dz$$

is possible, but

$$\int_0^1 \int_0^{1-x} \int_0^{1-y^2z^2} f(x, y, z)\,dx\,dy\,dz$$

is not (because the x variable has been integrated out by the time one integrates over y), and

$$\int_0^z \int_0^{1-x} \int_0^{1-y^2z^2} f(x, y, z)\,dx\,dy\,dz$$

is not (because z may not be a limit of an integral over z).

5. In some cases, it is necessary to divide a region into pieces in order to compute an integral. Here is an example.

   Let S be the region bounded by the x and y axes and the lines $x + y = 1$, $x + y = 2$. (See Figure 3.5.) We compute the integral of f over S where $f(x, y) = x + y$. We shall write it as a double integral,

$$\iint_S f(x, y)\,dy\,dx$$

   When $0 \le x \le 1$, y varies from $1 - x$ to $2 - x$. When $1 \le x \le 2$, however, y varies from 0 to $2 - x$. Thus we need to break S into two pieces S_1 and S_2 (see Figure 3.6):

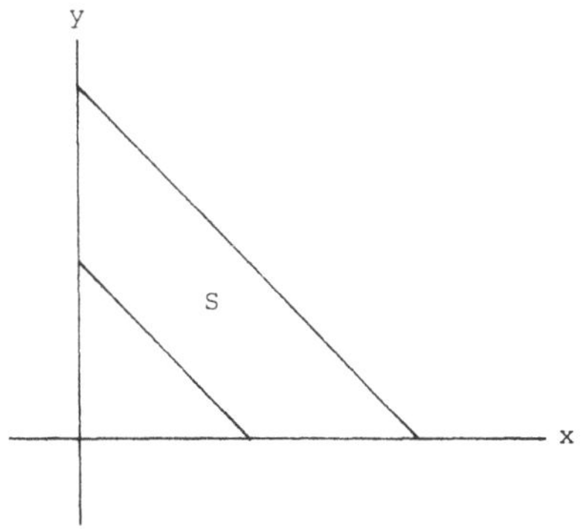

Figure 3.5

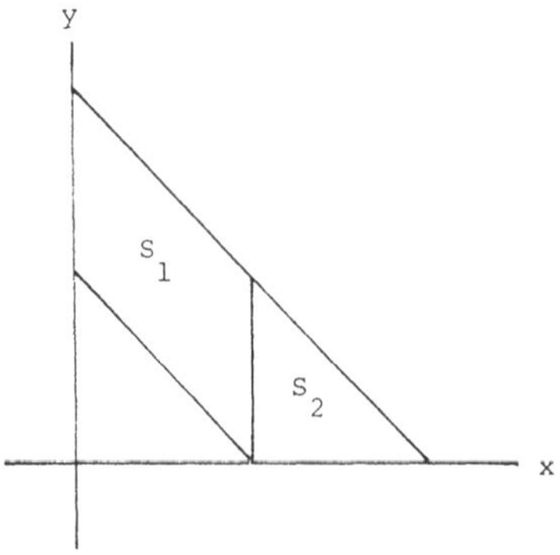

Figure 3.6

$$\begin{aligned}
\int_S f &= \int_0^1 \int_{1-x}^{2-x} (x+y)\,dy\,dx + \int_1^2 \int_0^{2-x} (x+y)\,dy\,dx \\
&= \int_0^1 \left(xy + \frac{y^2}{2}\right)\Bigg|_{y=1-x}^{y=2-x} dx + \int_1^2 \left(xy + \frac{y^2}{2}\right)\Bigg|_{y=0}^{y=2-x} dx \\
&= \int_0^1 \frac{3}{2}\,dx + \int_1^2 \frac{1}{2}(4 - x^2)\,dx \\
&= \frac{3}{2} + \left(2x - \frac{x^3}{6}\right)\Bigg|_1^2 = \frac{7}{3}
\end{aligned}$$

We could, of course, try this as an iterated integral in the other order; however it would still be necessary to divide S into two pieces. (See Exercise 3.)

Exercises

1. Sketch the region S and compute the integral of f over S where
   (a) $f(x, y) = x^2 - y^2 + xy - 3$, $S = \{(x, y) : 1 \le x \le 2, 4 \le y \le 5\}$.
   (b) $f(x, y) = x^2 + xy$, S is the region in $\mathbf{R}^2$ defined by the curves $y = x^2$ and $y = x^3$.
   (c) $f(x, y) = x \sin y$, S is the region in $\mathbf{R}^2$ defined by the curves $y = x^2$, $y = 2x^2$, and $x = 2$.
   (d) $f(x, y) = x$, $S = \{(x, y) : x^2 + y^2 \le 1\}$.
   (e) $f(x, y) = x \cos y$, S is the region in the first quadrant of $\mathbf{R}^2$ defined by the curves $x = \sqrt{\pi/2}$ and $y = x^2$.
   (f) $f(x, y) = \ln y$, S the region in the first quadrant of $\mathbf{R}^2$ defined by the lines $x = \frac{1}{2}$, $y = 1$, and $y = x$.

(g) $f(x, y) = x/\sqrt{x^2 + y^2}$, S is the region in the first quadrant of $\mathbf{R}^2$ defined by the curves $y = x$, $x = 1$, and $x = 2$.

2. Compute the integral of f over S, where
   (a) $f(x, y, z) = x + y + z$, S is the region in the first octant with $x + y + z \leq 1$.
   (b) $f(x, y, z) = x^2 + xyz$, S is the region defined in 2(a).
   (c) $f(x, y, z) = x + y^2 - xz$, S is the region bounded by the xy plane, the plane $z = 2$, and the cylinder $x^2 + y^2 = 1$.
   (d) $f(x, y, z) = z$, S is the region in the first octant bounded by $x^2 + y^2 + z^2 = 4$.
   (e) $f(x, y, z) = 2$, S is the region in $\mathbf{R}^3$ defined by the surfaces $z = x^2 + y^2$ and $z = 27 - 2x^2 - 2y^2$.
   (f) $f(x, y, z) = 1$, S is the region in $\mathbf{R}^3$ defined by the surfaces $z^2 = 8x$, $x^2 + y^2 = 4x$.
   (g) $f(x, y, z) = x^2 + y^2$, S is the region in $\mathbf{R}^3$ defined by the surfaces $x^2 + z^2 = a^2$ and $y^2 + z^2 = a^2$.
3. Compute the integral in Example 5 of the text by integrating in the other order.
4. Find the volume under the graph of $f(x, y) = 2x + 3y + 6$ over the region defined by the curves $x = 4 - y^2$ and $x = 1$.
5. Find the volume under the graph of $f(x, y) = 2x^2 + xy$ over the triangular region with vertices $(1, 1)$, $(2, 3)$, $(-1, 2)$.
6. Let B be the region obtained by cutting a hole of radius $\sqrt{r^2 - a^2}$ through a ball of radius r:

$$B = \{(x, y, z) \in \mathbf{R}^3 : r^2 - a^2 \leq x^2 + y^2 + z^2 \leq r^2\}$$

   Compute the volume of B.
7. Find the volume of the region defined by the two cylinders

$$x^2 + y^2 = 9 \qquad \text{and} \qquad y^2 + z^2 = 9$$

8. Find the volume of the region defined by the parabolic cylinders $y = x^2$ and $y = z^2$ and the plane $y = 1$.
9. Find the volume of the region defined by the three cylinders in $\mathbf{R}^3$ defined by $x^2 + y^2 = 1$, $x^2 + z^2 = 1$, and $y^2 + z^2 = 1$.
10. Suppose that the density of a body in a region $S \subset \mathbf{R}^3$ is given by a function $d(\mathbf{v})$. The *mass* of the body is

$$m = \int_S d(\mathbf{v})$$

   and the *center of mass* is the point $(\bar{x}, \bar{y}, \bar{z})$, where

$$\bar{x} = \frac{1}{m}\int_S x\,d(\mathbf{v}) \qquad \bar{y} = \frac{1}{m}\int_S y\,d(\mathbf{v}) \qquad \bar{z} = \frac{1}{m}\int_S z\,d(\mathbf{v})$$

   Let S be the region bounded by the three coordinate planes and the plane $x + y + 2z = 4$. Find the center of mass of a body in the region S whose density is given by
   (a) $d(\mathbf{v}) = 1$
   (b) $d(x, y, z) = x + z$

11. Find the center of mass of the region defined by $x^2 + y^2 + z^2 \leq 1$, $z \geq 0$, if the density is given by $d(x, y, z) = 3z$.
12. Find the center of mass of the region bounded by the cylinder $y^2 + z^2 = 16$ and the planes $x = 0$ and $x = 5$ if the density is given by $d(x, y, z) = 2x$.

4. The Proof of Fubini's Theorem

This section is devoted to the proof of Theorems 1.2 and 2.1. For simplicity of notation, we shall deal only with the case $n = 2$; the proof in the general case is quite similar.

We begin with Theorem 1.2. We have seen (Lemma 1.1) that if $g(x) = \int_{a_2}^{b_2} f(x, y)\,dy$, then g is continuous. Since we are assuming that $n = 2$, we need to prove that

$$\int_S f = \int_{a_1}^{b_1} g(x)\,dx = \int_{a_1}^{b_1} \left[\int_{a_2}^{b_2} f(x, y)\,dy \right] dx \tag{4.1}$$

We shall show that for any partition P of $S = R(\mathbf{a}, \mathbf{b})$, we have

$$L(f, P) \leq \int_{a_1}^{b_1} g(x)\,dx \leq U(f, P) \tag{4.2}$$

As a result,

$$\int_S f = \text{lub}\, L(f, P) \leq \int_{a_1}^{b_1} g(x)\,dx \leq \text{glb}\, U(f, P) = \int_S f$$

and equation (4.1) will be proved.

Let $P = P_1 \times P_2$ be any partition of S, where $P = \{t_0, t_1, \ldots, t_p\}$ and $P_2 = \{s_0, s_1, \ldots, s_q\}$. Define numbers M_{ij} and m_{ij}, $1 \leq i \leq p$, $1 \leq j \leq q$, by

$$M_{ij} = \text{lub}\,\{f(x, y) : x \in [t_{i-1}, t_i],\ y \in [s_{j-1}, s_j]\}$$

$$m_{ij} = \text{glb}\{f(x, y) : x \in [t_{i-1}, t_i],\ y \in [s_{j-1}, s_j]\}$$

Then

$$U(f, P) = \sum_{i=1}^{p} \sum_{j=1}^{q} M_{ij}(t_i - t_{j-1})(s_j - s_{j-1})$$

$$L(f, P) = \sum_{i=1}^{p} \sum_{j=1}^{q} m_{ij}(t_i - t_{i-1})(s_j - s_{j-1})$$

For each $x \in [a_1, t_1]$ define the function $f_x : [a_2, b_2] \to \mathbf{R}$ by $f_x(y) = f(x, y)$. Then, by the definition of the integral,

$$g(x) = \int_{a_2}^{b_2} f_x(y)\,dy \leq U(f_x, P_2)$$

The plan now is to prove that $U(g, P_1) \le U(f, P)$ so that

$$\int_{a_1}^{b_1} g(x)\,dx \le U(g, P_1) \le U(f, P)$$

which is one of the inequalities (4.2). We do this by looking at the definition of the upper sums. If $x \in [t_{i-1}, t_i]$, we have

$$f_x(y) = f(x, y) \le M_{ij}$$

for all $y \in [s_{j-1}, s_j]$. Thus, for all $x \in [t_{i-1}, t_i]$,

$$\text{lub}\,\{ f_x(y) : y \in [s_{j-1}, s_j]\} \le M_{ij}$$

Multiply this inequality by $s_j - s_{j-1}$ and sum over j:

$$g(x) \le U(f_x, P_2) \le \sum_{j=1}^{q} M_{ij}(s_j - s_{j-1})$$

for all $x \in [t_{i-1}, t_i]$. Let $N_i = \text{lub}\,\{g(x) : x \in [t_{i-1}, t_i]\}$; then

$$N_i \le \sum_{j=1}^{q} M_{ij}(s_j - s_{j-1})$$

and

$$\text{(4.3)} \qquad \begin{aligned} U(g, P_1) &= \sum_{i=1}^{p} N_i(t_i - t_{i-1}) \\ &\le \sum_{i=1}^{p} \left[\sum_{j=1}^{q} M_{ij}(s_j - s_{j-1}) \right](t_i - t_{i-1}) \\ &= U(f, P) \end{aligned}$$

We know, of course, that

$$\text{(4.4)} \qquad \int_{a_1}^{b_1} g(x)\,dx \le U(g, P_1)$$

Combining (4.3) and (4.4), we get

$$\int_{a_1}^{b_1} g(x)\,dx \le U(f, P)$$

which is half of (4.2). The same sort of reasoning proves that $L(g, P_1) \ge L(f, P)$, so that

$$\int_{a_1}^{b_1} g(x)\,dx \ge L(g, P_1) \ge L(f, P)$$

Therefore (4.2) is proved and the theorem follows.

The proof of Theorem 2.1 is identical with the proof of Theorem 1.2, except that the first sentence (concerning the continuity of g) should be omitted. From there on, the argument requires no change.

Exercises

1. Give the proof of Theorem 1.2 for arbitrary n.
2. Justify the remark after Theorem 1.2 by proving that theorem with g defined by integrating with respect to x instead of y.

*3. Let $S = R((a_1, a_2), (b_1, b_2))$, $f: S \to \mathbf{R}$ an integrable function, and define $h_x : [a_2, b_2] \to \mathbf{R}$ by

$$h_x(y) = f(x, y)$$

Suppose h_x is integrable for all but a finite number of values $x_1, \ldots, x_r$, and let $h_x : [a_2, b_2] \to \mathbf{R}$ by

$$\tilde{h}_x(y) = \begin{cases} 0 & \text{if } x = x_j,\ 1 \leq j \leq r \\ h_x(y) & \text{otherwise} \end{cases}$$

Prove that $\tilde{h}_x$ is integrable and that

$$\int_S f = \int_{a_1}^{b_1} \left[\int_{a_2}^{b_2} \tilde{h}_x(y)\, dy \right] dx$$

(The proof is similar to the proof of Theorem 2.1.)

4. State and prove the analogue of Exercise 3 for functions of n variables.

5. Differentiating under the Integral Sign

Let $S = R(\mathbf{a}, \mathbf{b})$ be a rectangle in $\mathbf{R}^2$ [as before, $\mathbf{a} = (a_1, a_2)$ and $\mathbf{b} = (b_1, b_2)$] and let U be an open set in $\mathbf{R}^2$ containing S. If $f: U \to \mathbf{R}$ is a continuously differentiable function, we can get a new function g, defined at least on the interval $a_1 \leq x \leq b_1$, by integrating f against the y variable:*

$$g(x) = \int_{a_2}^{b_2} f(x, y)\, dy$$

For example, if $f(x, y) = x^2 y^3$ on the rectangle given by (0, 0) and (1, 1), then

$$g(x) = \int_0^1 x^2 y^2\, dy = \frac{x^2}{4}$$

In this case, g is differentiable, and $g'(x) = \frac{x}{2}$. Notice also that

*As usual, we use (x, y) for the coordinates in $\mathbf{R}^2$.

$$\int_0^1 \frac{\partial f}{\partial x}(x, y)\, dy = \int_0^1 2xy^3\, dy = \frac{x}{2} = g'(x)$$

The main result of this section is the statement that this last equality holds quite generally.

Theorem 5.1 *Let f and g be as above. Then g is continuously differentiable on the interval (a_1, b_1), and*

$$g'(x) = \int_{a_2}^{b_2} \frac{\partial f}{\partial x}(x, y)\, dy$$

Proof. The quickest proof is indirect; we do not differentiate $g(x)$, but rather integrate $g'(x)$. Define $h : [a_1, b_1] \to \mathbf{R}$ by

$$h(t) = \int_{a_1}^{t} \left[\int_{a_2}^{b_2} \frac{\partial f}{\partial x}(x, y)\, dy \right] dx \tag{5.1}$$

The fundamental theorem of calculus (Theorem 2.3, Chapter 10) states that

$$h'(x) = \int_{a_2}^{b_2} \frac{\partial f}{\partial x}(x, y)\, dy \tag{5.2}$$

Note that h' is continuous, by Lemma 1.1.

We can evaluate $h(t)$ another way. Let S_t be the rectangle given by (a_1, a_2) and (t, b_2). (See Figure 5.1.) Then

$$h(t) = \int_{S_t} \frac{\partial f}{\partial x}(x, y)$$

by Theorem 1.2, since (5.1) is just an iterated integral over S_t. Integrate in the

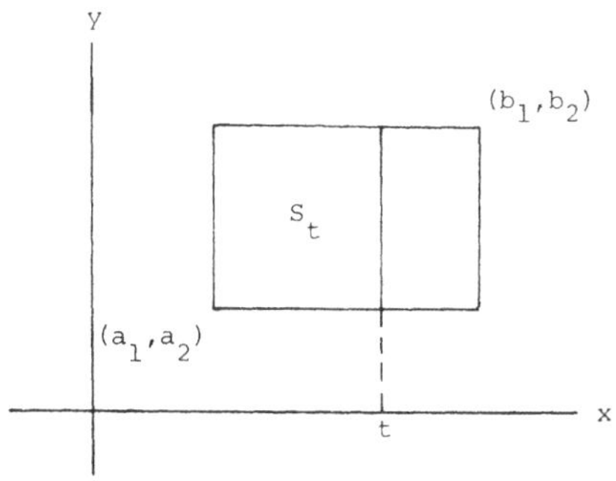

Figure 5.1 The rectangle S_t.

other order:

$$h(t) = \int_{a_2}^{b_2} \left[\int_{a_1}^{t} \frac{\partial f}{\partial x}(x, y)\, dx \right] dy$$

$$= \int_{a_2}^{b_2} \left[f(x, y) \Big|_{x=a_1}^{x=t} \right] dy$$

(again by the fundamental theorem of calculus)

$$= \int_{a_2}^{b_2} [f(t, y) - f(a_1, y)]\, dy = g(t) - g(a_1)$$

Therefore g and h differ by a constant, and so $g'(x) = h'(x)$. Now formula (5.2) completes the proof.

We have proved this theorem for functions in $\mathbf{R}^2$ because this case includes all of the usual situations which arise. Suppose, for instance, that $f: \mathbf{R}^n \to \mathbf{R}^1$ is continuously differentiable, and that $g: \mathbf{R}^{n-1} \to \mathbf{R}^1$ is defined by

$$g(x_1, \ldots, x_{n-1}) = \int_{a_n}^{b_n} f(x_1, \ldots, x_n)\, dx_n$$

Then g is continuously differentiable, and

$$\frac{\partial g}{\partial x_j} = \int_{a_n}^{b_n} \frac{\partial f}{\partial x_j}(x_1, \ldots, x_n)\, dx_n$$

The easiest way to verify this claim is to regard f as a function of x_j and x_n alone, fixing all the other coordinates. Then Theorem 5.1 applies directly. Since the partial derivatives of g are continuous, Proposition 6.1 of Chapter 4 implies that g is continuously differentiable.

We give some other variations on this theme in the exercises.

Exercises

1. Compute $g'(x)$ in two ways, where

(a) $g(x) = \int_0^1 \sin xy\, dy$

(b) $g(x) = \int_0^1 \frac{e^{xy}}{x}\, dy$

(c) $g(x) = \int_2^3 \frac{dy}{y + x}$

(d) $g(x) = \int_0^1 x \cos xy\, dy$

2. Let $f\colon \mathbf{R}^2 \to \mathbf{R}^1$ be continuously differentiable, and let $u, v\colon \mathbf{R}^1 \to \mathbf{R}^1$ be continuously differentiable functions with $u(x) < v(x)$, for all x. Define $h : \mathbf{R}^1 \to \mathbf{R}^1$ by

$$h(x) = \int_{u(x)}^{v(x)} f(x, y)\, dy$$

Show that $h(x)$ is differentiable, and that

$$h'(x) = \int_{u(x)}^{v(x)} \left[\frac{\partial f}{\partial x}(x, y)\right] dy + f(x, v(x))v'(x) - f(x, u(x))u'(x)$$

[*Hint*: What if $u(x)$ or $v(x)$ were constant? Use the chain rule.]

3. Let $f\colon \mathbf{R}^3 \to \mathbf{R}^1$ be continuously differentiable, and let

$$h(x) = \int_{a_2}^{b_2} \left[\int_{a_3}^{b_3} f(x, y, z)\, dy\right] dz$$

Show that h is differentiable and that $h'(x) = \int_{a_2}^{b_2} [\int_{a_3}^{b_3} (\partial f(x, y, z)/\partial x)\, dy]\, dz$.

*4. Given that $\int_0^1 e^{xy^2t^2}\, dt = 1$, compute dy/dx.

*5. Given that $\int_{-y}^{x} e^{t^3}\, dt = 2$, compute dy/dx.

*6. Given that $\int_{-y^2}^{x^2} e^{xyt}\, dt = 3$, compute dy/dx.

*7. Given that $\int_1^2 \ln(x + y + z + t)\, dt = 2$, compute $\partial z/\partial x$ and $\partial z/\partial y$.

6. The Change of Variable Formula

We now discuss the so-called change of variable formula. This formula describes the behavior of the integral of a function when we change coordinates and will be useful to us later in this chapter and the next. The proof is quite difficult; we give only a brief outline here.

In the course of outlining the proof of the change of variable formula, we introduce the notion of a Riemann sum and relate it to the integral. This will also be useful to us later on.

One of the standard techniques of integration, so-called substitution, states that if $f\colon [a, b] \to \mathbf{R}$ and $h : [c, d] \to [a, b]$ are differentiable functions with h surjective and $h'(x) \neq 0$ for x in $[c, d]$, then

$$\int_{h(c)}^{h(d)} f(x)\, dx = \int_c^d f(h(x))\, h'(x)\, dx \tag{6.1}$$

The proof of this formula involves the fundamental theorem of calculus and the clain rule. Explicitly, if $F(x)$ is an antiderivative for $f(x)$, then $F(h(x))$ is an antiderivative for $f(h(x))\, h'(x)$. Thus, using the fundamental theorem of calculus, we have

$$\int_{h(c)}^{h(d)} f(x)\,dx = F(h(d)) - F(h(c))$$
$$= \int_c^d f(h(x))\,h'(x)\,dx$$

We wish to generalize formula (6.1) to the integral of a function of n variables. To do this, we first reinterpret it.

First of all, we note that $h : [c, d] \to [a, b]$ is a diffeomorphism; that is, h is a differentiable function, h is invertible, and h^{-1} is differentiable. (See Theorem 2.2, Chapter 9.) If $h'(x) > 0$ for x in $[c, d]$, then $h(c) = a$ and $h(d) = b$ and (6.1) becomes

$$\int_a^b f(x)\,dx = \int_c^d f(h(x))\,h'(x)\,dx \tag{6.2}$$

If $h'(x) < 0$, then $h(c) = b$ and $h(d) = a$ and (6.1) becomes

$$\int_b^a f(x)\,dx = \int_c^d f(h(x))\,h'(x)\,dx$$

or

$$\int_a^b f(x)\,dx = \int_c^d f(h(x))\,(-h'(x))\,dx \tag{6.3}$$

Combining (6.2) and (6.3), we have the desired restatement of (6.1).

Proposition 6.1 *Let f: $[a, b] \to \mathbf{R}$ be a differentiable function and let $h : [c, d] \to [a, b]$ be a diffeomorphism. Then*

$$\int_a^b f(x)\,dx = \int_c^d f(h(x))\,|\,h'(x)\,|\,dx \tag{6.4}$$

Proposition 6.1 does not quite make sense as it stands for functions of several variables; we need an appropriate meaning for $|h'(x)|$. It turns out that the right generalization is to use the absolute value of the *determinant* of the derivative, $|\text{Det}(Dh)|$. (See Appendix A.) When $h : \mathbf{R}^2 \to \mathbf{R}^2$, we may write

$$h(\mathbf{v}) = (h_1(\mathbf{v}), h_2(\mathbf{v}))$$

and

$$Dh = \begin{bmatrix} \dfrac{\partial h_1}{\partial x_1} & \dfrac{\partial h_2}{\partial x_1} \\ \dfrac{\partial h_1}{\partial x_2} & \dfrac{\partial h_2}{\partial x_2} \end{bmatrix}$$

then $$|\text{Det}(Dh)| = \left| \frac{\partial h_1}{\partial x_1}\frac{\partial h_2}{\partial x_2} - \frac{\partial h_1}{\partial x_2}\frac{\partial h_2}{\partial x_1} \right|$$

When $h: \mathbf{R}^n \to \mathbf{R}^n$, $n > 2$, the formula is more complicated; in particular, the formula for the determinant rapidly becomes very long. (See Appendix A.)

At any rate, we can now state the appropriate generalization of Proposition 6.1.

Theorem 6.2 *Suppose that S and S' are compact subsets of $\mathbf{R}^n$, $f: S \to \mathbf{R}$ is a differentiable function, and $h: S' \to S$ a diffeomorphism. Then*

$$\int_S f = \int_{S'} (f \circ h)\,|\text{Det}(Dh)|$$

The integrand on the right side of the equation is the function taking a vector $\mathbf{v} \in S'$ into the number

$$f(h(\mathbf{v}))\,|\text{Det}(D_{\mathbf{v}} h)|$$

As we mentioned above, the proof is difficult, and we give only a sketch of the idea here,* with another approach to the integral.

Let S be a rectangle in $\mathbf{R}^n$, $f: S \to \mathbf{R}$ a continuous function, and P a partition of S into subrectangles $S_1, \ldots, S_m$. For each j, $1 \leq j \leq m$, we choose a point $\mathbf{v}_j$ in S_j. We can then form the sum

(6.5) $$\sum_{j=1}^{m} f(\mathbf{v}_j)\mu(S_j)$$

Such an expression is called a *Riemann sum for the function f relative to the partition P*. Of course, (6.5) depends on the choice of the points $\mathbf{v}_1, \ldots, \mathbf{v}_m$.

If we fix the partition P, then for any choice of points $\mathbf{v}_1, \ldots, \mathbf{v}_m$, the Riemann sum (6.5) lies between the upper sum and lower sum for f relative to P. Thus the Riemann sums can be thought of as approximations for the integral of f over S. In fact, we have the following.

Proposition 6.3 *Let S be a rectangle in $\mathbf{R}^n$ and $f: S \to \mathbf{R}$ a continuous function. Suppose $P_1, P_2, \ldots, P_m, \ldots$ is a sequence of partitions for S with*

$$\lim_{j \to \infty} (\text{mesh } P_j) = 0$$

Then, if R_j is a Riemann sum for f relative to the partition P_j, $j = 1, 2, \ldots$, we have

*A full proof is given in the article by J. T. Schwartz, "The Formula for Change of Variables in a Multiple Integral," *American Mathematical Monthly*, vol. 61, 1954, pp. 81–85. See also L. Corwin and R. Szczarba, *Calculus in Vector Spaces*, Marcel Dekker, New York, 1979, Section 12.7.

$$\lim_{j \to \infty} R_j = \int_S f$$

We leave the proof of this result as an exercise for the reader.

Returning to the proof of Theorem 6.2, we suppose that $n = 2$, S' is a rectangle, and $h : S' \to S$ is the restriction of a linear transformation. Let P be a partition of S' consisting of subrectangles $S'_1, \ldots, S'_m$ and let $\mathbf{w}_1, \ldots, \mathbf{w}_m$ be points of S' with $\mathbf{w}_j$ in S'_j, $1 \leq j \leq m$. We can then form the Riemann sum

$$\sum_{j=1}^{m} f(h(\mathbf{w}_j))\,|\mathrm{Det}(D_{\mathbf{w}_j} h)|\,\mu(S'_j) \tag{6.6}$$

approximating the integral

$$\int_{S'} (f \circ h)\,|\,\mathrm{Det}\,(Dh)|$$

The image of the rectangle S' under the linear transformation h is a parallelogram, as are the images of the rectangles S'_j. (See Figure 6.1.) Moreover, one effect of applying h to a rectangle is to multiply its area by $|\mathrm{Det}\,(h)|$:

$$\mu(h(S'_j)) = |\mathrm{Det}(h)|\,\mu(S'_j) = |\mathrm{Det}(D_{\mathbf{w}_j} h)|\,\mu(S'_j)$$

(since $D_{\mathbf{w}} h = h$). If we now think of the sets $S_j = h(S'_j)$ as forming a "partition" for the parallelogram $h(S') = S$, we can rewrite (6.6) as

$$\sum_{j=1}^{m} f(\mathbf{v}_j)\,\mu(S_j) \tag{6.7}$$

where $\mathbf{v}_j = h(\mathbf{w}_j)$, $1 \leq j \leq m$. Even though the sets S and $S_1, \ldots, S_m$ are not rectangles, it can be shown that sums of the form (6.7) do converge to the integral

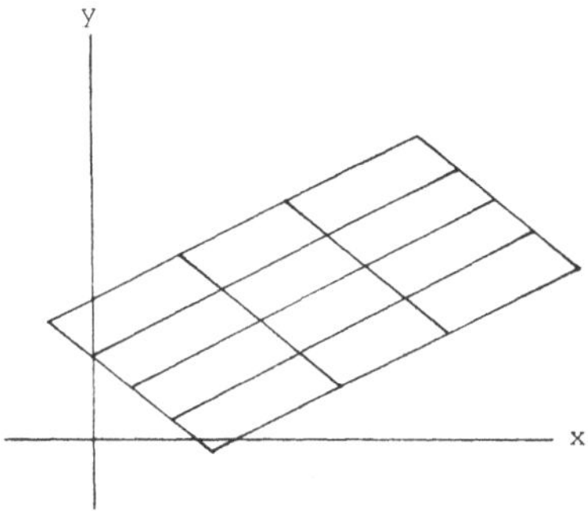

Figure 6.1 The image of S under h.

$$\int_S f$$

Since (6.7) is equal to (6.6) and sums of the form (6.6) converge to the integral (6.7), we have Theorem 6.2 in this case.

For general $h : S' \to S$, we proceed as above by picking points $\mathbf{w}_1, \ldots, \mathbf{w}_m$ in the subrectangles of a partition P for S'. The idea then is to approximate h by its derivative at the points $\mathbf{w}_j$ and use the special case treated above. For details see the references cited above.

Here are two examples of Theorem 6.2.

1. Let $h : \mathbf{R}^2 \to \mathbf{R}^2$ be defined by

$$(u, v) = h(x, y) = (x + y, x - y)$$

and let S be the unit disk:

$$S = \{(x, y) : x^2 + y^2 \le 1\}$$

The map h is certainly a diffeomorphism, since

$$h^{-1}(u, v) = \left(\frac{u + v}{2}, \frac{u - v}{2}\right)$$

Also, $\quad S' = h(S) = \{(x, y) : x^2 + y^2 \le 2\}$

since $\quad (x + y)^2 + (x - y)^2 = 2(x^2 + y^2)$

Thus $\|h(x,y)\|^2 = 2\|(x,y)\|^2$, and one effect of h is to multiply the norm of every vector by $\sqrt{2}$.

We need one other calculation:

$$D_{(x,y)}h = \begin{bmatrix} 1 & 1 \\ 1 & -1 \end{bmatrix}$$

and so $\quad \operatorname{Det} D_{(x,y)}h = -2 \qquad |\operatorname{Det}_{(x,y)}h| = 2$

Now let $f: S' \to \mathbf{R}$ be continuous. Then Theorem 6.2 says that

$$\int_{S'} f = \int_S 2(f \circ h)$$

or (as iterated integrals)

$$\int_{-\sqrt{2}}^{\sqrt{2}} \int_{-\sqrt{2-u^2}}^{\sqrt{2-u^2}} f(u, v)\, dv\, du = \int_{-1}^{1} \int_{-\sqrt{1-x^2}}^{\sqrt{1-x^2}} 2 f(x + y, x - y)\, dx\, dy$$

For instance, let $f(u,v) = uv$. Then

$$f(x + y, x - y) = (x + y)(x - y) = x^2 - y^2$$

and we get

$$\int_{-\sqrt{2}}^{\sqrt{2}} \int_{-\sqrt{2-u^2}}^{\sqrt{2-u^2}} uv\,dv\,du = \int_{-1}^{1} \int_{-\sqrt{1-x^2}}^{\sqrt{1-x^2}} (x^2 - y^2)\,dx\,dy$$

The reader should verify that the two sides are indeed equal (both, in fact, are 0).

2. Polar coordinates are defined in $\mathbf{R}^2$ by

$$(x, y) = h(r, \theta) = (r\cos\theta, r\sin\theta)$$

Since

$$h'(r, \theta) = \begin{bmatrix} \cos\theta & \sin\theta \\ -r\sin\theta & r\cos\theta \end{bmatrix}$$

we have

$$|\text{Det}\, h'(r, \theta)| = |r|$$

The map h is not a diffeomorphism of $\mathbf{R}^2$ to $\mathbf{R}^2$; for one thing, h' is not invertible whenever $r = 0$, and for another, $h(r, 0) = h(r, 2\pi)$. However, the theorem does hold on the set

$$U = \{(r, \theta) : r > 0, 0 < \theta < 2\pi\}$$

since h is a diffeomorphism from U onto

$$V = \{(x, y) : y \neq 0 \quad \text{or} \quad y = 0, x < 0\}$$

The set V is the space $\mathbf{R}^2$ with the positive x axis removed; because the set removed has content zero, we can generally ignore the fact that some points are missing. (We return to this matter below.)

Now let S' be a set in U and let $S = h(S')$. We can think of S as a set in the plane and S' as the same set in polar coordinates. For instance, if S' is the piece of the annulus shown in Figure 6.2,

$$S' = \{(x, y) : 1 \leq x^2 + y^2 \leq 2, x \geq 0, y \geq 0\}$$

then the points in S' can be expressed in polar coordinates as

$$S = \left\{(r, \theta) : 1 \leq r^2 \leq 1, 0 \leq \theta \leq \frac{\pi}{2}\right\}$$

Theorem 6.2 now says that if f is continuous, then

$$\int_{S'} f = \int_{S} (f \circ h)|\text{Det}\, h|$$

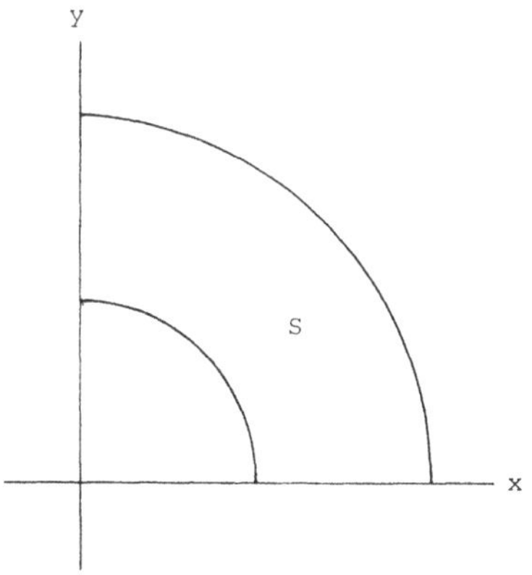

Figure 6.2 An annulus.

or
$$\int_{S'} f(x, y)\,dy\,dx = \int_S (f\circ h)(r, \theta)\,r\,dr\,d\theta$$

For instance, if $f(x, y) = xy$, then $(f\circ h)(r, \theta) = (r\cos\theta)(r\sin\theta) = r^2\sin\theta\cos\theta$, and the right-hand integral (for S, S' as above) is

$$\int_0^{\pi/2}\int_1^{\sqrt{2}} r^3\sin\theta\cos\theta\,dr\,d\theta = \int_0^{\pi/2} \frac{r^4}{4}\sin\theta\cos\theta\,d\theta\Bigg|_{r=1}^{r=\sqrt{2}}$$
$$= \int_0^{\pi/2} \sin\theta\cos\theta\,d\theta = \frac{\sin^2\theta}{2}\Bigg|_0^{\pi/2} = \frac{1}{2}$$

Computing the left-hand integral is harder, since it is not easy to write the limits of integration. (See Exercise 3.) Polar coordinates are often useful when symmetry about the origin is present in the problem.

The fact that $V \neq \mathbf{R}^2$ may seem to cause problems when one uses polar coordinates. In fact, it causes none at all. The missing part of $\mathbf{R}^2$ has content 0, and whether it is included or not does not affect the integral. For instance, if S' is the unit disk,

$$S' = \{(x, y) : x^2 + y^2 \leq 1\}$$

then $S' = h(S)$, where

$$S = \{(r, \theta) : 0 \leq r \leq 1,\ 0 \leq \theta \leq 2\pi\}$$

and the change of variables formula still applies:

$$\int_{S'} f(x, y) = \int_S f(r\cos\theta, r\sin\theta)$$

Exercises

1. Compute the integral of $f(x, y) = (x^2 + y^2)^{-3}$ over the region S, where
   (a) $S = \{(x, y) : 1 \leq x^2 + y^2 \leq 4\}$
   (b) $S = \{(x, y) : 1 \leq x^2 + y^2 \leq 4, x \geq 0, y \geq 0\}$
   (c) $S = \{(x, y) : 1 \leq x^2 + y^2 \leq 4, y \geq 0, \text{ and } y \leq x\}$
2. Compute the integral of $f(x, y) = x^2 y$ over the regions of Exercise 1.
3. Compute the integral of $f(x, y) = xy$ over the set
$$S = \{(x, y) : 1 \leq x^2 + y^2 \leq 2, x \geq 0, y \geq 0\}$$
   without transforming coordinates.
4. *Cylindrical coordinates* in $\mathbf{R}^3$ are defined by
$$(x, y, z) = h(r, \theta, z) = (r\cos\theta, r\sin\theta, z)$$
   (See Figure 6.3.) Show that $|\text{Det}(Dh)| = |r|$.
5. Use cylindrical coordinates to compute the integral of $f(x, y, z) = z(1 + x^2 + y^2)^{-1}$ over the region S, where
   (a) $S = \{(x, y, z) : 1 \leq x^2 + y^2 \leq 3, 1 \leq z \leq 5\}$
   (b) $S = \{(x, y, z) : 1 \leq x^2 + y^2 \leq 3, x \geq 0, y \geq 0, 1 \leq z \leq 5\}$
   (c) $S = \{(x, y, z) : 1 \leq x^2 + y^2 \leq 3, x \geq 0, y \geq x, 1 \leq z \leq 5\}$
6. Compute the integral of $f(x, y, z) = xz(1 + x^2 + y^2)^{-1}$ over the region S, where
   (a) $S = \{(x, y, z) : 1 \leq x^2 + y^2 \leq 3, 0 \leq z \leq 3\}$
   (b) $S = \{(x, y, z) : 1 \leq x^2 + y^2 \leq 3, x \geq 0, 0 \leq z \leq 3\}$
   (c) $S = \{(x, y, z) : 1 \leq x^2 + y^2 \leq 3, x\sqrt{3}/3 \leq y \leq x\sqrt{3}, 0 \leq z \leq 3\}$

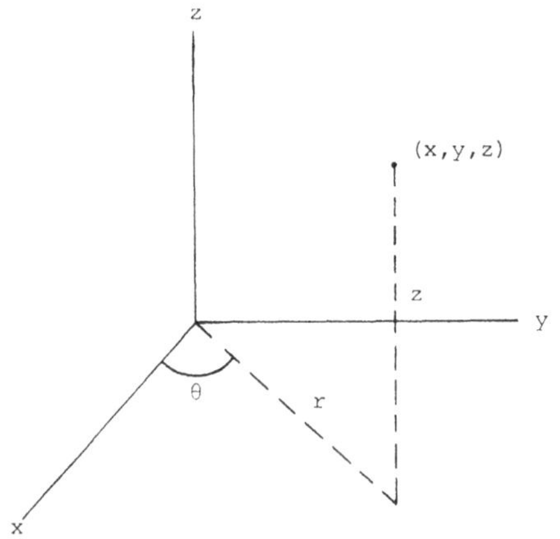

Figure 6.3 Cylindrical coordinates.

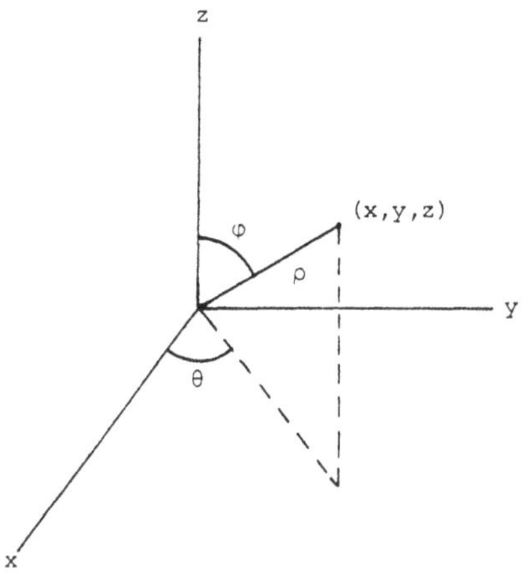

Figure 6.4 Spherical coordinates.

7. *Spherical coordinates* in $\mathbf{R}^3$ are defined by

$$(x, y, z) = h(\rho, \theta, \phi)$$
$$= (\rho \sin \phi \cos \theta, \rho \sin \phi \sin \theta, \rho \cos \phi)$$

(See Figure 6.4.) Show that $|\text{Det } h| = \rho^2 |\sin \phi|$.
8. Use spherical coordinates to compute the integral of $f(x, y, z) = x^2 + y^2 + z^2$ over the region S, where
   (a) $S = \{(x, y, z) : 1 \leq x^2 + y^2 + z^2 \leq 5, z \geq 0\}$
   (b) $S = \{(x, y, z) : 1 \leq x^2 + y^2 + z^2 \leq 5, z \geq 0, x\sqrt{3}/3 \leq y \leq x\sqrt{3}\}$
   (c) $S = \{(x, y, z) : 1 \leq x^2 + y^2 + z^2 \leq 5, z \geq 0, z^2 \geq x^2 + y^2\}$
9. Compute the integral of $f(x, y, z) = xy(x^2 + y^2 + z^2)^{-1}$ over the regions of Exercise 7.

12

Line Integrals

In this chapter, we develop the idea of the line integral of a function and of a vector field along a curve. As we shall see in this chapter and the next, these ideas are important in the physical sciences. The key result is Green's theorem, which relates the line integral of a vector field around a closed curve in $\mathbf{R}^2$ to the ordinary integral of a related function over the region enclosed by the curve.

1. Curves

In order to discuss line integrals, we need the notion of a curve in $\mathbf{R}^n$. We begin with some definitions.

A parametrized curve in $\mathbf{R}^n$ is a continuously differentiable function $\alpha\colon [a, b] \to \mathbf{R}^n$. (Recall that this means that there is a continuously differentiable function $\tilde{\alpha}$, defined on an open interval containing $[a, b]$, whose restriction to $[a, b]$ is α. As a result, we can differentiate α even at the ends of the interval.) For example, if $\mathbf{u}$ and $\mathbf{v}$ are distinct points in $\mathbf{R}^n$, then the parametrized curve $\alpha\colon [0, 1] \to \mathbf{R}^n$ given by

$$\alpha(t) = (1 - t)\mathbf{u} + t\mathbf{v}$$

defines the *line segment from* $\mathbf{u}$ to $\mathbf{v}$. (See Figure 1.1.) In fact, if we let t vary over all of $\mathbf{R}$, this same equation defines the *line determined by* $\mathbf{u}$ *and* $\mathbf{v}$.

Another example of a parametrized curve is

$$\alpha(t) = (\cos t, \sin t)$$

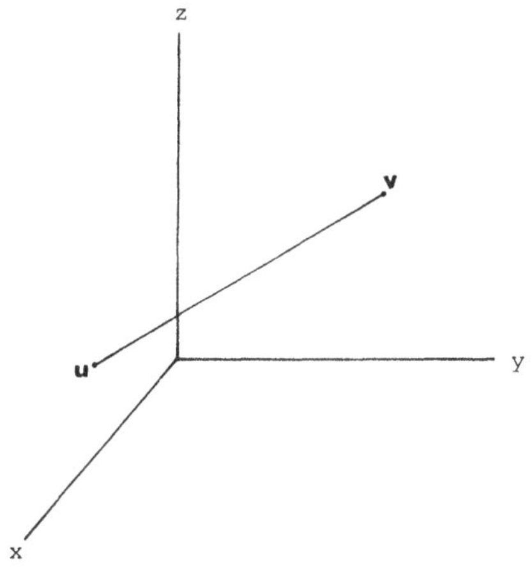

Figure 1.1 The line segment from **u** to **v**.

$t \in [0, 2\pi]$. The image set in this case is the *unit circle* in $\mathbf{R}^2$ with center the origin. (See Figure 1.2.)

Note that two different parametrized curves can have the same image set. For instance, the function $\beta : [1, 2] \to \mathbf{R}^2$ defined by

$$\beta(s) = (\cos 2\pi s, \sin 2\pi s)$$

is also a parametrization for the unit circle with center at the origin. For many purposes, however, it is useful to regard these curves as the same. We therefore make the following definition.

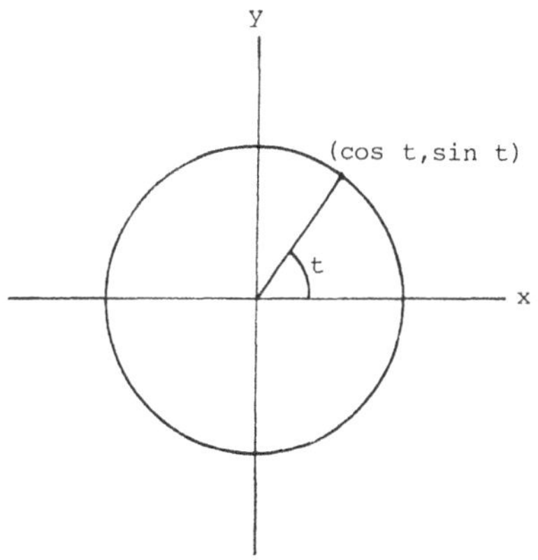

Figure 1.2 The unit circle.

Let $\alpha : [a, b] \to \mathbf{R}^n$, $\beta : [c, d] \to \mathbf{R}^n$ be two parametrized curves. We say that α is *equivalent* to β if there is a diffeomorphism $h : [c, d] \to [a, b]$ such that

$$\alpha(h(s)) = \beta(s)$$

for all s in $[c, d]$.

Since we have used the word *equivalent* in this definition, we should verify that this relation is an equivalence relation. (See Section 1.5.)

Proposition 1.1 *Let $\alpha : [a, b] \to \mathbf{R}^n$, $\beta : [c, d] \to \mathbf{R}^n$, and $\gamma : [e, f] \to \mathbf{R}^n$ be three parametrized curves. Then*

(a) *α is equivalent to α.*
(b) *If α is equivalent to β, then β is equivalent to α.*
(c) *If α is equivalent to β and β is equivalent to γ, then α is equivalent to γ.*

Thus equivalence as defined above is an equivalence relation.

We leave the proof as Exercise 5.

We now wish to define a *curve* as a parametrized curve or any other parametrized curve equivalent to it. As a matter of mathematical convenience and custom, we phrase the definition differently: *a curve is an equivalence class of parametrized curves with respect to the relation of equivalence given above.* (See Section 1.5 again.) Since any two equivalent parametrized curves clearly have the same image set (Exercise 6), it makes sense to speak of the image set of a curve. We denote the image set of the curve Γ by $\operatorname{Im} \Gamma$.

Remarks

1. The two parametrizations of the unit circle given above are equivalent, with $h : [1, 2] \to [0, 2\pi]$ given by $h(s) = 2\pi(s - 1)$.
2. Two different curves can have the same image set. For example, the parametrized curve $\gamma : [0, 2\pi] \to \mathbf{R}^2$ given by $\gamma(t) = (\sin t, \cos t)$ also has the unit circle as image set. However, γ is not equivalent to α (or β), as given above. One way to see this is to notice that the "endpoints" of α, $\alpha(0)$ and $\alpha(2\pi)$, are at $(1, 0)$, while those of γ are at $(0, 1)$. It is not hard to show that two equivalent parametrized curves have the same endpoints.
3. It is possible for the image set of a parametrized curve to have "corners." For example the $\alpha : [-1, 1] \to \mathbf{R}^2$ is defined by

$$\alpha(t) = (t^2, t^3)$$

then the image set has a corner at $(0, 0)$. (This curve is said to have a *cusp*; see Figure 1.3.)

A parametrized curve $\alpha : [a, b] \to \mathbf{R}^n$ is said to be *regular* if $\alpha'(t) \neq 0$ for all t in $[a, b]$. Since a regular curve has a continuous nowhere-vanishing tangent

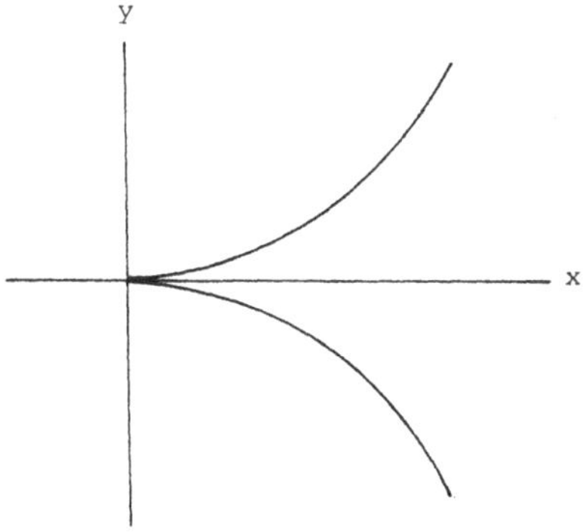

Figure 1.3 A cusp.

vector, it cannot have corners. In what follows, we deal only with regular curves.

Exercises

1. Find a parametrized curve in $\mathbf{R}^2$ whose image set is
   (a) The circle of radius 2 with center at the origin
   (b) The line segment from (2, 3) to (5, 8)
   (c) The piece of the parabola $y = \pm\sqrt{x}$, $0 \le x \le 1$
   (d) The unit circle with center (1, 0)
   (e) The ellipse $x^2/4 + y^2/9 = 1$
   (f) The graph of $y = x^3 - 7x^2 + 3x - 2$, $0 \le x \le 3$
2. Prove that the following pairs of parametrized curves are equivalent:
   (a) $\alpha(s) = (s, s + 1)$, $1 \le s \le 8$; $\beta(t) = (t^3, t^3 + 1)$, $1 \le t \le 2$
   (b) $\alpha(s) = (s, s^2)$, $2 \le s \le 3$; $\beta(t) = (\sqrt{t}, t)$, $4 \le t \le 9$
   (c) $\alpha(s) = (\sin s, \cos s)$, $0 \le s \le 2\pi$; $\beta(t) = (\cos t, \sin t)$, $\pi/2 \le t \le 5\pi/2$
   (d) $\alpha(s) = (2\cos s, 2\sin s)$, $0 \le s \le \pi/2$; $\beta(t) = ((2 - 2t^2)/(1 + t^2), 4t/(1 + t^2))$, $0 \le t \le 1$
3. Find a parametrized curve in $\mathbf{R}^3$ whose image set is
   (a) The line segment from (1, 1, 4) to (2, 3, 2)
   (b) The curve $y = x^2$ and $z = x^3$, $0 \le x \le 1$
   (c) The intersection of the planes $x + y + z = 1$ and $z = x - 2y$
   (d) The circle with radius 1 and center at the origin lying in the plane $z = 2x - 2y$
4. Prove that the following pairs of parametrized curves are equivalent:
   (a) $\alpha(s) = (s + 2, s + 3, 2s - 4)$; $\beta(t) = (t^2, t^2 + 1, 2t^2 - 8)$, $1 \le t \le 2$
   (b) $\alpha(s) = (e^s, e^{2s}, e^{3s})$, $0 \le s \le 1$; $\beta(t) = (t + 1, (t + 1)^2, (t + 1)^3)$, $0 \le t \le e - 1$
   (c) $\alpha(s) = (\sin s, \cos s, s)$, $0 \le s \le 7\pi/2$; $\beta(t) = (\cos t, \sin t, 4 - t)$, $\pi/2 \le t \le 4\pi$
5. Prove Proposition 1.1.
6. Prove that equivalent curves have the same image set.

2. Line Integrals of a Function

In this section, we show how to integrate a function along a curve. Besides being important in their own right, the notions we develop in this section will help when we deal with more general sorts of line integrals in the next section.

Suppose that U is an open subset of $\mathbf{R}^n$ and Γ is the curve defined by the parametrized curve $\alpha: [a, b] \to U$. If $f: U \to \mathbf{R}$ is a continuous function, we would like to integrate f along Γ. One way of doing this would be to pick a rectangle S containing $\operatorname{Im} \alpha$ and define $F: S \to \mathbf{R}$ by

$$F(\mathbf{v}) = \begin{cases} f(\mathbf{v}) & \text{if } \mathbf{v} \in \operatorname{Im} \alpha \\ 0 & \text{if } \mathbf{v} \notin \operatorname{Im} \alpha \end{cases}$$

and integrate F over S as described in Chapter 10. However, if $n > 1$, $\operatorname{Im} \alpha$ has content zero in $\mathbf{R}^n$. (See Exercise 7.) It follows easily that this integral is zero in these cases.

In order to motivate the proper definition of the integral of f along the curve Γ, we consider the problem of determining the mass of an inhomogeneous piece of wire. We imagine Γ to be a piece of wire and f as giving the density (mass per unit length) of the wire. We can get an approximation for the total mass of wire by choosing successive points $\mathbf{v}_0, \mathbf{v}_1, \ldots, \mathbf{v}_m$ on Γ and constructing the polygonal curve Γ' with these vertices. (See Figure 2.1.) If we assign the density $f(\mathbf{v}_i)$ to the segment of Γ' from $\mathbf{v}_{i-1}$ to $\mathbf{v}_i$, then the total mass of the wire represented by Γ' is

$$(2.1) \qquad \sum_{i=1}^{m} f(\mathbf{v}_i)\, \|\mathbf{v}_i - \mathbf{v}_{i-1}\|$$

If the points $\mathbf{v}_0, \ldots, \mathbf{v}_m$ are close together, then the expression (2.1) ought to be close to the mass of the wire represented by the curve Γ.

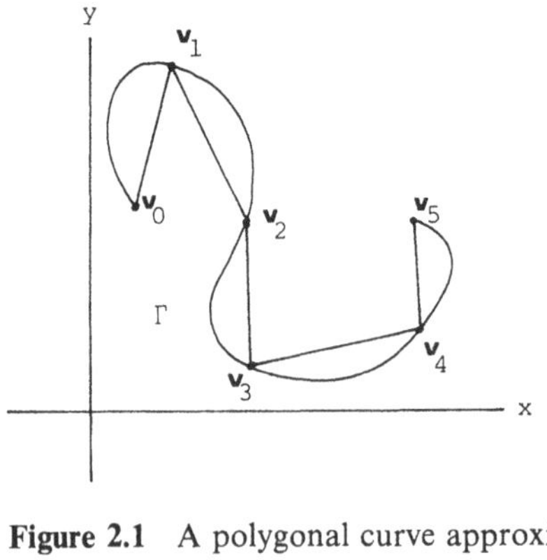

Figure 2.1 A polygonal curve approximating Γ.

We can rewrite (2.1) in terms of the parametrization for Γ as follows. Let $a = t_0 < \cdots < t_m = b$ be a partition of $[a, b]$ with the property that $\alpha(t_i) = \mathbf{v}_i$. Assume that the wire is in $\mathbf{R}^3$, so that we can write

$$\alpha(t) = (\alpha_1(t), \alpha_2(t), \alpha_3(t))$$

Then we can rewrite (2.1) as

$$\sum_{i=1}^{m} f(\alpha(t_i)) \left\{ \sum_{j=1}^{3} [\alpha_j(t_i) - \alpha_j(t_{i-1})]^2 \right\}^{1/2} \tag{2.2}$$

According to the mean value theorem, we can find points c_i, c_i', c_i'' in $[t_{i-1}, t_i]$ such that

$$\alpha_1(t_i) - \alpha_1(t_{i-1}) = \alpha_1'(c_i)(t_i - t_{i-1})$$
$$\alpha_2(t_i) - \alpha_2(t_{i-1}) = \alpha_2'(c_i')(t_i - t_{i-1})$$
$$\alpha_3(t_i) - \alpha_3(t_{i-1}) = \alpha_3'(c_i'')(t_i - t_{i-1})$$

Thus, (2.2) becomes

$$\sum_{i=1}^{m} f(\alpha(t_i)) \{ [\alpha_1'(c_i)]^2 + [\alpha_2'(c_i')]^2 + [\alpha_3'(c_i'')]^2 \}^{1/2} (t_i - t_{i-1})$$

This is approximately equal to

$$\sum_{i=1}^{m} f(\alpha(t_i)) \, \|\alpha'(t_i)\| \, (t_i - t_{i-1}) \tag{2.3}$$

since t_i, c_i, c_i', and c_i'' all lie in $[t_{i-1}, t_i]$.

As the partition of $[a, b]$ gets finer, we expect that the expression (2.3) should "converge" to the total mass of the wire. On the other hand, the expression (2.3) is a Riemann sum for the function $f(\alpha(t)) \, \|\alpha'(t)\|$, and so converges to the integral

$$\int_a^b f(\alpha(t)) \, \|\alpha'(t)\| \, dt \tag{2.4}$$

by Proposition 6.3 of Chapter 11. [Here, $\alpha'(t) = (\alpha_1'(t), \alpha_2'(t), \alpha_3'(t))$.] Thus the total mass of the wire is given by the integral (2.4).

This discussion may explain the definition which follows. If Γ is a curve in the open subset U of $\mathbf{R}^n$, if $\alpha: [a, b] \to U$ is a parametrization of Γ, and if $f: U \to \mathbf{R}$ is a continuous function, we define the *line integral* of f along Γ by

$$\int_\Gamma f = \int_a^b f(\alpha(t)) \, \|\alpha'(t)\| \, dt$$

[In this case, $\alpha'(t) = (\alpha_1'(t), \ldots, \alpha_n'(t))$.]

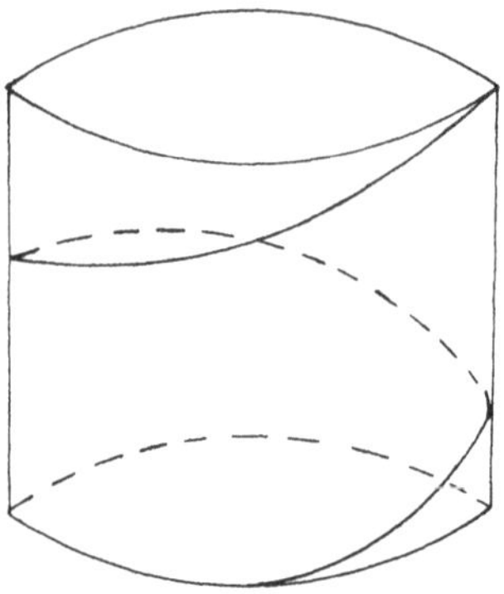

Figure 2.2 A circular helix.

For example, let Γ be the piece of the *circular helix* in $\mathbf{R}^3$ (see Figure 2.2) defined by

$$\alpha(t) = (\cos t, \sin t, t) \qquad 0 \le t \le \pi/2$$

and $f: \mathbf{R}^3 \to \mathbf{R}$ the function

$$f(x, y, z) = x^2 + y^2 + z^2$$

Then
$$\begin{aligned}\int_\Gamma f &= \int_0^{\pi/2} f(\alpha(t))\, \|\alpha'(t)\|\, dt \\ &= \int_0^{\pi/2} (\cos^2 t + \sin^2 t + t^2)(\sin^2 t + \cos^2 t + 1)^{1/2}\, dt \\ &= \int_0^{\pi/2} (1 + t^2)\sqrt{2}\, dt = \sqrt{2}\left(\frac{\pi}{2} + \frac{\pi^3}{24}\right)\end{aligned}$$

As we have defined it, the line integral seems to depend upon the parametrization chosen. The next result shows that this is not the case.

Proposition 2.1 *Let $\alpha: [a, b] \to U$, $\beta: [c, d] \to U$ be equivalent parametrized curves, and let $f: U \to \mathbf{R}$ a continuous function. Then*

$$\int_a^b f(\alpha(t))\, \|\alpha'(t)\|\, dt = \int_c^d f(\beta(t))\, \|\beta'(t)\|\, dt$$

Proof. The idea behind the proof is the change of variables formula, or the method of integration by substitution. Let $h: [c, d] \to [a, b]$ be the diffeomorphsim with $\alpha(h(s)) = \beta(s)$ for all s in $[c, d]$. Then

$$\int_c^d f(\beta(s))\,\|\beta'(s)\|\,ds = \int_c^d f(\alpha(h(s))\,\|\alpha'(h(s))\,h'(s)\|\,ds \tag{2.5}$$

since $\beta'(s) = \alpha'(h(s))\,h'(s)$ (by the chain rule). We know that $h'(s)$ is never 0 (since h is a diffeomorphism); therefore, $h'(s)$ is either always >0 or always <0. If $h'(s) > 0$, then $h(c) = a$ and $h(d) = b$, and the second integral in (2.5) is

$$\int_c^d f(\alpha(h(s)))\,\|\alpha'(h(s))\|\,h'(s)\,ds \tag{2.6}$$

If $F(t)$ is an antiderivative for the function $f(\alpha(t))\,\|\alpha'(t)\|$, then $F(h(s))$ is an antiderivative for $f(\alpha(h(s)))\,\|\alpha'(h(s))\|\,h'(s)$. (To see this, use the chain rule.) It follows from the fundamental theorem of calculus that the integral in (2.6) is equal to

$$\int_{h(c)}^{h(d)} f(\alpha(t))\,\|\alpha'(t)\|\,dt = \int_a^b f(\alpha(t))\,\|\alpha'(t)\|\,dt$$

If $h(s) < 0$, then $h(c) = b$ and $h(d) = a$ and a similar argument shows that

$$\int_c^d f(\beta(s))\,\|\beta'(s)\|\,ds = -\int_{h(c)}^{h(d)} f(\alpha(t))\,\|\alpha'(t)\|\,dt$$

$$= \int_a^b f(\alpha(t))\,\|\alpha'(t)\|\,dt$$

Thus Proposition 2.1 is proved.

Remarks

1. If f is the constant function 1, the line integral of f along Γ is simply the arc length of Γ.
2. We note that, as a consequence of Proposition 2.1, the line integral of a function f along a curve Γ does not depend on which direction we traverse Γ; we can start at either end. This is *not* true of the line integral discussed in the next section.
3. Finally, we observe that the requirement that the function α be continuously differentiable can be relaxed. Suppose $\alpha : [a, b] \to \mathbf{R}^n$ is a function and that there are points $a = a_0 < a_1 < \cdots < a_m = b$ such that the restriction of α to each of the intervals $[a_{i-1}, a_i]$ is continuously differentiable, $i = 1, \ldots, m$. In this case, α is called a *piecewise continuously differentiable parametrized curve.* We can define the line integral of a function F along the "curve" Γ represented by α to be the sum of the line integrals of F along the curves Γ_i defined by the restriction of α to the intervals $[a_{i-1}, a_i]$, $i = 1, \ldots, m$.

We close this section with a list of some of the properties of the line integral.

Proposition 2.2 *Let U be an open subset of $\mathbf{R}^n$, $\alpha: [a, b] \to U$ a parametrization for the curve Γ, f, $g: U \to \mathbf{R}$ continuous functions and c, d real numbers. Then*
(a) $\int_\Gamma (cf + dg) = c\int_\Gamma f + d\int_\Gamma g$
(b) If $f \geq g$ on U, then

$$\int_\Gamma f \geq \int_\Gamma g$$

(c) *Let c be a real number, $a < c < b$. Let $\alpha_1: [a, c] \to U$, $\alpha_2: [c, b] \to U$ be the restrictions of α and Γ_1, Γ_2, the corresponding curves. Then*

$$\int_\Gamma f = \int_{\Gamma_1} f + \int_{\Gamma_2} f$$

These results follow immediately from the corresponding properties of the standard integral. For example,

$$\begin{aligned}
\int_\Gamma (cf + dg) &= \int_a^b (cf + dg)(\alpha(t)) \, \|\alpha'(t)\| \, dt \\
&= \int_a^b [cf(\alpha(t)) + dg(\alpha(t))] \, \|\alpha'(t)\| \, dt \\
&= c\int_a^b f(\alpha(t)) \, \|\alpha'(t)\| \, dt + d\int_a^b g(\alpha(t)) \, \|\alpha'(t)\| \, dt \\
&= c\int_\Gamma f + d\int_\Gamma g
\end{aligned}$$

The remainder of the proof is left as Exercise 6.

Exercises

1. Compute the line integral of $f: \mathbf{R}^2 \to \mathbf{R}$ over the curve Γ with parametrization $\alpha: [a, b] \to \mathbf{R}^2$, where
   (a) $f(x, y) = 2xy$ and $\alpha(t) = (t, t + 1)$, $a = 1$, $b = 8$
   (b) $f(x, y) = 2xy$ and $\alpha(t) = (-t^3, 1 - t^3)$, $a = -2$, $b = -1$
   (c) $f(x, y) = x(1 + 4y)$ and $\alpha(t) = (t, t^2)$, $a = 2$, $b = 3$
   (d) $f(x, y) = x(1 + 4y)$ and $\alpha(t) = (\sqrt{t}, t)$, $a = 4$, $b = 9$
   (e) $f(x, y) = 3x + y$ and $\alpha(t) = (t - t^3/3, t^2 - 1)$, $a = 0$, $b = 3$
   (f) $f(x, y) = 2x + y$ and $\alpha(t) = (\cos t, \sin t)$, $a = 0$, $b = \pi$
2. Compute the line integral of $f: \mathbf{R}^3 \to \mathbf{R}$ over the curve Γ with parametrization $\alpha: [a, b] \to \mathbf{R}^3$, where

(a) $f(x, y, z) = xyz$ and $\alpha(t) = (t + 2, t + 3, 2t - 4)$, $a = -1$, $b = 2$
(b) $f(x, y, z) = xyz$ and $\alpha(t) = (t^2, t^2 + 1, 2t^2 - 8)$, $a = 1$, $b = 2$
(c) $f(x, y, z) = x^2y + y^2z$ and $\alpha(t) = (e^t, e^{2t}, 2e^{3t}/3)$, $a = 0$, $b = 1$
(d) $f(x, y, z) = x^2y + y^2z$ and $\alpha(t) = (t + 1, (t + 1)^2, \frac{2}{3}(t + 1)^3)$, $a = 0$, $b = e - 1$
(e) $f(x, y, z) = x$ and $\alpha(t) = (t^2 - t, 2 - t^2, \frac{4}{3}t^3 - t^2)$, $a = 0$, $b = 1$
(f) $f(x, y, z) = 2x - y + z - 1$ and $\alpha(t) = (\cos t, \sin t, t)$, $a = 0$, $b = \pi$

3. Find a parametrization for the curve Γ and evaluate the line integral of the function $f: \mathbf{R}^2 \to \mathbf{R}$ over Γ, where $f(x, y) = y$ and
(a) Im Γ is the line segment from (0, 1) to (2, 3).
(b) Im Γ is the piece of the parabola $y = x^2$ between (0, 0) and (2, 4).
(c) Im Γ is the piece of the circle $x^2 + y^2 = 4$ between (2, 0) and $(-\sqrt{2}, \sqrt{2})$.
(d) Im Γ is the polygon whose successive vertices are (1, 0), (1, 1), and (2, 2).

4. Find the total weight of a piece of wire 7 inches long if
(a) The density is proportional to the distance from the center of the wire.
(b) The density is proportional to the square of the distance from one end of the wire.

5. Find a parametrization for the curve Γ and evaluate the line integral of the function $f: \mathbf{R}^3 \to \mathbf{R}$ over Γ, where $f(x, y, z) = xyz$ and
(a) Im Γ is the line segment from (0, 1, 1) to (1, 1, 0).
(b) Im Γ is the polygon whose successive vertices are (0, 0, 1), (0, 1, 1), and (1, 2, 3).
(c) Im Γ is the unit circle in the plane $z = 1$ with center (0, 0, 1).
(d) Im Γ is the piece of the unit circle in the plane $z = 0$ between the points (1, 0, 0) and (0, 1, 0) together with the line segment from (0, 1, 0) to (1, 1, 1).

6. Prove statements (b) and (c) of Proposition 2.2.

*7. Let $\alpha: [a, b] \to \mathbf{R}^n$ be a regular parametrized curve, $n > 1$. Prove that Im α has content zero.

*8. Let $\alpha: [a, b] \to \mathbf{R}^n$ be a parametrized curve. We say that α is *parametrized by arc length* if, for each $t_0 \in [a, b]$, the length of the piece of α between $t = a$ and $t = t_0$ is exactly $t_0 - a$.
(a) Prove that α is parametrized by arc length if and only if $\|\alpha'(t)\| = 1$.
(b) Prove that, if α is parametrized by arc length, then $\langle \alpha'(t), \alpha''(t) \rangle = 0$ for all $t \in [a, b]$. [Here, $\alpha''(t) = (\alpha_1''(t), \ldots, \alpha_n''(t))$.]
(c) Let α be a regular curve. Prove that there is an equivalent curve β which is parametrized by arc length.

3. Line Integrals of Vector Fields

We now define another line integral, namely the integral of a vector field along a curve.

Let U be an open subset of $\mathbf{R}^n$. A *vector field* on U is a continuously differentiable function $\mathbf{F}: U \to \mathbf{R}^n$. We think of a vector field* as assigning to each point of U a vector based at the point. (See Figure 3.1.) In physics, for

*In this and later chapters, we shall denote vector fields by boldface.

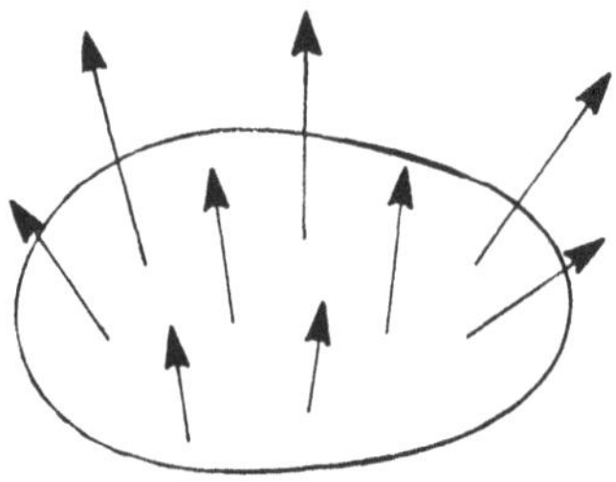

Figure 3.1 A vector field on U.

instance, a vector field could represent the force experienced by a particle in U. (In this case, a vector field is sometimes called a *force field*.)

For example, if we place a positive charge at the origin in $\mathbf{R}^2$, a negative charge at the point $\mathbf{v} \neq \mathbf{0}$ in $\mathbf{R}^2$ will feel a force directed toward the origin with magnitude

$$\frac{q}{\|\mathbf{v}\|^2}$$

where q is a positive constant. The corresponding force field is given by

$$\mathbf{F}: U \to \mathbf{R}^2$$

where U is the complement of the origin in $\mathbf{R}^2$ and

$$\mathbf{F}(x, y) = \left(\frac{-qx}{(x^2 + y^2)^{3/2}}, \frac{-qy}{(x^2 + y^2)^{3/2}} \right) \tag{3.1}$$

(See Figure 3.2.) (We exclude the origin because $q/\|\mathbf{v}\|^2$ is not defined when $\mathbf{v} = \mathbf{0}$.)

Suppose now that $\mathbf{F}$ is a vector field on the open set $U \subset \mathbf{R}^2$ and $\alpha: [a, b] \to U$ is a parametrized curve in U. We think of $\mathbf{F}$ as a force field and ask for the work done by the field in moving a particle from $\alpha(a)$ to $\alpha(b)$ along Γ. If the force is constant and the curve is a straight line, then the work is the product of the distance traveled and the force along the line:

$$W = \langle \mathbf{F}, \alpha(b) - \alpha(a) \rangle$$

In general, the curve is not a line and the force varies. In this case, we follow a procedure like that used in the last section for finding the total weight of a piece of wire. We first choose successive points along Γ,

$$\alpha(a) = \mathbf{v}_0, \mathbf{v}_1, \ldots, \mathbf{v}_m = \alpha(b)$$

and construct the polygon Γ' approximating Γ. (See Figure 2.1.) The work done in moving a particle along Γ' is approximately

$$\sum_{i=1}^{m} \langle \mathbf{F}(\mathbf{v}_i), \mathbf{v}_i - \mathbf{v}_{i-1} \rangle \tag{3.2}$$

[Here we are approximating the force along the segment of Γ' from $\mathbf{v}_{i-1}$ to $\mathbf{v}_i$ by $\mathbf{F}(\mathbf{v}_i)$.]

We can rewrite the expression (3.2) by choosing a partition for the interval $[a, b]$,

$$a = t_0 < t_1 < \cdots < t_m = b$$

with the property that $\alpha(t_i) = \mathbf{v}_i$, $0 \leq i \leq m$. Then (3.2) becomes

$$\sum_{i=1}^{m} \langle \mathbf{F}(\alpha(t_i)), \alpha(t_i) - \alpha(t_{i-1}) \rangle \tag{3.3}$$

We can now apply the mean value theorem just as in Section 2 to show that the expression (3.3) is approximately equal to

$$\sum_{i=1}^{m} \langle \mathbf{F}(\alpha(t_i)), \alpha'(t_i) \rangle (t_i - t_{i-1}) \tag{3.4}$$

As the partition gets finer, we expect (3.4) to converge to the work done in moving the particle from $\alpha(a)$ to $\alpha(b)$ along Γ. On the other hand, (3.4) is a Riemann sum for the function $\langle \mathbf{F}(\alpha(t)), \alpha'(t) \rangle$; so as the partition gets finer, (3.4) converges to the integral

$$\int_a^b \langle \mathbf{F}(\alpha(t)), \alpha'(t) \rangle \, dt \tag{3.5}$$

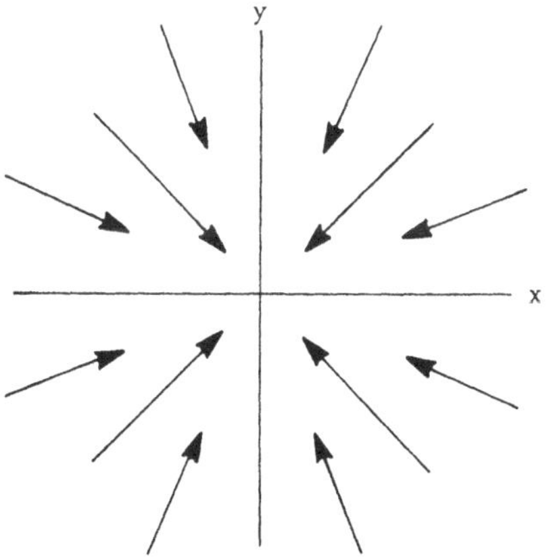

Figure 3.2 The force field from a charged particle.

by Proposition 6.3 of Chapter 11. Therefore, this integral gives the work done in moving a particle along Γ in the force field $\mathbf{F}$.*

For example, let $\mathbf{F}\colon U \to \mathbf{R}^2$ be the force field of (3.1) arising from the positive charge at the origin and $\alpha\colon [0, \pi/2] \to \mathbf{R}^2$ the piece of the unit circle given by

$$\alpha(t) = (\cos t, \sin t)$$

The work done in moving a negative charge along this curve from (1, 0) to (0, 1) is zero, since

$$\langle \mathbf{F}(\alpha(t)), \alpha'(t)\rangle = \langle(-q \cos t, \ -q \sin t), (-\sin t, \cos t)\rangle = 0$$

and therefore

$$\int_a^b \langle \mathbf{F}(\alpha(t)), \alpha'(t))\rangle\, dt = 0$$

Suppose we try another path from (1, 0) to (0, 1). Let $\beta\colon [0, 1] \to U$ describe the straight line segment from (1, 0) to (0, 1):

$$\beta(t) = (1 - t, t)$$

Then $\beta'(t) = (-1, 1)$ and the work done in moving the particle along this curve is given by

$$\begin{aligned}
&\int_0^1 \langle \mathbf{F}(\beta(t)), \beta'(t)\rangle\, dt \\
&\quad = q \int_0^1 \left\langle \left(\frac{-(1-t)}{[(1-t)^2 + t^2]^{3/2}}, \frac{-t}{[(1-t)^2 + t^2]^{3/2}}\right), (-1, 1) \right\rangle dt \\
&\quad = q \int_0^1 \frac{1 - 2t}{(1 - 2t + 2t^2)^{3/2}}\, dt \\
&\quad = q(1 - 2t + 2t^2)^{-1/2}\Big|_0^1 = 0
\end{aligned}$$

Therefore the work done in moving a particle from (1, 0) to (0, 1) in this force field is the same for these two paths. We shall see in the next section that this is no accident.

We use (3.5) to define the integral of a vector field $\mathbf{F}$ along a parametrized curve. Let $\alpha\colon [a, b] \to \mathbf{R}^n$ be the curve, let $\mathbf{F}$ be defined on the open set $U \subset \mathbf{R}^n$ with $\alpha([a, b]) \subset U$. The *line integral of* $\mathbf{F}$ along Γ is defined to be

*Often one wants to compute the work required to move the particle along Γ in the force field $\mathbf{F}$. This is $\int_a^b \langle -\mathbf{F}(\alpha(t)), \alpha'(t)\rangle\, dt$, since the force that must be applied to the particle must counteract the force field (and must therefore be opposite in direction to $\mathbf{F}$).

$$(3.6) \qquad \int_a^b \langle \mathbf{F}(\alpha(t)), \alpha'(t)\rangle \, dt$$

As the above examples show, the actual computation of (3.6) boils down to computing an ordinary integral.

We shall give various properties of line integrals in the next section. For the moment, we give only one result concerning the line integrals for different parametrizations of a curve. It is *not* necessarily true that all equivalent parametrizations give the same value for the line integral. In fact, it is not hard to see that the work required to move a particle along a path depends on the direction in which we traverse the path. By an argument like that used to prove Proposition 2.1, we can show that the integral (3.5) changes sign if we move along Γ in the opposite direction. This fact may help to explain the following definition.

The parametrized curve α: $[a, b] \to \mathbf{R}^n$ is *O-equivalent* (or *oriented-equivalent*) to the parametrized curve β: $[c, d] \to \mathbf{R}^n$ if there is a diffeomorphism $h : [c, d] \to [a, b]$ with $h'(s) > 0$ and $\alpha(h(s)) = \beta(s)$ for all s in $[c, d]$. Roughly speaking, α is O-equivalent to β if α is equivalent to β *and both* α *and* β *are traversed in the same direction* (as t goes from a to b and s from c to d). The relation of being O-equivalent is easily seen to be an equivalence relation (see Exercise 8); as in Section 1, an *oriented curve* is defined to be an equivalence class of parametrized curves under this relation.

For instance, the parametrized curves $\alpha, \beta : [0, 2\pi] \to \mathbf{R}^2$ given by

$$\alpha(t) = (\cos t, \sin t)$$

$$\beta(t) = (\cos t, -\sin t)$$

both define the unit circle in $\mathbf{R}^2$ with center the origin and are equivalent with $\alpha(h(t)) = \beta(t)$, $h(t) = 2\pi - t$, $h'(t) = -1$. However, α is *not* O-equivalent to β; in fact, α corresponds to traversing the unit circle in a counterclockwise direction and β corresponds to traversing the unit circle in a clockwise direction. Thus α and β define distinct oriented curves.

We say that $\alpha : [a, b] \to \mathbf{R}^n$ is a *parametrization for the oriented curve* Γ or a *parametrization for* Γ *as an oriented curve* if α is in the equivalence class of parametrized curves defining the oriented curve Γ. For example, if Γ is the oriented curve defined by $\alpha(t) = (\cos t, \sin t)$, $t \in [0, 2\pi]$, then $\beta(t) = (\cos t, -\sin t)$, $t \in [0, 2\pi]$, is not a parametrization of Γ as an oriented curve.

Suppose now that Γ is an oriented curve in the open subset $U \subset \mathbf{R}^n$ and $\mathbf{F} : U \to \mathbf{R}^n$ is a vector field on U. The line integral of $\mathbf{F}$ *along* Γ is defined by

$$\int_\Gamma \mathbf{F} = \int_a^b \langle \mathbf{F}(\alpha(t)), \alpha'(t)\rangle \, dt$$

where $\alpha : [a, b] \to U$ is a parametrization for Γ *as an oriented curve*. The next result shows that this integral does not depend on the parametrization chosen for Γ.

Proposition 3.1 *Let $\alpha : [a, b] \to U$, $\beta : [c, d] \to U$ be two parametrizations for the oriented curve Γ and $\mathbf{F} : U \to \mathbf{R}^n$ a vector field on U. Then*

$$\int_a^b \langle \mathbf{F}(\alpha(t)), \alpha'(t) \rangle \, dt = \int_c^d \langle \mathbf{F}(\beta(s)), \beta'(s) \rangle \, ds$$

Here $\alpha'(t) = (\alpha_1'(t), \ldots, \alpha_n'(t))$, where $\alpha(t) = (\alpha_1(t), \ldots, \alpha_n(t))$.

Proof. This proof is analogous to that of Proposition 2.1; we give only an outline here.

Suppose $\alpha(t) = \beta(h(t))$, where $h : [a, b] \to [c, d]$ is a diffeomorphism with $h'(t) > 0$, $t \in [a, b]$. Then

$$\begin{aligned}\int_a^b \langle \mathbf{F}(\alpha(t)), \alpha'(t) \rangle \, dt &= \int_a^b \langle \mathbf{F}(\beta(h(t))), \beta'(h(t)) h'(t) \rangle \, dt \\ &= \int_a^b \langle \mathbf{F}(\beta(h(t))), \beta'(h(t)) \rangle h'(t) \, dt \\ &= \int_c^d \langle \mathbf{F}(\beta(s)), \beta'(s) \rangle \, ds\end{aligned}$$

just as in the proof of Proposition 2.1. [In the last step, we let $s = h(t)$.]

There is another notation for the line integral of a vector field. Let $\alpha(t) = (\alpha_1(t), \ldots, \alpha_n(t))$ be a parametrization for the oriented curve Γ in U and $\mathbf{F} : U \to \mathbf{R}^n$ the vector field given by $\mathbf{F}(\mathbf{v}) = (f_1(\mathbf{v}), \ldots, f_n(\mathbf{v}))$. Then the line integral of $\mathbf{F}$ along Γ is given by

$$\int_a^b \sum_{i=1}^n f_i(\alpha(t)) \alpha_i'(t) \, dt = \sum_{i=1}^n \int_a^b f_1(\alpha(t)) \alpha_i'(t) \, dt$$

If we think of x_i as $\alpha_i(t)$, then we can set $dx_i = \alpha_i'(t) \, dt$ and write

$$\int_\Gamma \mathbf{F} = \int_\Gamma f_1 \, dx_1 + \cdots + f_n \, dx_n$$

This notation will make more sense when we deal with the integration of differential forms in Chapter 15.

Exercises

1. Evaluate the line integral of the vector field $\mathbf{F} : \mathbf{R}^2 \to \mathbf{R}^2$, where $\mathbf{F}(x, y) = (-y, x)$, along each of the curves of Exercise 1 of Section 2.

2. Find a parametrization for the curve Γ and determine the work done on a particle moving along Γ in $\mathbf{R}^3$ through the force field $\mathbf{F}: \mathbf{R}^3 \to \mathbf{R}^3$, where $\mathbf{F}(x, y, z) = (1, -x, z)$ and
   (a) Im Γ is the line segment from (0, 0, 0) to (1, 2, 1).
   (b) Im Γ is the polygonal curve with successive vertices (1, 0, 0), (0, 1, 1), and (2, 2, 2).
   (c) Im Γ is the unit circle in the plane $z = 1$ with center (0, 0, 1), beginning and ending at (1, 0, 1).
   (d) Im Γ is the piece of the unit circle in the plane $z = 0$ between (1, 0, 0) and (0, 1, 0) together with the line segment from (0, 1, 0) to (1, 1, 1).
3. Evaluate the line integral of the vector field $\mathbf{F}: \mathbf{R}^3 \to \mathbf{R}^3$ along the curve Γ, where
   (a) $\mathbf{F}(x, y, z) = (x^2, xy, 1)$ and Γ is defined by $\alpha: [0, 1] \to \mathbf{R}^3$, $\alpha(t) = (t, t^2, 1)$.
   (b) $\mathbf{F}(x, y, z) = (\cos z, e^x, e^y)$ and Γ is defined by $\alpha: [0, 2] \to \mathbf{R}^3$, $\alpha(t) = (1, t, e^t)$.
   (c) $\mathbf{F}(x, y, z) = (2x, z, y)$ and Γ is the *twisted cubic* defined by $\alpha: [0, 1] \to \mathbf{R}^3$, $\alpha(t) = (t, t^2, t^3)$.
   (d) $\mathbf{F}(x, y, z) = (\sin z, \cos z, -(xy)^{1/3})$ and Γ is defined by $\alpha: [0, \pi/2] \to \mathbf{R}^3$, $\alpha(t) = (\cos^3 t, \sin^3 t, t)$.
4. Compute $\int_\Gamma x\,dx + xy\,dy$, where Γ is given by
   (a) $\alpha(t) = (\sin t, \cos t)$, $0 \le t \le 2\pi$
   (b) $\alpha(t) = (t^2, t^3)$, $1 \le t \le 2$
   (c) $\alpha(t) = (e^{2t}, e^{3t})$, $0 \le t \le \ln 2$
   (d) $\alpha(t) = (t^2 - 1, t^2 + 1)$, $1 \le t \le 5$
5. Compute $\int_\Gamma y\,dx - z\,dy + x\,dz$, where Γ is given by
   (a) $\alpha(t) = (t, t^2, t^3)$, $0 \le t \le 3$
   (b) $\alpha(t) = (\sin t, \cos t, 2\cos t)$, $0 \le t \le \pi$
   (c) $\alpha(t) = (2t, 1 - t^2, 1 + t^2)$, $-1 \le t \le 1$
6. Let Γ be a parametrization of the unit circle centered at the origin, starting and ending at (1, 0) and traversed counterclockwise. Compute $\int_\Gamma x\,dx + y\,dy$, $\int_\Gamma y\,dx + x\,dy$, and $\int_\Gamma (y^3 - 2xy)\,dx + (3xy^2 - x^2)\,dy$.
*7. Let Γ be a curve in $\mathbf{R}^n$ with parametrization $\alpha: [a, b] \to \mathbf{R}^n$ and let $\mathbf{F}: U \to \mathbf{R}^n$ be a vector field defined on an open set containing Γ. Suppose that Γ is one-to-one and that, for every $u \in \Gamma$ with $u = \alpha(t)$, we have $\mathbf{F}(u) = \alpha'(t)$. Show that $\int_\Gamma \mathbf{F}$ is the arc length of Γ.
8. Prove that the relation "α is O-equivalent to β" is an equivalence relation.
9. Give a detailed proof of Proposition 3.1.
10. Let Γ be the boundary of the unit square in $\mathbf{R}^2$, and let S be the unit square. Show that if $\mathbf{F}$ is a continuously differentiable vector field in $\mathbf{R}^2$,

$$\mathbf{F}(\mathbf{v}) = (f_1(\mathbf{v}), f_2(\mathbf{v}))$$

then

$$\int_\Gamma \mathbf{F} = \int_S \left(\frac{\partial f_2}{\partial x} - \frac{\partial f_1}{\partial y}\right)$$

(the second integral is an ordinary area integral). Here, Γ goes around the square counterclockwise. (*Hint*: Compute the line integral along each line segment; show that the two horizontal ones give $-\int_S \partial f_1/\partial y$ and that the two vertical ones give $\int_S \partial f_2/\partial x$.)

4. Conservative Vector Fields

In this section, we study an important class of vector fields, the *conservative* vector fields. These are the vector fields whose line integrals do not depend on the particular curve chosen, only on its endpoints.

We begin with some of the properties of line integrals.

Proposition 4.1 *Let U be an open subset in $\mathbf{R}^n$ and α: $[a, b] \to U$ a parametrized curve defining the oriented curve Γ. Let $\mathbf{F}, \mathbf{G} : U \to \mathbf{R}^n$ be vector fields and c, d real numbers. Then*

$$\int_\Gamma (c\mathbf{F} + d\mathbf{G}) = c \int_\Gamma \mathbf{F} + d \int_\Gamma \mathbf{G}$$

Furthermore, let e lie between a and b; let α_1: $[a, e] \to U$, α_2: $[e, b] \to U$ the restrictions of α. (See Figure 4.1.) *Then, if Γ_1 and Γ_2 are the oriented curves defined by α_1 and α_2, we have*

$$\int_\Gamma \mathbf{F} = \int_{\Gamma_1} \mathbf{F} + \int_{\Gamma_2} \mathbf{F}$$

Finally, if $-\Gamma$ represents the curve Γ with its opposite orientation, then

$$\int_{-\Gamma} \mathbf{F} = -\int_\Gamma \mathbf{F}$$

The proofs are not hard. We shall prove the last part, leaving the rest as Exercise 3.

A parametrization for $-\Gamma$ is easily obtained by "running backward along Γ"; if α: $[a, b] \to \mathbf{R}^n$ parametrizes Γ, then $\beta : [-b, -a] \to \mathbf{R}^n$, given by

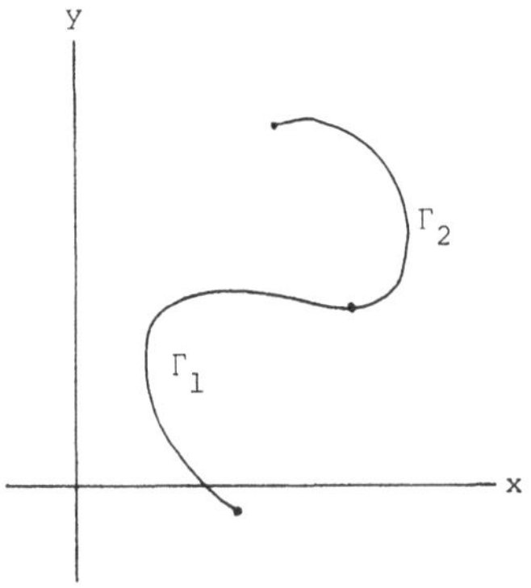

Figure 4.1

$$\beta(t) = \alpha(-t)$$

parametrizes $-\Gamma$. Thus

$$\begin{aligned}\int_{-\Gamma} \mathbf{F} &= \int_{-b}^{-a} \langle \mathbf{F}(\beta(t)), \beta'(t)\rangle \, dt \\ &= \int_{-b}^{-a} \langle \mathbf{F}(\alpha(-t)), -\alpha'(-t)\rangle \, dt \\ &\quad \text{(by the chain rule and the definition of } \beta) \\ &= \int_{b}^{a} - \langle \mathbf{F}(\alpha(s)), -\alpha'(s)\rangle \, ds \\ &\quad \text{(letting } s = -t) \\ &= \int_{a}^{b} \langle \mathbf{F}(\alpha(s)), -\alpha'(s)\rangle \, ds = -\int_{\Gamma} \mathbf{F}\end{aligned}$$

as claimed.

Recall that in the previous section we computed line integrals of a function $\mathbf{F}$ along two different curves connecting the same points and found that the answers agreed. This is not always the case. For instance, suppose that $\mathbf{F}: \mathbf{R}^2 \to \mathbf{R}^2$ is defined by

$$\mathbf{F}(x, y) = (y, x)$$

and let Γ be the first quadrant of the unit circle from $(1, 0)$ to $(0, 1)$. Then Γ is parametrized by $\alpha_1: [0, \pi/2] \to \mathbf{R}^2$, where

$$\alpha_1(t) = (\cos t, \sin t)$$

and
$$\begin{aligned}\int_{\Gamma_1} \mathbf{F} &= \int_{0}^{\pi/2} \langle(\sin t, -\cos t), (-\sin t, \cos t)\rangle \, dt \\ &= \int_{0}^{\pi/2} -(\sin^2 t + \cos^2 t) \, dt = \int_{0}^{\pi/2} -1 \, dt = -\frac{\pi}{2}\end{aligned}$$

If, however, Γ_2 is the straight line from $(1, 0)$ to $(0, 1)$, so that Γ_2 is parametrized by $\alpha_2: [0, 1] \to \mathbf{R}^2$, where

$$\alpha_2(t) = (1 - t, t)$$

then
$$\begin{aligned}\int_{\Gamma_2} \mathbf{F} &= \int_{0}^{1} \langle(t, t - 1), (-1, 1)\rangle \, dt \\ &= \int_{0}^{1} -1 \, dt = -1 \neq -\frac{\pi}{2}\end{aligned}$$

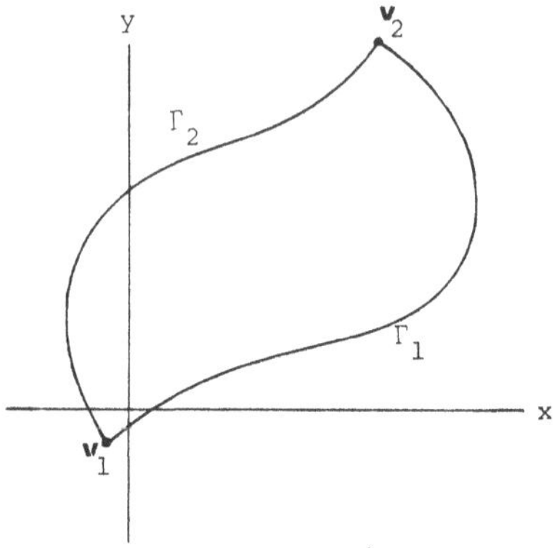

Figure 4.2 Two curves from $\mathbf{v}_1$ to $\mathbf{v}_2$.

We now consider the question of when the line integral of **F** along Γ depends only on the endpoints of Γ. A vector field **F** with this property is called *conservative*. Such fields are of importance in physics because of the law of conservation of evergy. For example, we have seen that the work done in moving a particle along a curve Γ in a force field **F** is $\int_\Gamma \mathbf{F}$. If Γ_1 and Γ_2 are two curves connecting the same points $\mathbf{v}_1$ and $\mathbf{v}_2$ (as in Figure 4.2), then the work done by the field in going from $\mathbf{v}_1$ to $\mathbf{v}_2$ along Γ_1 and back to $\mathbf{v}_1$ along Γ_2 is (we traverse Γ_2 in the reverse direction) is

$$\int_{\Gamma_1} \mathbf{F} + \int_{-\Gamma_2} \mathbf{F} = \int_{\Gamma_1} \mathbf{F} - \int_{\Gamma_2} \mathbf{F}$$

by Proposition 4.1. The law of conservation of energy says that when we return to the original state, the total work done should be 0; that is,

$$\int_{\Gamma_1} \mathbf{F} - \int_{\Gamma_2} \mathbf{F} = 0$$

or

$$\int_{\Gamma_1} \mathbf{F} = \int_{\Gamma_2} \mathbf{F}$$

That is, the law of conservation of energy requires force fields to be conservative fields.*

Let **F** be a vector field on the open set $U \subset \mathbf{R}^n$. A function $V : U \to \mathbf{R}$ of class C^2 is called a *potential function* for the vector field **F** if **F** is the gradient of V,

$$\mathbf{F} = \text{grad } V$$

*The law of conservation of energy may fail for some physical system because of friction, or because energy is being added. In such cases, force fields need not be conservative.

For example, let U be the complement of the origin in $\mathbf{R}^2$ and $\mathbf{F}: U \to \mathbf{R}^2$ the force field arising from a positive charge at the origin, given by (3.1). Then $V: U \to \mathbf{R}$ defined by

$$V(x, y) = \frac{q}{(x^2 + y^2)^{1/2}}$$

is a potential function for $\mathbf{F}$. On the other hand, if $\mathbf{F}: \mathbf{R}^2 \to \mathbf{R}^2$ is given by

$$\mathbf{F}(x, y) = (y, -x)$$

then $\mathbf{F}$ has no potential function V. For if $\mathbf{F} = \operatorname{grad} V$, then

$$\frac{\partial V}{\partial x} = y \qquad \frac{\partial V}{\partial y} = -x$$

Thus
$$\frac{\partial^2 V}{\partial y \partial x} = \frac{\partial}{\partial y} y = 1$$

and
$$\frac{\partial^2 V}{\partial x \partial y} = \frac{\partial}{\partial x}(-x) = -1$$

which contradicts Theorem 1.1 of Chapter 6.

Proposition 4.2 *Suppose that* $\mathbf{F}$ *is a vector field on the open set* $U \subset \mathbf{R}^n$ *with a potential function* V. *Then* $\mathbf{F}$ *is conservative.*

Proof. Let $\mathbf{v}_0$ and $\mathbf{v}_1$ be points in U, and let Γ be an oriented curve with parametrization $\alpha: [a, b] \to U$, $\alpha(a) = \mathbf{v}_0$, $\alpha(b) = \mathbf{v}_1$. Write

$$\alpha(t) = (\alpha_1(t), \ldots, \alpha_n(t))$$

We compute:

$$\begin{aligned}
\int_\Gamma \mathbf{F} &= \int_a^b \langle \mathbf{F}(\alpha(t)), \alpha'(t) \rangle \, dt \\
&= \int_a^b \langle (\operatorname{grad} V)(\alpha(t)), \alpha'(t) \rangle \, dt \\
&= \int_a^b \left[\sum_{j=1}^n \frac{\partial V}{\partial x_j}(\alpha(t)) \frac{d\alpha_j}{dt} \right] dt \\
&= \int_a^b \left[\frac{d}{dt} V(\alpha(t)) \right] dt \qquad \text{by the chain rule} \\
&= V(\alpha(a)) - V(\alpha(b)) \qquad \text{by the fundamental theorem of calculus} \\
&= V(\mathbf{v}_1) - V(\mathbf{v}_0)
\end{aligned}$$

Thus, if the vector field $\mathbf{F}$ has a potential function, the line integral of $\mathbf{F}$

along Γ depends only on the endpoints of Γ, not on the curve itself. Hence, $\mathbf{F}$ is conservative, as we claimed.

Proposition 4.2 has an almost complete converse, as we now see.

Theorem 4.3 *Let U be an open subset of $\mathbf{R}^n$ with the property that any two points in U can be connected by a curve. (For instance, U could be an open ball.) A vector field $\mathbf{F}$ on U is conservative if and only if it has a potential function.*

A subset U of $\mathbf{R}^n$ satisfying the condition of Theorem 4.3 (any two points can be connected by a curve) is called *pathwise connected.**

Proof. We have already seen that a vector field is conservative if it has a potential function. We prove the converse.

Let $\mathbf{v}_0$ be a fixed point of U. Define $V : U \to \mathbf{R}$ by

$$V(\mathbf{v}) = \int_\Gamma \mathbf{F}$$

where Γ is a curve on U from $\mathbf{v}_0$ to $\mathbf{v}$. The curve Γ exists by hypothesis and V is well defined (that is, independent of the curve Γ) since $\mathbf{F}$ is conservative (see Proposition 4.1). We now compute $\partial V/\partial x_i$.

Let $\mathbf{v} = (x_1, \ldots, x_n)$ and Γ_h the curve with parametrization $\alpha_h : [0, 1] \to U$,

$$\alpha_h(t) = (x_1, \ldots, x_i + th, \ldots, x_n)$$

If Γ is a curve from $\mathbf{v}_0$ to $\mathbf{v}$, we let Γ_h' be the curve from $\mathbf{v}_0$ to the point $\mathbf{v}_h = \alpha_h(1) = (x_1, \ldots, x_i + h, \ldots, x_n)$ obtained by first traversing Γ and then traversing Γ_h; Γ_h' has the parametrization $\beta_h : [a, b + 1] \to U$ given by

$$\beta_h(t) = \begin{cases} \alpha(t) & a \le t \le b \\ \alpha_h(t - b) & b \le t \le b + 1 \end{cases}$$

(See Figure 4.3.) Then

$$\begin{aligned} \frac{\partial V}{\partial x_i}(\mathbf{v}) &= \lim_{h \to 0} \frac{1}{h}[V(\mathbf{v}_h) - V(\mathbf{v})] \\ &= \lim_{h \to 0} \frac{1}{h}\left(\int_{\Gamma_h'} \mathbf{F} - \int_\Gamma \mathbf{F}\right) \\ &= \lim_{h \to 0} \frac{1}{h}\int_{\Gamma_h} \mathbf{F} \end{aligned}$$

*The usual definition of pathwise connected only requires the existence of a continuous curve between any two points of the set. (See Section 7.6.) Here, we require the curve to be continuously differentiable. However, it can be shown that these two definitions are equivalent for open sets.

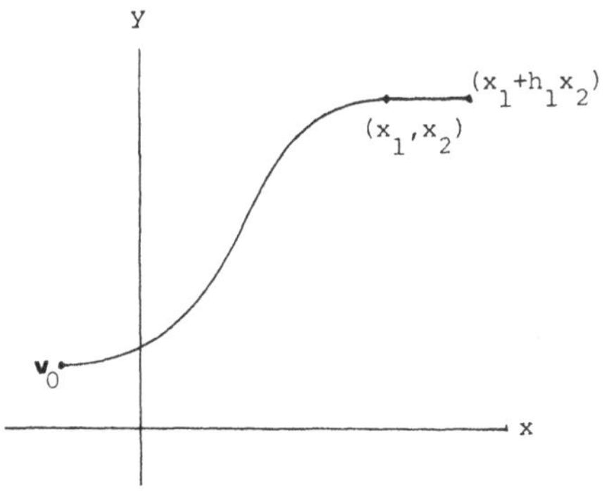

Figure 4.3

by Proposition 4.1. Thus

$$\begin{aligned}\frac{\partial V}{\partial x_i}(\mathbf{v}) &= \lim_{h\to 0}\frac{1}{h}\int_0^1 \langle \mathbf{F}(\alpha_h(t)),\, \alpha_h'(\mathbf{t})\rangle\, dt \\ &= \lim_{h\to 0}\frac{1}{h}\int_0^1 \langle \mathbf{F}(\alpha_h(\mathbf{t})),\, h\mathbf{e}_i\rangle\, dt \\ &= \lim_{h\to 0}\int_0^1 F_i(x_i, \ldots, x_i + th, \ldots, x_n)\, dt\end{aligned}$$

where $\mathbf{e}_i$ is the ith basis vector in $\mathbf{R}^n$ and F_i is the ith component of $\mathbf{F}$. Now

$$\int_0^1 F_i(x_1, \ldots, x_i + th, \ldots, x_n)\, dt$$

is a continuous function of h. (See Lemma 1.1, Chapter 11.) Hence

$$\begin{aligned}\frac{\partial V}{\partial x_i}(\mathbf{v}) &= \lim_{h\to 0}\int_0^1 F_i(x_1, \ldots, x_i + th, \ldots, x_n)\, dt \\ &= \int_0^1 F_i(x_1, \ldots, x_n)\, dt \\ &= F_i(x_1, \ldots, x_n)\end{aligned}$$

Therefore grad $V = \mathbf{F}$, and Theorem 4.3 is proved.

We still need a criterion for determining when a force field is conservative. For the first results along those lines, however, we need Green's theorem, the subject of the next section.

Exercises

1. Find a potential function for each of the following force fields and determine the work required to move a particle from the point $\mathbf{v}_0$ to the point $\mathbf{v}_1$, where
   (a) $\mathbf{F}(x, y) = (2xy^3, 3x^2y^2)$, $\mathbf{v}_0 = (1, 0)$, $\mathbf{v}_1 = (2, 1)$.
   (b) $\mathbf{F}(x, y) = (y \cos xy, x \cos xy)$, $\mathbf{v}_0 = (0, 1)$, $\mathbf{v}_1 = (\pi/2, 1)$.
   (c) $\mathbf{F}(x, y, z) = (2x + y, x, 1)$, $\mathbf{v}_0 = (0, 0, 0)$, $\mathbf{v}_1 = (1, 2, -1)$.
   (d) $\mathbf{F}(x_1, \ldots, x_n) = (2x_1, \ldots, 2x_n)$, $\mathbf{v}_0 = (0, \ldots, 0)$, $\mathbf{v}_1 = (1, \ldots, 1)$.
   (e) $\mathbf{F}(x,y,z) = (yz \cos xyz, xz \cos xyz, xy \cos xyz)$, $\mathbf{v}_0 = (1,1,0)$, $\mathbf{v}_1 = (\frac{1}{3}, \frac{1}{2}, 2\pi)$.
   (f) $\mathbf{F}(x, y) = (2xye^{x^2y}, x^2e^{x^2y})$, $\mathbf{v}_0 = (0, 0)$, $\mathbf{v}_1 = (2, 1)$.
   (g) $\mathbf{F}(x, y) = (x^2 + 4xy + 4y^2, 2x^2 + 8xy + 8y^2)$, $\mathbf{v}_0 = (2, -1)$, $\mathbf{v}_1 = (-4, 2)$.
   (h) $\mathbf{F}(x, y, z) = (2xy, x^2 + z, y)$, $\mathbf{v}_0 = (1, 3, 2)$, $\mathbf{v}_1 = (2, 3, 1)$.
   (The work done is $-\int_\Gamma \mathbf{F}$, since one pushes *against* the force.)
2. Show that if $\mathbf{F} = (f_1, \ldots, f_n)$ is a conservative force field on $\mathbf{R}^n$, then, for all i, j with $1 \leq i \leq j \leq n$,
$$\frac{\partial f_i}{\partial x_j} = \frac{\partial f_j}{\partial x_i}$$
3. Prove the first two assertions of Proposition 4.1.
4. Let U be an open subset of $\mathbf{R}^n$ and $\mathbf{F}: U \to \mathbf{R}^n$ a vector field. Prove that $\mathbf{F}$ is conservative if and only if the line integral around any closed curve is zero.

5. Green's Theorem

In Chapter 10, we defined the integral of a function of n variables over a region in $\mathbf{R}^n$; in Section 3, we defined the line integral of a vector field over a curve in $\mathbf{R}^n$. In this section, we show that, when $n = 2$, these two integrals are closely

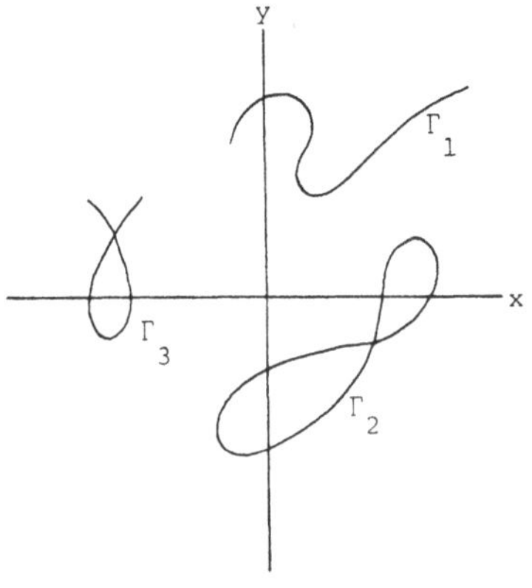

Figure 5.1 Three curves.

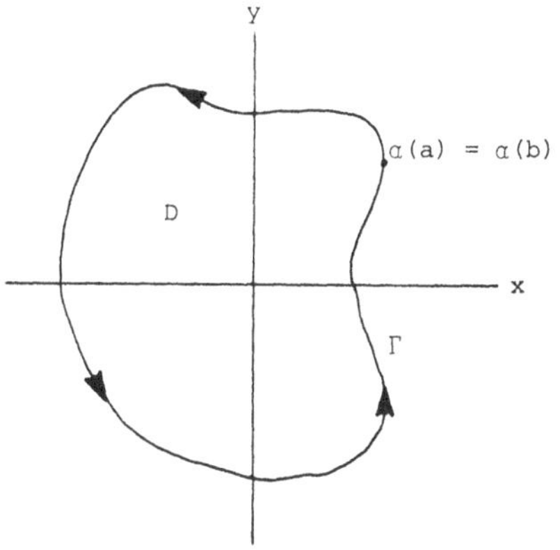

Figure 5.2 A simple closed curve bounding D.

related. The precise result is called *Green's theorem*. Its proof will be given in the next section; extending it to $\mathbf{R}^n$ will occupy us for the next three chapters.

Let $\alpha : [a, b] \to \mathbf{R}^n$ be a parametrization for a curve Γ. We say that Γ is *simple* if it does not intersect itself; that is, if α is injective. For instance, the curve Γ_1 in Figure 5.1 is simple, but Γ_2 and Γ_3 are not. The curve Γ is said to be *closed* if $\alpha(a) = \alpha(b)$, so that Γ does indeed close up on itself. In Figure 5.1, Γ_2 is the only closed curve.

It is clearly impossible for α to be injective and for $\alpha(a)$ to equal $\alpha(b)$; that is, for Γ to be simple and closed. Nevertheless, we define Γ to be a *simple closed curve* if Γ is closed and if α is injective on the interval $[a, b)$, as in Figure 5.2.

We can now state our main theorem for this section.

Theorem 5.1 (Green's Theorem) *Let Γ be an oriented piecewise continuously differentiable simple closed curve in $\mathbf{R}^2$ bounding a region D.* Suppose that Γ is oriented so that the region D is on the left as we traverse Γ.* (See Figure 5.2.) *Then, if $\mathbf{F}\colon D \to \mathbf{R}^2$ is a vector field on D, $\mathbf{F}(\mathbf{v}) = (f_1(\mathbf{v}), f_2(\mathbf{v}))$, we have*

$$\int_\Gamma \mathbf{F} = \int_D \left(\frac{\partial f_2}{\partial x} - \frac{\partial f_1}{\partial y} \right) \tag{5.1}$$

As mentioned above, the proof will be given in the next section.

*A theorem (known as the *Jordan curve theorem*) says that a simple closed curve divides the plane into two regions, only one of which is bounded. This result seems fairly obvious. Proving it, however, turns out to be surprisingly difficult.

Example Let $\mathbf{F}(x, y) = (y, x)$, and let Γ be the unit circle in $\mathbf{R}^2$, traversed counterclockwise. We can compute $\int_\Gamma \mathbf{F}$ quite easily with the help of Green's theorem. Let D be the unit disk. Then, from Green's theorem,

$$\int_\Gamma \mathbf{F} = \int_D \left(\frac{\partial}{\partial x} x - \frac{\partial}{\partial y} y \right) = 0$$

We use the idea of this example to derive a criterion for a vector field to be conservative. Suppose that U is a pathwise connected open subset of $\mathbf{R}^2$ and that $\mathbf{F}$ is a conservative vector field on $U, \mathbf{F}(\mathbf{v}) = (f_1(\mathbf{v}), f_2(\mathbf{v}))$. We know from Proposition 4.3 that $\mathbf{F}$ has a potential function V. If $\mathbf{F}$ has a continuous derivative, then

$$\frac{\partial f_1}{\partial y} = \frac{\partial^2 V}{\partial y \partial x} = \frac{\partial^2 V}{\partial x \partial y} = \frac{\partial f_2}{\partial x}$$

by Theorem 1.1 of Chapter 8. Thus if $\mathbf{F}$ is conservative and continuously differentiable, then $\partial f_1/\partial y = \partial f_2/\partial x$.

The converse, unfortunately, is not true in general. For example, let U be the complement of the origin in $\mathbf{R}^2$ and define $\mathbf{F}: U \to \mathbf{R}^2$ by

$$\mathbf{F}(x, y) = \left(\frac{-y}{x^2 + y^2}, \frac{x}{x^2 + y^2} \right)$$

Then

$$f_1(x, y) = \frac{-y}{x^2 + y^2} \qquad f_2(x, y) = \frac{x}{x^2 + y^2}$$

and

$$\frac{\partial f_1}{\partial y} = \frac{y^2 - x^2}{x^2 + y^2} = \frac{\partial f_2}{\partial x}$$

However, if Γ is the closed curve defined by the parametric curve $\alpha: [0, 2\pi] \to U$,

$$\alpha(t) = (\cos t, \sin t)$$

then

$$\begin{aligned} \int_\Gamma \mathbf{F} &= \int_0^{2\pi} \langle \mathbf{F}(\alpha(t)), \alpha'(t) \rangle \, dt \\ &= \int_0^{2\pi} \langle (-\sin t, \cos t), (-\sin t, \cos t) \rangle \, dt \\ &= \int_0^{2\pi} dt = 2\pi \end{aligned}$$

Thus $\mathbf{F}$ is not conservative.

In spite of this example, we can prove that a vector field in $\mathbf{R}^2$ is conservative whenever $\partial f_1/\partial y = \partial f_2/\partial x$ if we suitably restrict the open set U. We say that a pathwise connected set $X \subset \mathbf{R}^2$ is *simply connected* if the region enclosed by any simple closed curve in X is entirely contained in X. For example, any (open or closed) ball in $\mathbf{R}^2$ is simply connected. However, the set

$$U = \{(x, y) \in \mathbf{R}^2 : (x, y) \neq (0, 0)\}$$

is not simply connected; the region enclosed by the unit circle is the unit ball and is not contained in U.

We can now state the result, promised last section, which gives a criterion for a vector field on $\mathbf{R}^2$ to be conservative.

Proposition 5.2 *Let* $\mathbf{F} = (f_1, f_2)$ *be a vector field on the simply connected open set* $U \subset \mathbf{R}^2$. *Then* $\mathbf{F}$ *is conservative if and only if*

$$\frac{\partial f_1}{\partial y} = \frac{\partial f_2}{\partial x} \tag{5.2}$$

Proof. We have proved above that equation (5.2) holds if $\mathbf{F}$ is conservative. For the converse, we use Green's theorem.

Suppose $\mathbf{F}$ is a vector field on the simply connected open set U satisfying (5.2), and let Γ' and Γ'' by two curves joining $\mathbf{v}_0$ to $\mathbf{v}_1$. We must show that

$$\int_{\Gamma'} \mathbf{F} = \int_{\Gamma} \mathbf{F}$$

Let Γ be the closed curve obtained by first traversing Γ' from $\mathbf{v}_0$ to $\mathbf{v}_1$ and then traversing $-\Gamma''$ from $\mathbf{v}_1$ to $\mathbf{v}_0$. Then

$$\int_{\Gamma} \mathbf{F} = \int_{\Gamma'} \mathbf{F} - \int_{\Gamma''} \mathbf{F}$$

Thus we need to show that the line integral of $\mathbf{F}$ over Γ vanishes whenever Γ is a closed curve.

Suppose, first of all, that Γ is a simple closed curve enclosing the region D on its left as in Figure 5.3. Then $D \subset U$, since U is simply connected, and we can apply Green's theorem to obtain the equation

$$\int_{\Gamma} \mathbf{F} = \int_{D} \left(\frac{\partial f_2}{\partial x} - \frac{\partial f_1}{\partial y} \right)$$

However, the integrand is 0, by equation (5.2). Therefore $\int_\Gamma \mathbf{F} = 0$, as desired.

If Γ is not a simple closed curve we break Γ into a number of simple closed curves and apply the argument above to each of these. For example, if Γ is the closed curve of Figure 5.3, then

$$\int_{\Gamma} \mathbf{F} = \int_{\Gamma_1} \mathbf{F} + \int_{\Gamma_2} \mathbf{F} + \int_{\Gamma_3} \mathbf{F} \tag{5.3}$$

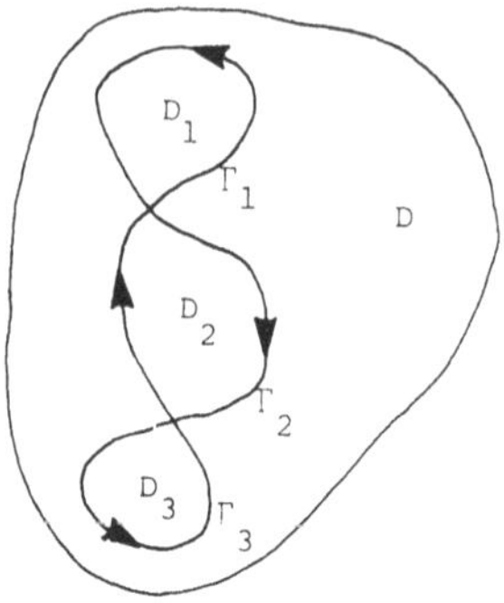

Figure 5.3 A nonsimple closed curve.

Now each of the regions D_1, D_2, D_3 is contained in U since U is simply connected and Green's theorem gives us the equations

$$\int_{\Gamma_1} \mathbf{F} = \int_{D_1} \left(\frac{\partial f_2}{\partial x} - \frac{\partial f_1}{\partial y} \right)$$

$$\int_{\Gamma_2} \mathbf{F} = - \int_{D_2} \left(\frac{\partial f_2}{\partial x} - \frac{\partial f_1}{\partial y} \right)$$

$$\int_{\Gamma_3} \mathbf{F} = - \int_{D_2} \left(\frac{\partial f_2}{\partial x} - \frac{\partial f_1}{\partial y} \right)$$

However, each of these integrals in zero, since equation (5.2) holds. Thus

$$\int_{\Gamma} \mathbf{F} = 0$$

by equation (5.3).

This reasoning clearly proves the proposition in the case where Γ intersects itself finitely many times. It is possible for Γ to intersect itself infinitely often. The proposition remains true, but the proof requires some extra work. We shall omit this case here.

Exercises

1. Let Γ be the boundary curve of the rectangle $R(\mathbf{a}, \mathbf{b})$ [traversed so that $R(\mathbf{a}, \mathbf{b})$ is on the left], where $\mathbf{a} = (0, 0)$, $\mathbf{b} = (2, 3)$. Use Green's theorem to compute the integral of the vector field $\mathbf{F}$ over Γ, where
   (a) $\mathbf{F}(x, y) = (0, x)$

(b) $\mathbf{F}(x, y) = (x + y, y^2)$
(c) $\mathbf{F}(x, y) = (xy^2, 2x - y)$
(d) $\mathbf{F}(x, y) = \left(\sin \frac{\pi}{2} xy, 2x\right)$
(e) $\mathbf{F}(x, y) = (xy, x + y)$
(f) $\mathbf{F}(x, y) = \left(x \sin y, \frac{1}{2} x^2 \cos y\right)$

2. Determine which of the following vector fields $\mathbf{F}$ on $\mathbf{R}^2$ are conservative. If $\mathbf{F}$ is conservative, find a potential function for $\mathbf{F}$.
(a) $\mathbf{F}(x, y) = (x^2 - y^2, 2xy)$
(b) $\mathbf{F}(x, y) = (x^2, 2y^3)$
(c) $\mathbf{F}(x, y) = (x \sin y, y \cos x)$
(d) $\mathbf{F}(x, y) = (x^2 + y^2, 2xy)$
(e) $\mathbf{F}(x, y) = (xe^{xy}, ye^{xy})$
(f) $\mathbf{F}(x, y) = (y^2 - 2x, 2xy + 6y^2)$
3. A subset X in $\mathbf{R}^n$ is said to be *convex* if the line segment converting any two points of X is entirely contained in X; that is, whenever $\mathbf{v}_1, \mathbf{v}_2 \in X$, then $t\mathbf{v}_1 + (1 - t)\mathbf{v}_2$ is in X for all $t \in [0, 1]$.
(a) Prove that any convex subset of $\mathbf{R}^n$ is connected.
(b) Prove that any convex subset of $\mathbf{R}^2$ is simply connected.
4. Let U be a convex open subset of $\mathbf{R}^2$ and $\mathbf{F}$: $U \to \mathbf{R}^2$ a vector field, $\mathbf{F}(\mathbf{v}) = (f_1(\mathbf{v}), f_2(\mathbf{v}))$, with

$$\frac{\partial f_2}{\partial x} = \frac{\partial f_1}{\partial y}$$

Prove that the function $V: U \to \mathbf{R}$ defined by

$$V(\mathbf{u}) = (x - a) \int_0^1 f_1(t\mathbf{u} + (1 - t)\mathbf{u}_0)\, dt + (y - b) \int_0^1 f_2(t\mathbf{u} + (1 - t)\mathbf{u}_0)\, dt$$

[where $\mathbf{u}_0 = (a, b)$ is a fixed point of U and $\mathbf{u} = (x, y)$] is a potential function for $\mathbf{F}$. Where is the fact that U is convex used?
5. (a) Let Γ be a simple closed curve in $\mathbf{R}^2$ bounding a region U. Show that the area of U is $-\int_\Gamma y\, dx$, where Γ is oriented so that U is to the left of Γ.
(b) Show that the area of U is also $\int_\Gamma x\, dy$.

6. The Proof of Green's Theorem

We now give the proof of Green's theorem. The proof involves five cases. The first two deal with the result when the region D has a simple form. We give fairly complete details in these cases. Two others are similar to the first two and are left as Exercises 2 and 3. The last is the general case and we give only an indication of the proof here.

Case 1 Suppose D is the rectangle defined by the points (a, b) and (c, d) (see Figure 6.1):

$$D = \{(x, y) : a \leq x \leq c, b \leq y \leq d\}$$

We prove that

$$\int_\Gamma f_1\, dx = -\int_D \frac{\partial f_1}{\partial y} \tag{6.1}$$

and leave the proof of the equality

$$\int_\Gamma f_2\, dy = \int_D \frac{\partial f_2}{\partial x}$$

to the reader.

Recall that if Γ_1 is parametrized by α: $[a, b] \to \mathbf{R}^2$, where $\alpha(t) = (\alpha_1(t), \alpha_2(t))$, then

$$\int_{\Gamma_1} f_1\, dx = \int_a^b f_1(\alpha(t))\, \alpha_1'(t)\, dt$$

If Γ_1 is a vertical line segment, then $\alpha_1(t)$ is constant and

$$\int_{\Gamma_1} f_1\, dx = 0$$

Thus

$$\int_\Gamma f_1\, dx$$

is equal to the line integrals over the horizontal line segments:

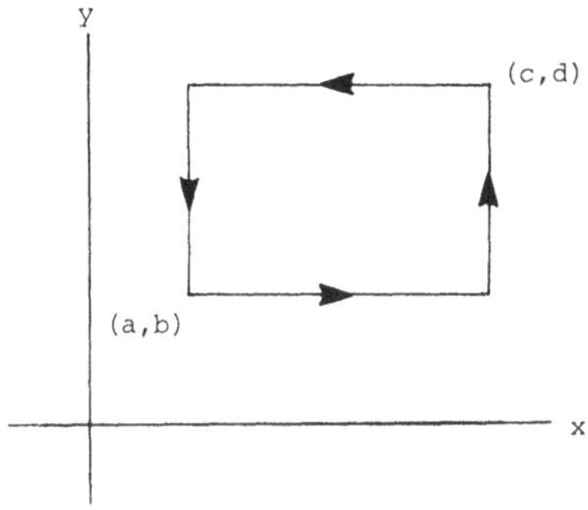

Figure 6.1 A rectangle in $\mathbf{R}^2$.

$$\begin{aligned}\int_\Gamma f_1\,dx &= \int_a^c f_1(x, b)\,dx + \int_c^a f_1(x, d)\,dx\\ &= \int_a^c [f_1(x, b) - f_1(x, d)]\,dx\\ &= \int_a^c \left(-\int_b^d \frac{\partial f_1}{\partial y}\,dy\right)dx\\ &= -\int_D \frac{\partial f_1}{\partial y}\end{aligned}$$

by Theorem 1.2 of Chapter 11. {Here, we have parametrized the line from (a, b) to (c, b) by $\alpha: [a, c] \to \mathbf{R}^2$ such that $\alpha(t) = (t, b)$. The parametrization for the other line is similar.}

Case 2 In this case we assume that D is defined by

$$D = \{(x, y) : a \le x \le b,\ 0 \le y \le h(x)\}$$

where $h : [a, b] \to \mathbf{R}$ is a positive differentiable function. (See Figure 6.2.) Let S be the rectangle in $\mathbf{R}^2$ defined by

$$S = \{(x, y) : a \le x \le b,\ 0 \le y \le 1\}$$

and define $\varphi: S \to D$ by

$$\varphi(x, y) = (x, yh(x))$$

Then φ is bijective and differentiable. Furthermore, if $\alpha: [c, d] \to \mathbf{R}^2$ is a parametrization of the boundary curve Δ to S such that S is on the left of Δ, then $\beta = \varphi \circ \alpha$ is a parametrization of Γ with D on the left. Thus

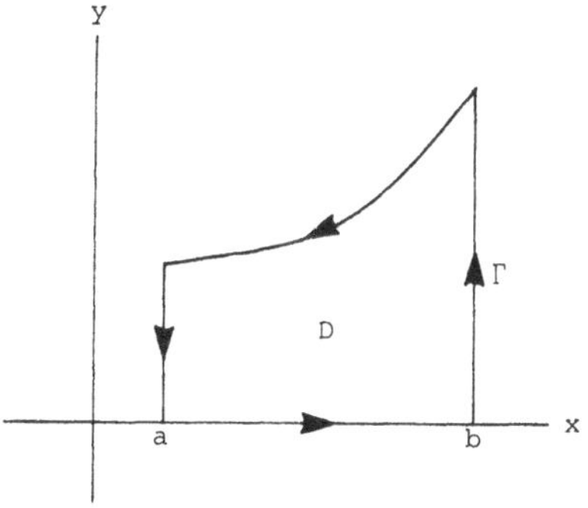

Figure 6.2

(6.2) $$\int_\Gamma f_1\,dx + f_2\,dy = \int_c^d \langle f(\beta(t)), \beta'(t)\rangle\,dt$$
$$= \int_c^d \langle f(\varphi(t))), (D_{\alpha(t)}\varphi)\,\alpha'(t)\rangle\,dt$$

Since $\varphi(x, y) = (x, yh(x))$, the linear transformation $D_{\mathbf{v}}\varphi$ has matrix

$$\begin{bmatrix} 1 & 0 \\ yh'(x) & h(x) \end{bmatrix}$$

In particular, $D_{\mathbf{v}}\varphi$ is invertible; so φ^{-1} is differentiable by Theorem 2.3 of Chapter 9.

Now define $g: \Delta \to \mathbf{R}^2$, $g = (g_1(\mathbf{v}), g_2(\mathbf{v}))$, by

$$g(\mathbf{v}) = (D_{\mathbf{v}}\varphi)^*(f \circ \varphi(\mathbf{v}))$$

where the matrix for $(D_{\mathbf{v}}\varphi)^*$ is

$$\begin{bmatrix} 1 & yh'(x) \\ 0 & h(x) \end{bmatrix}$$

Notice that if $\mathbf{w}_1$ and $\mathbf{w}_2$ are any two vectors in $\mathbf{R}^2$, then

$$\langle (D_{\mathbf{v}}\varphi)^*\,\mathbf{w}_1, \mathbf{w}_2\rangle = \langle \mathbf{w}_1, (D_{\mathbf{v}}\varphi)\mathbf{w}_2\rangle$$

[to prove this, let $\mathbf{w}_1 = (x_1, y_1)$ and $\mathbf{w}_2 = (x_2, y_2)$; then compute]. So, from (6.2), we see that

$$\int_\Gamma f_1\,dx + f_2\,dy = \int_c^d \langle (D_{\alpha(t)}\varphi)^*(f \circ \varphi)(\alpha(t)), \alpha'(t)\rangle\,dt$$
$$= \int_\Delta g_1\,dx + g_2\,dy$$

From the matrix expression for $(D_{\mathbf{v}}\varphi)^*$, we can compute g_1 and g_2:

$$g_1(x, y) = f_1(x, yh(x)) + yh'(x)\,f_2(x, yh(x))$$
$$g_2(x, y) = h(x)\,f_2(x, yh(x))$$

Therefore

$$\frac{\partial g_1}{\partial y} = h(x)\frac{\partial f_1}{\partial y}(x, yh(x)) + h'(x)\,f_2(x, yh(x)) + yh'(x)\frac{\partial f_2}{\partial y}(x, yh(x))\,h(x)$$

$$\frac{\partial g_2}{\partial x} = h'(x)\,f_2(x, yh(x)) + h(x)\frac{\partial f_2}{\partial x}(x, yh(x)) + h(x)\frac{\partial f_2}{\partial y}(x, yh(x))\,yh'(x)$$

We can now apply case 1:

$$\begin{aligned}\int_\Delta g_1\,dx + g_2\,dy &= \int_S \left(\frac{\partial g_2}{\partial x} - \frac{\partial g_1}{\partial y}\right)\\ &= \int_S \left[\frac{\partial f_2}{\partial x}(x, yh(x)) - \frac{\partial f_1}{\partial y}(x, yh(x))\right]h(x)\\ &= \int_S \left(\frac{\partial f_2}{\partial x}\varphi - \frac{\partial f_1}{\partial y}\varphi\right)\operatorname{Det}(D\varphi)\\ &= \int_D \left(\frac{\partial f_2}{\partial x} - \frac{\partial f_1}{\partial y}\right)\end{aligned}$$

(The last equality is by the change of variable formula, Theorem 6.2, Chapter 11; recall that $h > 0$.) This finishes the proof of case 2.

Case 3 Suppose now that Γ is an arbitrary oriented closed piecewise continuously differentiable curve parametrized by the function $\alpha(t) = (\alpha_1(t), \alpha_2(t))$, $a \le t \le b$. We partition the interval $[a, b]$ into a finite number of subintervals,

$$a = t_0 < t_1 < \cdots < t_m = b$$

and let Γ_j be the curve obtained by restricting α to the interval $[t_{j-1}, t_j]$, $1 \le j \le m$. We construct this partition so that the following are true.

(i) The function α is differentiable on $[t_{j-1}, t_j]$ for $1 \le j \le m$.
(ii) For each j, $1 \le j \le m$, we have either $\alpha_1'(t) \ne 0$ for all $t \in (t_{j-1}, t_j)$ or $\alpha_2'(t) \ne 0$ for all $t \in (t_{j-1}, t_j)$. (See Figure 6.3.)

We can then divide the region D into subregions whose boundary curves are either pieces of Γ or horizontal or vertical line segments. Each of these subregions is, up to a rotation of $\pm 90°$ or $180°$, the type considered in case 2 (see also Exercise 3); so

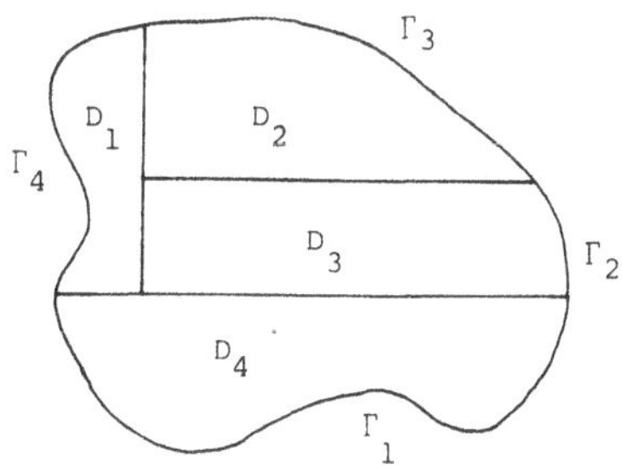

Figure 6.3

$$\int_{\Delta_j} f_1\,dx + f_2\,dy = \int_{D_j} \left(\frac{\partial f_2}{\partial x} - \frac{\partial f_1}{\partial y}\right)$$

where Δ_j is the bounding curve of D_j. Since each of the interior horizontal or vertical line segments occurs as the boundary of exactly two regions, one sees easily that their contributions cancel out and we have

$$\begin{aligned}\int_{\Gamma} f_1\,dx + f_2\,dy &= \sum_{j=1}^{m} \int_{\Gamma_j} f_1\,dx + f_2\,dy \\ &= \sum_{j=1}^{m} \int_{D_j} \left(\frac{\partial f_2}{\partial x} - \frac{\partial f_1}{\partial y}\right) \\ &= \int_{D} \left(\frac{\partial f_2}{\partial x} - \frac{\partial f_1}{\partial y}\right)\end{aligned}$$

This completes our account of the proof of Green's theorem.

Note. The reason that this discussion does not constitute a proof is that we have not explicitly described how to subdivide the region D. It is easy to convince oneself that D can be split into regions of the desired shape, but harder to describe a procedure for doing the splitting. Such a description leads us to considerations rather far removed from the main topics of our text.

Exercises

1. Complete the other half of case 1 of the proof of Theorem 4.1, namely that

$$\int_{\Gamma} f_2\,dy = \int_{D} \frac{\partial f_2}{\partial x}$$

2. Give a direct proof of Theorem 5.1 for regions defined by a right triangle with two sides parallel to the coordinate axes.
3. Some of the regions in Figure 6.3, like D_6, are distorted triangles; that is, they are (up to a rotation) regions like those in Figure 6.2, except that $h(a) = 0$ [and $h(t) > 0$ for $a < t \le b$]. Prove Green's theorem for such regions. (*Hint*: Modify the proof of case 2 and use Exercise 2.)
4. Show that if $U \in \mathbf{R}^n$ is a connected open set, then any two points $\mathbf{v}, \mathbf{w} \in U$ can be connected by a continuous piecewise differentiable path. (*Hint*: Use a polygonal path.)

13

Surface Integrals

In this chapter, we begin the task of extending the results of the previous section to higher dimensions. The first step is defining surface integrals for surfaces in $\mathbf{R}^3$. As with line integrals, there are two sorts of surface integrals: integrals of functions and integrals of vector fields. We next relate the surface integral of a vector field to the line integral of a related vector field along the boundary of the surface. This result, Stokes' theorem, is an analogue of Green's theorem. We state it here and give an application to hydrodynamics. The proof is given in Chapter 15.

As will be seen, surface integrals are considerably more difficult to work with than line integrals. For further generalizations, we need to introduce differential forms. This, however, must wait until the next chapter.

1. The Cross Product in $\mathbf{R}^3$

In general, it is not possible to introduce an interesting multiplication in $\mathbf{R}^n$, one where the product of two vectors is again a vector. It is possible in $\mathbf{R}^1$, since we can multiply real numbers; it is also possible in $\mathbf{R}^2$, since the point $(a, b) \in \mathbf{R}^2$ can be regarded as the complex number $a + bi$. In this section, we introduce a multiplication in $\mathbf{R}^3$ with properties strikingly different from ordinary multiplication. This multiplication will be useful in our study of surface integrals.

Let $\mathbf{u} = (a_1, a_2, a_3)$ and $\mathbf{v} = (b_1, b_2, b_3)$ be vectors in $\mathbf{R}^3$. The *cross product of* $\mathbf{u}$ *with* $\mathbf{v}$ is defined to be the vector

(1.1) $$\mathbf{u} \times \mathbf{v} = (a_2b_3 - a_3b_2, a_3b_1 - a_1b_3, a_1b_2 - a_2b_1)$$

For example, if $\mathbf{e}_1$, $\mathbf{e}_2$, and $\mathbf{e}_3$ are the standard basis vectors in $\mathbf{R}^3$, with $\mathbf{e}_1 = (1, 0, 0)$, $\mathbf{e}_2 = (0, 1, 0)$, and $\mathbf{e}_3 = (0, 0, 1)$, then

(1.2) $$\mathbf{e}_1 \times \mathbf{e}_2 = \mathbf{e}_3 \qquad \mathbf{e}_2 \times \mathbf{e}_3 = \mathbf{e}_1 \qquad \mathbf{e}_3 \times \mathbf{e}_1 = \mathbf{e}_2$$

In fact, if $\mathbf{u}$ and $\mathbf{v}$ are as above, then one can write $\mathbf{u} \times \mathbf{v}$ as the determinant of the matrix

(1.3) $$\begin{bmatrix} \mathbf{e}_1 & \mathbf{e}_2 & \mathbf{e}_3 \\ a_1 & a_2 & a_3 \\ b_1 & b_2 & b_3 \end{bmatrix}$$

[This determinant is meant "symbolically": if one expands in the usual way, then (1.3) becomes (1.1).]

The elementary properties of the cross product are contained in the following proposition.

Proposition 1.1 *Let a be a real number and $\mathbf{u}$, $\mathbf{v}$, $\mathbf{w}$ vectors in $\mathbf{R}^3$. Then*

(a) $a(\mathbf{u} \times \mathbf{v}) = (a\mathbf{u}) \times \mathbf{v} = \mathbf{u} \times (a\mathbf{v})$
(b) $(\mathbf{u} + \mathbf{w}) \times \mathbf{v} = \mathbf{u} \times \mathbf{v} + \mathbf{w} \times \mathbf{v}$
(c) $\mathbf{u} \times \mathbf{v} = -\mathbf{v} \times \mathbf{u}$
(d) $\langle \mathbf{u} \times \mathbf{v}, \mathbf{u} \rangle = \langle \mathbf{u} \times \mathbf{v}, \mathbf{v} \rangle = 0$

Proof. These properties follow by direct computation. For example, to prove $a(\mathbf{u} \times \mathbf{v}) = (a\mathbf{u}) \times \mathbf{v}$, we use the definition:

$$\begin{aligned} a(\mathbf{u} \times \mathbf{v}) &= a(a_2b_3 - a_3b_2, a_3b_1 - a_1b_3, a_1b_2 - a_2b_1) \\ &= (a(a_2b_3) - a(a_3b_2), a(a_3b_1) - a(a_1b_3), a(a_1b_2) - a(a_2b_1)) \\ &= ((aa_2)b_3 - (aa_3)b_2, (aa_3)b_1 - (aa_1)b_3, (aa_1)b_2 - (aa_2)b_1) \\ &= (a\mathbf{u}) \times \mathbf{v} \end{aligned}$$

The remainder of the proof is left to the reader.

Corollary *For any vector $\mathbf{v}$ in $\mathbf{R}^3$, $\mathbf{v} \times \mathbf{v} = \mathbf{0}$.*

Proof. By (c) above, $\mathbf{v} \times \mathbf{v} = -\mathbf{v} \times \mathbf{v}$, and $\mathbf{0}$ is clearly the only vector with this property.

Another useful property of the cross product is the following "converse" to the corollary above.

Proposition 1.2 *Let $\mathbf{u}$ and $\mathbf{v}$ be vectors in $\mathbf{R}^3$ with $\mathbf{u} \times \mathbf{v} = \mathbf{0}$. Then either $\mathbf{u}$ is a multiple of $\mathbf{v}$ or $\mathbf{v}$ is a multiple of $\mathbf{u}$.*

Proof. Let $\mathbf{u} = (a_1, a_2, a_3)$ and $\mathbf{v} = (b_1, b_2, b_3)$. By definition, if $\mathbf{u} \times \mathbf{v} = \mathbf{0}$, then

$$a_2b_3 - a_3b_2 = a_3b_1 - a_1b_3 = a_1b_2 - a_2b_1 = 0$$

If either vector is zero, then it is a multiple of the other. Suppose $\mathbf{v} \neq \mathbf{0}$, and for simplicity, assume $b_1 \neq 0$. Using the equations above, we see that

$$a_1 = \frac{a_1}{b_1}b_1 \qquad a_2 = \frac{a_1}{b_1}b_2 \qquad a_3 = \frac{a_1}{b_1}b_3$$

so that $\mathbf{u} = (a_1/b_1)\mathbf{v}$. A similar argument holds if $b_1 = 0$ and some other component of $\mathbf{v}$ is nonzero.

Proposition 1.1 shows that the cross product does not satisfy the commutative law: $\mathbf{u} \times \mathbf{v} \neq \mathbf{v} \times \mathbf{u}$ generally. [In fact, (c) of Proposition 1.1 is often called the *anticommutative law*.] The associative law also fails for cross products: in general,

$$\mathbf{u} \times (\mathbf{v} \times \mathbf{w}) \neq (\mathbf{u} \times \mathbf{v}) \times \mathbf{w}$$

For instance,

$$\mathbf{e}_1 \times (\mathbf{e}_1 \times \mathbf{e}_2) = \mathbf{e}_1 \times \mathbf{e}_3 = -(\mathbf{e}_3 \times \mathbf{e}_1) = -\mathbf{e}_2$$

while

$$(\mathbf{e}_1 \times \mathbf{e}_1) \times \mathbf{e}_2 = \mathbf{0} \times \mathbf{e}_2 = \mathbf{0}$$

Instead of the associative law, a more complicated law (the *Jacobi identity*) holds for cross products. See Exercises 5 and 6.

We conclude with a geometric description of the cross product.

Proposition 1.3 *Let* $\mathbf{u}$ *and* $\mathbf{v}$ *be vectors in* $\mathbf{R}^3$ *and let* θ *be the angle between* $\mathbf{u}$ *and* $\mathbf{v}$, $0 \leq \theta \leq \pi$. *Then*

$$\|\mathbf{u} \times \mathbf{v}\| = \|\mathbf{u}\|\,\|\mathbf{v}\| \sin\theta$$

(This is exactly the area of the parallelogram spanned by $\mathbf{u}$ and $\mathbf{v}$.)

Proof. Let $\mathbf{u} = (a_1, a_2, a_3)$ and $\mathbf{v} = (b_1, b_2, b_3)$. Then

$$\begin{aligned}\|\mathbf{u} \times \mathbf{v}\|^2 &= (a_1^2 + a_2^2 + a_3^2)(b_1^2 + b_2^2 + b_3^2) - (a_1b_1 + a_2b_2 + a_3b_3)^2 \\ &= \|\mathbf{u}\|^2\,\|\mathbf{v}\|^2 - \langle \mathbf{u}, \mathbf{v}\rangle^2\end{aligned}$$

by direct computation. Now

$$\langle \mathbf{u}, \mathbf{v}\rangle = \|\mathbf{u}\|\,\|\mathbf{v}\| \cos\theta$$

so that

$$\|\mathbf{u} \times \mathbf{v}\|^2 = \|\mathbf{u}\|^2\,\|\mathbf{v}\|^2(1 - \cos^2\theta) = \|\mathbf{u}\|^2\,\|\mathbf{v}\|^2 \sin^2\theta$$

This result leads to another characterization of the cross product $\mathbf{u} \times \mathbf{v}$. Let $\mathbf{n}$ be the unit vector orthogonal to the plane of $\mathbf{u}$ and $\mathbf{v}$. We orient $\mathbf{n}$ so that turning a right-hand screw along the axis of $\mathbf{n}$ from $\mathbf{u}$ to $\mathbf{v}$ moves in the direction of $\mathbf{n}$. Then

$$\text{(1.4)} \qquad \mathbf{u} \times \mathbf{v} = (\|\mathbf{u}\| \, \|\mathbf{v}\| \sin \theta)\mathbf{n}$$

Why? Recall [Proposition 1.1(d)] that $\mathbf{u} \times \mathbf{v}$ is orthogonal to $\mathbf{u}$ and $\mathbf{v}$; in addition, we know the length of $\mathbf{u} \times \mathbf{v}$ from Proposition 1.3. From these facts, we see that

$$\mathbf{u} \times \mathbf{v} = \pm(\|\mathbf{u}\| \, \|\mathbf{v}\| \sin \theta)\mathbf{n}$$

To check the sign, we use the fact that $\mathbf{u} \times \mathbf{v}$ varies continuously with $\mathbf{u}$ and $\mathbf{v}$ and notice that the formula works for the cases in (1.2).

Exercises

1. Compute the cross product $\mathbf{u} \times \mathbf{v}$, where
   (a) $\mathbf{u} = (0, 1, 1)$, $\mathbf{v} = (1, 1, 0)$
   (b) $\mathbf{u} = (2, 1, 0)$, $\mathbf{v} = (1, 1, 0)$
   (c) $\mathbf{u} = (2, 1, -1)$, $\mathbf{v} = (1, 3, -4)$
   (d) $\mathbf{u} = (1, 3, 2)$, $\mathbf{v} = (2, 1, 3)$
   (e) $\mathbf{u} = (4, -2, -2)$, $\mathbf{v} = (5, 3, 4)$
   (f) $\mathbf{u} = (2, -5, 1)$, $\mathbf{v} = (7, -7, 4)$
2. Let $\mathbf{u} = (a_1, a_2, a_3)$, $\mathbf{v} = (b_1, b_2, b_3)$, and $\mathbf{w} = (c_1, c_2, c_3)$ be vectors in $\mathbf{R}^3$.
   (a) Prove that $\langle \mathbf{u}, \mathbf{v} \times \mathbf{w} \rangle$ is the determinant of the matrix

$$\begin{bmatrix} a_1 & a_2 & a_3 \\ b_1 & b_2 & b_3 \\ c_1 & c_2 & c_3 \end{bmatrix}$$

   (b) Prove that $\langle \mathbf{u}, \mathbf{v} \times \mathbf{w} \rangle = \langle \mathbf{v}, \mathbf{w} \times \mathbf{u} \rangle = \langle \mathbf{w}, \mathbf{u} \times \mathbf{v} \rangle$.
3. Let $\mathbf{u} = (1, 2, -1)$, $\mathbf{v} = (2, 0, 3)$, $\mathbf{w} = (-3, 1, 5)$, $\mathbf{x} = (1, -2, 2)$, and compute
   (a) $\langle \mathbf{u}, \mathbf{v} \times \mathbf{w} \rangle$
   (b) $\langle \mathbf{v}, \mathbf{w} \times \mathbf{v} \rangle$
   (c) $\langle \mathbf{v} + \mathbf{x}, \mathbf{w} \times \mathbf{u} \rangle$
   (d) $(\mathbf{v} + \mathbf{x}) \times (\mathbf{u} - \mathbf{v})$
   (e) $\mathbf{v} \times (\mathbf{u} - 2\mathbf{v} + 7\mathbf{x})$
   (f) $(\mathbf{v} \times \mathbf{w}) \times (\mathbf{u} \times \mathbf{v})$
4. Prove properties (b), (c), and (d) of Proposition 1.1.
5. Prove $(\mathbf{u} \times \mathbf{v}) \times \mathbf{w} + (\mathbf{v} \times \mathbf{w}) \times \mathbf{u} + (\mathbf{w} \times \mathbf{u}) \times \mathbf{v} = \mathbf{0}$. (This is called the *Jacobi identity*.)
6. Prove $(\mathbf{u} \times \mathbf{v}) \times \mathbf{w} = \mathbf{u} \times (\mathbf{v} \times \mathbf{w})$ if and only if $(\mathbf{u} \times \mathbf{w}) \times \mathbf{v} = \mathbf{0}$.

*7. Let S be the rectangle in $\mathbf{R}^2$ defined by the points (a_1, a_2) and (b_1, b_2), and let $T : \mathbf{R}^2 \to \mathbf{R}^3$ be a linear transformation. Prove that the area of $T(S)$ is given by

$$(b_1 - a_1)(b_2 - a_2) \, \| T(\mathbf{e}_1) \times T(\mathbf{e}_2) \|$$

*8. Let $T: \mathbf{R}^3 \to \mathbf{R}^3$ be a linear transformation.
   (a) Show by example that $T(\mathbf{u} \times \mathbf{v})$ is not necessarily equal to $T(\mathbf{u}) \times T(\mathbf{v})$.
   (b) Suppose that $\langle T\mathbf{u}, T\mathbf{v}\rangle = \langle \mathbf{u}, \mathbf{v}\rangle$ for all $\mathbf{u}, \mathbf{v} \in \mathbf{R}^3$. (Such a T is called an *orthogonal transformation.*) Prove that

$$T(\mathbf{u} \times \mathbf{v}) = \varepsilon T(\mathbf{u}) \times T(\mathbf{v})$$

where $\varepsilon = \pm 1$. (In fact, ε is the determinant of the matrix of T.)

2. Surfaces

We now give the definition of a surface as well as some examples.

Let S be a closed subset of $\mathbf{R}^2$ with nonempty interior. A *parametrized surface* is a continuously differentiable function $\alpha: S \to \mathbf{R}^n$.

For example, let $S = \mathbf{R}^2$ and $\mathbf{u}$, $\mathbf{v}$, $\mathbf{w}$ vectors in $\mathbf{R}^3$, with $\mathbf{v}$ and $\mathbf{w}$ independent. Then

(2.1) $$\alpha(x, y) = \mathbf{u} + x\mathbf{v} + y\mathbf{w}$$

defines a parametrized surface whose image is a *plane.* (See Figure 2.1.) If $\mathbf{v}'$ and $\mathbf{w}'$ are any two linearly independent vectors in the span of $\mathbf{v}$ and $\mathbf{w}$, then $\beta(x, y) = \mathbf{u} + x\mathbf{v}' + y\mathbf{w}'$ defines the same plane.

Recall that in Section 5.3 we defined a plane in $\mathbf{R}^3$ in terms of a point on the plane and a normal to the plane. If the image of α defined in (2.1) is to be a plane P, then the endpoint of $\mathbf{u} = (c_1, c_2, c_3)$ should be on P and $\mathbf{v} \times \mathbf{w}$ should be a normal. Let $\mathbf{v} = (a_1, a_2, a_3)$ and $\mathbf{w} = (b_1, b_2, b_3)$. Then the equation of the plane containing $\mathbf{u}$ with normal $\mathbf{v} \times \mathbf{w}$ is

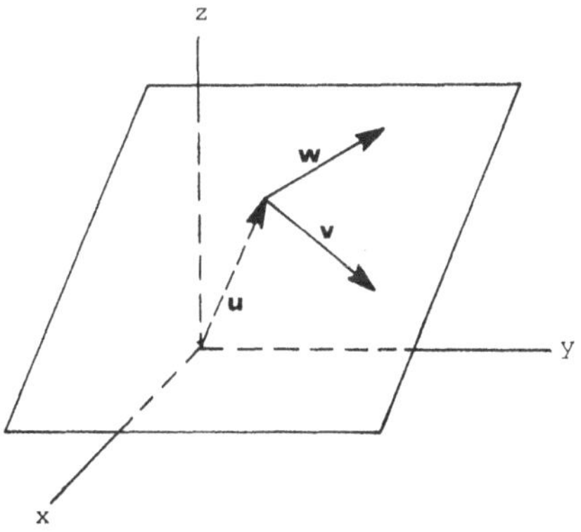

Figure 2.1 A plane in $\mathbf{R}^3$.

$$(2.2) \qquad (a_2b_3 - a_3b_2)(x_1 - c_1) + (a_3b_1 - a_1b_3)(x_2 - c_2) + (a_1b_2 - a_2b_1)(x_3 - c_3) = 0$$

Now, any point in the image of α has the form

$$\begin{aligned}\mathbf{u} + x\mathbf{v} + y\mathbf{w} &= (c_1, c_2, c_3) + x(a_1, a_2, a_3) + y(b_1, b_2, b_3) \\ &= (c_1 + xa_1 + yb_1, c_2 + xa_2 + yb_2, c_3 + xa_3 + yb_3)\end{aligned}$$

for some $(x, y) \in \mathbf{R}^2$. A straightforward computation shows that any such point satisfies equation (2.2).

If we restrict the functions α of equation (2.1) to a rectangle $S = R(\mathbf{a}, \mathbf{b})$, we get a *parallelogram*. (See Figure 2.2.)

In general, if $\alpha: S \to \mathbf{R}^3$ is a parametrized surface and $\mathbf{z}_0 \in S$ is a point at which $(D_{\mathbf{z}_0}\alpha)\mathbf{e}_1$ and $(D_{\mathbf{z}_0}\alpha)\mathbf{e}_2$ are independent, we define the *tangent plane* at $\mathbf{u} = \alpha(\mathbf{z}_0)$ to be the plane

$$(2.3) \qquad \beta(x, y) = \mathbf{u} + (D_{\mathbf{z}_0}\alpha)\mathbf{e}_1 + y(D_{\mathbf{z}_0}\alpha)\mathbf{e}_2$$

(See Figure 2.3.) One sees easily that this notion of tangent plane corresponds to the one defined in Section 5.7. (See Exercise 6.)

For a more interesting example of a surface, let $S = B_1(\mathbf{0})$ be the closed ball about the origin of radius 1 in $\mathbf{R}^2$ and

$$(2.4) \qquad \alpha(x, y) = \left(\frac{2x}{1 + x^2 + y^2}, \frac{2y}{1 + x^2 + y^2}, \frac{1 - x^2 - y^2}{1 + x^2 + y^2}\right)$$

[Here (x, y) are coordinates in $\mathbf{R}^2$.] The image of α is a unit upper hemisphere in $\mathbf{R}^3$. (See Figure 2.4.) This mapping is obtained by *stereographic projection*;

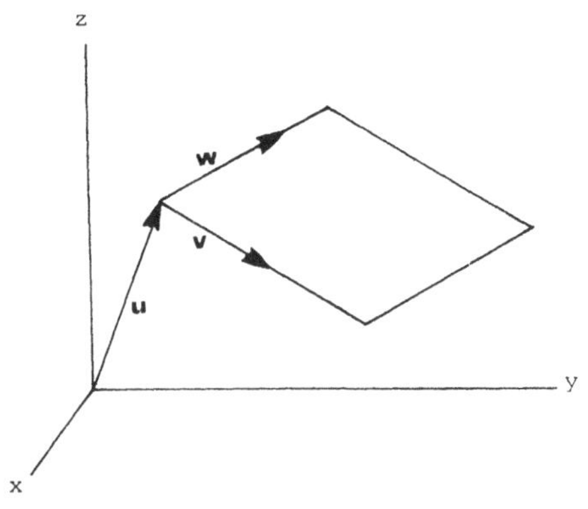

Figure 2.2 A parallelogram in $\mathbf{R}^3$.

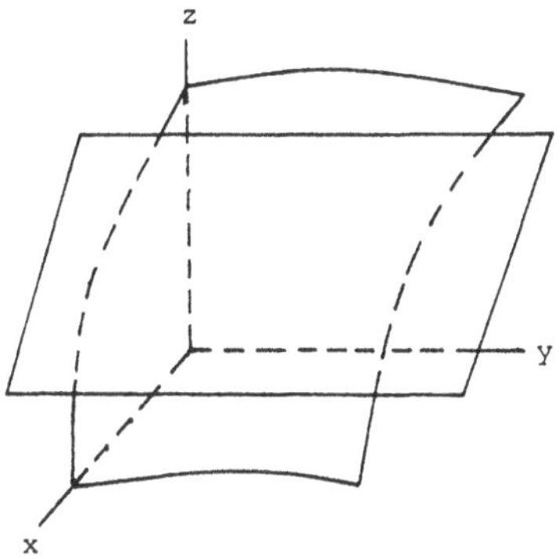

Figure 2.3 A tangent plane.

$\alpha(x, y)$ is the point of intersection of the upper hemisphere with the line through the points $(0, 0, -1)$ and $(x, y, 0)$. Similarly,

$$\beta(x, y) = \left(\frac{2x}{1 + x^2 + y^2}, \frac{2y}{1 + x^2 + y^2}, \frac{x^2 + y^2 - 1}{1 + x^2 + y^2} \right)$$

defines the lower hemisphere.

In fact, we can consider both α and β above as defined on all of $\mathbf{R}^2$. If we do, then the image of α is the complement of the "south pole" $(0, 0, -1)$ and the image of β is the complement of the north pole $(0, 0, 1)$. (See Figure 2.5.) Computing the tangent plane is not hard. If $\mathbf{v} = (x, y)$, we see from (2.4) that $D_{\mathbf{v}}\alpha$ has matrix

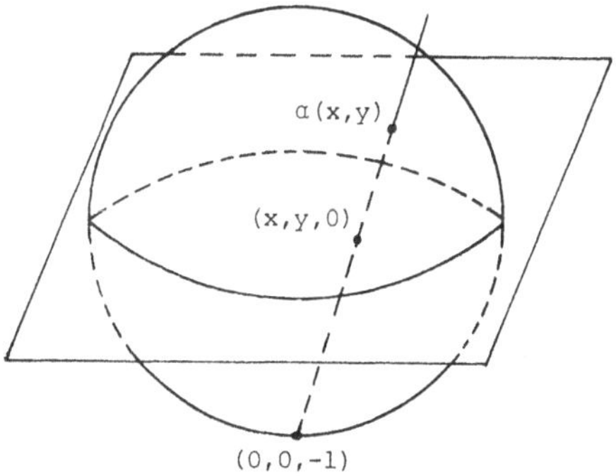

Figure 2.4 Stereographic projection of the upper hemispheres.

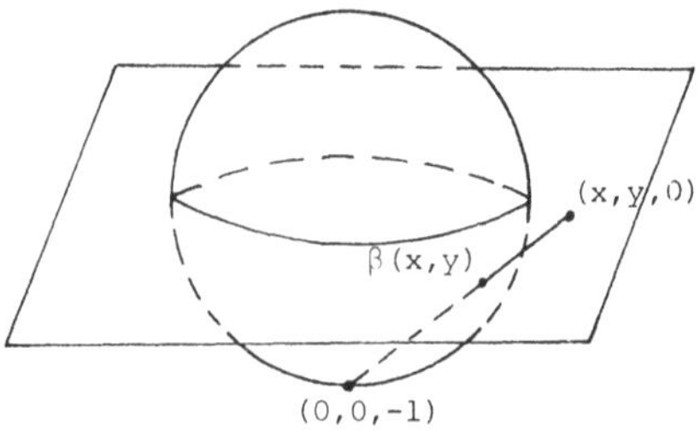

Figure 2.5 Stereographic projection of the lower hemisphere.

(2.5)
$$\begin{bmatrix} \dfrac{2(1-x^2+y^2)}{d^2} & \dfrac{-4xy}{d^2} \\ \dfrac{-4xy}{d^2} & \dfrac{2(1+x^2-y^2)}{d^2} \\ \dfrac{-4x}{d^2} & \dfrac{-4y}{d^2} \end{bmatrix}$$

where $d = 1 + x^2 + y^2$.

So, for example, if $x = y = \frac{1}{2}$, so that $\alpha(x, y) = (\frac{2}{3}, \frac{2}{3}, \frac{1}{3})$, then $D_{\mathbf{v}}\alpha$ has matrix

$$\begin{bmatrix} \frac{8}{9} & \frac{-4}{9} \\ \frac{-4}{9} & \frac{8}{9} \\ \frac{-8}{9} & \frac{-8}{9} \end{bmatrix}$$

and $\quad (D_{\mathbf{v}}\alpha)\mathbf{e}_1 = \left(\frac{8}{9}, \frac{-4}{9}, \frac{-8}{9}\right) \qquad (D_{\mathbf{v}}\alpha)\mathbf{e}_2 = \left(\frac{-4}{9}, \frac{8}{9}, \frac{-8}{9}\right)$

Therefore the tangent plane is given by

$$\begin{aligned} \gamma(x,y) &= \left(\frac{2}{3}, \frac{2}{3}, \frac{1}{3}\right) + \left(\frac{8}{9}, \frac{-4}{9}, \frac{-8}{9}\right)x + \left(\frac{-4}{9}, \frac{8}{9}, \frac{-8}{9}\right)y \\ &= \frac{1}{9}(6 + 8x - 4y, 6 - 4x + 8y, 3 - 8x - 8y) \end{aligned}$$

An easier point at which to visualize the tangent plane is $\mathbf{v} = (0, 0, 1) = \alpha(0, 0)$. There, $D_{\mathbf{v}}\alpha$ has matrix

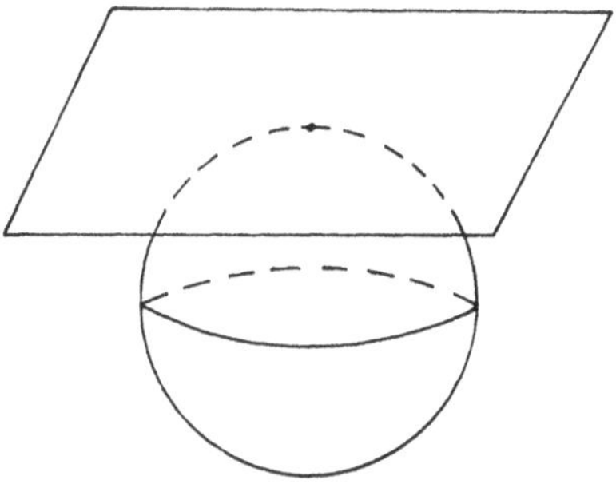

Figure 2.6 Tangent plane to the sphere at the north pole.

$$\begin{bmatrix} 2 & 0 \\ 0 & 2 \\ 0 & 0 \end{bmatrix}$$

and $\qquad (D_{\mathbf{v}}\alpha)\mathbf{e}_1 = (2, 0, 0) \qquad (D_{\mathbf{v}}\alpha)\mathbf{e}_2 = (0, 2, 0)$

Thus the tangent plane is

$$\delta(x, y) = (0, 0, 1) + (2, 0, 0)x + (0, 2, 0)y = (2x, 2y, 1)$$

This is the plane through $(0, 0, 1) = \alpha(0, 0)$ parallel to the x, y plane, as it should be. (See Figure 2.6.)

Just as in the case of curves, we say that two parametrized surfaces $\alpha : S \to \mathbf{R}^n$, $\beta : S' \to \mathbf{R}^n$ are *equivalent* if there is a diffeomorphism $h : S' \to S$ such that $\alpha(h(\mathbf{v})) = \beta(\mathbf{v})$ for all $\mathbf{v}$ in S'. This is easily seen to be an equivalence relation. A *surface** is an equivalence class Σ of parametrized surfaces under this relation. We say that $\alpha : S \to \mathbf{R}^n$ is a *parametrization* for Σ if α is in the equivalence class defining Σ.

To illustrate, let $\mathbf{u}, \mathbf{v}, \mathbf{w}$ be three vectors in $\mathbf{R}^n$ with $\mathbf{v}$ and $\mathbf{w}$ independent, and define $\beta : \mathbf{R}^2 \to \mathbf{R}^n$,

$$\beta(x, y) = \mathbf{u} + x\mathbf{v} + y\mathbf{w}$$

to be the corresponding plane. Define $\alpha : \mathbf{R}^2 \to \mathbf{R}^n$ by

$$\alpha(s, t) = \mathbf{u} + s\mathbf{v} + t(\mathbf{w} + \mathbf{v})$$

Then α defines the same plane. In fact, α is equivalent to β with $h : \mathbf{R}^2 \to \mathbf{R}^2$ defined by $h(x, y) = (x - y, y)$, since

*We shall need a more general notion of surface in the next chapter. It will involve piecing together surfaces of the kind considered here.

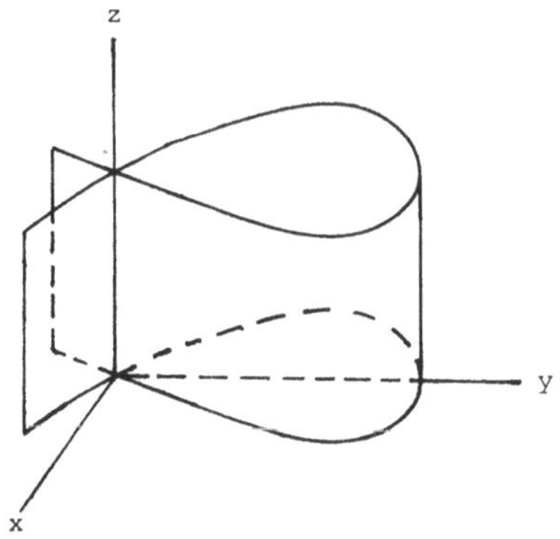

Figure 2.7 A nonsimple surface.

$$(\alpha \circ h)(x, y) = \alpha(x - y, y) = \mathbf{u} + (x - y)\mathbf{v} + y(\mathbf{w} + \mathbf{v})$$
$$= \mathbf{u} + x\mathbf{v} + y\mathbf{w} = \beta(x, y)$$

and h has a differentiable inverse $[h^{-1}(x, y) = (x + y, y)]$.

A surface is said to be *simple* if it has a parametrization which is injective. All of the examples given above are simple surfaces. An example of a surface which is not simple is the one with parametrization $\alpha : S \to \mathbf{R}^3$, where S is the rectangle

$$S = \left\{(x, y) : -\frac{\pi}{2} \leq x \leq \frac{\pi}{2}, 0 \leq y \leq 1\right\}$$

and $\alpha(x, y) = (\sin x \cos 2x, \cos x \cos 2x, y)$. (See Figure 2.7.) Note that $\alpha(-\pi/4, y) = \alpha(\pi/4, y)$ for all y.

As with curves, a surface can have corners. For example, the surface with parametrization

$$\beta : S' \to \mathbf{R}^3$$

where S' is the rectangle

$$S' = \{(x, y) : -1 \leq x \leq 1, 0 \leq y \leq 1\}$$

and $\alpha(x, y) = (x^3, x^2, y)$ has a corner along the z axis. (See Figure 2.8.) A surface is said to be *regular* if it has a parametrization $\alpha : S \to \mathbf{R}^n$ with Jacobian of rank 2 at each point of S.* A regular surface cannot have corners, since it has a well defined tangent plane at each of its points. In the future, we shall generally restrict attention to regular surfaces.

*Recall that the Jacobian is of rank 2 if its two columns (regarded as vectors in $\mathbf{R}^n$) are linearly independent.

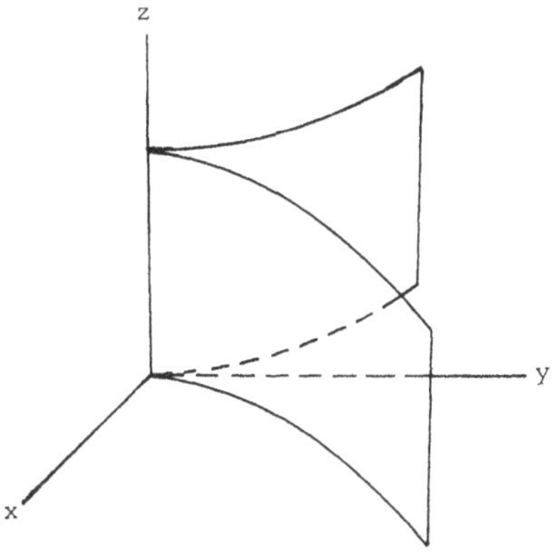

Figure 2.8 A nonregular surface.

Exercises

1. (a) Find a parametrization for the plane through (1, 0, 0) and orthogonal to $\mathbf{v}_0 = (1, -2, 2)$.
   (b) Find a parametrization for the plane through (1, 3, 4) and orthogonal to $\mathbf{v}_0 = (2, 1, -1)$.
   (c) Find a parametrization for the plane containing the points (1, 3, 2), (2, −5, 4), and (1, 0, 1).
   (d) Find a parametrization for the plane containing the points (1, −1, 1), (3, −2, 3), and (4, 0, 7).
   (e) Find a parametrization for the parallelogram two of whose sides are the lines connecting (1, −2, 1) with (1, 4, 3) and (1, −2, 1) with (2, 3, −1).
   (f) Find a parametrization for the parallelogram two of whose sides are the lines connecting (3, 2, 1) with (4, 7, 5) and (3, 2, 1) with (5, −2, −3).
   (g) Find a parametrization for the points on the sphere of radius 2 centered about the origin.
   (h) Find a parametrization for the points on the ellipsoid $x^2/4 + y^2/9 + z^2/16 = 1$.
   (i) Find a parametrization for the triangle with vertices at (1, 0, 0), (0, 1, 0), and (0, 0, 1).
   (j) Find a parametrization for the points on the cylinder $x^2 + y^2 = \frac{1}{4}$ lying inside the unit sphere.
2. Show that the following parametrizations are equivalent:
   (a) $\alpha(s, t) = (s^2, t^2, s^2t^2)$, $8 \le s, t \le 27$; $\beta(u,v) = (u^3, v^3, u^3v^3)$, $4 \le u,v \le 9$.
   (b) $\alpha(s, t) = (s, t, (1 - s^2 - t^2)^{1/2})$, $s^2 + t^2 \le \frac{1}{4}$; $\beta(u,v) = (v, u, (1 - u^2 - v^2)^{1/2})$, $u^2 + v^2 \le \frac{1}{4}$.
   (c) $\alpha(s, t) = (s^2 - t^2, 2s, s + t)$, $s^2 + t^2 \le 1$; $\beta(u,v) = (uv, u + v, v)$, $u^2 + v^2 \le 2$.
   (d) $\alpha(s, t) = (e^s, e^{s+2t}, e^{t+2s})$, $0 \le s \le 1, -1 \le t \le 0$; $\beta(u,v) = (v, u^2v, v^2u)$, $1/e \le u \le 1$, $1 \le v \le e$.
3. Write the equations of the parametrization $\beta\colon U \to \mathbf{R}^3$ for the upper unit

hemisphere in $\mathbf{R}^3$ in which U is the unit disk and $F(\mathbf{v})$ is the point where the line from $(0, 0, 2)$ through $\mathbf{v}$ meets the hemisphere.

4. Let $U = \{\mathbf{v} = (x, y) : 1 \leq \|\mathbf{v}\| \leq 2, y \geq 0\}$, and let a parametrization for the surface Σ be given by $\alpha\colon U \to \mathbf{R}^3\colon \alpha(x, y) = (x, y, x^2 + y^2)$. Find an equivalent parametrization $\beta\colon V \to \mathbf{R}^3$, where $V = \{(s, t) : 1 \leq t \leq 2, 0 \leq t \leq \pi\}$.
5. Let $\alpha : \mathbf{R}^2 \to \mathbf{R}^3$ be given by $\alpha(x, y) = (x, y, x^2 + y^2)$. Compute the tangent plane to the surface Σ parametrized by α at (1,2).
6. Let $f\colon \mathbf{R}^2 \to \mathbf{R}$ be a differentiable function, and define $\alpha(x, y) = (x, y, f(x, y))$. Compute the tangent plane to the surface Σ parametrized by α at (x_0, y_0).

3. Surface Area

A natural approach in attempting to define the area of a simple surface is to approximate it by surfaces whose areas we can compute and then pass to some kind of limit. For example, suppose the simple surface Σ is defined by the injective mapping $\alpha\colon S \to \mathbf{R}^3$, where S is the rectangle

$$S = \{(x, y) : a_1 \leq x \leq a_2,\ b_1 \leq y \leq b_2\}$$

The points $\alpha(a_1, b_1)$, $\alpha(a_2, b_1)$, and $\alpha(a_1, b_2)$ span a planar triangle in $\mathbf{R}^2$, as do the points $\alpha(a_2, b_1)$, $\alpha(a_1, b_2)$, and $\alpha(a_2, b_2)$. (See Figure 3.1.) These two triangles fit together to form a polygonal surface which can be considered to be an approximation to the surface Σ, admittedly a rather poor one.

We can improve this approximation as follows. Let P be a partition of S and S_{ij} one of the subrectangles of P. Just as we did above, we form a pair of planar triangles from the vertices of S_{ij}. Repeating this construction for each of the subrectangles of P, we obtain a polygonal surface which approximates Σ. We expect that the approximation gets better as the partition gets finer and that the surface area of Σ should be the least upper bound over all partitions of S of the areas of these polygonal surfaces. Unfortunately, this does not work. Examples of smooth surfaces can be given for which the least upper bound of the areas of the polygonal approximations does not exist.*

To give the proper definition of surface area, for surfaces in $\mathbf{R}^3$, we use the fact that the derivative of α at $\mathbf{v} \in S$ approximates the function α near $\mathbf{v}$.

Let Σ be the surface defined by the function $\alpha : S \to \mathbf{R}^3$, where $S = R(\mathbf{a}, \mathbf{b})$ is a rectangle, and let P be a partition of S defined by

$$a_1 = t_0 < t_1 < \cdots < t_m = a_2$$

$$b_1 = s_0 < s_1 < \cdots < s_n = b_2$$

Let S_{ij} be the subrectangle given by

*Such an example can be found in the article "What is Surface Area?" by T. Radó, *Amer. Math. Monthly*, vol. 50, 1943, pp. 139–141.

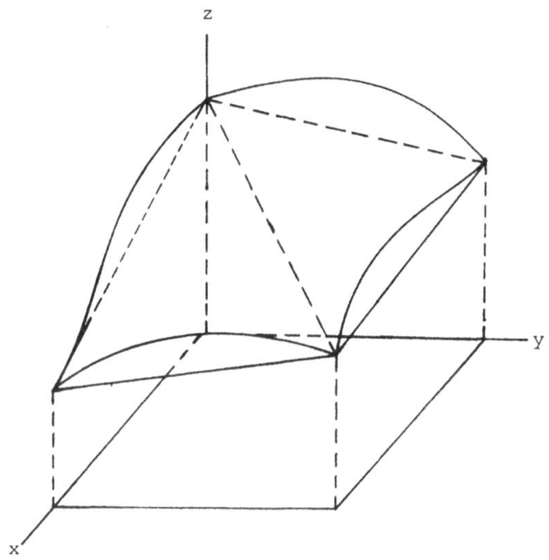

Figure 3.1 An approximating polygonal surface.

$$S_{ij} = \{(x, y) : t_{i-1} \le x \le t_i,\ s_{j-1} \le y \le s_j\}$$

and let $\mathbf{c}_{ij}$ be any point in S_{ij}. If $\beta : S_{ij} \to \mathbf{R}^3$ is defined by

$$\beta(\mathbf{v}) = (D_{\mathbf{c}_{ij}}\,\alpha)(\mathbf{v} - \mathbf{c}_{ij}) + \alpha(\mathbf{c}_{ij})$$

then β is a good approximation for α near $\mathbf{c}_{ij}$. Consequently, if the partition P is fine enough, the area of the image of S_{ij} under β should be approximately the area of the image of S_{ij} under α. However, the image of S_{ij} under β is a parallelogram with area

$$(t_i - t_{i-1})(s_j - s_{j-1})\,\|(D_{\mathbf{c}_{ij}}\,\alpha)(\mathbf{e}_1) \times (D_{\mathbf{c}_{ij}}\,\alpha)(\mathbf{e}_2)\|$$

(See Exercise 7, Section 1.) Thus, the expression

$$\sum_{i=1}^{m}\ \sum_{j=1}^{n}\ (t_i - t_{i-1})(s_j - s_{j-1})\,\|(D_{\mathbf{c}_{ij}}\,\alpha)(\mathbf{e}_1) \times (D_{\mathbf{c}_{ij}}\,\alpha)(\mathbf{e}_2)\|$$

should converge to the area of the surface Σ as the partition P gets finer. This expression is a Riemann sum for the function taking $\mathbf{v}$ into $\|(D_{\mathbf{v}}\alpha)(\mathbf{e}_1) \times (D_{\mathbf{v}}\alpha)(\mathbf{e}_2)\|$, so it converges to the integral

$$\int_S \|(D\alpha)(\mathbf{e}_1) \times (D\alpha)(\mathbf{e}_2)\| \tag{3.1}$$

as the partition P gets finer, by Proposition 6.3 of Chapter 11 (provided that the function $\|(D\alpha)(\mathbf{e}_1) \times (D\alpha)(\mathbf{e}_2)\| : S \to \mathbf{R}$ is integrable). We therefore define the *area* of the simple surface Σ to be the integral (3.1) when it exists. [Note that this definition applies only to surfaces in $\mathbf{R}^3$; indeed, (3.1) makes sense only in $\mathbf{R}^3$, since the cross product is defined only in $\mathbf{R}^3$.]

For example, let Σ be the piece of the cylinder with parametrization $\alpha: S \to \mathbf{R}^3$, where

$$S = \left\{(x, y) : 0 \leq x \leq \frac{\pi}{2}, 0 \leq y \leq 1\right\}$$

and

$$\alpha(x, y) = (\cos x, \sin x, y)$$

(See Figure 3.2.)
The Jacobian matrix of α is

$$J_\alpha(x, y) = \begin{bmatrix} -\sin x & 0 \\ \cos x & 0 \\ 0 & 1 \end{bmatrix}$$

Thus, if $\mathbf{v} = (x, y)$,

$$(D_\mathbf{v}\alpha)\mathbf{e}_1 = (-\sin x, \cos x, 0)$$
$$(D_\mathbf{v}\alpha)\mathbf{e}_2 = (0, 0, 1)$$

and

(3.2) $$(D_\mathbf{v}\alpha)\mathbf{e}_1 \times (D_\mathbf{v}\alpha)\mathbf{e}_2 = (\cos x, \sin x, 0)$$

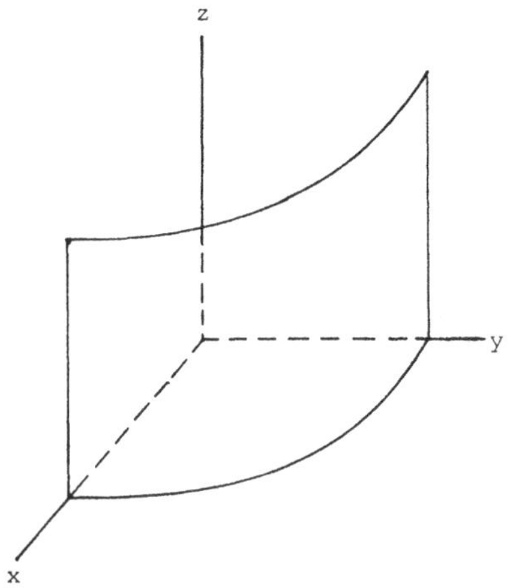

Figure 3.2 A cylinder.

The surface area of Σ is then

$$\int_S \|(\cos x, \sin x, 0)\| = \int_S 1 = \frac{\pi}{2}$$

As a more complicated example, we compute the surface area of a hemisphere. As we saw in Section 2, we can parametrize the hemisphere by $\alpha: S \to \mathbf{R}^3$, where S is the closed ball of radius 1 about the origin and

$$\alpha(x, y) = \left(\frac{2x}{1 + x^2 + y^2}, \frac{2y}{1 + x^2 + y^2}, \frac{1 - x^2 - y^2}{1 + x^2 + y^2}\right)$$

We see from (2.5) that, if $\mathbf{v} = (x, y)$, then

$$(D_{\mathbf{v}}\alpha)(\mathbf{e}_1) = \mathbf{u} \qquad (D_{\mathbf{v}}\alpha)(\mathbf{e}_2) = \mathbf{w}$$

where
$$\mathbf{u} = \frac{2}{(1 + x^2 + y^2)^2}(1 - x^2 + y^2, -2xy, -2x)$$
$$\mathbf{w} = \frac{2}{(1 + x^2 + y^2)^2}(-2xy, 1 + x^2 - y^2, -2y)$$

Thus

$$\mathbf{u} \times \mathbf{w} = \frac{4}{(1 + x^2 + y^2)^3}(2x, 2y, 1 - x^2 - y^2) \tag{3.3}$$

and

$$\|\mathbf{u} \times \mathbf{w}\| = \frac{4}{(1 + x^2 + y^2)^3}[(2x)^2 + (2y)^2 + (1 - x^2 - y^2)^2]^{1/2} = \frac{4}{(1 + x^2 + y^2)^2} \tag{3.4}$$

Hence the area of the surface is

$$\int_S \frac{4}{(1 + x^2 + y^2)^2}$$

This integral is hard to evaluate in this form, but in polar coordinates it becomes easy. On the surface S, r varies from 0 to 1 and θ from 0 to 2π, so the area of the hemisphere is given by

$$\int_0^1 \int_0^{2\pi} \frac{4}{(1 + r^2)^2} r\, dr\, d\theta = \int_0^1 \frac{8\pi r}{(1 + r^2)^2} dr = \left.\frac{-4\pi}{(1 + r^2)}\right|_0^1 = 2\pi$$

(See the example at the end of Section 11.6.) Thus the area of the sphere of radius 1 is 4π.

We may note one odd fact about this computation, which serves as a partial check. Notice that $\mathbf{u} \times \mathbf{w}$ [given in (3.3)] is a multiple of $\alpha(x, y)$. We know that $\mathbf{u}$ and $\mathbf{w}$ are vectors generating the tangent plane to the surface at $\alpha(x, y)$; since $\mathbf{u} \times \mathbf{w}$ is orthogonal to $\mathbf{u}$ and to $\mathbf{w}$, $\mathbf{u} \times \mathbf{w}$ is orthogonal to the tangent plane. That is, $\mathbf{u} \times \mathbf{w}$ points in the direction of the normal to the hemisphere at (x, y). But the radius line to a point on a sphere is normal to the sphere—that is, $\alpha(x, y)$ is normal to the surface. Therefore $\mathbf{u} \times \mathbf{w}$ and $\alpha(x, y)$ should be proportional, and they are.

The next result shows that the area of a simple surface Σ does not depend on how we parametrize it.

Proposition 3.1 *Let $\alpha : S \to \mathbf{R}^3$, $\beta : S' \to \mathbf{R}^3$ be two parametrizations for simple surface Σ. Then*

$$\int_S \|(D\alpha)(\mathbf{e}_1) \times (D\alpha)(\mathbf{e}_2)\| = \int_{S'} \|(D\beta)(\mathbf{e}_1) \times (D\beta)(\mathbf{e}_2)\|$$

Proof. Let $h : S \to S'$ be a diffeomorphism with $\beta(h(\mathbf{v})) = \alpha(\mathbf{v})$ for all $\mathbf{v}$ in S. Let $f : S \to \mathbf{R}$, $g : S' \to \mathbf{R}$ be defined by

$$f(\mathbf{v}) = \|(D_{\mathbf{v}}\alpha)\mathbf{e}_1 \times (D\alpha)\mathbf{e}_2\|$$

$$g(\mathbf{w}) = \|(D_{\mathbf{w}}\beta)\mathbf{e}_1 \times (D_{\mathbf{w}}\beta)\mathbf{e}_2\|$$

We must show $\int_S f = \int_{S'} g$.

Now

$$\begin{aligned} f(\mathbf{v}) &= \|(D_{\mathbf{v}}\alpha)\mathbf{e}_1 \times (D_{\mathbf{v}}\alpha)\mathbf{e}_2\| \\ &= \|(D_{\mathbf{v}}(\beta \circ h))\mathbf{e}_1 \times (D_{\mathbf{v}}(\beta \circ h))\mathbf{e}_2\| \\ &= \|(D_{h(\mathbf{v})}\beta) \circ (D_{\mathbf{v}}h)(\mathbf{e}_1) \times (D_{h(\mathbf{v})}\beta) \circ (D_{\mathbf{v}}h)(\mathbf{e}_2)\| \end{aligned}$$

by the chain rule. (See Section 4.4.) The matrix of $D_{\mathbf{v}}h$ is the Jacobian of h,

$$J_h = \begin{bmatrix} a_{11} & a_{12} \\ a_{21} & a_{22} \end{bmatrix}$$

where the a_{ij} are the partial derivatives of the components of h. Thus, setting $T = D_{h(\mathbf{v})}\beta$, we have

$$\begin{aligned} f(\mathbf{v}) &= \|T(a_{11}\mathbf{e}_1 + a_{21}\mathbf{e}_2) \times T(a_{12}\mathbf{e}_1 + a_{22}\mathbf{e}_2)\| \\ &= \|a_{11}a_{12}(T\mathbf{e}_1 \times T\mathbf{e}_1) + a_{11}a_{22}(T\mathbf{e}_1 \times T\mathbf{e}_2) \\ &\quad + a_{21}a_{12}(T\mathbf{e}_2 \times T\mathbf{e}_1) + a_{21}a_{22}(T\mathbf{e}_2 \times T\mathbf{e}_2)\| \\ &= \|(a_{11}a_{22} - a_{21}a_{12})(T\mathbf{e}_1 \times T\mathbf{e}_2)\| \\ &= |\text{Det } J_h| \, \|(D_{h(\mathbf{v})}\beta)\mathbf{e}_1 \times (D_{h(\mathbf{v})}\beta)\mathbf{e}_2\| \\ &= |\text{Det } J_h| \, g(h(\mathbf{v})) \end{aligned}$$

It follows that

$$\int_S f = \int_S (g \circ h)\,|\text{Det } J_h| = \int_{S'} g$$

by the change of variable formula, Theorem 6.2 of Chapter 11. This proves the proposition.

Let $\alpha\colon S \to \mathbf{R}^3$ be a parametrization for the regular surface Σ so that $D_{\mathbf{v}}\alpha\colon \mathbf{R}^2 \to \mathbf{R}^3$ has rank 2 for each $\mathbf{v} \in S$. Then $(D_{\mathbf{v}}\alpha)(\mathbf{e}_1)$ and $(D_{\mathbf{v}}\alpha)(\mathbf{e}_2)$ are independent and span the tangent plane to the surface Σ at $\alpha(\mathbf{v})$. Thus, $\mathbf{n}(\mathbf{v}) = (D_{\mathbf{v}}\alpha)(\mathbf{e}_1) \times (D_{\mathbf{v}}\alpha)(\mathbf{e}_2)$ is nonzero and orthogonal to the tangent plane to Σ at $\alpha(\mathbf{v})$. (We saw an example of this earlier, when considering the hemisphere.) For any surface Σ, regular or not, we call the function $\mathbf{n}\colon S \to \mathbf{R}^3$ given by

$$\mathbf{n}(\mathbf{v}) = (D_{\mathbf{v}}\alpha)(\mathbf{e}_1) \times (D_{\mathbf{v}}\alpha)(\mathbf{e}_2)$$

the normal field to Σ *defined by* α. (See Figure 3.3.) Thus, the surface area of Σ is given by $\int_S \|\mathbf{n}\|$.

We will sometimes write $\mathbf{n}_\alpha(\mathbf{v})$ for $\mathbf{n}(\mathbf{v})$ to emphasize the dependence of $\mathbf{n}$ on α.

Exercises

1. (a) Let S be a closed set in $\mathbf{R}^2$ and $f\colon S \to \mathbf{R}$ a continuously differentiable function. Show that the graph of f,

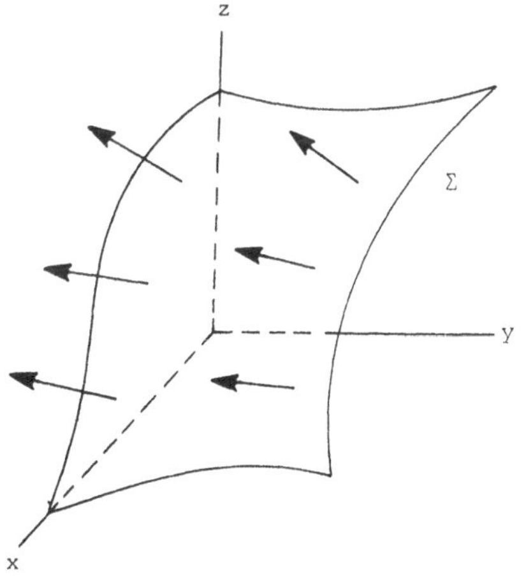

Figure 3.3 A normal field to Σ.

$$G(f) = \{(x, y, z) \in \mathbf{R}^3 : z = f(x, y)\}$$

is the image of a parametrized surface.

(b) Show that the area of the surface $G(f)$ is given by

$$\int_S \left[1 + \left(\frac{\partial f}{\partial x}\right)^2 + \left(\frac{\partial f}{\partial y}\right)^2\right]^{1/2}$$

2. Find the area of the surface $x^2 + y^2 + 2z - 4 = 0$ over the region S in the xy plane, where

$$S = \{(x, y, z) : 1 \leq x^2 + y^2 \leq 2, z = 0\}$$

3. Find the area of the surface $z = xy$ over the unit disk in the xy plane with center the origin.
4. A *surface of revolution* is obtained by rotating a curve $z = f(x)$, $0 \leq a \leq x \leq b$, in the xz plane about the z axis.
   (a) Show that this surface has the equation $z = f(r)$ in cylindrical coordinates. (See Exercise 4 of Section 11.6.)
   (b) Prove that the area of the surface is given by

$$2\pi \int_a^b [1 + (f'(r))^2]^{1/2}\, r\, dr$$

   (c) Find the area of the surface obtained by rotating the curve $z = 2x + 3$, $0 \leq x \leq 4$, about the z axis.
5. Let $S = \{(x, y) : 0 \leq x \leq 2\pi,\ 0 \leq y \leq \pi\}$ and define the parametrized surface $\alpha : S \to \mathbf{R}^3$ by

$$\alpha(x, y) = (\cos x \sin y, \sin x \sin y, \cos y)$$

Show that the image of α is the unit 2 sphere,

$$\{(x, y, z) \in \mathbf{R}^3 : x^2 + y^2 + z^2 = 1\}$$

and use this parametrization to find the surface area of the unit 2 sphere.
6. Find the area of the surface of the sphere of radius r.
7. Find the area of the piece of the paraboloid $z = x^2 + y^2$ which is cut out by the region between the cylinder $x^2 + y^2 = 2$ and the cylinder $x^2 + y^2 = 6$.

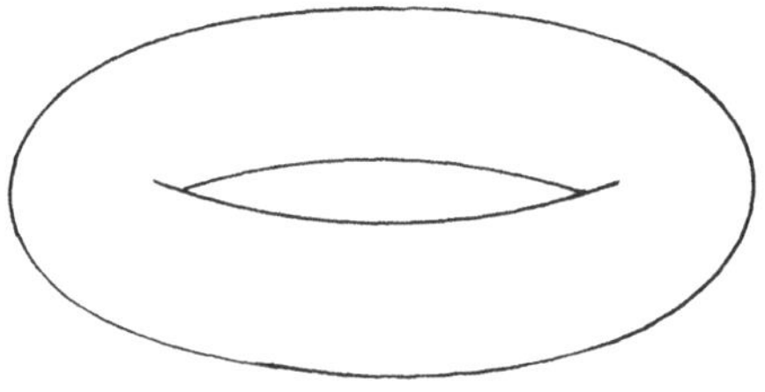

Figure 3.4 A torus.

8. Set up the iterated integral for the surface area of the ellipsoid

$$\frac{x^2}{a^2} + \frac{y^2}{b^2} + \frac{z^2}{c^2} = 1$$

9. Find the area of the surface obtained by rotating the circle $z^2 + (x - a)^2 = r^2$, $a > r$, about the z axis. (This surface is called a *torus*. See Figure 3.4.)

4. Surface Integrals

Suppose that Σ is a surface in the open subset U of $\mathbf{R}$ and $f\colon U \to \mathbf{R}^3$ is a continuous function. In analogy with the situation for line integrals, we define the *surface integral of f over Σ* to be

$$(4.1) \qquad \int_\Sigma f = \int_S (f \circ \alpha)\, \|(D\alpha)\mathbf{e}_1 \times (D\alpha)\mathbf{e}_2\| = \int_S (f \circ \alpha)\, \|\mathbf{n}\|$$

where $\alpha : S \to \mathbf{R}^3$ is a parametrization for the surface Σ and the integrals on the right are the ordinary integrals of the function

$$\mathbf{v} \to f(\alpha(\mathbf{v}))\, \|\mathbf{n}_\alpha(\mathbf{v})\| = f(\alpha(\mathbf{v}))\, \|(D_\mathbf{v}\alpha)\mathbf{e}_1 \times (D_\mathbf{v}\alpha)\mathbf{e}_2\|$$

over the region $S \subset \mathbf{R}^2$. If we think of the surface as representing a thin sheet of material and f (restricted to the surface) as giving the density (weight per unit area), then (4.1) gives the total weight of the sheet.

As an example, let Σ be the unit hemisphere, and let $f(x, y, z) = z$. Then we may use the parametrization of (2.1). From (3.4), we get

$$\|\mathbf{n}(x, y)\| = \frac{4}{(1 + x^2 + y^2)^2}$$

while

$$(f \circ \alpha)(x, y) = \frac{1 - x^2 - y^2}{(1 + x^2 + y^2)}$$

Thus

$$\int_\Sigma f = \int_S \frac{4(1 - x^2 - y^2)}{(1 + x^2 + y^2)^3}$$

where S is the unit disk. Again, this integral is easier in polar coordinates:

$$\int f = \int_0^1 \int_0^{2\pi} \frac{(1 - r^2)}{(1 + r^2)^3} r\, d\theta\, dr = 8\pi \int_0^1 \frac{(1 - r^2)}{(1 + r^2)^3} r\, dr$$

Let $u = r^2$; the integral becomes

$$4\pi \int_0^1 \frac{1-u}{(1+u)^3}\,du = 4\pi \int_0^1 \left(\frac{2}{(1+u)^3} - \frac{1}{(1+u)^2}\right) du$$
$$= 4\pi\left(\frac{1}{1+u} - \frac{1}{(1+u)^2}\right)\Bigg|_0^1 = \pi$$

As in the case of surface area, the integral (4.1) does not depend upon the parametrization of the surface. Explicitly:

Proposition 4.1 *Let $\alpha: S \to \mathbf{R}^3$ and $\beta: S' \to \mathbf{R}^3$ be two parametrizations for the surface Σ. Then*

$$\int_S (f\circ\alpha)\,\|(D\alpha)\mathbf{e}_1 \times (D\alpha)\mathbf{e}_2\| = \int_{S'} (f\circ\beta)\,\|(D\beta)\mathbf{e}_1 \times (D\beta)\mathbf{e}_2\|$$

The proof is almost exactly the same as the proof of Proposition 3.1. We leave the details to the reader.

The surface integral of a function has the usual properties of integrals.

Proposition 4.2 *Let Σ be a surface in the open subset U in $\mathbf{R}^3$, $f, g: U \to \mathbf{R}$ continuous functions, and a, b real numbers. Then*

(a) $\displaystyle\int_\Sigma (af + bg) = a\int_\Sigma f + b\int_\Sigma g$

(b) *If $f(\mathbf{v}) \geq g(\mathbf{v})$ for all $\mathbf{v}$ in U, then*

$$\int_\Sigma f \geq \int_\Sigma g$$

(c) *Let $\alpha: S \to \mathbf{R}^3$ be a parametrization of Σ, where*

$$S = \{(x, y) : a \leq x \leq b,\ c \leq y \leq d\}$$

and let

$$S_1 = \{(x, y) : a \leq x \leq e,\ c \leq y \leq d\}$$
$$S_2 = \{(x, y) : e \leq x \leq b,\ c \leq y \leq d\}$$

where $a < e < b$. Let $\alpha_1: S_1 \to \mathbf{R}^3$, $\alpha_2: S_2 \to \mathbf{R}^3$ be the restrictions of α, and Σ_1 and Σ_2 the corresponding surfaces. Then

$$\int_\Sigma f = \int_{\Sigma_1} f + \int_{\Sigma_2} f$$

This proposition follows from the properties of the standard integral. For example, to prove (a), we have

$$\int_\Sigma (af+bg) = \int_S [(af+bg)\circ\alpha]\,\|\mathbf{n}\|$$
$$= \int_S [a(f\circ\alpha) + b(g\circ\alpha)]\,\|\mathbf{n}\|$$
$$= a\int_S (f\circ\alpha)\,\|\mathbf{n}\| + b\int_S (g\circ\alpha)\,\|\mathbf{n}\|$$
$$= a\int_\Sigma f + b\int_\Sigma g$$

The remainder of the proof is left as an exercise.

Suppose now that U is an open subset of $\mathbf{R}^3$ and $\alpha: S \to U$ is a parametrization for the surface Σ. Let $\mathbf{F}: U \to \mathbf{R}^3$ be a vector field on U. *The surface integral of the vector field* $\mathbf{F}$ *along* Σ (or simply the *integral of* $\mathbf{F}$ *along* Σ) is defined by

(4.2) $$\int_\Sigma \mathbf{F} = \int_S \langle \mathbf{F}\circ\alpha, \mathbf{n}_\alpha\rangle = \int_S \langle \mathbf{F}\circ\alpha, (D\alpha)\mathbf{e}_1 \times (D\alpha)\mathbf{e}_2\rangle$$

Here are two examples. In both cases, we let $\mathbf{F}(x, y, z) = (x, y, z)$.

1. Let Σ be the piece of the cylinder with parametrization

$$\alpha(x, y) = (\cos x, \sin x, y)$$

$0 \le x \le \pi/2$, $0 \le y < 1$. (See Figure 4.1.) We computed $\mathbf{n}_\alpha$ in Section 3

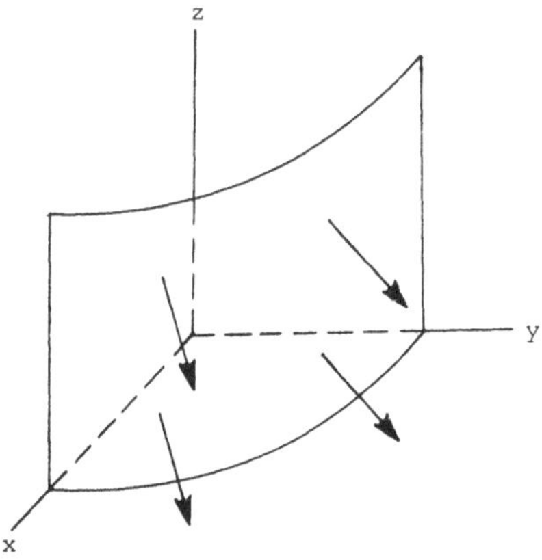

Figure 4.1 A normal field to a cylinder.

[see equation (3.2)]:

$$\mathbf{n}_\alpha(x, y) = (\cos x, \sin x, 0)$$

Furthermore,

$$(\mathbf{F}\circ\alpha)(x, y) = (\cos x, \sin x, y)$$

Thus
$$\int_\Sigma \mathbf{F} = \int_S (\cos^2 x + \sin^2 x + 0y)$$
$$= \int_0^1 \int_0^{\pi/2} 1\, dx\, dy = \frac{\pi}{2}$$

2. Let Σ be the unit hemisphere, parametrized as in (2.4). Then

$$(\mathbf{F}\circ\alpha)(x, y) = \left(\frac{2x}{1 + x^2 + y^2}, \frac{2y}{1 + x^2 + y^2}, \frac{1 - x^2 - y^2}{1 + x^2 + y^2}\right)$$

and [from (3.3)]

$$\mathbf{n}_\alpha = \frac{4}{(1 + x^2 + y^2)^2}(\mathbf{F}\circ\alpha)(x, y)$$

Thus

$$\langle(\mathbf{F}\circ\alpha)(x, y), \mathbf{n}_\alpha(x, y)\rangle = \frac{4}{(1 + x^2 + y^2)^2}\langle\mathbf{F}\circ\alpha(x, y), \mathbf{F}\circ\alpha(x, y)\rangle$$
$$= \frac{4}{(1 + x^2 + y^2)^2}$$

since $\|(\mathbf{F}\circ\alpha)(x, y)\| = 1$. Hence

$$\int_\Sigma \mathbf{F} = \int_S \frac{4}{(1 + x^2 + y^2)^2}$$

This is the same integral we evaluated in Section 3; thus

$$\int_\Sigma \mathbf{F} = 2\pi$$

In order to determine the dependence of this integral on the parametrization, we need the notion of an *oriented surface.*

Let D_1 and D_2 be two subsets of $\mathbf{R}^n$. A diffeomorphism $h : D_1 \to D_2$ is said to be *orientation preserving* if $\operatorname{Det}(J_h(\mathbf{v})) > 0$ for all $\mathbf{v}$ in D_1 and *orientation reversing* if $\operatorname{Det}(J_h(\mathbf{v})) < 0$ for all $\mathbf{v}$ in D_1. Two simple parametrized surfaces $\alpha : S \to \mathbf{R}^n$, $\beta : S' \to \mathbf{R}^n$ are *O-equivalent* if there is an orientation-preserving diffeomorphism $h : S' \to S$ such that $\alpha(h(\mathbf{v})) = \beta(\mathbf{v})$ for all $\mathbf{v}$ in S'. This is easily seen to be an equivalence relation. We define an *oriented surface* Σ to be an equivalence class of simple parametrized surfaces under the relation of O-equivalence. (This is like the definition for curves given in Section 12.3.)

We say that α is a *parametrization for the oriented surface* Σ if α is one of the parametrized surfaces in the equivalence class determining Σ as an oriented surface.

Suppose that $\alpha: S \to \mathbf{R}^n$ is a parametrization for the oriented surface Σ, where

$$S = \{(x, y) : a \leq x \leq b, c \leq y \leq d\}$$

Define $\tilde{\alpha} : S \to \mathbf{R}^n$ by

$$\tilde{\alpha}(x, y) = \alpha(a + b - x, y)$$

Then, if $h : S \to S$ is defined by

$$h(x, y) = (a + b - x, y)$$

we have $\tilde{\alpha}(x, y) = \alpha(h(x, y))$ so that α and $\tilde{\alpha}$ are equivalent. However,

$$J_h(x, y) = \begin{bmatrix} -1 & 0 \\ 0 & 1 \end{bmatrix}$$

so Det $J_h(x, y) = -1$, and it follows that α and $\tilde{\alpha}$ are not O-equivalent and represent different oriented surfaces. We let $-\Sigma$ denote the surface determined by $\tilde{\alpha}$.

We can now prove that the integral of a vector field along an *oriented* surface is independent of which parametrization we choose. (This is the analogue of Proposition 4.1.)

Proposition 4.3 *Let U be an open subset of $\mathbf{R}^3$ and let $\alpha : S \to U, \beta : S' \to U$ be two parametrizations for the oriented surface Σ. Then, if $\mathbf{F} : U \to \mathbf{R}^3$ is a vector field on U, we have*

$$\int_S \langle \mathbf{F} \circ \alpha, (D\alpha)\mathbf{e}_1 \times (D\alpha)\mathbf{e}_2 \rangle = \int_{S'} \langle \mathbf{F} \circ \beta, (D\beta)\mathbf{e}_1 \times (D\beta)\mathbf{e}_2 \rangle$$

Proof. Since this proof is similar to the proof of Proposition 4.1, we give only a sketch here. Define $f: S \to \mathbf{R}$, $g: S' \to \mathbf{R}$ by

$$f(\mathbf{v}) = \langle (\mathbf{F} \circ \alpha)(\mathbf{v}), (D_{\mathbf{v}}\alpha)\mathbf{e}_1 \times (D_{\mathbf{v}}\alpha)\mathbf{e}_2 \rangle$$

$$g(\mathbf{w}) = \langle (\mathbf{F} \circ \beta)(\mathbf{w}), (D_{\mathbf{w}}\beta)\mathbf{e}_1 \times (D_{\mathbf{w}}\beta)\mathbf{e}_2 \rangle$$

We must show that

(4.3) $$\int_S f = \int_{S'} g$$

Suppose now that $h : S \to S'$ is an orientation-preserving diffeomorphism with Det $J_h > 0$ and $(\beta \circ h)(\mathbf{v}) = \alpha(\mathbf{v})$. Then

$$f(\mathbf{v}) = \langle(\mathbf{F}(\beta \circ h)\mathbf{v}, D_{\mathbf{v}}(\beta \circ h)\mathbf{e}_1 \times D_{\mathbf{v}}(\beta \circ h)\mathbf{e}_2\rangle$$

But the calculation in Proposition 3.1 shows that

$$D_{\mathbf{v}}(\beta \circ h)\mathbf{e}_1 \times D_{\mathbf{v}}(\beta \circ h)\mathbf{e}_2 = \operatorname{Det}(D_{\mathbf{v}} h)((D_{h(\mathbf{v})}\beta)\mathbf{e}_1 \times (D_{h(\mathbf{v})}\beta)\mathbf{e}_2)$$

and therefore that

$$\begin{aligned} f(\mathbf{v}) &= \operatorname{Det}(D_{\mathbf{v}} h)\langle \mathbf{F} \circ \beta(h(\mathbf{v})), (D_{h(\mathbf{v})}\beta)\mathbf{e}_1 \times (D_{h(\mathbf{v})}\beta)\mathbf{e}_2\rangle \\ &= \operatorname{Det}(D_{\mathbf{v}} h)\, g\,(h(\mathbf{v})) \end{aligned}$$

[Notice that because we are not taking norms, we end up with $\operatorname{Det}(D_{\mathbf{v}} h)$ rather than $|\operatorname{Det}(D_{\mathbf{v}} h)|$. This explains the need for oriented surfaces.] Equation (4.3) now follows from the change of variable formula, Theorem 6.2 of Chapter 11.

It is not difficult to show that properties analogous to (a) and (c) of Proposition 4.2 hold for the line integral of a vector field along a curve. We leave the explicit statement of these properties and their proofs to the reader.

Let $\alpha : S \to \mathbf{R}^3$ and $\beta : S' \to \mathbf{R}^3$ be two parametrizations for the oriented surface Σ and let $h : S \to S'$ be an orientation-preserving diffeomorphism with $(\beta \circ h)(\mathbf{v}) = \alpha(\mathbf{v})$. Just as in the proof of Proposition 3.1, we see that

$$(D_{\mathbf{v}}\alpha)(\mathbf{e}_1) \times (D_{\mathbf{v}}\alpha)(\mathbf{e}_2) = \operatorname{Det}(D_{\mathbf{v}}h)\,[(D_{h(\mathbf{v})}\beta)(\mathbf{e}_1) \times (D_{h(\mathbf{v})}\beta)(\mathbf{e}_2)]$$

Equivalently,

$$\mathbf{n}_\alpha(\mathbf{v}) = (\operatorname{Det}(D_{\mathbf{v}}h))\mathbf{n}_\beta(h(\mathbf{v}))$$

Since $\operatorname{Det}(D_{\mathbf{v}}h) > 0$, we see that *any two parametrizations for the same regular oriented surface define normal fields that point in the same direction.* Thus we can speak of the two "sides" of an oriented surface.

We note that not all surfaces have two sides. The Möbius band, pictured in Figure 4.2, is an example of a surface with only one side. Thus, the Möbius band cannot be considered as an oriented surface. (Of course, it is not a simple surface in our sense.)

Exercises

1. Compute the surface integral of the function $f(x_1, x_2, x_3) = x_1x_3 + x_2^2$ over the surface Σ, where

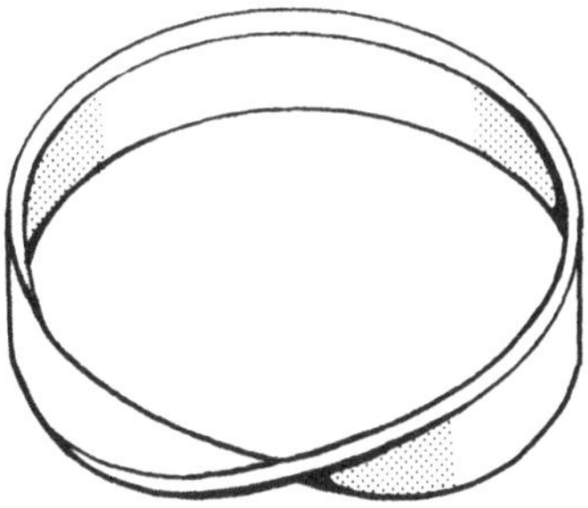

Figure 4.2 A Möbius band.

(a) Σ is the cylinder with parametrization $\alpha(x, y) = (\cos x, \sin x, y)$, $0 \le x \le \pi/2$, $0 \le y \le 2$.
(b) Σ has the parametrization $\alpha(x, y) = (x^2, y^2/2, xy)$, $0 \le x \le 1$, $0 \le y \le 2$.
(c) Σ has the parametrization $\alpha\colon S \to \mathbf{R}^3$, $S = \{(x, y) : 0 \le x, y \le 1\}$ and $\alpha(x, y) = (x, x^2, y)$.
(d) Σ is the parallelogram with vertices (0, 0, 0), (1, 0, 2), (0, 1, 1), and (1, 1, 3). (Find a parametrization for Σ.)

2. Compute the surface integral of the vector field $\mathbf{F}(x_1, x_2, x_3) = (x_1 - x_2, x_3, 1)$ over the surfaces Σ of Exercise 1.
3. Let $h_1, h_2\colon S' \to S$ be orientation-preserving diffeomorphisms. Prove that the composite $h_1 \circ h_2$ is also orientation preserving. What can be said about $h_1 \circ h_2$ if h_1 and h_2 are orientation reversing? If h_1 is orientation preserving and h_2 orientation reversing?
4. Prove that the relation between surfaces "α is O-equivalent to β" is an equivalence relation.
5. Prove Proposition 4.1.
6. Prove Proposition 4.2.
7. Give a detailed proof of Proposition 4.3.
8. Let Σ be an oriented surface in the open set U of $\mathbf{R}^3$ and $f\colon U \to \mathbf{R}^3$ a vector field. Prove that

$$\int_{-\Sigma} \mathbf{F} = -\int_{\Sigma} \mathbf{F}$$

9. State and prove the analogue of Proposition 4.2 for the surface integral of a vector field.

5. Stokes' Theorem

We now state an analogue of Green's theorem known as *Stokes' theorem* and give an application to hydrodynamics. The proof of Stokes' theorem is given in Chapter 15.

In order to state Stokes' theorem, we need to introduce the *curl* of a vector field.

Let $\mathbf{F}$ be a vector field on the open set $U \subset \mathbf{R}^3$, $\mathbf{F}(\mathbf{v}) = (f_1(\mathbf{v}), f_2(\mathbf{v}), f_3(\mathbf{v}))$. The *curl* of $\mathbf{F}$ is the vector field curl $\mathbf{F}: U \to \mathbf{R}^3$ defined by

$$\operatorname{curl} \mathbf{F} = \left(\frac{\partial f_3}{\partial x_2} - \frac{\partial f_2}{\partial x_3}, \frac{\partial f_1}{\partial x_3} - \frac{\partial f_3}{\partial x_1}, \frac{\partial f_2}{\partial x_1} - \frac{\partial f_1}{\partial x_2}\right)$$

For example, if $\mathbf{F}(x, y, z) = (xy^2, x + 3, yz)$, then

$$(\operatorname{curl} \mathbf{F})(x, y, z) = (z - 1, 0, 1 - 2xy)$$

Note the similarity between the formula for the curl and the formula for the cross product. In fact, if we use ∇, the symbol for the gradient, as a "symbolic vector,"

$$\nabla = \left(\frac{\partial}{\partial x}, \frac{\partial}{\partial y}, \frac{\partial}{\partial z}\right)$$

then

$$\operatorname{curl} \mathbf{F} = \nabla \times \mathbf{F}$$

as one can verify by expanding the formula for the cross product. (Of course, this formula is purely formal; ∇ is not really a vector.)

We can now state Stokes' theorem. As we see, it relates line integrals to surface integrals.

Theorem 5.1 (Stokes' Theorem) *Let U be an open subset of $\mathbf{R}^3$, S a rectangle in $\mathbf{R}^2$, and $\alpha: S \to U$ a parametrization for the oriented surface Σ. The restriction of α to the boundary of S defines a piecewise differentiable oriented curve which we denote by Γ. If $\mathbf{F}: U \to \mathbf{R}^3$ is a vector field on U, we have*

$$\int_\Gamma \mathbf{F} = \int_\Sigma \operatorname{curl} \mathbf{F}$$

As we indicated above, the proof of this result is given in Chapter 15; we give an interpretation of it here.

Suppose now that U is an open subset of $\mathbf{R}^3$ containing a fluid which is in motion and $\mathbf{F}$ is the vector field on U which assigns to each point $\mathbf{v} \in U$ the velocity of the fluid at $\mathbf{v}$; $\mathbf{F}$ is called a *velocity field*. In this case, curl $\mathbf{F}$ measures the angular velocity or rotation of the fluid. For example, if $U = \mathbf{R}^3$ and

$$\mathbf{F}(x, y, z) = (0, 1, 0)$$

the fluid is flowing uniformly in the positive y direction and has no rotation. (See Figure 5.1.) This is reflected in the fact that curl $\mathbf{F} = (0, 0, 0)$.

On the other hand, if $U = \mathbf{R}^3$ and

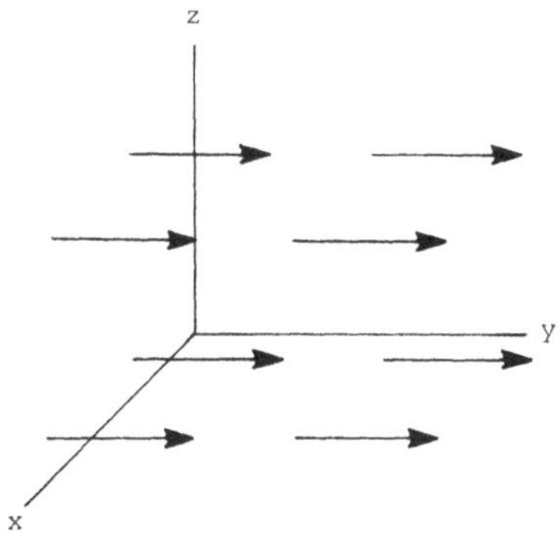

Figure 5.1 Uniform fluid flow.

$$\mathbf{F}(x, y, z) = (y, -x, 0)$$

we have a "whirlpool" along the z axis. (See Figure 5.2.) This is reflected in the fact that

$$\text{curl } \mathbf{F} = (0, 0, -2)$$

The fact that the nonzero entry in curl $\mathbf{F}$ is in the z coordinate indicates that the rotation is about the z axis. The size of the entry indicates something about the rate of rotation. In general, the curl varies from point to point, because the field rotates at different rates near different points.

A velocity field $\mathbf{F}$ is said to be *irrotational* if curl $\mathbf{F} = \mathbf{0}$.

Let Γ be a closed curve in the open set U. We define the *circulation of the*

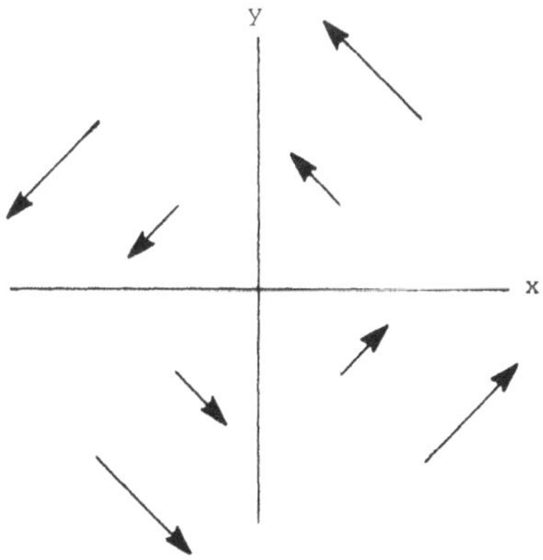

Figure 5.2 A whirlpool—looking down along the z axis.

velocity field $\mathbf{F}$ *around* Γ to be the line integral of $\mathbf{F}$ around Γ. For instance, if $\mathbf{F}$ is the velocity field for the whirlpool described above,

$$\mathbf{F}(x, y, z) = (y, -x, 0)$$

and if Γ is defined by the parametrized curve $\alpha : [0, 2\pi] \to U$,

$$\alpha(t) = (\cos t, \sin t, 0)$$

the circulation of $\mathbf{F}$ around Γ is given by

$$\int_{\Gamma} \mathbf{F} = \int_0^{2\pi} \langle \mathbf{F}(\alpha(t)), \alpha'(t) \rangle \, dt = -2\pi$$

We can use this notion to interpret the curl.

Let $\mathbf{n}$ be a unit vector (a "direction") at the point $\mathbf{v}_0 \in U$. It is not difficult to show that, for any sufficiently small $r > 0$, a parametrized surface $\alpha_r : B_r \to U$ can be found whose image is the flat disk of radius r, center $\mathbf{v}_0$, and normal

$$(D_{\mathbf{v}}\alpha_r)(\mathbf{e}_1) \times (D_{\mathbf{v}}\alpha_r)(\mathbf{e}_2) = \mathbf{n}$$

for all $\mathbf{v} \in B_r$. (See Figure 5.3.) [Here $B_r = \bar{B}_r(\mathbf{0})$ is the closed disk of radius r about $\mathbf{0}$ in $\mathbf{R}^2$.] Let Σ_r denote the image surface and Γ_r its boundary curve. Then, by Stokes' theorem, the circulation of $\mathbf{F}$ around Γ_r is given by

$$\int_{\Gamma_r} \mathbf{F} = \int_{\Sigma_r} \operatorname{curl} \mathbf{F} = \int_{B_r} \langle \operatorname{curl} \mathbf{F}, \mathbf{n} \rangle$$

According to the mean value theorem for integrals (see Exercise 6, Section 10.5), we can find a point $\mathbf{v}_r \in B_r$ such that

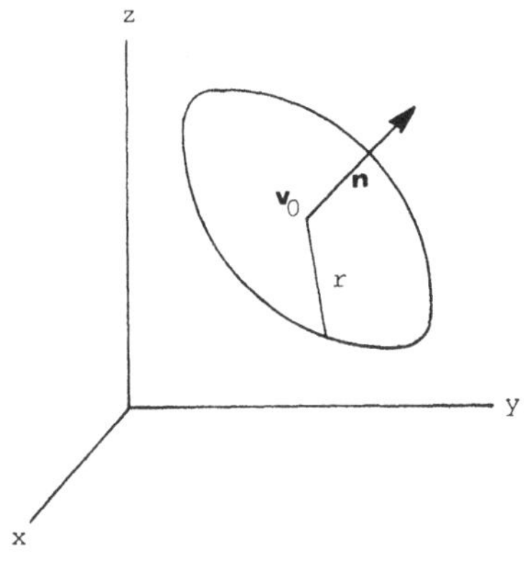

Figure 5.3 A disk with normal $\mathbf{n}$.

$$\int_{B_r} \langle \operatorname{curl} \mathbf{F}, \mathbf{n} \rangle = \mu(B_r) \langle (\operatorname{curl} \mathbf{F})(\mathbf{v}_r), \mathbf{n} \rangle$$

where $\mu(B_r) = \pi r^2$ is the area of B_r. Thus,

$$\langle (\operatorname{curl} \mathbf{F})(\mathbf{v}_r), \mathbf{n} \rangle = \frac{1}{\mu(B_r)} \int_{\Gamma_r} \mathbf{F}$$

Now, as r approaches zero, $\mathbf{v}_r$ approaches $\mathbf{v}_0$, and we have

$$\langle (\operatorname{curl} \mathbf{F})(\mathbf{v}_0), \mathbf{n} \rangle = \lim_{r \to 0} \frac{1}{\mu(B_r)} \int_{\Gamma_r} \mathbf{F}$$

Therefore, the quantity $\langle (\operatorname{curl} \mathbf{F})(\mathbf{v}_0), \mathbf{n} \rangle$ is the limiting value of the circulation of $\mathbf{F}$ per unit area on a surface orthogonal to $\mathbf{n}$ at $\mathbf{v}_0$. Because this quantity is clearly maximized when

$$\mathbf{n} = \frac{(\operatorname{curl} \mathbf{F})(\mathbf{v}_0)}{\|(\operatorname{curl} \mathbf{F})(\mathbf{v}_0)\|}$$

(that is, when the circulation is measured in the direction of curl $\mathbf{F}$), curl $\mathbf{F}$ is called the *vorticity (vector) field of the velocity field* $\mathbf{F}$. (Note that *vortex* means "whirlpool.")

Exercises

1. Let U be an open subset of $\mathbf{R}^3$, $\mathbf{F}$, $\mathbf{G}$: $U \to \mathbf{R}^3$ vector fields, and a, b real numbers. Prove that

$$\operatorname{curl}(a\mathbf{F} + b\mathbf{G}) = a \operatorname{curl} \mathbf{F} + b \operatorname{curl} \mathbf{G}$$

2. Let U be an open subset of $\mathbf{R}^3$ and f: $U \to \mathbf{R}$ a C^2 function. Prove that curl (grad f) = 0.
3. Compute the vorticity field of the following velocity fields.
   (a) $\mathbf{F}(x, y, z) = (xy, yz, xyz)$
   (b) $\mathbf{F}(x, y, z) = (x^2 + y^2, xz - y, 2xyz^2)$
   (c) $\mathbf{F}(x, y, z) = (z \sin x, \sin x, x + y + z)$

*4. A subset $X \subset \mathbf{R}^3$ is said to be *simply connected* if, whenever Γ is a closed curve in X, there is a surface Σ in X whose boundary curve is Γ. Let U be an open subset of $\mathbf{R}^3$ which is both pathwise connected and simply connected and let $\mathbf{F}$ be a vector field on U. Prove that U has a potential function if and only if curl $\mathbf{F} = \mathbf{0}$.

*5. Let $\mathbf{n}$ be a unit vector at $\mathbf{v}_0 \in \mathbf{R}^3$ and $B_r = \bar{B}_r(\mathbf{0})$ the closed ball of radius r about $\mathbf{0}$ in $\mathbf{R}^2$. Prove that there is a parametrized surface α_r: $B_r \to \mathbf{R}^3$ whose image is the planar disk of radius r with center $\mathbf{v}_0$ and whose normal $(D_{\mathbf{v}_0} \alpha_r)(\mathbf{e}_1) \times (D_{\mathbf{v}_0} \alpha_r)(\mathbf{e}_2)$ is $\mathbf{n}$.

14

Differential Forms

In this chapter and the next we show how to generalize line and surface integrals to more general sorts of integrals on $\mathbf{R}^n$. At the end, we shall give a far-reaching generalization of Green's theorem and Stokes' theorem. Before then, however, we need to develop a good deal of material, much of it formal and not obviously connected with integration. This chapter is concerned with the more formal aspects of the theory; in the next chapter we show how to give it substance. As a result, this chapter may seem rather abstract and artificial.

The subject matter of this chapter is algebraic operations on differential forms. We define differential forms, show how to add, multiply, and differentiate them, and discuss the relationships among these operations.

1. The Algebra of Differential Forms

In this section, we introduce the notion of a differential form on an open subset of Euclidean space. Eventually it will be clear what the relevance of this section, and those that follow, is to Green's and Stokes' theorems, but for some time the connection will be mysterious. The best procedure for the moment is simply to regard differential forms as completely new mathematical objects and to learn how to compute with them, using the rules for algebraic manipulations presented here. From time to time, we shall give illustrations of the relation between differential forms and the notions of the past few chapters.

Let U be an open subset of $\mathbf{R}^n$. A *differential* 0 *form* is a differentiable function $f: U \to \mathbf{R}$. A *differential* 1 *form* on U is an expression

$$f_1\, dx_1 + f_2\, dx_2 + \cdots + f_n\, dx_n$$

where each f_i is a differentiable function, $f_i: U \to \mathbf{R}$, $1 \le i \le n$. For example,

$$xy\, dx + 2dy$$

is a differential 1 form on $\mathbf{R}^2$,

$$\frac{1}{z}\, dy + 2 \sin x\, dz$$

is a differential 1 form on the open set $U = \{(x, y, z) : z \neq 0\}$ in $\mathbf{R}^3$, and

$$x_1\, dx_2 + x_4^2\, dx_7$$

is a differential 1 form on $\mathbf{R}^7$.

For convenience, we drop the word *differential* from "differential form" and speak simply of *forms*. For example, a differential 0 form will be called an 0 *form* and a differential 1 form will be called a 1 *form.*

Notice that the expressions we integrated when working with line integrals (see the end of Section 13.3) were 1 forms. The reason for introducing differential forms is, in fact, to allow us to generalize line integrals to higher dimensional surfaces.

We can add two 1 forms on U and multiply a 1 form on U by an 0 form on U in the obvious ways:

$$\left(\sum_{i=1}^{n} f_i\, dx_i\right) + \left(\sum_{i=1}^{n} g_i\, dx_i\right) = \sum_{i=1}^{n} (f_i + g_i)\, dx_i$$

$$g\left(\sum_{i=1}^{n} f_i\, dx_i\right) = \sum_{i=1}^{n} g f_i\, dx_i$$

These operations satisfy the usual formal properties (for example, the commutative and distributive laws).

A 2 *form* ω *on* U is an expression of the kind

$$\omega = \sum_{i,j=1}^{n} f_{ij}\, dx_i\, dx_j \tag{1.1}$$

where each f_{ij} is a differentiable function on U. Here we think of the term $dx_i\, dx_j$ as the "product" of the 1 form dx_i with the 1 form dx_j. This multiplication is defined to have the property

$$dx_i\, dx_j = -dx_j\, dx_i \tag{1.2}$$

for all $i, j = 1, \ldots, n$. In particular, $dx_i\, dx_i = -dx_i\, dx_i$, so that $dx_i\, dx_i = 0$ for $i = 1, \ldots, n$.

We note that, using (1.2), we can write the 2 form ω of (1.1) as

$$\omega = \sum_{1 \le i < j \le n} g_{ij}\, dx_i\, dx_j$$

where $g_{ij} = f_{ij} - f_{ji}$. For instance,

$$\sum_{i,j=1}^{3} x_i^2 x_j^3\, dx_i\, dx_j = (x_1^2 x_2^3 - x_1^3 x_2^2)\, dx_1\, dx_2 + (x_1^2 x_3^3 - x_1^3 x_3^2)\, dx_1\, dx_3 + (x_2^2 x_3^3 - x_2^3 x_3^2)\, dx_2\, dx_3$$

since the $dx_1 dx_1$, $dx_2 dx_2$, and $dx_3 dx_3$ terms are all 0.

We define the *product* of two 1 forms by

$$\left(\sum_{i=1}^{n} f_i\, dx_i\right)\left(\sum_{j=1}^{n} g_j\, dx_j\right) = \sum_{i,j=1}^{n} f_i g_j\, dx_i\, dx_j = \sum_{1 \le i < j \le n} (f_i g_j - f_j g_i)\, dx_i\, dx_j$$

Here are two examples:

1. If $\omega = x\, dy + xy\, dz$ and $\eta = 2dx - 4 \sin z\, dy + x^2 dx$, then

$$\omega\eta = x\, dy\, dx - 4x \sin z\, dy\, dy + x^3 dy\, dz + 2xy\, dz\, dx - 4xy \sin z\, dz\, dy + x^3 y\, dz\, dz$$
$$= 2x\, dy\, dx + x^3 dy\, dz + 2xy\, dz\, dx - 4xy \sin z\, dz\, dy$$

(since $dy\, dy = dz\, dz = 0$)

$$= -2x\, dx\, dy - 2xy\, dx\, dz + (x^3 + 4xy \sin z)\, dy\, dz$$

(since $dy\, dx = -dx\, dy$, $dz\, dx = -dx\, dz$, and $dz\, dy = -dy\, dz$)

2. If

$$\omega = \sum_{i=1}^{3} a_i\, dx_i \qquad \eta = \sum_{i=1}^{3} b_j\, dx_j$$

are 1 forms on $\mathbf{R}^3$, then

$$\omega\eta = (a_1 b_2 - a_2 b_1)\, dx_1\, dx_2 + (a_1 b_3 - a_3 b_1)\, dx_1\, dx_3 + (a_2 b_3 - a_3 b_2)\, dx_2\, dx_3$$

Note the similarity here with the cross product of the vectors (a_1, a_2, a_3) and (b_1, b_2, b_3). (See Section 13.1.)

It is easy to see that the multiplication of 1 forms defined above satisfies all the usual properties of multiplication, except that

$$\omega\eta = -\eta\omega$$

for any two 1 forms ω and η. (See Exercise 3. To make sense of this exercise, we need to be able to add 2 forms and to multiply them by 0 forms. The definitions should be obvious; in any case we shall give them soon.)

More generally, if p is any positive integer, we define a p form on U to be an expression of the kind

$$\omega = \Sigma f_{i_1 i_2 \cdots i_p} dx_{i_1} dx_{i_2} \cdots dx_{i_p} \tag{1.3}$$

where the sum ranges over all sets of p integers $i_1, \ldots, i_p$ with $1 \le i_1, i_2, \ldots, i_p \le n$, and where each $f_{i_1 i_2 \cdots i_p}$ is a differentiable real-valued function on U. For example,

$$x_1 x_7 \, dx_1 \, dx_3 \, dx_4 - \cos x_3 \, dx_3 \, dx_5 \, dx_8$$

is a 3 form on $\mathbf{R}^8$ and

$$(\ln x_4) \, dx_1 \, dx_3 \, dx_4 \, dx_5 + (x_2 - x_4^2) \, dx_1 \, dx_2 \, dx_3 \, dx_4$$

is a 4 form on the set $U = \{(x_i) \in \mathbf{R}^5 : x_4 > 0\}$.

As above, we think of the terms $dx_{i_1} \cdots dx_{i_p}$ as the product of the 1 forms $dx_{i_1}, \ldots, dx_{i_p}$, again with the condition that $dx_i dx_j = -dx_j \, dx_i$. We can also add two p forms, multiply a p form by an 0 form, and multiply a p form and a q form to obtain a $(p+q)$ form. For addition, we simply sum all the terms of the two forms: if

$$\omega_1 = \Sigma f_{i_1 \cdots i_p} dx_{i_1} \cdots dx_{i_p}$$

and

$$\omega_2 = \Sigma g_{i_1 \cdots i_p} dx_{i_1} \cdots dx_{i_p}$$

then

$$\omega_1 + \omega_2 = \Sigma (f_{i_1 \cdots i_p} + g_{i_1 \cdots i_p}) \, dx_{i_1} \cdots dx_{i_p}$$

If we multiply ω_1 by the 0 form h, we obtain

$$\Sigma \, h f_{i_1 \cdots i_p} dx_{i_1} \cdots dx_{i_p}$$

To multiply a p form by a q form, we multiply each term of the p form by each term of the q form, remembering the rule that $dx_i dx_j = -dx_j dx_i$, and add the products together. Here is an example. If

$$\omega = x_1 \, dx_1 dx_3 - \sin x_2 \, dx_2 dx_5$$

and

$$\eta = 2x_3^2 \, dx_1 \, dx_3 + 2x_4 \, dx_2 \, dx_3 - (x_1 + x_2) \, dx_4 \, dx_5$$

then

$$\begin{aligned} \omega\eta = {} & 2x_1 x_3^2 \, dx_1 \, dx_3 \, dx_1 \, dx_3 + 2x_1 x_4 \, dx_1 \, dx_3 \, dx_2 \, dx_3 \\ & - x_1(x_1 + x_2) \, dx_1 \, dx_3 \, dx_4 \, dx_5 \\ & - 2x_3^2 \sin x_2 \, dx_2 \, dx_5 \, dx_1 \, dx_3 \\ & - 2x_4 \sin x_2 \, dx_2 \, dx_5 \, dx_2 \, dx_3 \\ & + (x_1 + x_2) \sin x_2 \, dx_2 \, dx_5 \, dx_4 \, dx_5 \end{aligned}$$

Now

$$2x_1x_3^2\,dx_1\,dx_3\,dx_1\,dx_3 = -2x_1x_3^2\,dx_1\,dx_1\,dx_3\,dx_3$$

since $dx_1dx_3 = -dx_3dx_1$. However, $dx_1dx_1 = 0$, so this term vanishes. Similarly,

$$2x_1x_4\,dx_1\,dx_3\,dx_2\,dx_3 = 0$$

$$2x_4\sin x_2\,dx_2\,dx_5\,dx_2\,dx_3 = 0$$

and

$$(x_1 + x_2)\sin x_2\,dx_2\,dx_5\,dx_4\,dx_5 = 0$$

(More generally, any term which repeats a dx_j factor is 0. See Exercise 3.) Thus

$$\begin{aligned}\omega\eta &= -x_1(x_1+x_2)\,dx_1\,dx_3\,dx_4\,dx_5 - 2x_3^2\sin x_2\,dx_2\,dx_5\,dx_1\,dx_3\\ &= -x_1(x_1+x_2)\,dx_1\,dx_3\,dx_4\,dx_5 + 2x_3^2\sin x_2\,dx_1\,dx_2\,dx_3\,dx_5\end{aligned}$$

since $dx_2\,dx_5\,dx_1\,dx_3 = -dx_1\,dx_2\,dx_3\,dx_5$. Three interchanges are needed to get $dx_1\,dx_2\,dx_3\,dx_5$ from $dx_2\,dx_4\,dx_1\,dx_3$:

$$\begin{aligned}dx_2\,dx_5\,dx_1\,dx_3 &= -dx_2\,dx_1\,dx_5\,dx_3\\ &= dx_1\,dx_2\,dx_5\,dx_3\\ &= -dx_1\,dx_2\,dx_3\,dx_5\end{aligned}$$

Exercises

1. Let $\omega_1, \omega_2, \omega_3$ be the 1 forms on $\mathbf{R}^3$ given by

$$\begin{aligned}\omega_1 &= x_1\,dx_2 - dx_3\\ \omega_2 &= 2x_3^2dx_1\\ \omega_3 &= dx_1 - x_2x_3dx_2\end{aligned}$$

   Compute the following.
   (a) $x_2\omega_1 + \omega_2$
   (b) $x_3\omega_2 - x_1\omega_3$
   (c) $\omega_1\omega_3$
   (d) $(2\omega_1 - x_2\omega_3)\omega_2$
   (e) $\omega_1\omega_2\omega_3$
   (f) $(\omega_1 - x_1\omega_2)(2\omega_3 - \omega_1)(x_2^2\omega_2 - \omega_3)$

2. Let $\omega_1, \omega_2, \omega_3$, and ω_4 be the 1 forms on $\mathbf{R}^4$ given by

$$\begin{aligned}\omega_1 &= x_2dx_1 + \sin x_1\,dx_3 + 2dx_4\\ \omega_2 &= x_3^2dx_2 + x_2x_4dx_3\\ \omega_3 &= (x_1 - x_4)\,dx_1 + x_2dx_4\\ \omega_4 &= x_2dx_2 - x_1x_4dx_3\end{aligned}$$

Compute the following.
(a) $x_2\omega_1 + 3\omega_2$
(b) $2x_4\omega_2 - x_3^2\omega_4$
(c) $\omega_1\omega_3$
(d) $(2\omega_1 + x_4\omega_2)\omega_4$
(e) $\omega_1\omega_2\omega_3$
(f) $(x_4\omega_1 + \omega_3)(2\omega_1 - \omega_2)\omega_4$
(g) $(3\omega_2 - \omega_1)(\omega_2 + \omega_3)(\omega_1 - x_2\omega_4)$
(h) $\omega_1\omega_2\omega_3\omega_4$
(i) $(\omega_1 - x_2\omega_3)(\omega_2 + \omega_3)\omega_4(\omega_2 - \omega_1)$

3. Let $\omega = f\,dx_{i_1} \cdots dx_{i_p}$ and $\eta = g\,dx_{j_1} \cdots dx_{j_q}$. Show that if $i_k = j_l$ for any k, l, then $\omega\eta = 0$. (For example,

$$(f\,dx_1\,dx_4\,dx_5)(g\,dx_2\,dx_3\,dx_4) = 0$$

because $i_2 = 4 = j_3$.)
4. Prove that $\omega\eta = -\eta\omega$ for any 1 forms ω and η.
5. (a) Prove that $\omega^2 = \omega\omega = 0$ for any 1 form ω.
(b) Show that there is a 2 form η on $\mathbf{R}^4$ with $\eta^2 \neq 0$.

2. Basic Properties of the Sum and Product of Forms

We now derive some of the basic properties of the addition and multiplication of differential forms. Our first result states that the associative law and distributive laws hold for these operations.

Proposition 2.1 *Let ω_1 and ω_2 be p forms, η_1 and η_2 q forms, and γ an r form. Then*

(a) $(\omega_1 + \omega_2)\eta_1 = \omega_1\eta_1 + \omega_2\eta_1$
(b) $\omega_1(\eta_1 + \eta_2) = \omega_1\eta_1 + \omega_1\eta_2$
(c) $\omega_1(\eta_1\gamma) = (\omega_1\eta_1)\gamma$

Proof. This result is an immediate consequence of the definitions; we give the proof of (a), leaving the proofs of (b) and (c) as exercises.

Suppose

$$\omega_1 = \sum_I f_{i_1 \cdots i_p}\,dx_{i_1} \cdots dx_{i_p}$$
$$\omega_2 = \sum_I g_{i_1 \cdots i_p}\,dx_{i_1} \cdots dx_{i_p}$$
$$\eta_1 = \sum_J h_{j_1 \cdots j_q}\,dx_{j_1} \cdots dx_{j_q}$$

where $\sum_I$ denotes the sum over all $i_1, \ldots, i_p$ with $1 \le i_1, \ldots, i_p \le n$ and $\sum_J$ denotes the sum over all $j_1, \ldots, j_q$ with $1 \le j_1, \ldots, j_q \le n$. Then

$$
\begin{aligned}
(\omega_1+\omega_2)\eta_1 &= \left(\textstyle\sum_I f_{i_1\cdots i_p}\,dx_{i_1}\cdots dx_{i_p} + \sum_I g_{i_1\cdots i_p}\,dx_{i_1}\cdots dx_{i_p}\right)\\
&\quad \times \textstyle\sum_J h_{j_1\cdots j_q}\,dx_{j_1}\cdots dx_{j_q}\\
&= \left[\textstyle\sum_I (f_{i_1\cdots i_p}+g_{i_1\cdots i_p})\,dx_{i_1}\cdots dx_{i_p}\right]\\
&\quad \times \textstyle\sum_J h_{j_1\cdots j_q}\,dx_{j_1}\cdots dx_{j_q}\\
&\quad \text{(by the definition of the sum of forms)}\\
&= \textstyle\sum_I \sum_J (f_{i_1\cdots i_p}+g_{i_1\cdots i_p})h_{j_1\cdots j_q}\,dx_{i_1}\cdots dx_{i_p}\,dx_{j_1}\cdots dx_{j_q}\\
&\quad \text{(by the definition of the product of forms)}\\
&= \textstyle\sum_I \sum_J (f_{i_1\cdots i_p}h_{j_1\cdots j_q}\\
&\quad + g_{i_1\cdots i_p}h_{j_1\cdots j_q})\,dx_{i_1}\cdots dx_{i_p}\,dx_{j_1}\cdots dx_{j_q}\\
&= \textstyle\sum_I \sum_J f_{i_1\cdots i_p}h_{j_1\cdots j_q}\,dx_{i_1}\cdots dx_{i_p}\,dx_{j_1}\cdots dx_{j_q}\\
&\quad + \textstyle\sum_I \sum_J g_{i_1\cdots i_p}h_{j_1\cdots j_q}\,dx_{i_1}\cdots dx_{i_p}\,dx_{j_1}\cdots dx_{j_q}\\
&\quad \text{(by the definition of the sum of forms)}\\
&= \omega_1\eta_1+\omega_2\eta_1
\end{aligned}
$$

(by the definition of the product of forms). Thus, statement (a) of Proposition 2.1 is proved.

The next result is a version of the commutative law for the multiplication of forms (sometimes called the *anticommutative law*).

Proposition 2.2 *Let ω be a p form on U and η a q form on U. Then*

$$\omega\eta = (-1)^{pq}\,\eta\omega \tag{2.1}$$

Proof. Suppose $\omega = f\,dx_{i_1}\cdots dx_{i_p}$ and $\eta = g\,dx_{j_1}\cdots dx_{j_q}$. Then

$$\omega\eta = fg\,dx_{i_1}\cdots dx_{i_p}\,dx_{j_1}\cdots dx_{j_q}$$

and

$$\eta\omega = fg\,dx_{j_1}\cdots dx_{j_q}\,dx_{i_1}\cdots dx_{i_p}$$

To get from $\omega\eta$ to $\eta\omega$ clearly involves interchanging pq of the dx_{i_l} and dx_{j_k}, since each of the q dx_{j_k}'s must pass p dx_i's. Each of these interchanges introduces a factor of -1. Therefore equation (2.1) holds in this case.

If ω and η are arbitrary forms, then each is a sum of terms of the kind considered above, and equation (2.1) follows from Proposition 2.1.

We now show that p forms on $\mathbf{R}^n$ for $p \geq n$ have a particularly simple expression.

Proposition 2.3 *Let ω be an n form on the open set U of $\mathbf{R}^n$. Then*

$$\omega = f\,dx_1\,dx_2 \cdots dx_n$$

If ω is a p form on U with $p > n$, then $\omega = 0$.

Proof. As is not hard to see, any p form ω can be written

$$\omega = \textstyle\sum_I f_{i_1 \cdots i_p}\,dx_{i_1} \cdots dx_{i_p}$$

where $1 \le i_1 < i_2 < \cdots < i_p < n$. If $p = n$, it follows that $i_1 = 1$, $i_2 = 2$, $\ldots$, $i_n = n$, and the sum consists of exactly one term, $f_{1 \cdots n}\,dx_1 \cdots dx_n$.

If ω is a p form with $p > n$, then, in each of the summands of ω, at least one of the dx_i must occur twice. Thus $\omega = 0$, since $dx_i\,dx_i = 0$.

Our final result expresses the determinant of an $n \times n$ matrix in terms of the product of 1 forms. (See Appendix A for a discussion of determinants.)

Proposition 2.4 *Let $A = (a_{ij})$ be an $n \times n$ matrix and define 1 forms $\omega_1, \ldots, \omega_n$ by*

$$\omega_j = \sum_{j=1}^{n} a_{ij}\,dx_i$$

Then

$$\omega_1\omega_2 \cdots \omega_n = \text{Det } A\,dx_1\,dx_2 \cdots dx_n$$

Proof. We give a sketch of the proof, leaving the details as Exercise 4. For any $n \times n$ matrix $A = (a_{ij})$, we define a number $\Delta(A)$ by the equation

$$\omega_1\omega_2 \cdots \omega_n = \Delta(A)\,dx_1\,dx_2 \cdots dx_n$$

where $\omega_1, \ldots, \omega_n$ are defined as in Proposition 2.4. Using the properties of addition and multiplication of forms, we see that the function Δ (from matrices to real numbers) satisfies the following:

(2.2) If any two columns of A are equal, then $\Delta(A) = 0$.

(2.3) If all but one of the columns of A is held fixed, then Δ is a linear function of the remaining column. For example, if $n = 2$, and a, $b \in \mathbf{R}$,

$$\Delta\begin{bmatrix} aa_{11} + ba'_{11} & a_{12} \\ aa_{21} + ba'_{21} & a_{22} \end{bmatrix} = a\Delta\begin{bmatrix} a_{11} & a_{12} \\ a_{21} & a_{22} \end{bmatrix} + b\Delta\begin{bmatrix} a'_{11} & a_{12} \\ a'_{21} & a_{22} \end{bmatrix}$$

(2.4) If I_n is the $n \times n$ identity matrix, then $\Delta(I_n) = 1$.

However, it is shown in Appendix A that the only function from $n \times n$

matrices to real numbers satisfying (2.2), (2.3), and (2.4) is the determinant function. Thus Proposition 2.4 is proved.

Exercises

1. Prove statement (b) of Proposition 3.1.
2. Prove statement (c) of Proposition 2.1.
3. Verify Proposition 2.4 directly for (a) 2×2 matrices; (b) 3×3 matrices.
4. Give the details of the proof of Proposition 2.4.

3. The Exterior Differential

In this section, we describe differentiation of differential forms. Given a p form ω, we produce a $(p+1)$ form $d\omega$, called the *exterior differential* of ω.

To begin with, recall that a real-valued function f defined on an open subset U in $\mathbf{R}^n$ is said to be a C^k *function* (or of *class* C^k) if all partial derivatives of f of order k exist and are continuous. (Here, $1 \leq k < \infty$.) We say f is a C^∞ function if f is C^k for all k. Similarly, the p form ω of (1.3) is called a C^k *form* if each of the functions $f_{i_1 \cdots i_p}$ is C^k, $1 \leq k \leq \infty$. We assume all forms in this section to be C^k for some $k \geq 2$.

Let f be a C^2 function on the open set U in $\mathbf{R}^n$. The *exterior differential* of f, df, is the 1 form defined by

$$df = \sum_{j=1}^{n} \frac{\partial f}{\partial x_j} dx_j$$

For example, if $f: \mathbf{R}^3 \to \mathbf{R}$ is given by $f(x_1, x_2, x_3) = x_1^2 + x_2 x_3$, then

$$df = 2x_1 dx_1 + x_3 dx_2 + x_2 dx_3$$

In defining 1 forms in the previous section, we introduced the formal expressions dx_i, $1 \leq i \leq n$. In fact, dx_i can be identified with the exterior differential of the ith coordinate function. More explicitly, let $x_i : \mathbf{R}^n \to \mathbf{R}$ be the function that assigns to any point in $\mathbf{R}^n$ its ith coordinate: $x_i(y_1, \ldots, y_n) = y_i$. Then

$$d(x_i) = \sum_{j=1}^{n} \frac{\partial x_i}{\partial x_j} dx_j = dx_i$$

since $\partial x_i / \partial x_j$ is 1 if $i = j$ and 0 otherwise.

If ω is a p form,

$$\omega = \textstyle\sum_I f_{i_1 \cdots i_p} dx_{i_1} \cdots dx_{i_p}$$

with $1 \leq i_1, \ldots, i_p \leq n$, then $d\omega$ is the $(p+1)$ form defined by

$$d\omega = \textstyle\sum_I \; d(f_{i_1 \cdots i_p})\, dx_{i_1} \cdots dx_{i_p}$$
$$= \textstyle\sum_I \; \sum_j \; \dfrac{\partial f_{i_1 \cdots i_p}}{\partial x_j} dx_j dx_{i_1} \cdots dx_{i_p}$$

This last sum is over all $j, i_1, \ldots, i_p$ with $1 \leq j \leq n$ and $1 \leq i_1, \ldots, i_p \leq n$.

To illustrate, let ω be the 1 form on $\mathbf{R}^4$ given by

$$\omega = x_2 x_4 \, dx_1 + x_1^2 \, dx_3$$

Then
$$\begin{aligned} d\omega &= d(x_2 x_4)\, dx_1 + d(x_1^2)\, dx_3 \\ &= (x_4\, dx_2 + x_2\, dx_4)\, dx_1 + (2x_1\, dx_1)\, dx_3 \\ &= -\, x_4\, dx_1\, dx_2 + 2x_1\, dx_1\, dx_3 - x_2\, dx_1\, dx_4 \end{aligned}$$

Another example: let ω be the 2 form on $\mathbf{R}^4$ given by

$$\omega = x_4\, dx_1\, dx_2 + x_2 x_3\, dx_1\, dx_3$$

Then
$$\begin{aligned} d\omega &= dx_4\, dx_1\, dx_2 + (x_3\, dx_2 + x_2\, dx_3)\, dx_1\, dx_3 \\ &= dx_1\, dx_2\, dx_4 - x_3\, dx_1\, dx_2\, dx_3 \end{aligned}$$

since $dx_3\, dx_1\, dx_3 = dx_1\, dx_3\, dx_3 = 0$.

We derive some of the properties of the exterior differential in the next section. We conclude this section with a discussion of the relationship between the exterior differential and certain important operations on vector fields and functions.

Let U be an open subset of $\mathbf{R}^n$, $f: U \to \mathbf{R}$ a differentiable function, and $\mathbf{F}: U \to \mathbf{R}^n$ a vector field, $\mathbf{F}(\mathbf{v}) = (f_1(\mathbf{v}), \ldots, f_n(\mathbf{v}))$. The gradient of f and the curl of $\mathbf{F}$ (when $n = 3$) are vector fields on U defined earlier:

$$\text{grad } f = \left(\frac{\partial f}{\partial x_1}, \ldots, \frac{\partial f}{\partial x_n} \right)$$

$$\text{curl } \mathbf{F} = \left(\frac{\partial f_3}{\partial x_2} - \frac{\partial f_2}{\partial x_3}, \frac{\partial f_1}{\partial x_3} - \frac{\partial f_3}{\partial x_1}, \frac{\partial f_2}{\partial x_1} - \frac{\partial f_1}{\partial x_2} \right)$$

We define the *divergence* of $\mathbf{F}$ to be the function div $\mathbf{F}: U \to \mathbf{R}$ given by

$$\text{div } \mathbf{F} = \frac{\partial f_1}{\partial x_1} + \frac{\partial f_2}{\partial x_2} + \cdots + \frac{\partial f_n}{\partial x_n}$$

We have already seen how the gradient and curl arise in applications; we shall give an interpretation of the divergence in the next chapter. Notice that (formally) the divergence of $\mathbf{F}$ behaves like an inner product of $\nabla = (\partial/\partial x_1, \ldots, \partial/\partial x_n)$ with $\mathbf{F} = (f_1, \ldots, f_n)$, and the curl of $\mathbf{F}$ like the cross product $\nabla \times \mathbf{F}$.

Let U be an open subset of $\mathbf{R}^n$ and $\Omega^p(U)$ the set of all C^2 p forms on U. We can add the elements of $\Omega^p(U)$ and multiply them by real numbers. Indeed, these operations satisfy analogues of properties (2.3) through (2.10) of Chapter 2. (A set equipped with an addition and scalar multiplication satisfying these properties is called a *vector space*.)

Let f be an element of $\Omega^0(U)$; then $f: U \to \mathbf{R}$ is a C^2 function. We define a mapping ψ from $\Omega^0(U)$ into $\Omega^n(U)$ which takes the 0 form f into the n form

$$\psi(f) = f\,dx_1 \cdots dx_n$$

This mapping is easily seen to be injective and Proposition 2.3 tells us that it is also surjective. Therefore, this mapping defines a one-to-one correspondence between $\Omega^0(U)$ and $\Omega^n(U)$.

The mapping ψ has an addition useful property; it preserves addition and scalar multiplication. That is, if $a, b \in \mathbf{R}$ and $f, g \in \Omega^0(U)$, then

$$\psi(af + bg) = a\psi(f) + b\psi(g)$$

since $\quad (af + bg)\,dx_1 \cdots dx_n = a(f\,dx_1 \cdots dx_n) + b(g\,dx_1 \cdots dx_n)$

A bijective mapping satisfying this property is called an *isomorphism* and $\Omega^0(U)$ and $\Omega^n(U)$ are said to be *isomorphic*.

Similarly, we define a mapping from $\Omega^1(U)$ into $\Omega^{n-1}(U)$ by taking the 1 form

$$\omega = \sum_{i=1}^{n} f_i\,dx_i$$

into the $(n-1)$ form

$$\tilde{\omega} = \sum_{i=1}^{n} (-1)^{i+1} f_i\,dx_1 \cdots dx_{i-1}\,dx_{i+1} \cdots dx_n$$

[The factor $(-1)^{i+1}$ is included for convenience; this will be explained below.] This mapping is easily seen to be an isomorphism between $\Omega^1(U)$ and $\Omega^{n-1}(U)$. (See Exercise 7.)

For example, the 1 form

$$x_2^3\,dx_1 + 3x_1x_2\,dx_4$$

in $\mathbf{R}^4$ corresponds to the 3 form

$$x_2^3\,dx_2\,dx_3\,dx_4 - 3x_1x_2\,dx_1\,dx_2\,dx_3$$

under this isomorphism.

In the same way, it can be shown that $\Omega^p(U)$ and $\Omega^{n-p}(U)$ are isomorphic for $1 \le p \le n$. (See Exercise 8.)

Now let $\Phi(U)$ be the set of all vector fields on U (with the obvious addition and scalar multiplication). Define a mapping from $\Phi(U)$ to $\Omega^1(U)$ by assigning the 1 form

$$\sum_{i=1}^{n} f_i \, dx_i$$

to the vector field $\mathbf{F} = (f_1, \ldots, f_n)$. This mapping is an isomorphism (Exercise 9), and as a result, we can identify $\Omega^1(U)$, $\Omega^{n-1}(U)$, and $\Phi(U)$. Under this identification, the vector field

$$\mathbf{F} = (f_1, \ldots, f_n) \tag{3.1}$$

the 1 form

$$\omega = \sum_{i=1}^{n} f_i \, dx_i \tag{3.2}$$

and the $(n-1)$ form

$$\tilde{\omega} = \sum_{i=1}^{n} (-1)^{i+1} f_i \, dx_1 \cdots dx_{i-1} \, dx_{i+1} \cdots dx_n \tag{3.3}$$

correspond to one another.

We can now reinterpret the divergence, gradient, and curl in terms of operations on differential forms.

Suppose, first of all, that f is a differentiable function on U. Then grad f is the vector field

$$\text{grad } f = \left(\frac{\partial f}{\partial x_1}, \ldots, \frac{\partial f}{\partial x_n}\right)$$

which corresponds to the 1 form

$$\sum_{i=1}^{n} \frac{\partial f}{\partial x_i} dx_i$$

under the identification described above. However, this 1 form is exactly df, the exterior differential of f. Thus we see that grad f *is the vector field corresponding to the* 1 *form* df.

Next, let $\mathbf{F} = (f_1, \ldots, f_n)$ and let $\tilde{\omega}$ be the $(n-1)$ form corresponding to $\mathbf{F}$ [given in (3.3)]. Then $d\tilde{\omega}$ is an n form:

$$\begin{aligned} d\tilde{\omega} &= \sum_{i,j=1}^{n} (-1)^{i+1} \frac{\partial f_i}{\partial x_j} dx_j \, dx_1 \cdots dx_{i-1} \, dx_{i+1} \cdots dx_n \\ &= \sum_{i=1}^{n} (-1)^{i+1} \frac{\partial f_i}{\partial x_i} dx_i \, dx_1 \cdots dx_{i-1} \, dx_{i+1} \cdots dx_n \qquad \text{(see Exercise 5)} \\ &= \left(\sum_{i=1}^{n} \frac{\partial f_i}{\partial x_i}\right) dx_1 \cdots dx_n \qquad (\text{since } dx_i \, dx_j = -dx_j \, dx_i) \\ &= (\text{div } \mathbf{F}) \, dx_1 \cdots dx_n \end{aligned}$$

Thus div **F** *is the function corresponding to the exterior differential of the* $(n-1)$ *form* $\tilde{\omega}$ *corresponding to* **F**.

Finally, let **F** be a vector field on the open subset U of $\mathbf{R}^3$ and let ω be the corresponding 1 form [given in (3.2)]. Then a straightforward computation shows that

$$d\omega = \left(\frac{\partial f_2}{\partial x_1} - \frac{\partial f_1}{\partial x_2}\right) dx_1\, dx_2 + \left(\frac{\partial f_3}{\partial x_1} - \frac{\partial f_1}{\partial x_3}\right) dx_1\, dx_3 + \left(\frac{\partial f_3}{\partial x_2} - \frac{\partial f_2}{\partial x_3}\right) dx_2\, dx_3$$

which is exactly the 2 form corresponding to the vector field curl **F**. (Note that $n-1=2$ here since $n=3$.) Therefore, curl **F** *is the vector field corresponding the exterior differential of the* 1 *form* ω *corresponding to* **F**.

We note that the factor $(-1)^{i+1}$ which occurred in defining the correspondence between 1 forms and $(n-1)$ forms was needed in reinterpreting both div **F** and curl **F**.

Exercises

1. Compute $d\omega$, where ω is the form on $\mathbf{R}^3$ given by
   (a) $\omega = xy + z^2$
   (b) $\omega = x\,dy - y\,dz$
   (c) $\omega = xy\,dx + (x-2z)y\,dy + xyz^2\,dz$
   (d) $\omega = 7xz\,dx\,dy + \sin xy\,dx\,dz - 2z\,dy\,dz$
   (e) $\omega = \cos xy\,dx\,dy - yz^2\,dx\,dz + (x^2y - yz^2)\,dy\,dz$
2. Compute $d\omega$, where ω is the form on $\mathbf{R}^4$ given by
   (a) $\omega = x_1x_3 - x_2^2x_4$
   (b) $\omega = x_2\,dx_1 + x_1x_3\,dx_2 - x_2x_4^2\,dx_4$
   (c) $\omega = \sin x_1x_2\,dx_1 + x_3\cos x_4\,dx_3 - (x_1^2 + x_2^2)\,dx_4$
   (d) $\omega = x_2^2x_4\,dx_1\,dx_4 + \sin x_1x_2\,dx_2dx_3$
   (e) $\omega = (x_3^2 - x_1x_4)\,dx_1\,dx_3 + 7\cos(x_2^2 + x_3^2)\,dx_2dx_3$
   (f) $\omega = x_1x_3^2\,dx_1\,dx_2\,dx_4 - \cos x_2x_3\,dx_1dx_3\,dx_4$
3. Verify that $d(d\omega) = 0$ for the forms in Exercises 2 and 3.
4. Let U be an open subset of $\mathbf{R}^n$, **F**, **G**: $U \to \mathbf{R}^n$ vector fields on U, and a, b real numbers. Prove that

$$\operatorname{div}(a\mathbf{F} + b\mathbf{G}) = a \operatorname{div} \mathbf{F} + b \operatorname{div} \mathbf{G}$$

5. Verify that if $\omega = (-1)^{j+1} dx_1 \cdots dx_{j-1}\,dx_{j+1} \cdots dx_n$, then

$$(dx_k)\omega = \begin{cases} dx_1 \cdots dx_n & \text{if } j = k \\ 0 & \text{otherwise} \end{cases}$$

6. Verify that for $n = 3$, if the vector field **F** corresponds to the 1 form ω and **G** corresponds to η, then $\mathbf{F} \times \mathbf{G}$ corresponds to $\omega \times \eta$ and curl **F** corresponds to $d\omega$.

7. Verify that the mapping taking the 1 form

$$\omega = \sum_{i=1}^{n} f_i\, dx_i$$

into the $(n-1)$ form

$$\tilde{\omega} = \sum (-1)^{i+1} f_i\, dx_1 \cdots dx_{i-1}\, dx_{i+1} \cdots dx_n$$

is an isomorphism between $\Omega^1(U)$ and $\Omega^{n-1}(U)$ for U open in $\mathbf{R}^n$.
8. Prove that $\Omega^p(U)$ and $\Omega^{n-p}(U)$ are isomorphic.
9. Verify that the mapping taking the vector field $\mathbf{F} = (f_1, \ldots, f_n)$ into the 1 form

$$\omega = \sum_{i=1}^{n} f_i\, dx_i$$

is an isomorphism of $\varphi(U)$ onto $\Omega^1(U)$.

4. Basic Properties of the Exterior Differential

We now derive some of the important properties of the exterior differential. We begin by showing that it preserves the sum of forms.

Proposition 4.1 *If ω_1 and ω_2 are p forms on the open set U of $\mathbf{R}^n$, then*

$$d(\omega_1 + \omega_2) = d\omega_1 + d\omega_2$$

Proof. Let ω_1 and ω_2 be given by

$$\omega_1 = \sum_I f_{i_1 \cdots i_p}\, dx_{i_1} \cdots dx_{i_p}$$

$$\omega_2 = \sum_I g_{i_1 \cdots i_p}\, dx_{i_1} \cdots dx_{i_p}$$

Then

$$d(\omega_1 + \omega_2) = d\left(\sum_I f_{i_1 \cdots i_p}\, dx_{i_1} \cdots dx_{i_p} + g_{i_1 \cdots i_p}\, dx_{i_1} \cdots dx_{i_p}\right)$$

$$= d\left(\sum_I (f_{i_1 \cdots i_p} + g_{i_1 \cdots i_p})\, dx_{i_1} \cdots dx_{i_p}\right)$$

$$= \sum_I d(f_{i_1 \cdots i_p} + g_{i_1 \cdots i_p})\, dx_{i_1} \cdots dx_{i_p}$$

$$= \sum_I (df_{i_1 \cdots i_p} + dg_{i_1 \cdots i_p})\, dx_{i_1} \cdots dx_{i_p}$$

since the partial derivative of the sum of two functions is the sum of the partial derivatives. This last expression is just $d\omega_1 + d\omega_2$.

An easy induction proves the following extension of Proposition 4.1.

Corollary *Let $\omega_1, \ldots, \omega_k$ be p forms on the open set U of $\mathbf{R}^n$. Then*

$$d\left(\sum_{i=1}^{k} \omega_i\right) = \sum_{i=1}^{k} d\omega_i$$

We now determine the behavior of the exterior differential on the product of two forms.

Proposition 4.2 *Let ω be a p form on U and η a q form on U. Then*

$$d(\omega\eta) = (d\omega)\eta + (-1)^p \omega\, d\eta$$

Proof. Note first of all that, for 0 forms f and g,

$$d(fg) = g\,df + f\,dg$$

by the product rule for differentiation. Thus, if

$$\omega = f\,dx_{i_1} \cdots dx_{i_p}$$

and

$$\eta = g\,dx_{j_1} \cdots dx_{j_q}$$

then

$$\begin{aligned} d(\omega\eta) &= d(fg\,dx_{i_1} \cdots dx_{i_p}\,dx_{j_1} \cdots dx_{j_q}) \\ &= (g\,df + f\,dg)(dx_{i_1} \cdots dx_{i_p}\,dx_{j_1} \cdots dx_{j_q}) \\ &= (df)\,dx_{i_1} \cdots dx_{i_p}\,g\,dx_{j_1} \cdots dx_{j_q} \\ &\quad + f\,dg(dx_{i_1} \cdots dx_{i_p}\,dx_{j_1} \cdots dx_{j_q}) \end{aligned}$$

Now dg is a 1 form and $dx_{i_1} \cdots dx_{i_p}$ a p form; so

$$dg(dx_{i_1} \cdots dx_{i_p}) = (-1)^p (dx_{i_1} \cdots dx_{i_p})\,dg$$

by Proposition 2.2. Thus

$$\begin{aligned} d(\omega\eta) &= [(df)\,dx_{i_1} \cdots dx_{i_p}]g\,dx_{j_1} \cdots dx_{j_q} \\ &\quad + (-1)^p (f\,dx_{i_1} \cdots dx_{i_p})\,dg\,dx_{j_1} \cdots dx_{j_q} \\ &= (d\omega)\eta + (-1)^p \omega\,d\eta \end{aligned}$$

Since any p form and any q form are the sum of terms of the type treated above, the proof of Proposition 4.2 for general ω and η follows from Proposition 4.1 and the special case above.

The final result of this section describes the result of applying the exterior differential twice.

Proposition 4.3 *For any p form ω, $d(d\omega) = 0$.*

Sketch of Proof. It is easily seen that Proposition 4.3 for an arbitrary p

form will follow if we prove $d(df) = 0$, where f is a 0 form. In this case, direct computation shows that

$$d(df) = \sum_{1 \le i < j \le n} \left(\frac{\partial^2 f}{\partial x_i \partial x_j} - \frac{\partial^2 f}{\partial x_j \partial x_i} \right) dx_i\, dx_j$$

which vanishes by Theorem 1.1 of Chapter 6 (since f is assumed to be of class C^2).

Exercises

1. Prove the corollary to Proposition 4.1.
2. Give the details of the proof of Proposition 4.3 when $p > 0$. (The problem is to keep track of the subscripts.)
3. Let U be an open subset of $\mathbf{R}^3$, $f\colon U \to \mathbf{R}$ a function, and $\mathbf{F}\colon U \to \mathbf{R}^3$ a vector field. Use Proposition 4.3 to prove

$$\text{curl (grad } f) = 0$$

$$\text{div (curl } \mathbf{F}) = 0$$

5. The Action of Differentiable Functions on Forms

Suppose that U is an open subset of $\mathbf{R}^n$, V an open subset of $\mathbf{R}^m$, and $\varphi : U \to V$ a differentiable function of class C^2. (For the definition of class C^2, see Section 3.) In coordinates,

$$\varphi(\mathbf{v}) = (\varphi_1(\mathbf{v}), \ldots, \varphi_m(\mathbf{v}))$$

We now associate to each p form ω on V a p form $\varphi^*\omega$ on U.

If f is an 0 form on V, φ^*f is defined to be the composite $f \circ \varphi$. If $y_1, \ldots, y_m$ are coordinates in $\mathbf{R}^m$, we define $\varphi^* dy_i$ by

$$\varphi^* dy_i = \sum_{j=1}^{n} \frac{\partial \varphi_i}{\partial x_j} dx_j \tag{5.1}$$

For an arbitrary 1 form ω on U,

$$\omega = f_1 dy_1 + \cdots + f_m dy_m$$

we define $\varphi^*\omega$ by

$$\begin{aligned} \varphi^*\omega &= \sum_{i=1}^{m} \varphi^*(f_i)\varphi^* dy_i \\ &= \sum_{i=1}^{m} \sum_{j=1}^{n} (f_i \circ \varphi) \frac{\partial \varphi}{\partial x_j} dx_j \end{aligned} \tag{5.2}$$

Finally, if

$$\omega = \sum f_{i_1 \cdots i_p} dy_{i_1} \cdots dy_{i_p}$$

is a p form, then

$$\varphi^*\omega = \sum (\varphi^*(f_{i_1 \cdots i_p}))\varphi^*(dy_{i_1}) \cdots \varphi^*(dy_{i_p}) \tag{5.3}$$

We call φ^* the *mapping induced by* φ *on forms* or simply the *mapping induced by* φ.

The definition of φ^* given above may seem somewhat artificial. The following remarks may help motivate it.

If ω is a 0 form, say $\omega = f : V \to \mathbf{R}$, then defining $\varphi^*\omega$ to be the composite $f \circ \varphi$ is quite natural. The definition of $\varphi^*\omega$ when ω is a p form, $p > 0$, is dictated by the desire to have φ^* satisfy the following:

$$d(\varphi^*\omega) = \varphi^* d\omega \tag{5.4}$$

$$\varphi^*(\omega + \omega') = \varphi^*\omega + \varphi^*\omega' \tag{5.5}$$

$$\varphi^*(\omega\eta) = (\varphi^*\omega)(\varphi^*\eta) \tag{5.6}$$

Here ω and ω' are p forms and η is a q form. (We prove that these properties do actually hold in the next section.)

For example, if $x_j : \mathbf{R}^n \to \mathbf{R}$, $1 \leq j \leq n$, are the coordinate functions in $\mathbf{R}^n$ and $y_i : \mathbf{R}^m \to \mathbf{R}$, $1 \leq i \leq m$, the coordinate functions in $\mathbf{R}^m$, then, if (5.4) is to hold, we must have

$$\begin{aligned}
\varphi^*(dy_i) &= \varphi^*(d(y_i)) \\
&= d(\varphi^*(y_i)) \\
&= d(y_i \circ \varphi) \\
&= d\varphi_i \\
&= \sum_{j=1}^{n} \frac{\partial \varphi_i}{\partial x_j} dx_j
\end{aligned}$$

This is exactly equation (5.1). To define φ^* on an arbitrary 1 form, we use properties (5.5) and (5.6) to obtain (5.2).

The same reasoning shows that equation (5.3), which defines φ^* on an arbitrary p form, is also a consequence of properties (5.4), (5.5), and (5.6). (See Exercise 9.)

Here are some examples.

1. Let $\varphi : \mathbf{R} \to \mathbf{R}$ and let $\omega = f\,dy$ be a 1 form on $\mathbf{R}$. Then

$$\varphi^*\omega = \varphi^*(f)\varphi^*(dy) = (f \circ \varphi)\frac{d\varphi}{dx} dx$$

2. *Let* $\varphi : \mathbf{R}^2 \to \mathbf{R}^2$ be given by

$$\varphi(x_1, x_2) = (ax_1 + bx_2, cx_1 + ex_2)$$

and let $\omega = dx_1\, dx_2$. Then

$$\varphi^* dx_1 = \frac{\partial \varphi_1}{\partial x_1} dx_1 + \frac{\partial \varphi_1}{\partial x_2} dx_2 = a\,dx_1 + b\,dx_2$$

$$\varphi^* dx_2 = \frac{\partial \varphi_2}{\partial x_1} dx_1 + \frac{\partial \varphi_2}{\partial x_2} dx_2 = c\,dx_1 + e\,dx_2$$

Thus

$$\begin{aligned} \varphi^*\omega &= (\varphi^* dx_1)(\varphi^* dx_2) \\ &= (a\,dx_1 + b\,dx_2)(c\,dx_1 + e\,dx_2) \\ &= (ac - be)\,dx_1\,dx_2 \end{aligned}$$

We generalize this example in Exercise 7.

3. Let $\varphi : \mathbf{R}^2 \to \mathbf{R}^3$ be defined by

$$\varphi(x_1, x_2) = (x_1 x_2, x_1^2 - x_2^2, 2x_1 + 3x_2)$$

and set

$$\omega = y_1\, dy_1\, dy_2 - y_2^2\, dy_2 dy_3$$

Then

$$\begin{aligned} \varphi^*(y_1) &= y_1 \circ \varphi = x_1 x_2 \\ \varphi^*(y_2) &= x_1^2 - x_2^2 \\ \varphi^*(y_3) &= 2x_1 + 3x_2 \\ \varphi^*(dy_1) &= x_2\, dx_1 + x_1\, dx_2 \\ \varphi^*(dy_2) &= 2x_1\, dx_1 - 2x_2\, dx_2 \end{aligned}$$

and

$$\varphi^*(dy_3) = 2dx_1 + 3dx_2$$

Therefore

$$\begin{aligned} \varphi^*\omega &= x_1 x_2 (x_2\, dx_1 + x_1\, dx_2)(2x_1\, dx_1 - 2x_2\, dx_2) \\ &\quad - (x_1^2 - x_2^2)^2 (2x_1\, dx_1 - 2x_2\, dx_2)(2dx_1 + 3dx_2) \\ &= -2[x_1 x_2^3 + x_1^3 x_2 + 3x_1(x_1^2 - x_2^2) + 2x_2(x_1^2 - x_2^2)^2]\, dx_1\, dx_2 \end{aligned}$$

We give the basic properties of the mapping induced by φ in the next section.

Exercises

1. Let $\varphi : \mathbf{R}^2 \to \mathbf{R}^2$ be given by $\varphi(x, y) = (x^2 + y^2, 2xy)$ and determine $\varphi^*\omega$, where
   (a) $\omega = st^2$ (Here, s and t are the coordinates in the second $\mathbf{R}^2$.)
   (b) $\omega = t\,ds$

(c) $\omega = t^2\, ds - s\, dt$
(d) $\omega = ds\, dt$
(e) $\omega = \sin st\, ds\, dt$

2. Let $\varphi : \mathbf{R}^3 \to \mathbf{R}^2$ be given by $\varphi(x, y, z) = (xy, yz)$ and determine $\varphi^*\omega$ for the forms ω of Exercise 1.
3. Let $\varphi : \mathbf{R}^3 \to \mathbf{R}^3$ be given by $\varphi(x_1, x_2, x_3) = (2x_1 + x_2, x_1x_3, x_2 - x_3)$ and determine $\varphi^*\omega$, where
   (a) $\omega = y_1y_2y_3$ (Here, y_1, y_2, and y_3 are the coordinates in the second $\mathbf{R}^3$.)
   (b) $\omega = dy_2$
   (c) $\omega = y_2\, dy_1 - y_1\, dy_3$
   (d) $\omega = dy_1\, dy_3$
   (e) $\omega = y_1y_3^2\, dy_1\, dy_2 - y_1^3\, dy_1\, dy_3 + \sin y_1\, dy_2\, dy_3$
   (f) $\omega = dy_1\, dy_2 dy_3$
   (g) $\omega = (y_1y_2 - y_3^2)\, dy_1\, dy_2\, dy_3$
4. Let $\varphi : \mathbf{R}^3 \to \mathbf{R}^3$ be given by $\varphi(x_1, x_2, x_3) = (x_1^2x_3, x_2x_3, x_2^3)$ and determine $\varphi^*\omega$ for the forms in Exercise 3.
5. Show that $\varphi^*d\omega$ and $d\varphi^*\omega$ are equal for φ and ω as in Exercise 1.
6. Show that $\varphi^*d\omega$ and $d\varphi^*\omega$ are equal for φ and ω as in Exercise 3.
7. Let U and V be open subsets of $\mathbf{R}^n$, $\varphi : U \to V$ a C^2 function, and $\omega = g\, dx_1 \cdots dx_n$ be an n form on V. Prove that

$$\varphi^*\omega = (g \circ \varphi)\,[\mathrm{Det}\,(D\varphi)]\, dx_1 \cdots dx_n$$

   (*Hint*: Recall Proposition 2.4.)
8. Let $\beta : \mathbf{R}^2 \to \mathbf{R}^3$, $\mathbf{F} : \mathbf{R}^3 \to \mathbf{R}^3$ be C^2 functions, where $\mathbf{F}(\mathbf{v}) = (f_1(\mathbf{v}), f_2(\mathbf{v}), f_3(\mathbf{v}))$. Let ω be the 1 form defined by

$$\omega = f_1\, dx_1 + f_2\, dx_2 + f_3\, dx_3$$

   and prove that

$$\beta^*d\omega = \langle (\mathrm{curl}\ \mathbf{F}) \circ \beta, (D\beta)\mathbf{e}_1 \times (D\beta)\mathbf{e}_2 \rangle\, dx\, dy$$

   where x and y are coordinates in $\mathbf{R}^2$.
9. Show that, given the definition of φ^* on 0 forms and 1 forms, the definition (5.3) of φ^* on p forms, $p > 1$, follows from (5.4), (5.5), and (5.6).

6. Basic Properties of the Induced Mapping

We give here some of the important properties of the induced mapping on forms.

Proposition 6.1 *Let U be an open subset of $\mathbf{R}^n$, V an open subset of $\mathbf{R}^m$, and $\varphi : U \to V$ a differentiable function. Then*

$$\varphi^*(\omega + \omega') = \varphi^*\omega + \varphi^*\omega' \tag{6.1}$$

and

(6.2) $$\varphi^*(\omega\eta) = (\varphi^*\omega)(\varphi^*\eta)$$

for any p forms ω, ω' and q form η on V. Furthermore, if W is an open subset of $\mathbf{R}^k$ and $\psi: V \to W$ a differentiable function, then

(6.3) $$(\psi \circ \varphi)^* = \varphi^* \circ \psi^*$$

Proof. The first two assertions of the proposition are immediate consequences of the definition of φ^* and are left as Exercises 1 and 2. We prove the third.

If μ is the zero form $f: W \to \mathbf{R}$, then

$$\begin{aligned}(\psi \circ \varphi)^* f &= f \circ \psi \circ \varphi \\ &= (f \circ \psi) \circ \varphi \\ &= \varphi^*(f \circ \psi) \\ &= \varphi^*(\psi^* \circ f)\end{aligned}$$

Let $z_1, \ldots, z_k$ be coordinates in $\mathbf{R}^k$ and suppose that

$$\varphi(\mathbf{v}) = (\varphi_1(\mathbf{v}), \ldots, \varphi_m(\mathbf{v}))$$

and

$$\psi(\mathbf{w}) = (\psi_1(\mathbf{w}), \ldots, \psi_k(\mathbf{w}))$$

Then

$$\begin{aligned}(\psi \circ \varphi)^* dz_i &= \sum_{j=1}^{n} \frac{\partial(\psi_i \circ \varphi)}{\partial x_j} dx_j \\ &= \sum_{j=1}^{n} \sum_{l=1}^{m} \left(\frac{\partial \psi_i}{\partial y_l} \circ \varphi\right) \frac{\partial \varphi_l}{\partial x_j} dx_j\end{aligned}$$

by the chain rule. Therefore,

$$\begin{aligned}(\psi \circ \varphi)^* dz_i &= \sum_{l=1}^{m} \left(\frac{\partial \psi_i}{\partial y} \circ \varphi\right) \sum_{j=1}^{n} \frac{\partial \varphi_l}{\partial x_j} dx_j \\ &= \sum_{l=1}^{m} \varphi^* \frac{\partial \psi_i}{\partial y_l} \varphi^*(dy_l) \\ &= \varphi^* \sum_{l=1}^{m} \frac{\partial \psi_i}{\partial y} dy_l \\ &= \varphi^*(\psi^* dz_i) \\ &= (\varphi^* \circ \psi^*)\, dz\end{aligned}$$

Thus, if $\mu = f_1\, dz_1 + \cdots + f_k\, dz_k$, we have

$$\begin{aligned}(\psi \circ \varphi)^*\mu &= \sum_{i=1}^{k} ((\psi \circ \varphi)^* f_i)(\psi \circ \varphi)^* dz_i \\ &= \sum_{i=1}^{k} (\varphi^* \circ \psi^* f_i)(\varphi^* \circ \psi^* dz_i) \\ &= \varphi^* \sum_{i=1}^{k} (\psi^* f_i)(\psi^* dz_i) \\ &= (\varphi^* \circ \psi^*)\mu\end{aligned}$$

The verification of (6.3) when μ is a p form now follows from the definition of φ^*, formulas (6.1) and (6.2), and the cases $p = 0, 1$ above. The details are left as Exercise 3.

The next proposition shows that the action of φ^* and the exterior differential commute.

Proposition 6.2 *Let U be an open subset of $\mathbf{R}^n$, V an open subset of $\mathbf{R}^m$, and $\varphi : U \to V$ a differentiable function. Then*

$$d(\varphi^*\omega) = \varphi^*(d\omega)$$

for any p form ω on V.

Proof. Let $x_1, \ldots, x_n$ be coordinates in $\mathbf{R}^n$, $y_1, \ldots, y_m$ coordinates in $\mathbf{R}^m$, and $\varphi(\mathbf{v}) = (\varphi_1(\mathbf{v}), \ldots, \varphi_m(\mathbf{v}))$. The proof breaks up into four cases. We proceed directly by computing both $\varphi^*(d\omega)$ and $d(\varphi^*\omega)$.

Case 1 $\omega = f$ is a 0 form.

$$\begin{aligned}\varphi^* d\omega &= \varphi^* \sum_{i=1}^{m} \frac{\partial f}{\partial y_i} dy_i \\ &= \sum_{i=1}^{m} \varphi^*\left(\frac{\partial f}{\partial y_i}\right)\varphi^*(dy_i) \\ &= \sum_{i=1}^{m} \left(\frac{\partial f}{\partial y_i} \circ \varphi\right) \sum_{j=1}^{n} \frac{\partial \varphi_i}{\partial x_j} dx_j\end{aligned}$$

and

$$\begin{aligned}d(\varphi^*\omega) &= d(f \circ \varphi) \\ &= \sum_{j=1}^{n} \frac{\partial (f \circ \varphi)}{\partial x_j} dx_j \\ &= \sum_{j=1}^{n} \left(\sum_{i=1}^{m} \frac{\partial f}{\partial y_i} \circ \varphi\right) \frac{\partial \varphi_i}{\partial x_j} dx_j\end{aligned}$$

by the chain rule. Thus the formula holds in this case.

Case 2 $\omega = dy_k$.

First of all, $\varphi^* d\omega = 0$ since $d\omega = d(dy_k) = 0$ by Proposition 4.3. On the other hand,

$$\begin{aligned} d\varphi^*\omega &= d \sum_{j=1}^{n} \frac{\partial \varphi_k}{\partial x_j} dx_j \\ &= \sum_{i,j=1}^{n} \frac{\partial^2 \varphi_k}{\partial x_i \partial x_j} dx_i\, dx_j \\ &= \sum_{1 \le i < j \le n} \left(\frac{\partial^2 \varphi_k}{\partial x_i \partial x_j} - \frac{\partial^2 \varphi_k}{\partial x_j \partial x_i} \right) dx_i\, dx_j = 0 \end{aligned}$$

by Theorem 1.1 of Chapter 6, since the functions φ_k have continuous second partial derivatives.

Case 3 $\omega = f\, dy_{i_1} \cdots dy_{i_p}$.

To begin with,

$$\begin{aligned} \varphi^*(d\omega) &= \varphi^*((df)\, dy_{i_1} \cdots dy_{i_p}) \\ &= (\varphi^* df)\varphi^* dy_i \cdots \varphi^* dy_{i_p} \qquad \text{(by Proposition 6.1)} \\ &= (d\varphi^* f)\varphi^* dy_{i_1} \cdots \varphi^* dy_{i_p} \end{aligned}$$

by case 1. Thus we need to show that

$$(6.4) \qquad d(\varphi^* f)\varphi^* dy_{i_1} \cdots \varphi^* dy_{i_p} = d((\varphi^* f)(\varphi^* dy_{i_1} \cdots (\varphi^* dy_{i_p}))$$

Using Proposition 4.2, we see that

$$(6.5) \quad d((\varphi^* f)(\varphi^* dy_{i_1}) \cdots (\varphi^* dy_{i_p})) = (d(\varphi^* f))(\varphi^* dy_{i_1} \cdots \varphi^* dy_{i_p})$$

$$+ \sum_{j=1}^{p} (-1)^{j-1} \varphi^* f \varphi^* dy_{i_1} \cdots \varphi^* dy_{i_{j-1}} d(\varphi^* dy_{i_j}) \varphi^* dy_{i_{j+1}} \cdots \varphi^* dy_{i_p}$$

However, case 2 shows that

$$d(\varphi^* dy_{i_j}) = \varphi^* d(dy_{i_j}) = 0$$

and thus only the first term on the left-hand side of (6.5) remains. Therefore (6.4) does hold, and case 3 is done.

Case 4 ω an arbitrary p form.

Since any p form is the sum of terms of the kind treated in case 3, this case is a consequence of Proposition 4.1 and case 3. The proposition is therefore proved.

Exercises

1. Prove equation (6.1) of Proposition 6.1.
2. Prove equation (6.2) of Proposition 6.1.
3. Prove that equation (6.3) of Proposition 6.1 holds for p forms, $p > 1$, assuming that it holds for $p = 0$ and 1.
4. Give the details of case 4 of the proof of Proposition 6.2.

15

Integration of Differential Forms

In this chapter, we give meaning to the various notions of the previous chapter. We show first how to integrate a differential p form over a so-called singular p chain. The result is an integral which includes both line and surface integrals of vector fields as special cases. We then derive a general Stokes' theorem for this integral and show that both Green's theorem and Stokes' theorem are special cases of it. As another special case, we prove Gauss' theorem, which relates certain surface integrals in $\mathbf{R}^3$ to volume integrals, and give an application to hydrodynamics.

1. Integration of Forms

We now define the integral of a p form over a singular p cube in $\mathbf{R}^n$. This notion generalizes both line and surface integrals.

To begin with, let ω be a p form on an open subset $U \subset \mathbf{R}^p$ and S a closed subset of U. Then ω can be written $\omega = f\,dx_1 \cdots dx_p$ and we define the *integral of ω over S* by

$$\int_S \omega = \int_S f$$

where the second integral is the usual integral f over S, the one defined in Chapter 10.

In order to define the integral of a p form in $\mathbf{R}^n$, we need the notion of a singular p cube. The *standard p cube* is the subset I^p of $\mathbf{R}^p$ defined by

$$I^p = \{(x_1, \ldots, x_p) : 0 \le x_i \le 1,\ 1 \le i \le p\}$$

if $p > 0$. We set $I^0 = \{0\}$.

For $p > 0$, I^p is simply the rectangle in $\mathbf{R}^p$ defined by the origin and the point $(1, 1, \ldots, 1)$. (See, for example, Figure 1.1.) A *singular p cube* on a subset B of $\mathbf{R}^n$ is a differentiable mapping $T\colon I^p \to B$. For example, we can consider a parametrized curve $\alpha\colon I^1 \to \mathbf{R}^n$ to be a singular 1 cube and a parametrized surface $\alpha\colon I^2 \to \mathbf{R}^n$ to be a singular 2 cube. In general, however, singular cubes can be quite degenerate; the mapping $T\colon I^p \to \mathbf{R}^n$, $T(\mathbf{v}) = \mathbf{0}$, is an extreme example.

Let U be an open subset of $\mathbf{R}^n$, ω a p form on U, and $T\colon I^p \to U$ a singular p cube. We define the *integral of ω over T* by

$$\int_T \omega = \int_{I^p} T^*\omega$$

where the integral on the right is the integral of the p form $T^*\omega$ over the subset $I^p \subset \mathbf{R}^p$ defined above. Here are some examples.

1. Let $p = 1$, $\omega = f_1\,dx_1 + \cdots + f_n\,dx_n$, and $\alpha\colon I \to \mathbf{R}^n$ a singular 1 cube, $\alpha(t) = (\alpha_1(t), \ldots, \alpha_n(t))$. (We write I for I^1.) Then

$$\int_\alpha \omega = \int_I \alpha^*\omega$$

$$= \int_I (f_1\alpha_1)\alpha_1' + \cdots + (f_n\alpha_n)\alpha_n'$$

$$= \int_I \langle \mathbf{F}\circ\alpha, \alpha' \rangle$$

which coincides with our earlier definition of the integral of the vector field $\mathbf{F} = (f_1, \ldots, f_n)$ along the parametrized curve α.

2. Let $p = 2$ and $n = 3$. Then ω is a 2 form,

$$\omega = f_1\,dx_2\,dx_3 - f_2\,dx_1\,dx_3 + f_3\,dx_1\,dx_2$$

(note the minus sign; see Section 14.3 for an explanation) and $\alpha\colon I^2 \to \mathbf{R}^3$, $\alpha(\mathbf{v}) = (\alpha_1(\mathbf{v}), \alpha_2(\mathbf{v}), \alpha_3(\mathbf{v}))$, is a differentiable singular 2 cube. Suppose we let (x, y) be coordinates in $\mathbf{R}^2$ and (x_1, x_2, x_3) be coordinates in $\mathbf{R}^3$. Then

$$\alpha^*\omega = (f_1\circ\alpha)\alpha^*(dx_2)\alpha^*(dx_3) - (f_2\circ\alpha)\alpha^*(dx_1)\alpha^*(dx_3) + (f_3\circ\alpha)\alpha^*(dx_1)\alpha^*(dx_2)$$

where $$\alpha^*(dx_1) = \frac{\partial\alpha_1}{\partial x}dx + \frac{\partial\alpha_1}{\partial y}dy$$

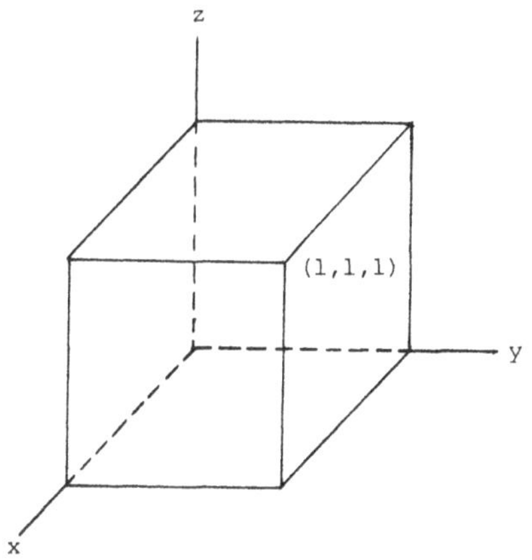

Figure 1.1 The standard 3 cube.

$$\alpha^*(dx_2) = \frac{\partial \alpha_2}{\partial x} dx + \frac{\partial \alpha_2}{\partial y} dy$$

and

$$\alpha^*(dx_3) = \frac{\partial \alpha_3}{\partial x} dx + \frac{\partial \alpha_3}{\partial y} dy$$

It follows that

$$\alpha^*\omega = \left[\left(f_1 \circ \alpha \right) \left(\frac{\partial \alpha_2}{\partial x} \frac{\partial \alpha_3}{\partial y} - \frac{\partial \alpha_2}{\partial y} \frac{\partial \alpha_3}{\partial x} \right) - \left(f_2 \circ \alpha \right) \left(\frac{\partial \alpha_1}{\partial x} \frac{\partial \alpha_3}{\partial y} - \frac{\partial \alpha_1}{\partial y} \frac{\partial \alpha_3}{\partial x} \right) \right.$$
$$\left. + \left(f_3 \circ \alpha \right) \left(\frac{\partial \alpha_1}{\partial x} \frac{\partial \alpha_2}{\partial y} - \frac{\partial \alpha_1}{\partial y} \frac{\partial \alpha_2}{\partial x} \right) \right] dx\, dy$$
$$= \langle \mathbf{F} \circ \alpha, (D\alpha)\mathbf{e}_1 \times (D\alpha)\mathbf{e}_2 \rangle \, dx\, dy$$

where $\mathbf{F}: U \to \mathbf{R}^3$ is the vector field given by $\mathbf{F}(\mathbf{v}) = (f_1(\mathbf{v}), f_2(\mathbf{v}), f_3(\mathbf{v}))$. Thus

$$\int_\alpha \omega = \int_{I^2} \alpha^*\omega = \int_{I^2} \langle \mathbf{F} \circ \alpha, (D\alpha)\mathbf{e}_1 \times (D\alpha)\mathbf{e}_2 \rangle$$

This coincides with our definition of the surface integral of $\mathbf{F}$ over the parametrized surface α.

3. Let ω be the 3 form on $\mathbf{R}^4$ defined by

$$\omega = x_3^2 x_4\, dx_1\, dx_3\, dx_4 + x_1 x_2\, dx_2\, dx_3\, dx_4$$

and let $T: I^3 \to \mathbf{R}^4$ be defined by

$$T(x, y, z) = (z, xy, x, y + z)$$

Then
$$T^*\omega = T^*(x_3^2x_4)\,T^*(dx_1)\,T^*(dx_3)\,T^*(dx_4) + T^*(x_1x_2)\,T^*(dx_2)\,T^*(dx_3)\,T^*(dx_4)$$

Now
$$T^*(x_3^2x_4) = x^2(y+z)$$
$$T^*(x_1x_2) = xyz$$
$$T^*(dx_1) = dz$$
$$T^*(dx_2) = y\,dx + x\,dy$$
$$T^*(dx_3) = dx$$

and
$$T^*(dx_4) = dy + dz$$

Thus

$$\begin{aligned} T^*\omega &= x^2(y+z)\,dz\,dx\,(dy+dz) + xyz(y\,dx + x\,dy)\,dx\,(dy+dz) \\ &= [x^2(y+z) - x^2yz]\,dx\,dy\,dz \end{aligned}$$

and
$$\begin{aligned} \int_T \omega &= \int_{I^3} x^2(y+z-yz) \\ &= \int_0^1\int_0^1\int_0^1 x^2(y+z-yz)\,dx\,dy\,dz \\ &= \int_0^1\int_0^1 \frac{1}{3}(y+z-yz)dy\,dz \\ &= \int_0^1 \frac{1}{3}\left(\frac{1}{2}+z-\frac{z}{2}\right)dz = \frac{1}{4} \end{aligned}$$

We shall have more to say about the integrals of n forms on singular n cubes in $\mathbf{R}^n$ in Section 3.

Exercises

1. Calculate the integral of the p form ω over the singular p cube T, with
   (a) $T: I^1 \to \mathbf{R}^3$, $T(s) = (s, s^2, s^3)$, and $\omega = dx + dz$.
   (b) T as in (a) and $\omega = xy\,dx - z\,dy$.
   (c) $T: I^2 \to \mathbf{R}^3$, $T(s, t) = (s, t, st)$, and $\omega = dxdz$.
   (d) T as in (c), $\omega = dy\,dz$.
   (e) T as in (c), $\omega = 2dx\,dz - x\,dy\,dz$.
   (f) T as in (c), $\omega = xy^2\,dx\,dy - 2yz\,dx\,dz + 4dy\,dz$.
   (g) $T: I^3 \to \mathbf{R}^4$, $T(s, t, u) = (st^2, tu, s, s+u)$, $\omega = dx_1\,dx_2\,dx_4$.
   (h) T as in (g), $\omega = x_2\,dx_1\,dx_3\,dx_4 - 3dx_2\,dx_3\,dx_4$.
   (i) T as in (g), $\omega = (x_2x_3 - x_1^2)\,dx_1\,dx_2\,dx_4 + x_3^2x_4\,dx_2\,dx_3\,dx_4$.
2. Calculate the integral of the 2 form ω over the singular 2 cube $T: I^2 \to \mathbf{R}^4$ given by $T(s, t) = (st, s^2, t^2, s+t)$ when

(a) $\omega = dx_1\, dx_3$.
(b) $\omega = x_2\, dx_1\, dx_2 - 3x_3^2\, dx_3\, dx_4$.
(c) $\omega = x_1x_3\, dx_1\, dx_3 - (x_2^2 + x_3x_4)\, dx_2\, dx_4$.
(d) $\omega = 2x_2x_3\, dx_2\, dx_3 + x_1x_2x_3\, dx_2\, dx_4$.

3. Calculate the integral of the p form ω over the singular p cube $T: I^3 \to \mathbf{R}^4$, where $T(s, t, u) = (stu, s^2t, t^2u, u^3)$ and
(a) $\omega = dx_1\, dx_3\, dx_4$.
(b) $\omega = x_2\, dx_1\, dx_2\, dx_4 - 2dx_2\, dx_3\, dx_4$.
(c) $\omega = x_2x_3^2\, dx_1\, dx_2\, dx_3 + x_1x_4\, dx_2\, dx_3\, dx_4$.
(d) $\omega = x_4\, dx_1\, dx_2\, dx_4 - 2x_2^2x_3\, dx_1\, dx_3\, dx_4$.

2. The General Stokes' Theorem

In this section, we state the general form of Stokes' theorem. This is an extremely important result. We derive Green's theorem (Theorem 5.1, Chapter 12) and Stokes' theorem (Theorem 5.1, Chapter 13) from it in Section 3 and give an application to hydrodynamics in Section 4. The proof of Stokes' theorem is given in Section 5.

We begin with some definitions.

A *singular p chain* on a subset $B \subset \mathbf{R}^n$ is a formal linear combination of singular p cubes on B:

$$c = a_1T_1 + a_2T_2 + \cdots + a_rT_r$$

where $a_1, \ldots, a_r$ are real numbers and $T_1, \ldots, T_r$ singular p cubes on B. We emphasize that we do not consider c to be a function, just a formal expression. We shall give a geometric interpretation of certain chains in the next section. (See also the definition of ∂T below.)

If U is an open subset of $\mathbf{R}^n$, ω a p form on U, and c a singular p chain on U, we define the integral of ω over c by

$$\int_c \omega = \sum_{j=1}^{r} a_j \int_{T_j} \omega$$

We can add p chains and multiply them by real numbers in the obvious way; if $c = a_1T_1 + \cdots + a_rT_r$, $c' = b_1T_1' + \cdots + b_mT_m'$, and a is a real number, then

$$c + c' = a_1T_1 + \cdots + a_rT_r + b_1T_1' + \cdots + b_mT_m'$$

$$ac = (aa_1)T_1 + \cdots + (aa_r)T_r$$

Our first result of this section gives some of the elementary properties of the integral of a p form over a p chain.

Proposition 2.1 *Let a be a real number, ω and ω' p forms on $U \subset \mathbf{R}^n$, and c, c' singular p chains on U. Then*

$$\int_c (\omega + \omega') = \int_c \omega + \int_c \omega'$$

$$\int_{c+c'} \omega = \int_c \omega + \int_{c'} \omega$$

$$\int_{ac} \omega = a \int_c \omega = \int_c a\omega$$

The proofs are straightforward; for example, if $c = a_1 T_1 + \cdots + a_r T_r$, we have

$$\begin{aligned}
\int_c (\omega + \omega') &= \sum_{i=1}^{r} a_i \int_{T_i} (\omega + \omega') \\
&= \sum_{i=1}^{r} a_i \int_{I^p} T_i^*(\omega + \omega') \\
&= \sum_{i=1}^{r} a_i \int_{I^p} (T_i^*\omega + T_i^*\omega') \\
&= \sum_{i=1}^{r} a_i \left(\int_{I^p} T_i^*\omega + \int_{I^p} T_i^*\omega' \right) \\
&= \sum_{i=1}^{r} a_i \int_{I^p} T_i^*\omega + \sum_{i=1}^{r} a_i \int_{I^p} T_i^*\omega' \\
&= \int_c \omega + \int_c \omega'
\end{aligned}$$

We leave the verification of the remaining two equations to the reader.

For each positive integer p, define mappings

$$\partial_i, \partial_i' : I^{p-1} \to I^p \qquad 1 \le i \le p$$

by

$$\partial_i(t_1, \ldots, t_{p-1}) = (t_1, \ldots, t_{i-1}, 0, t_i, \ldots, t_{p-1})$$

$$\partial_i'(t_1, \ldots, t_{p-1}) = (t_1, \ldots, t_{i-1}, 1, t_i, \ldots, t_{p-1})$$

These functions correspond to the inclusion of I^{p-1} into the various faces of I^p. For instance, if $p = 2$, ∂_2' takes I^1 onto the edge of I^2 between the vertex (0, 1) and the vertex (1, 1). If $T: I^p \to \mathbf{R}^n$ is a singular p cube, the *boundary of T* is defined to be the singular $(p - 1)$ chain

$$\partial T = \sum_{i=1}^{p} (-1)^i (T \circ \partial_i - T \circ \partial_i')$$

where $T \circ \partial_i$ and $T \circ \partial_i'$ are the composite functions. This chain is essentially the restriction of T to the boundary of I^p. If $p = 2$,

$$\partial T = T \circ \partial_2 + T \circ \partial'_1 - T \circ \partial'_2 - T \circ \partial_1$$

which is the sum, with signs, of the restrictions of T to the four edges of I^2. In fact, if we interpret the terms $-T \circ \partial'_2$ and $-T \circ \partial'_1$ as corresponding to traversing these edges in the negative direction (from 1 to 0), then ∂T corresponds to the closed curve obtained by restricting T to the boundary of I^2, oriented so that the set I^2 is on the left. (See Figure 2.1.)

If $c = a_1 T_1 + \cdots + a_r T_r$ is an arbitrary singular p chain, we define ∂c by

$$\partial c = \sum_{i=1}^{r} a_i(\partial T_i)$$

Proposition 2.2 *If c and c' are singular p chains and a, b are real numbers, then*

$$\partial(ac + bc') = a\partial c + b\partial c'$$

Furthermore, $\partial\partial c = 0$ for any singular p chain c.

The proof of the first part of the proposition is trivial and left as an exercise.

The proof of the second part of the proposition is somewhat complicated. However, since we do not need this result in what follows, we also leave its proof as an exercise. (See Exercises 8 to 11.)

We can now state the general form of Stokes' theorem. The proof is given in Section 5.

Theorem 2.3 *Let U be an open subset of $\mathbf{R}^n$, ω a $p - 1$ form on U, and c a singular p chain on U. Then*

$$\int_{\partial c} \omega = \int_c d\omega$$

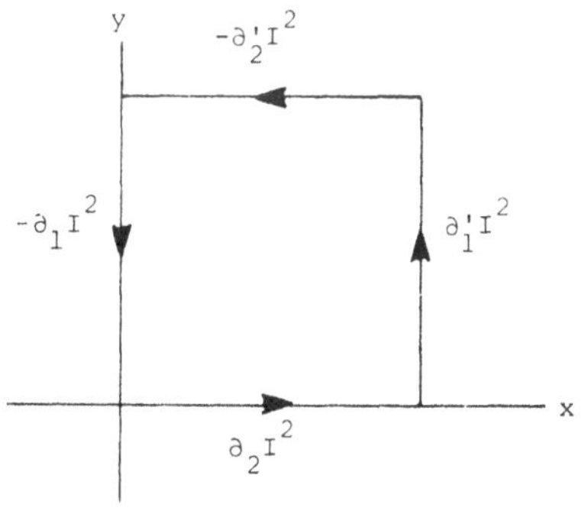

Figure 2.1 The boundary of the standard 2 cube.

In the next section, we shall see that this result is indeed a generalization of Green's and Stokes' theorems.

Exercises

1. Let $T_1, T_2, T_3\colon I^2 \to \mathbf{R}^3$ be given by
$$T_1(s, t) = (s, t, s + t)$$
$$T_2(s, t) = (st, s^2, t^2)$$
$$T_3(s, t) = (s - t, s + t, st)$$
and let $\omega = y\,dx\,dz$ be a 2 form on $\mathbf{R}^3$. Compute the integral of ω over the singular 2 chain c, where
(a) $c = T_1 - T_2$
(b) $c = 2T_1 + T_2 + T_3$
(c) $c = T_3 - 4T_1$
(d) $c = 3T_1 - 2T_2 + 5T_3$
2. Let $\omega = 2xy\,dx\,dy - y^2\,dy\,dz$ and compute the integral of ω over the singular 2 chains of Exercise 1.
3. Let $T_1, T_2, T_3\colon I^3 \to \mathbf{R}^4$ be given by
$$T_1(s, t, u) = (s, t + u, t, s - u)$$
$$T_2(s, t, u) = (st, su, tu, s + t + u)$$
$$T_3(s, t, u) = (s - u, t^2, tu, st)$$
and let $\omega = dx_1\,dx_2\,dx_4$. Compute the integral of ω over the singular 3 chains c defined by the formulas of (a) through (d) of Exercise 1.
4. Let $\omega = x_1 x_3\,dx_1\,dx_2\,dx_3 + x_3^2\,dx_2\,dx_3\,dx_4$ and compute the integral of ω over the singular 3 chains of Exercise 1.
5. Prove the second and third assertions of Proposition 2.1.
6. Prove the first assertion of Proposition 2.2.
7. Show that
$$(\partial_i)^*\,dx_j = (\partial_i')^*\,dx_j = \begin{cases} dx_j & \text{if } j < i \\ 0 & \text{if } j = i \\ dx_{j-1} & \text{if } j > i \end{cases}$$
8. Let T be a singular p cube. Prove that
$$\partial(\partial T) = \sum_{i=1}^{p} (-1)^i \sum_{j=1}^{p-1} (-1)^j (T\circ\partial_i\circ\partial_j - T\circ\partial_i\circ\partial_j' + T\circ\partial_i'\circ\partial_j - T\circ\partial_i'\circ\partial_j')$$
9. Prove that, for any i, j, $1 \le i \le j \le p - 1$, we have
$$\partial_i\circ\partial_j = \partial_{j+1}\circ\partial_i$$
$$\partial_i\circ\partial_j' = \partial_{j+1}'\circ\partial_i$$
$$\partial_i'\circ\partial_j = \partial_{j+1}\circ\partial_i'$$
$$\partial_i'\circ\partial_j' = \partial_{j+1}'\circ\partial_i'$$

10. Let T be a singular p chain. Prove that

$$\sum_{i=1}^{p}\sum_{j=1}^{p-1}(-1)^{i+j}\,T\circ\partial_i\circ\partial_j = 0$$

$$\sum_{i=1}^{p}\sum_{j=1}^{p-1}(-1)^{i+j}(T\circ\partial_i\circ\partial_j' + T\circ\partial_i'\circ\partial_j) = 0$$

$$\sum_{i=1}^{p}\sum_{j=1}^{p-1}(-1)^{i+j}\,T\circ\partial_i'\circ\partial_j' = 0$$

11. Let c be a singular p chain. Prove that $\partial(\partial c) = 0$.

3. Green's Theorem and Stokes' Theorem

We now derive both Green's theorem (Theorem 5.1, Chapter 12) and Stokes' theorem (Theorem 5.1, Chapter 13). We begin with an interpretation of the integral of an n form in $\mathbf{R}^n$ over a particular kind of singular n cube.

Let $T\colon I^n \to \mathbf{R}^n$ be a singular n cube which also happens to be an orientation-preserving diffeomorphism onto its image B. Suppose that ω is an n form on an open set containing B, $\omega = f\,dx_1 \cdots dx_n$. Then

$$\int_T \omega = \int_{I^n} T^*\omega$$

$$= \int_{I^n} (f\circ T)(\text{Det } DT)\,dx_1 \cdots dx_n$$

(by Proposition 2.4, Chapter 14; see also Exercise 4, Section 14.3)

$$= \int_{I^n} (f\circ T)(\text{Det } DT)$$

(by the definition of the integral of an n form over a subset of $\mathbf{R}^n$). However

$$\int_{I^n} (f\circ T)(\text{Det } DT) = \int_B f$$

by Theorem 6.2 of Chapter 12. Thus

$$\int_T \omega = \int_B f \tag{3.1}$$

A similar calculation shows that if T reverses orientation, then

$$\int_T \omega = -\int_B f$$

Because of this dependence on orientation, we refer to the integral of ω over T as an *oriented integral.*

We can use this observation to derive Green's theorem from Theorem 2.3. Let Γ be a simple closed curve in $\mathbf{R}^2$ enclosing a region B in such a way that Γ is oriented so that B is on the left. Let $T: I^2 \to \mathbf{R}^2$ be a singular 2 cube which is an orientation-preserving diffeomorphism of I^2 onto B. (Such diffeomorphisms can be shown to exist.) Then ∂T can be thought of as defining a piecewise differentiable parametrization of Γ consistent with its orientation. (See Figure 2.1.) Now, if

$$\omega = f_1\, dx_1 + f_2\, dx_2$$

is a 1 form on B, Theorem 2.3 gives us the equation

$$\int_{\partial T} \omega = \int_T d\omega \tag{3.2}$$

If $\mathbf{F} = (f_1, f_2): B \to \mathbf{R}^2$ is such that $\mathbf{F}$ corresponds to ω (see Section 14.3), then $d\omega = (\partial f_2/\partial x_1 - \partial f_1/\partial x_2)\, dx_1\, dx_2$, and (3.2) becomes

$$\int_\Gamma \mathbf{F} = \int_B \left(\frac{\partial f_2}{\partial x_1} - \frac{\partial f_1}{\partial x_2}\right) \tag{3.3}$$

which is Green's theorem.

Suppose now that $\beta: I^2 \to \mathbf{R}^3$ is a simple parametrized surface, which we think of as a singular 2 cube. Let

$$\omega = f_1\, dx_1 + f_2\, dx_2 + f_3\, dx_3$$

be a 1 form defined on an open set containing the image of β. Then

$$\int_{\partial\beta} \omega = \int_\beta d\omega$$

by Theorem 2.3. As we saw earlier, $\partial\beta$ can be considered as defining an oriented curve Γ, and by definition, we have

$$\int_{\partial\beta} \omega = \int_\Gamma \mathbf{F} \tag{3.4}$$

where $\mathbf{F}(v) = (f_1(v), f_2(v), f_3(v))$. Furthermore,

$$\int_\beta d\omega = \int_{I^2} \beta^* d\omega$$

and

$$\beta^* d\omega = \beta^*\left[\left(\frac{\partial f_2}{\partial x_1} - \frac{\partial f_1}{\partial x_2}\right)dx_1 dx_2 + \left(\frac{\partial f_3}{\partial x_1} - \frac{\partial f_1}{\partial x_3}\right)dx_1 dx_3 + \left(\frac{\partial f_3}{\partial x_2} - \frac{\partial f_2}{\partial x_3}\right)dx_2 dx_3\right]$$

$$= \sum_{1\le i<j\le 3}\left(\frac{\partial f_j}{\partial x_i}\circ\beta - \frac{\partial f_i}{\partial x_j}\circ\beta\right)\left(\frac{\partial \beta_i}{\partial x}\frac{\partial \beta_j}{\partial y} - \frac{\partial \beta_i}{\partial y}\frac{\partial \beta_j}{\partial x}\right)dx\,dy$$

$$= \langle(\text{curl }\mathbf{F})\circ\beta, (D\beta)\mathbf{e}_1 \times (D\beta)\mathbf{e}_2\rangle\, dx\,dy$$

(See also Section 14.3, Exercise 8.) Here (x, y) are coordinates in I^2. Thus

(3.5) $$\int_\beta d\omega = \int_{I^2} \langle(\text{curl }\mathbf{F})\beta, (D\beta)\mathbf{e}_1 \times (D\beta)\mathbf{e}_2\rangle$$

This is just the surface integral of the vector field curl $\mathbf{F}$ over the parametrized surface β.

Combining equations (3.4) and (3.5), we have

(3.6) $$\int_\Gamma \mathbf{F} = \int_\beta \langle(\text{curl }\mathbf{F})\circ\beta, (D\beta)\mathbf{e}_1 \times (D\beta)\mathbf{e}_2\rangle$$

where Γ is the boundary curve of the parametrized surface β and $\mathbf{F}: U \to \mathbf{R}^3$ is a differentiable function defined on the open set U containing the image of β. This is Stokes' theorem (Theorem 5.1, Chapter 13).

4. The Gauss Theorem and Incompressible Fluids

The Gauss theorem is a special case of the general Stokes' theorem when $n = p = 3$. It relates the surface integral of a vector field $\mathbf{F}$ over a closed surface with the ordinary integral of the divergence of $\mathbf{F}$ over the region enclosed by the surface.

Suppose that $T: I^3 \to \mathbf{R}^3$ is a singular 3 cube which is also an orientation-preserving diffeomorphism onto its image $D \subset \mathbf{R}^3$. Let ω be a 2 form on D,

$$\omega = f_1\, dx_2\, dx_3 - f_2\, dx_1\, dx_3 + f_3\, dx_1\, dx_2$$

(Note the minus sign.) Theorem 2.3 states that

(4.1) $$\int_{\partial T} \omega = \int_T d\omega$$

Set $\mathbf{F} = (f_1, f_2, f_3): D \to \mathbf{R}^3$. Then, using equation (3.1) and the discussion at the end of Section 14.3, we have

$$\int_T d\omega = \int_D \operatorname{div} \mathbf{F}$$

By definition, the left side of (4.1) is

$$\sum_{i=1}^{3} \int_{T\partial_i} \omega - \sum_{i=1}^{3} \int_{T\partial_i'} \omega$$

The functions $T\partial_i, T\partial_i'$: $I^2 \to \mathbf{R}^3$ can be considered as defining oriented surfaces $\sum_i$, $\sum_i'$, $1 \leq i \leq 3$, and

$$\int_{T\partial_i} \omega = \int_{\Sigma_i} \mathbf{F} \qquad \int_{T\partial_i'} \omega = \int_{\Sigma_i'} \mathbf{F}$$

In fact, these surfaces fit together to make up the boundary surface Σ of the region D. (See Figure 4.1.) Furthermore, the signs are arranged so that the orientations of the surfaces $\sum_i$ and $\sum_i'$ define an orientation of Σ. Thus, it makes sense to write the left-hand side of equation (3.5) as $\int_\Sigma \mathbf{F}$.

We state the consequence of this discussion as a theorem.

Theorem 4.1 (The Gauss Theorem) *Let D be a region in $\mathbf{R}^3$ with boundary surface Σ oriented so that the normal to Σ points into D. Let $\mathbf{F} : D \to \mathbf{R}^3$ be a differentiable vector field. Then*

$$\int_\Sigma \mathbf{F} = \int_D \operatorname{div} \mathbf{F}$$

For an application of this result, we return to the study of

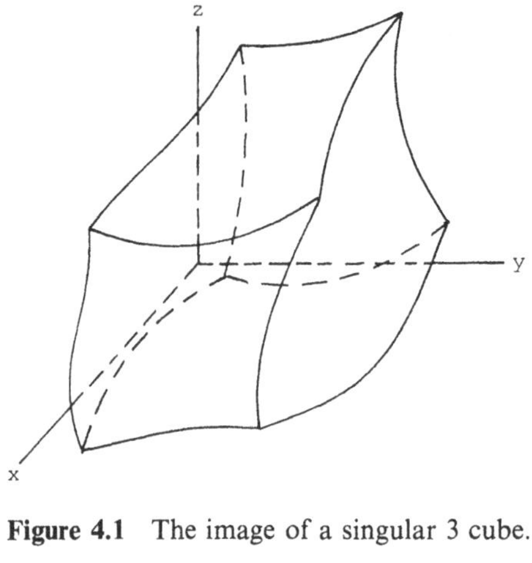

Figure 4.1 The image of a singular 3 cube.

hydrodynamics. (See Section 13.5.) We suppose that $U \subset \mathbf{R}^3$ is an open set in which a fluid is in motion and $\mathbf{F}: U \to \mathbf{R}^3$ is the corresponding velocity field.

Let Σ be the surface of a closed ball $B \subset U$. Then $\int_\Sigma \mathbf{F}$ represents the net flow of fluid through the surface Σ. We say that the fluid is *incompressible* if this integral is zero for any ball $B \subset U$. As a consequence of Theorem 4.1, we have

$$\int_B \operatorname{div} \mathbf{F} = 0$$

for any ball $B \subset U$. It follows that div $\mathbf{F} = 0$ on U. Thus the fluid is incompressible if and only if div $\mathbf{F} = 0$.

Suppose further that U is an open ball and that $\mathbf{F}$ is irrotational. (See Section 13.5.) Then $\mathbf{F} = \operatorname{grad} \varphi$ for some C^2 function $\varphi: U \to \mathbf{R}$. (See Exercise 4, Section 13.5.) If the fluid is also incompressible, we have

$$\operatorname{div} \mathbf{F} = \operatorname{div} \operatorname{grad} \varphi = \frac{\partial^2 \varphi}{\partial x^2} + \frac{\partial^2 \varphi}{\partial y^2} + \frac{\partial^2 \varphi}{\partial z^2} = 0$$

This equation is called the *Laplace equation* and any function satisfying it is said to be *harmonic.*

Thus we see that the velocity field of an irrotational, incompressible fluid in a simply connected subset of $\mathbf{R}^3$ has a harmonic potential function.

The Gauss theorem can also be applied to electromagnetism. Let $\mathbf{E}$ and $\mathbf{H}$ be vector fields on the open set $U \subset \mathbf{R}^3$, $\mathbf{E}$ the *electric field* and $\mathbf{H}$ the *magnetic field.* We assume that these vector fields are (differentiable) functions of time t. These two vector fields satisfy the *Maxwell equations*:

$$\operatorname{div} \mathbf{E} = 4\pi\rho \qquad \operatorname{div} \mathbf{H} = 0$$

$$\operatorname{curl} \mathbf{E} = -\frac{1}{c}\frac{\partial \mathbf{H}}{\partial t} \qquad \operatorname{curl} \mathbf{H} = \frac{1}{c}\frac{\partial \mathbf{E}}{\partial t}$$

where c is a universal constant and ρ is the *charge density.* If $\mathbf{H} = \mathbf{0}$ (this is the electrostatic case), then $\mathbf{E}$ is independent of time and

$$\operatorname{curl} \mathbf{E} = \mathbf{0}$$

Thus, if U is simply connected,

$$\mathbf{E} = -\operatorname{grad} \varphi$$

for some function $\varphi: U \to \mathbf{R}$. We then have (from the first Maxwell equation)

$$\operatorname{div} \operatorname{grad} \varphi = -4\pi\rho$$

In particular, φ is harmonic if U is free of charge (that is, if $\rho = 0$).

Suppose now that D is a region in U with boundary surface Σ. The total charge contained in D is the integral

$$\int_D \rho = \frac{1}{4\pi}\int_D \operatorname{div} \mathbf{E}$$

(again using the first Maxwell equation). According to the Gauss theorem, this last integral is equal to

$$\frac{1}{4\pi}\int_\Sigma \mathbf{E}$$

This quantity is called the *flux across the surface* Σ. Thus we have proved that the *total charge contained in a region D is equal to the flux across the boundary of D*.

It is possible to apply Stokes' theorem to prove an analogue of Theorem 4.1 for regions D which are not the image of a singular 2 cube. To deal with this situation, we break D up into subregions, each of which is the image of a singular 3 cube. For example, if D is the *solid torus* (or doughnut) pictured in Figure 4.2, then the subregions D_1 and D_2 are images of singular 3 cubes $T_1, T_2 : I^3 \to \mathbf{R}^3$. Let $\sum_1$ and $\sum_2$ be the boundary surfaces of D_1 and D_2. Then, if $\mathbf{F} : D \to \mathbf{R}^3$ is a differentiable function, we can apply Theorem 4.1 to obtain

$$\int_{\Sigma_1} \mathbf{F} = \int_{D_1} \operatorname{div} \mathbf{F} \qquad \int_{\Sigma_2} \mathbf{F} = \int_{D_2} \operatorname{div} \mathbf{F}$$

Adding these two equations, we have

$$\int_{\Sigma_1} \mathbf{F} + \int_{\Sigma_2} \mathbf{F} = \int_{D_1} \operatorname{div} \mathbf{F} + \int_{D_2} \operatorname{div} \mathbf{F} \tag{4.2}$$

Now the right side of equation (4.2) is just the integral of div $\mathbf{F}$ over the region D. To determine the left side of equation (4.2), we note the surfaces $\sum_1$ and $\sum_2$ have the two pieces $\sum_0$ and $\sum_0'$ in common. (See Figure 4.2.) Furthermore, $\sum_0$ and $\sum_0'$ inherit orientations from both $\sum_1$ and $\sum_2$ and the orientation inherited from $\sum_1$ is the negative of that inherited from $\sum_2$. It follows that the

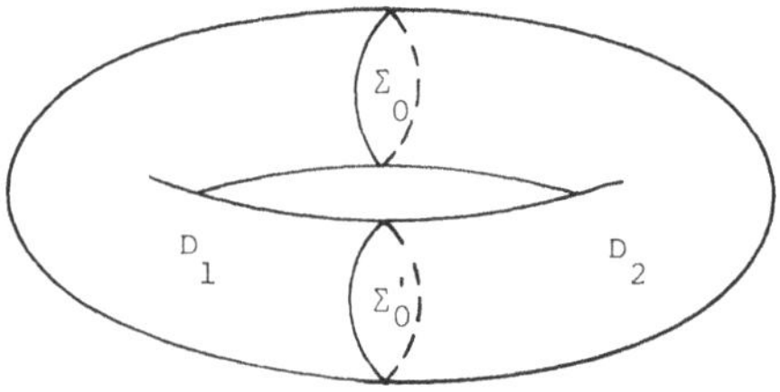

Figure 4.2 A solid torus.

integrals of **F** over $\sum_0$ and $\sum_0'$ cancel so that the right side of equation (4.2) can be identified with the integral of **F** over Σ. Thus we have

$$\int_\Sigma \mathbf{F} = \int_D \operatorname{div} \mathbf{F}$$

in this case also.

Any region U in $\mathbf{R}^3$ with a smooth boundary can be divided into subregions such that the new boundary surfaces created by the division cancel out (as $\sum_0$ and $\sum_0'$ did in the example above). Therefore the divergence theorem applies to such regions. Proving that such a division exists is not easy, however. We would need to prove that examples like the Möbius band (see Figure 3.4 of Chapter 13) cannot happen when we deal with open sets in $\mathbf{R}^3$ rather than surfaces. Such matters are studied in topology, and we shall not attempt any further discussion here.

Exercises

1. Let S be the surface of the sphere $x^2 + y^2 + z^2 = 9$. Evaluate

$$\int_S x^2\, dy\, dz + y^2\, dz\, dx + z^2\, dx\, dy$$

by using Gauss' theorem.

2. Let S be the surface of the sphere $(x-2)^2 + (y-3)^2 + (z-1)^2 = 25$. Evaluate

$$\int_S x^2\, dy\, dz + y^2\, dz\, dx + z^2\, dx\, dy$$

3. Let U be a compact region in $\mathbf{R}^3$, and suppose that Gauss' theorem can be applied to U and ∂U. Show that the volume of U is given by

$$\operatorname{Vol}(U) = \int_{\partial U} x\, dy\, dz = \int_{\partial U} y\, dz\, dx = \int_{\partial U} z\, dx\, dy$$

4. Let U be as in the previous problem. Show that the center of mass of U (see Exercise 10 of Section 11.3) is given by

$$\bar{x} = \frac{1}{2\operatorname{Vol}(U)} \int_{\partial U} x^2\, dy\, dz$$

$$\bar{y} = \frac{1}{\operatorname{Vol}(U)} \int_{\partial U} xy\, dy\, dz$$

$$\bar{z} = \frac{1}{\operatorname{Vol}(U)} \int_{\partial U} xz\, dy\, dz$$

(Note: The density of U is assumed to be 1 everywhere.)

*5. Prove *Green's identities*: If f and $g : \mathbf{R}^3 \to \mathbf{R}$ have two continuous derivatives in the region U (where U is as above), then

$$\int_U (f\,\Delta g + \langle \nabla f, \nabla g \rangle) = \int_{\partial U} f \nabla g$$

and

$$\int_U (f\,\Delta g - g\,\Delta f) = \int_{\partial U} (f \nabla g - g\,\nabla f)$$

Here

$$\Delta f = \frac{\partial^2 f}{\partial x^2} + \frac{\partial^2 f}{\partial y^2} + \frac{\partial^2 f}{\partial z^2}$$

is the Laplacian of f and $\nabla f = \text{grad } f$. (*Hint*: Let $F = f\nabla g$ in Gauss' theorem.)

6. (a) Let U be any compact region of $\mathbf{R}^3$ which does not include the origin, and let $\mathbf{F}(\mathbf{v}) = \mathbf{v}/\|\mathbf{v}\|^3$. Show that

$$\int_{\partial U} \mathbf{F} = 0$$

(b) Let V be a ball in $\mathbf{R}^3$ centered at the origin, and let $\mathbf{F}$ be as in (a). Show that

$$\int_{\partial V} \mathbf{F} = 4\pi$$

*7. (The heat equation) Suppose that a body occupies a region U of space, and let $T(x, y, z, t)$ be the temperature at the point $(x, y, z) \in U$ at time t. The *specific heat c* measures the amount of heat per unit mass the body can absorb: if V is a subset of U and ρ is the density of the body, then

$$\int_V c\rho \frac{\partial T}{dt}$$

is the amount of heat absorbed by V. (Both c and ρ may be functions of x, y, and z.)

Now fix a time t_0, and regard $T(x, y, z, t_0)$ as a function of x, y, and z only. The usual law of heat conduction says that the flow of heat through a surface S is given by

$$\int_S (-k\,\nabla T)$$

where k is a constant.

(a) Let V be a subset of U such that Gauss' theorem holds for V and ∂V. Show that

$$\int_V c\rho \frac{\partial T}{dt} - k \operatorname{div}(\nabla T) = 0$$

(Note: Heat entering V is flowing in the direction opposite to the outward normal.)

(b) Conclude (assuming that everything is continuous) that T satisfies the *heat equation*:

$$c\rho \frac{\partial T}{dt} = k\left(\frac{\partial^2 T}{\partial x^2} + \frac{\partial^2 T}{\partial y^2} + \frac{\partial^2 T}{\partial z^2}\right)$$

5. Proof of the General Stokes' Theorem

We now prove Theorem 2.3. The proof is divided into three cases, the first one being the crucial one.

Case 1 Suppose first of all that $n = p$, so that

$$\omega = \sum_{i=1}^{p} f_i\, dx_1 \cdots dx_{i-1}\, dx_{i+1} \cdots dx_p$$

Suppose further that c is the singular p chain represented by $I^p \subset \mathbf{R}^p$. [More precisely, $c = T\colon I^p \to \mathbf{R}^p$, where T is the *inclusion* mapping, $T(\mathbf{v}) = \mathbf{v}$.] Then

$$\begin{aligned}\int_c d\omega &= \int_c \sum_{i=1}^{p} (df_i)\, dx_1 \cdots dx_{i-1}\, dx_{i+1} \cdots dx_p \\ &= \sum_{i=1}^{p} (-1)^{i-1} \int_c \frac{\partial f_i}{\partial x_i} dx_1 \cdots dx_p \\ &= \sum_{i=1}^{p} (-1)^{i-1} \int_{I^p} \frac{\partial f_i}{\partial x_i}\end{aligned}$$

Using the Fubini theorem (Theorem 1.2, Chapter 11), we have

$$\int_{I^p} \frac{\partial f_i}{\partial x_i} = \int_{I^{p-1}} \left(\int_0^1 \frac{\partial f_i}{\partial x_i} dx_i\right) = \int_{I^{p-1}} (f_i \partial_i' - f_i \partial_i)$$

Now
$$(\partial_i)^* dx_j = (\partial_i')^* dx_j = \begin{cases} dx_j & \text{if } j < i \\ 0 & \text{if } j = i \\ dx_{j-1} & \text{if } j > i \end{cases}$$

(See Exercise 7, Section 2.) Thus

$$\begin{aligned}(5.1)\qquad \int_{\partial_i'} \omega &= \int_{I^{p-1}} (\partial_i')^* \omega \\ &= \int_{I^{p-1}} \sum_{j=1}^{n} (f_j \circ \partial_i'^* (dx_1 \cdots dx_{j-1}\, dx_{j+1} \cdots dx_p) \\ &= \int_{I^{p-1}} f_i \circ \partial_i'\end{aligned}$$

[If dx_i appears, then $\partial_i'^*(dx_1 \cdots dx_{j-1}\,dx_{j+1} \cdots dx_p) = 0$.] Similarly,

$$\int_{\partial_i} \omega = \int_{I^{p-1}} f_i \partial_i \tag{5.2}$$

Thus, from (5.1) and (5.2),

$$\int_c d\omega = \sum_{i=1}^{p} (-1)^{i-1} \int_{\partial_i' - \partial_i} \omega$$
$$= \int_{\Sigma(-1)^i(\partial_i - \partial_i')} \omega = \int_{\partial c} \omega$$

using Proposition 2.1.

The remaining cases are now direct consequences of case 1 and the definitions, though the computation may seem confusing.

Case 2 Suppose now that $c = T\colon I^p \to \mathbf{R}^n$ is an arbitrary singular p cube. Then

$$\int_T d\omega = \int_{I^p} T^*(d\omega) = \int_{I^p} d(T^*\omega)$$

by Proposition 6.2 of Chapter 14. By case 1 above, we have

$$\int_{I^p} d(T^*\omega) = \int_{\partial I^p} T^*\omega = \sum_{i=1}^{p} (-1)^i \int_{I^{p-1}} [\partial_i^*(T^*\omega) - \partial_i'^*(T^*\omega)]$$
$$= \sum_{i=1}^{p} (-1)^i \int_{I^{p-1}} [(T\partial_i)^*\omega - (T\partial_i')^*\omega]$$

(from Proposition 6.2, Chapter 14)

$$= \int_{\partial T} \omega$$

Case 3 If c is any singular p chain, then c is a linear combination of singular p cubes. Stokes' theorem in this case follows from case 2 (above) and Proposition 2.1.

16

Infinite Series

This chapter is devoted to the study of infinite series. We begin by defining what it means for an infinite series to converge. We then give various tests for determining convergence, first for series of nonnegative terms and afterwards for more general series.

1. Infinite Series

Suppose $a_1, a_2, \ldots, a_n, \ldots$ is an infinite sequence of real or complex numbers. Our first problem is to make sense of the sum of the *infinite series*

$$\sum_{j=1}^{\infty} a_j = a_1 + a_2 + \cdots + a_n + \cdots \tag{1.1}$$

To do this, we define the sequence of *partial sums*, $S_1, S_2, \ldots, S_n, \ldots$ by

$$S_k = a_1 + a_2 + \cdots + a_k = \sum_{j=1}^{k} a_j$$

If the sum (1.1) is to make sense, this sequence should clearly converge to the sum. Accordingly, we say that the infinite series (1.1) *converges with sum S* if the sequence of partial sums converges to S; that is, if

$$\lim_{k \to \infty} S_k = S$$

If the series (1.1) does not converge, we say that it *diverges*.

Of course, an infinite series need not begin with the term a_1; it could equally well begin with a_2, a_7, a_{100}, or any other term. We shall generally begin series with a_0 or a_1, however.

Here are some examples.

1. The series

$$\sum_{j=1}^{\infty} \frac{1}{2^j} = \frac{1}{2} + \frac{1}{4} + \frac{1}{8} + \cdots$$

converges to 1. The sequence of partial sums is

$$\left\{\frac{1}{2}, \frac{3}{4}, \frac{7}{8}, \ldots, 1 - \frac{1}{2^n}, \ldots\right\}$$

which clearly converges to 1.

2. The series

$$\sum_{j=1}^{\infty} \frac{1}{j(j+1)} = \frac{1}{2} + \frac{1}{6} + \frac{1}{12} + \frac{1}{20} + \cdots$$

also converges to 1. Notice that $1/j(j+1) = 1/j - 1/(j+1)$; thus the series is

$$\left(1 - \frac{1}{2}\right) + \left(\frac{1}{2} - \frac{1}{3}\right) + \left(\frac{1}{3} - \frac{1}{4}\right) + \cdots$$

and the sequence of partial sums is $\{1 - \frac{1}{2}, 1 - \frac{1}{3}, \ldots, 1 - 1/n, \ldots\}$, which has limit 1.

3. The series

$$\sum_{j=1}^{\infty} 1 = 1 + 1 + 1 + \cdots$$

diverges. The sequence of partial sums is $\{1, 2, 3, 4, \ldots\}$ which clearly has no limit.

4. The series

$$\sum_{j=0}^{\infty} (-1)^j = 1 - 1 + 1 - 1 + 1 - 1 + \cdots$$

also diverges. The sequence of partial sums is $\{1, 0, 1, 0, 1, 0, \ldots\}$ which does not have a limit.

5. The series

$$\sum_{j=1}^{\infty} \frac{1}{j}$$

(the *harmonic series*) also diverges. To prove this, notice that

$$1 > \frac{1}{2} \qquad \frac{1}{2} \geq \frac{1}{2}$$

$$\frac{1}{3} + \frac{1}{4} > \frac{1}{4} + \frac{1}{4} = \frac{1}{2}$$

$$\frac{1}{5} + \frac{1}{6} + \frac{1}{7} + \frac{1}{8} > \frac{1}{8} + \frac{1}{8} + \frac{1}{8} + \frac{1}{8} = \frac{1}{2}$$

and, more generally,

$$\frac{1}{2^n + 1} + \frac{1}{2^n + 2} + \cdots + \frac{1}{2^{n+1}} > \frac{1}{2^{n+1}} + \cdots + \frac{1}{2^{n+1}} \,(2^n \text{ terms}) = \frac{1}{2}$$

Thus if $S_k = \sum_{j=1}^{k} 1/j$ is the jth partial sum, then

$$\begin{aligned} S_{2^n} &= 1 + \frac{1}{2} + \left(\frac{1}{3} + \frac{1}{4}\right) + \cdots + \left(\frac{1}{2^{n-1} + 1} + \cdots + \frac{1}{2^n}\right) \\ &> \frac{1}{2} + \frac{1}{2} + \frac{1}{2} + \cdots + \frac{1}{2} \qquad (n + 1 \text{ terms}) \\ &= \frac{n+1}{2} \end{aligned}$$

It follows that the terms of the sequence $\{S_k\}$ are not bounded. According to Proposition 6.5 of Chapter 3, the sequence $\{S_k\}$ cannot converge, and the proof is complete.

The next proposition gives some basic properties of infinite series.

Proposition 1.1 *Let $\sum_{j=1}^{\infty} a_j$ and $\sum_{j=1}^{\infty} b_j$ be infinite series, and let c be a (real or complex) number. Then*

(a) *If $\sum_{j=1}^{\infty} a_j = s$ and $\sum_{j=1}^{\infty} b_j = t$, then $\sum_{j=1}^{\infty} (a_j + b_j) = s + t$.*
(b) *If $\sum_{j=1}^{\infty} a_j = s$ and $\sum_{j=1}^{\infty} b_j$ diverges, then $\sum_{j=1}^{\infty} (a_j + b_j)$ diverges.*
(c) *If $\sum_{j=1}^{\infty} a_j = s$, then $\sum_{j=1}^{\infty} ca_j = cs$.*
(d) *If $\sum_{j=1}^{\infty} a_j$ diverges and $c \neq 0$, then $\sum_{j=1}^{\infty} ca_j$ diverges.*

Proof. Once we use the definition of infinite series to see what we have to prove, then (a) becomes obvious. If $\{s_k\}$ and $\{t_k\}$ are the sequences of partial sums for $\sum_{j=1}^{\infty} a_j$ and $\sum_{j=1}^{\infty} b_j$, respectively, then $\{s_k + t_k\}$ is the sequence of partial sums for $\sum_{j=1}^{\infty} (a_j + b_j)$. From Proposition 6.2 of Chapter 3, we see that $\lim_{k \to \infty} (s_k + t_k) = s + t$, and (a) follows. The proof of (c) is similar, and is left as Exercise 7.

Parts (b) and (d) are also easy; we prove (d) here, leaving (b) as Exercise 8. If $\sum_{j=1}^{\infty} ca_j$ converges, then $\sum_{j=1}^{\infty} (1/c)(ca_j)$ would converge by part (c). This contradicts the assumption that $\sum_{j=1}^{\infty} a_j$ diverges.

Results about infinite series are generally attempts to answer one of two questions:

(a) Given a specific series, does it converge or not?
(b) How can one find an infinite series of functions which converges to a given function?

The answer to question (b) can have practical applications; for instance, it can provide a method of computing the values of familiar functions like sines and exponentials. Still, the fundamental question is (a), since the basic fact we need to know about an infinite series is whether it converges in the first place. We shall begin to provide answers in the next section.

We conclude this section with two simple results.

Proposition 1.2 *The convergence or divergence of the series $\sum_{j=1}^{\infty} a_j$ is not affected by adding or removing a finite number of terms.*

Note The proposition does not say that the sum does not change if finitely many terms are changed, but merely that such a change cannot transform a convergent series into a divergent one or vice versa. Thus (to go back to the earlier examples) the series $300 + \frac{1}{2} + \frac{1}{4} + \frac{1}{8} + \cdots$ (the nth term is $1/2^{n-1}$, except when $n = 1$) converges, while the series

$$\sum_{j=1,000,000}^{\infty} \frac{1}{j}$$

diverges.

Proof of Proposition 1.2. It suffices to prove the result in the special case where only the first term is added or removed, since mathematical induction can be used to prove the result generally. We prove it in the case where the first term is removed, leaving the other case (adding a term) as an exercise.

Notice first that $b_1 + b_2 + b_3 + \cdots$ converges if and only if $0 + b_1 + b_2 + b_3 + \cdots$ does also (the sequences of partial sums are the same, except that the second sequence starts with an extra 0). Now suppose that $\sum_{j=1}^{\infty} a_j$ converges. The series $-a_1 + 0 + 0 + 0 + \cdots$ also converges, and thus, by Proposition 1.1,

$$0 + a_2 + a_3 + \cdots + a_n + \cdots$$

converges. Therefore $\sum_{j=2}^{\infty} a_j$ converges, as claimed. The proof when $\sum_{j=1}^{\infty} a_j$ diverges is nearly identical.

Proposition 1.3 *If $\sum_{j=1}^{\infty} a_j$ converges, then $\lim_{j\to\infty} a_j = 0$.*

Proof. Let s_n be the nth partial sum of the series, and let $s_0 = 0$. Then $s_n - s_{n-1} = a_n$. By hypothesis, $\{s_n\}$ converges to some limit s. Then

$\lim_{n\to\infty} s_{n-1} = s$ as well, and therefore

$$0 = s - s = \lim_{n\to\infty} (s_n - s_{n-1}) = \lim_{n\to\infty} a_n$$

Remark. The converse of this result is false: there are divergent series whose terms tend to 0. Example 5 (the harmonic series) is such a series. This proposition can be used to prove that series diverge (if the terms do not go to 0, the series must diverge), but *not* to prove that a series converges. The error of assuming that the converse of Proposition 1.3 is true is very common, and most mathematicians are sick of seeing it. Try not to distress them further.

Exercises

1. A series of the form

$$(b_1 - b_2) + (b_2 - b_3) + \cdots + (b_n - b_{n+1}) + \cdots = \sum_{j=1}^{\infty} (b_j - b_{j+1})$$

is called a *telescoping series*. Show that the series converges exactly when the sequence $\{b_n\}$ converges, and that its sum is $b_1 - \lim_{n\to\infty} b_n$.

2. Suppose that $\sum_{j=1}^{\infty} \mathbf{v}_j$ is an infinite series of vectors in $\mathbf{R}^n$. We say that $\sum_{j=1}^{\infty} \mathbf{v}_j = \mathbf{v}$ if the sequence of partial sums converges to $\mathbf{v}$.

   Write $\mathbf{v}_j = (a_{ij}, a_{2j}, \ldots, a_{nj})$. Show that $\sum_{j=1}^{\infty} \mathbf{v}_j$ converges if and only if $\sum_{j=1}^{\infty} a_{ij}$ converges for $i = 1, 2, \ldots, n$.

3. A *geometric* series is one of the form $\sum_{j=0}^{\infty} ar^j (a \neq 0)$. Show that
   (a) $\sum_{j=0}^{n} ar^j = a(1 - r^{n+1})/(1 - r)$ (if $r \neq 1$).
   (b) The series converges if $|r| < 1$ and diverges otherwise.

4. Show that $\sum_{n=1}^{\infty} 1/(2n - 1) = 1 + \frac{1}{3} + \frac{1}{5} + \cdots$ diverges.

5. Let $\{s_n\}$ $(n = 1, 2, \ldots)$ be an arbitrary sequence of real numbers. Find a series $\sum_{j=1}^{\infty} a_j$ whose sequence of partial sums is $\{s_n\}$. (Thus any theorem about series can be converted to one about sequences, and vice versa.)

*6. Prove that $\sum_{n=2}^{\infty} 1/(n \ln n)$ diverges. (*Hint*: Use a proof like the one in Example 5.)

7. Prove (c) of Proposition 1.1.

8. Prove (b) of Proposition 1.1.

9. Give a detailed proof of Proposition 1.2.

*10. Prove that $\sum_{n=1}^{\infty} (-1)^{n-1}/n$ converges. (*Hint*: Show that the odd partial sums decrease to a limit and that the even partial sums increase to a limit. Then show that the two limits are the same.)

*11. Show that $e = \sum_{n=0}^{\infty} 1/n!$ is irrational. (*Hint*: If $e = p/q$, where p/q is a fraction in lowest terms, then $q!e$ is an integer. On the other hand, $q!e =$ an integer $+ \sum_{n=q+1}^{\infty} q!/n!$. Deduce a contradiction.)

12. We defined improper integrals of the form $\int_a^{\infty} f(x)\,dx$ in Exercise 15 of Section 10.4. This exercise (and some future ones) will deal further with these improper integrals. Here, we shall assume that the lower limit of integration is 0; this is purely arbitrary (and any other lower limit would serve as well).
   (a) Show that if $\int_0^{\infty} f$ and $\int_0^{\infty} g$ exist, then $\int_0^{\infty} (f + g)$ exists, with

$$\int_0^\infty (f+g) = \int_0^\infty f + \int_0^\infty g$$

(b) Show that if $\int_0^\infty f$ exists and $a \in \mathbf{R}$, then $\int_0^\infty af$ exists, with

$$\int_0^\infty af = a \int_0^\infty f$$

(c) Prove the analogues of (b) and (d) of Proposition 1.1 for improper integrals.

*(d) Show that it is possible for $\int_0^\infty f$ to exist even if $f(x)$ does not approach 0 as x approaches infinity. In fact, f need not be bounded. (Thus the analogue of Proposition 1.3 for integrals is *false*.)

2. Series of Positive Terms

In this section and the next, we determine some tests for telling whether a given infinite series converges or not. The most useful tests apply to those series where every term is positive (or at least nonnegative).

The first result that we need is a translation of an elementary result on increasing sequences of real numbers. (See Exercise 3, Section 7.3.)

Lemma 2.1 *Suppose $a_j \geq 0$ for all j. Then the series $\sum_{j=1}^\infty a_j$ converges if and only if the sequence of partial sums is bounded. If the sequence of partial sums is bounded, then the sum of the series is the least upper bound of this sequence.*

The proof involves observing that the sequence of partial sums is increasing and bounded and using the exercise mentioned above. We leave the details as Exercise 12.

This lemma has an extremely useful consequence.

Theorem 2.2 (The Comparison Test) *Let $\sum_{j=1}^\infty a_j$ and $\sum_{j=1}^\infty b_j$ be two infinite series with $b_j \geq a_j \geq 0$ for all j. Then*

(a) *The series $\sum_{j=1}^\infty a_j$ converges if $\sum_{j=1}^\infty b_j$ converges (moreover, $\sum_{j=1}^\infty a_j \leq \sum_{j=1}^\infty b_j$).*

(b) *The series $\sum_{j=1}^\infty b_j$ diverges if $\sum_{j=1}^\infty a_j$ diverges.*

Proof. (a) Let $\{s_n\}$ and $\{t_n\}$ be the partial sums of $\sum_{j=1}^\infty a_j$ and $\sum_{j=1}^\infty b_j$, respectively. By hypothesis, $t_n \geq s_n \geq 0$ for all n. If $\sum_{j=1}^\infty b_j = t$, then $\lim_{n\to\infty} t_n = t$, and $t \geq t_n$ for all n (since $\{t_n\}$ is nondecreasing). Therefore $t \geq s_n$ for all n; so $\{s_n\}$ is bounded, and Lemma 1.1 says that $\sum_{j=1}^\infty a_n$ converges to a number $\leq t$.

(b) This is simply the contrapositive of (a).

Corollary *The conclusion holds even if the hypothesis of (a) fails finitely often—that is, even if $b_j < a_j$ for finitely many j.*

Proof. We can, by Proposition 1.2, delete the terms for which the hypothesis fails from both series without affecting convergence.

Note Many of the theorems which follow have corollaries like the one given above. We shall usually refrain from stating these corollaries each time, though we shall use them freely.

Theorem 2.2 provides a useful method of finding out whether a series converges or diverges: find a convergent series of larger terms or a divergent series of smaller ones. All we need is a good list of series whose convergence or divergence is known. The simplest examples are geometric series (see Section 1, Exercise 3). A *geometric series* is one of the form

$$\sum_{j=0}^{\infty} ar^j \tag{2.1}$$

Since we are dealing with series of positive terms, we assume that $a > 0$ and $r \geq 0$. As the exercise states, the geometric series diverges if $r > 1$ and converges to $a/(1 - r)$ if $r < 1$.

There is a standard way of applying the comparison test with geometric series.

Theorem 2.3 (The Ratio Test) *Let $\sum_{j=1}^{\infty} a_j$ be a series of positive terms. Consider the ratios a_{n+1}/a_n of successive terms.*

(a) *If there is a number $r < 1$ such that $(a_{n+1}/a_n) \leq r$ for all n, then the series converges.*
(b) *If $(a_{n+1}/a_n) \geq 1$ for all n, then the series diverges.*
(c) *The conclusions of (a) and (b) hold even if the hypotheses [about (a_{n+1}/a_n)] do not hold for a finite number of values of n.*
(d) *Suppose that $\lim_{n\to\infty} (a_{n+1}/a_n) = r$. If $r < 1$, the series converges: if $r > 1$, it diverges.*

Before giving the proof, we note some facts about the theorem.

1. If $\lim_{n\to\infty} (a_{n+1}/a_n) = 1$, the theorem says nothing about convergence. The reason is that there are examples of series $\sum_{j=1}^{\infty} a_j$ where this limit is 1 and the series converges ($\sum_{j=1}^{\infty} 1/j^2$ is one, as we shall see in Section 3), and other examples (like $\sum_{j=1}^{\infty} 1/j$) where it diverges. Thus the test fails in this case.
2. Of course, $\lim_{n\to\infty} (a_{n+1}/a_n)$ need not exist at all. In such a case, we may be

able to use (a) or (b) of the theorem, but usually the ratio test simply fails then. In practice, one uses (d) rather than (a) or (b).

Proof of Theorem 2.3. (a) If $a_{n+1}/a_n \le r$ for all n, then

$$a_2 \le a_1 r \qquad a_3 \le a_2 r \le a_1 r^2$$

and, in general,

$$a_n \le a_1 r^{n-1}$$

(The proof of this last statement is an easy exercise in mathematical induction; see Exercise 13.) Since $\sum_{n=1}^{\infty} a_1 r^{n-1}$ converges (because $r < 1$), Theorem 2.1 implies that the given series converges.

(b) If $(a_{n+1}/a_n) \ge 1$, then any easy induction [like that in part (a)] shows that $a_n \ge a_1$ for all n. Therefore the terms a_n do not tend to 0, and the series diverges.

(c) We may remove finitely many terms from the series without affecting convergence or divergence. We simply remove enough so that the hypothesis of (a) or (b) holds for the remaining terms.

(d) Suppose first that $r < 1$. Then $(1 - r)/2 > 0$, and we can find an N such that

$$\left|\frac{a_{n+1}}{a_n} - r\right| < \frac{1 - r}{2}$$

when $n > N$. Then

$$\begin{aligned}\frac{a_{n+1}}{a_n} &= \left|\frac{a_{n+1}}{a_n} - r + r\right| \\ &< \left|\frac{a_{n+1}}{a_n} - r\right| + r < \frac{1 + r}{2} < 1\end{aligned}$$

when $n > N$. Thus the series converges by (a) and (c).

If $r > 1$, we can find N such that

$$\left|\frac{a_{n+1}}{a_n} - r\right| < r - 1$$

when $n > N$. In this case,

$$\begin{aligned}\frac{a_{n+1}}{a_n} &= \left|r + \left(\frac{a_{n+1}}{a_n} - r\right)\right| \\ &\ge r - \left|\frac{a_{n+1}}{a_n} - r\right| > 1\end{aligned}$$

when $n > N$. Now (b) and (c) imply that the series diverges.

This theorem is more useful for proving convergence of series than divergence; if $r > 1$, then $\lim_{n\to\infty} a_n \neq 0$, and it is often easy to tell by inspection that the series diverges. Here are some examples of applications of the theorem.

1. The series $\sum_{n=0}^{\infty} 1/n!$ converges. (*Note*: $0! = 1$.) Here, $a_n = 1/n!$ and

$$\frac{a_{n+1}}{a_n} = \frac{n!}{(n+1)!} = \frac{1}{n+1}$$

   Thus $\lim_{n\to\infty} (a_{n+1}/a_n) = 0$ and case (d) of Theorem 2.3 applies to give the result.
2. The series $\sum_{n=1}^{\infty} n^n/n!$ diverges. This time, $a_n = n^n/n!$ and

$$\frac{a_{n+1}}{a_n} = \frac{(n+1)^{n+1}}{(n+1)!}\frac{n!}{n^n} = \frac{(n+1)^{n+1}}{(n+1)n^n} = \left(\frac{n+1}{n}\right)^n$$

   Thus $(a_{n+1}/a_n) > 1$ for all n, and case (b) of Theorem 2.3 applies.

   There are various strengthenings of the ratio test; see, for instance, Exercise 5 of the next section.

Exercises

1. Check the following series for convergence or divergence.

   (a) $\sum_{n=1}^{\infty} \left(-\frac{2}{3}\right)^n$ (b) $\sum_{k=1}^{\infty} \left(\frac{5}{3}\right)^k$

   (c) $\sum_{n=1}^{\infty} \frac{n}{\sqrt{2^n}}$ (d) $\sum_{n=1}^{\infty} \frac{2^n}{n!}$

   (e) $\sum_{n=1}^{\infty} \frac{n^4}{n!}$ (f) $\sum_{n=1}^{\infty} \frac{n^3}{2^n}$

   (g) $\sum_{n=2}^{\infty} n^3 e^{-n^3}$ (h) $\sum_{n=2}^{\infty} n^2 e^{n^{-3}}$

   (i) $\sum_{k=1}^{\infty} \frac{n^k}{(2k)!}$ (j) $\sum_{n=1}^{\infty} \frac{\sqrt{n!}}{4^n}$

2. (Root test) Let $\sum_{n=1}^{\infty} a_n$ be an infinite series of nonnegative terms, and suppose that $r = \lim_{n\to\infty} \sqrt[n]{a_n}$ exists. Show that

   If $r < 1$, the series converges.
   If $r > 1$, the series diverges.

   (For a stronger form, see Exercise 14.)
3. Suppose that $\sum_{n=1}^{\infty} a_n$ and $\sum_{n=1}^{\infty} b_n$ are series of positive terms and that $\lim_{n\to\infty} a_n/b_n = K$ exists and is nonzero. Show that the series either both converge or both diverge.
4. Show that for both $\sum_{n=1}^{\infty} 1/n(n+1)$ and $\sum_{n=1}^{\infty} 1/n$, the limit ratio of succcessive terms is 1, but that one series converges and the other diverges.

*5. (a) Use the comparison test to prove that $\sum_{n=1}^{\infty} 1/2n(2n-1)$ converges. (See Example 2 of Section 1.)
(b) Show that the even partial sums of $\sum_{n=1}^{\infty} (-1)^n/n$ converge.
(c) Now prove that the series in (b) converges.

6. A repeating decimal, like .123123123 · · · , may be regarded as an infinite sum. Show that such a decimal represents a rational number.

7. Let $\sum_{n=1}^{\infty} a_n$ and $\sum_{n=1}^{\infty} b_n$ be convergent series of positive terms. Prove that $\sum_{n=1}^{\infty} a_n b_n$ converges.

*8. (a) Let $\{a_k\}$ be a sequence of positive numbers such that $\sum_{k=1}^{\infty} a_k$ diverges, and let $s_n = \sum_{k=1}^{n} a_k$. Prove that $\sum_{k=1}^{\infty} a_k/s_k$ diverges. (*Hint*: Show that if $s_n > 2s_m$, then $\sum_{k=m+1}^{n} a_k/s_k > \frac{1}{2}$.)
(b) Use this result (with an appropriate sequence $\{a_k\}$) to prove that the harmonic series diverges.

9. Show that if $\{a_n\}$ and $\{b_n\}$ are sequences of positive numbers such that $\sum_{n=1}^{\infty} a_n^2$ and $\sum_{n=1}^{\infty} b_n^2$ both converge, then $\sum_{n=1}^{\infty} a_n b_n$ converges.

10. Suppose that $\{a_k\}$ is a sequence of positive terms with $\lim_{n\to\infty} a_{n+1}/a_n = r$. For what values of $x(\geq 0)$ does $\sum_{k=1}^{\infty} a_k x^k$ certainly converge, and for what values does the series diverge?

*11. We have seen that the harmonic series diverges. Show, on the other hand, that the sum of the reciprocals of all integers which do not have a 9 in their decimal expansions converges. (*Hint*: Count the number of such integers between 10^n and 10^{n+1}.)

12. Prove Lemma 2.1.

13. Give the inductive proofs in Theorem 2.3.

*14. (Root test, stronger form) Show that if the series $\sum_{n=1}^{\infty} a_n$ of positive terms
(a) Converges if $\limsup \sqrt[n]{a_n} < 1$.
(b) Diverges if $\limsup \sqrt[n]{a_n} > 1$.
(See Exercise 12, Section 7.2, for the definition of lim sup.)

*15. Let $\sum_{n=1}^{\infty} a_n$ be an infinite series of positive decreasing terms; that is, $a_1 \geq a_2 \geq a_3 \geq \cdots$, and $a_n > 0$ for all n. Show that $\sum_{n=1}^{\infty} a_n$ converges if and only if $\sum_{n=1}^{\infty} 2^n a_{2^n}$ converges. (This result is called *Cauchy's condensation test.*) *Hint*: Use the idea of Example 5, Section 1.

3. The Integral Test

We now develop another useful test, the integral test. We need, first of all, to recall a few facts about improper integrals.

Suppose that $f:[a,\infty)\to \mathbf{R}$ is a continuous function. Then $\int_a^x f(t)\,dt$ makes sense whenever $x > a$. We define

$$\int_a^{\infty} f(t)\,dt = \lim_{x\to\infty} \int_a^x f(t)\,dt$$

whenever the limit exists and we say that the integral $\int_a^{\infty} f(t)\,dt$ converges; if the limit does not exist, the integral is said to diverge. (See Exercise 15 of Section 10.4.)

If f is always nonnegative, then $g(x) = \int_a^x f(t)\,dt$ increases (or, at least, never decreases). In this case, either $g(x)$ is unbounded as x approaches infinity, in which case $\int_a^\infty f(t)\,dt$ diverges, or $g(x)$ is bounded, in which case $\int_a^\infty f(t)\,dt = \text{lub}_{x>a}\, g(x)$. (See Exercise 3.)

Theorem 3.1 (The Integral Test) *Suppose that $f: [1, \infty) \to \mathbf{R}$ is continuous, nonnegative, and nonincreasing. Then*

(a) *If $\int_1^\infty f(t)\,dt$ converges, then $\sum_{n=1}^\infty f(n)$ converges, and*

$$\int_1^\infty f(t)\,dt \le \sum_{n=1}^\infty f(n) \le f(1) + \int_1^\infty f(t)\,dt$$

(b) *If $\int_1^\infty f(t)\,dt$ diverges, so does $\sum_{n=1}^\infty f(n)$.*

The proof follows momentarily. Here, first, are some examples.

1. Consider the series

$$\sum_{n=1}^\infty \frac{1}{n^2}$$

If we define $f: [1, \infty) \to \mathbf{R}$ by $f(x) = 1/x^2$, we have $1/n^2 = f(n)$ and

$$\begin{aligned}\int_1^\infty f(t)\,dt &= \lim_{x\to\infty} \int_1^x \frac{1}{t^2}\,dt \\ &= \lim_{x\to\infty} \left(\frac{-1}{t}\Bigg|_1^x \right) \\ &= \lim_{x\to\infty} \left(1 - \frac{1}{x}\right) = 1\end{aligned}$$

Thus

$$\sum_{n=1}^\infty \frac{1}{n^2} = \sum_{n=1}^\infty f(n)$$

converges with sum $\le 1 + f(1) = 2$.

2. We can use Theorem 3.1 to show that the harmonic series diverges, a result that we have already proved. Let $f: [1, \infty) \to \mathbf{R}$ be defined by $f(t) = 1/t$. Then

$$\begin{aligned}\int_1^\infty \frac{1}{t}\,dt &= \lim_{x\to\infty} \int_1^x \frac{1}{t}\,dt \\ &= \lim_{x\to\infty} \left(\ln t \Bigg|_1^x \right) \\ &= \lim_{x\to\infty} (\ln x)\end{aligned}$$

Since this limit does not exist, the series

$$\sum_{n=1}^{\infty} \frac{1}{n} = \sum_{n=1}^{\infty} f(n)$$

diverges.

Proof of Theorem 3.1. The key inequality we need is the following: for any integer $n \geq 1$,

$$f(n) \geq \int_n^{n+1} f(t)\,dt \geq f(n+1) \tag{3.1}$$

The reason should be obvious from Figure 3.1: compare the areas. For a formal proof, notice that

$$f(n) \geq f(t) \geq f(n+1) \tag{3.2}$$

for $n \leq t \leq n + 1$. Now integrate each term of (3.2) from n to $n + 1$, and the inequality (3.1) results.

If we sum (3.1) from 1 to N, we get

$$\sum_{n=1}^{N} f(n) \geq \sum_{n=1}^{N} \int_n^{n+1} f(t)\,dt \geq \sum_{n=1}^{N} f(n+1)$$

However,
$$\sum_{n=1}^{N} \int_n^{n+1} f(t)\,dt = \int_1^{N+1} f(t)\,dt$$

and
$$\sum_{n=1}^{N} f(n+1) = \sum_{n=2}^{N+1} f(n) \geq \sum_{n=2}^{N} f(n)$$

since $f(N + 1) \geq 0$. Thus

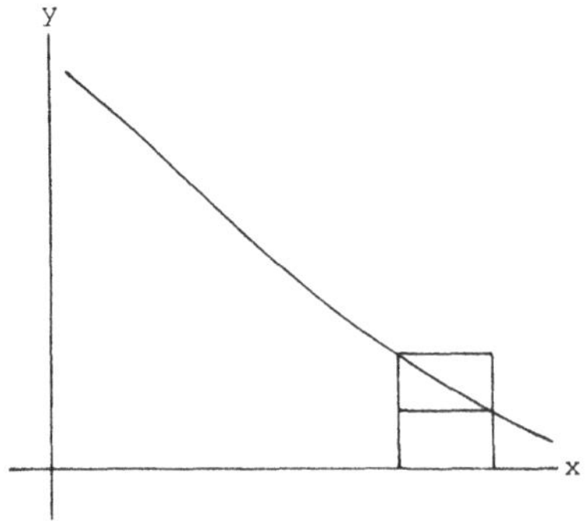

Figure 3.1

$$\sum_{n=1}^{N} f(n) \geq \int_{1}^{N+1} f(t)\,dt \geq \sum_{n=2}^{N} f(n) \tag{3.3}$$

If $\int_1^\infty f(t)\,dt$ converges, then

$$\begin{aligned} f(1) + \int_1^\infty f(t)\,dt &\geq f(1) + \int_1^{N+1} f(t)\,dt \\ &\geq f(1) + \sum_{n=2}^{N} f(n) \\ &= \sum_{n=1}^{N} f(n) \end{aligned}$$

Thus the partial sums of the series

$$\sum_{n=1}^{\infty} f(n)$$

are bounded and the series converges by Lemma 2.1. The inequality on the sum of this series follows from (3.3).

If $\int_1^\infty f(t)\,dt$ diverges, then

$$g(x) = \int_1^x f(t)\,dt$$

must be unbounded as x approaches infinity. Since

$$\sum_{n=1}^{N} f(n) \geq g(N+1)$$

by (3.3), the partial sums of the series

$$\sum_{n=1}^{\infty} f(n)$$

are unbounded. Thus this series cannot converge.

Remark. As usual, the hypotheses of Theorem 3.1 need only apply for large values of t; in particular, the function $f(t)$ need not be nonincreasing when t is small. In this case, of course, the estimate on the sum of the series needs to be adjusted.

We close this section with some additional examples of how the results of this section are applied to particular series.

1. Consider the series

(3.4) $$\sum_{n=1}^{\infty} \frac{1}{n^r}$$

where r is a positive real number. If $r \neq 1$, we have

$$\begin{aligned}\int_1^{\infty} \frac{1}{t^r}\,dt &= \lim_{x\to\infty} \int_1^x \frac{1}{t^r}\,dt \\ &= \lim_{x\to\infty}\left(\frac{t^{-r+1}}{-r+1}\bigg|_1^x\right) \\ &= \lim_{x\to\infty}\left(\frac{x^{-r+1}}{-r+1} - \frac{1}{-r+1}\right)\end{aligned}$$

This limit exists if $r > 1$ and does not if $r < 1$. Thus, using Theorem 3.1, we know that the series (3.4) converges if $r > 1$ and diverges if $r \leq 1$. [If $r = 1$, (3.4) is the harmonic series.]

2. We can combine Example 1 with Theorem 2.2 in a useful way. Consider the series

(3.5) $$\sum_{n=1}^{\infty} \frac{1}{n^3 + 3n + 7}$$

Now $$n^3 < n^3 + 3n + 7$$

for $n \geq 1$, so that

$$\frac{1}{n^3 + 3n + 7} < \frac{1}{n^3}$$

for $n \geq 1$. Since

$$\sum_{n=1}^{\infty} \frac{1}{n^3}$$

converges by Example 1, the series (3.5) converges by Theorem 2.2.

3. For a more complicated example of the same kind, consider the series

(3.6) $$\sum_{n=1}^{\infty} \frac{1}{2n^3 - n^2 - 7}$$

(This series has positive terms except for $n = 1$.)

If we can show that $n^3 > n^2 + 7$ for large n, we will have

$$2n^3 - n^2 - 7 > 2n^3 - n^3 = n^3$$

for large n so that

$$\frac{1}{2n^3 - n^2 - 7} < \frac{1}{n^3}$$

for large n. Thus the series (3.6) will converge by Theorem 2.2.

To see that $n^3 > n^2 + 7$ for large n, we show that the function $g(x) = x^3 - x^2 - 7$ is positive for large values of x. Indeed, $g'(x) = 3x^2 - 2x$ is positive when $x > \frac{3}{2}$; so $g(x)$ is increasing for $x \geq \frac{3}{2}$. However, $g(3) = 11 > 0$; so $g(x)$ is positive for $x \geq 3$. Thus $n^3 > n^2 + 7$ for $n \geq 3$, and the series converges.

Exercises

1. Check the following series for convergence or divergence.

(a) $\sum_{n=1}^{\infty} \frac{2n}{n^2+7}$

(b) $\sum_{n=1}^{\infty} \frac{n-2}{n^2-4n-6}$

(c) $\sum_{n=3}^{\infty} \frac{1}{n^3+7}$

(d) $\sum_{n=2}^{\infty} \frac{1}{n^3+n^2-7}$

(e) $\sum_{n=2}^{\infty} \frac{1}{n^4-n^3+7n}$

(f) $\sum_{k=1}^{\infty} \frac{k}{k^2+2}$

(g) $\sum_{k=2}^{\infty} \frac{k+3}{k^3-2k^2+1}$

(h) $\sum_{n=2}^{\infty} \frac{1}{n \ln n}$

(i) $\sum_{n=1}^{\infty} \frac{1}{\sqrt{n(n+1)}}$

(j) $\sum_{n=2}^{\infty} \frac{1}{\sqrt{n(n+1)(n-1)}}$

(k) $\sum_{n=1}^{\infty} (\sqrt{n^2+1} - n)$

(l) $\sum_{n=1}^{\infty} \frac{1}{3^n - 2^n}$

(m) $\sum_{n=1}^{\infty} \frac{\ln n}{n^2}$

(n) $\sum_{n=1}^{\infty} \frac{n \ln n}{n^3+3n^2-n-2}$

(o) $\sum_{n=1}^{\infty} \frac{n \ln n}{2^n}$

*(p) $\sum_{n=2}^{\infty} \frac{1}{\ln n^{\ln \ln n}}$

*(q) $\sum_{n=1}^{\infty} \sin \frac{1}{n}$

2. For which values of r does the series

$$\sum_{n=2}^{\infty} \frac{1}{n(\ln n)^r}$$

converge and for which does it diverge?

3. Let $f: [a, \infty) \to \mathbf{R}$ be a nonnegative integrable function and define $g: [a, \infty) \to \mathbf{R}$ by

$$g(x) = \int_a^x f(t)\,dt$$

If g is bounded, prove that

$$\int_a^\infty f(t)\,dt = \text{lub}\,\{g(x) : x \geq a\}$$

4. Let f satisfy the conditions of the integral test, and let $s_N = \sum_{n=1}^{N} f(n)$. Suppose that $\sum_{n=1}^{\infty} f(n)$ converges to s. Show that $|s - s_N| \leq \int_N^\infty f(t)\,dt$.
5. (Raabe's test) If the ratio in the ratio test is 1, the test fails. There are various more delicate tests to take care of this case. Here is one.

   Suppose that $\sum_{n=1}^{\infty} a_n$ is an infinite series of positive terms and that $\lim_{n\to\infty} n(a_{n+1}/a_n - 1) = r_1$ exists. If $r_1 < -1$, the series converges; if $r_1 > -1$ it diverges. (If $r_1 = -1$, the test fails.)

   (*Hint*: Compare the series with $\sum_{n=1}^{\infty} 1/n^k$ for appropriate k.)
6. Show that $\lim_{n\to\infty} \sum_{j=n+1}^{2n} 1/j = \ln 2$. (*Hint*: Use the estimation procedure of the integral test.)

*7. Show that $\gamma = \lim_{n\to\infty} (\sum_{j=1}^{n} 1/j - \ln n)$ exists and is between $\frac{1}{2}$ and 1. (*Hint*: Use the idea behind the integral test.) The number γ is often called *Euler's constant*.

*8. (a) Show that if $\sum_{n=1}^{\infty} a_n$ converges, where the numbers a_n are decreasing and positive, then $\lim_{n\to\infty} na_n = 0$. (*Hint*: Show that $2na_{2n} \leq 2\sum_{k=n+1}^{2n} a_k$ approaches zero. Show next that $(2n+1)a_{2n+1} \leq (2n+1)a_{2n}$ approaches zero.)

   (b) Find a divergent series $\sum_{n=1}^{\infty} a_n$ such that the numbers a_n are decreasing and positive, with $\lim_{n\to\infty} na_n = 0$. [Thus the converse of (a) does *not* hold.]

*9. For what values of a and b (>0) does

$$\frac{a}{b} + \frac{a(a+1)}{b(b+1)} + \frac{a(a+1)(a+2)}{b(b+1)(b+2)} + \cdots$$

converge? (See Problem 5.)

*10. (Infinite products) We define

$$\prod_{k=1}^{n} (1 + a_k) = (1 + a_1)(1 + a_2)\cdots(1 + a_n)$$

The expression $\prod_{k=1}^{\infty} (1 + a_k)$ is called an *infinite product*. (One usually assumes that $a_k \neq -1$ for all k, for obvious reasons.) If $\lim_{n\to\infty} \prod_{k=1}^{n} (1 + a_k)$ exists, then $\prod_{k=1}^{\infty} (1 + a_k)$ is said to *converge* to this limit; if the limit fails to exist, then $\prod_{k=1}^{\infty} (1 + a_k)$ is said to *diverge*. (Sometimes the product is said to diverge if the limit is 0; this definition makes it easier to state some theorems briefly. We shall not need to concern ourselves with this matter, however.)

Assume that $a_k \geq 0$ for all k. Show that

$$\prod_{k=1}^{\infty} (1 + a_k) \text{ converges} \Leftrightarrow \sum_{k=1}^{\infty} \ln(1 + a_k) \text{ converges}$$

$$\Leftrightarrow \sum_{k=1}^{\infty} a_k \text{ converges}$$

4. Absolute and Conditional Convergence

Let $\sum_{n=1}^{\infty} a_n$ be a series of terms, *not* necessarily all nonnegative. The series is said to be *absolutely convergent* if $\sum_{n=1}^{\infty} |a_n|$ converges, and *conditionally convergent* if $\sum_{n=1}^{\infty} a_n$ converges, but $\sum_{n=1}^{\infty} |a_n|$ does not.

Examples

1. The series $1 - \frac{1}{2} + \frac{1}{4} - \frac{1}{8} + \cdots = \sum_{n=0}^{\infty} (-\frac{1}{2})^n$ converges absolutely, since $\sum_{n=0}^{\infty} (\frac{1}{2})^n$ converges.
2. The series $1 - 1 + \frac{1}{2} - \frac{1}{2} + \frac{1}{3} - \frac{1}{3} + \cdots$ converges, since the partial sums, $1, 0, \frac{1}{2}, 0, \frac{1}{3}, 0, \ldots$, clearly converge to 0. But the series converges conditionally, since $1 + 1 + \frac{1}{2} + \frac{1}{2} + \cdots$ diverges.
3. The series $\sum_{n=1}^{\infty} (-1)^{n-1}/n = 1 - \frac{1}{2} + \frac{1}{3} - \frac{1}{4} + \cdots$ does not converge absolutely (see Example 5 of Section 1), but it does converge, as we shall see below. Therefore it converges conditionally.

Notice that we did not assume in our definition of absolute convergence that an absolutely convergent series converges. The reason is that it is not necessary.

Theorem 4.1 *An absolutely convergent series converges.*

Proof. Let $\sum_{n=1}^{\infty} a_n$ be the series. We split it into positive and negative terms as follows: let

$$b_n = \begin{cases} a_n & \text{if } a_n > 0 \\ 0 & \text{if } a_n \le 0 \end{cases}$$

$$c_n = \begin{cases} 0 & \text{if } a_n \ge 0 \\ -a_n & \text{if } a_n < 0 \end{cases}$$

The result of these definitions is that

$$(4.1) \qquad \begin{aligned} &0 \le b_n \le |a_n| \qquad 0 \le c_n \le |a_n| \qquad \text{for all } n \\ &b_n + c_n = |a_n| \qquad b_n - c_n = a_n \qquad \text{for all } n \end{aligned}$$

We know that $\sum_{n=1}^{\infty} |a_n|$ converges. Therefore $\sum_{n=1}^{\infty} b_n$ and $\sum_{n=1}^{\infty} c_n$ converge by (4.1) and Theorem 2.2. Now Proposition 1.1 says that $\sum_{n=1}^{\infty} a_n = \sum_{n=1}^{\infty} (b_n - c_n)$ converges, proving the theorem.

Remark. In addition, $|\sum_{n=1}^{\infty} a_n| \le \sum_{n=1}^{\infty} |a_n|$; see Exercise 3.

Conditionally convergent series are more difficult to work with than absolutely convergent ones. For one thing, they tend to converge more slowly. There are also more pitfalls in using them. The next result illustrates one.

Proposition 4.2 *Let $\sum_{n=1}^{\infty} a_n$ be a conditionally convergent series, and let K be any number. Then it is possible to rearrange the terms of the series so that they sum to K. It is also possible to rearrange the terms so that the series diverges.*

Proof. Define b_n and c_n as in the previous theorem. Notice that both b_n and c_n approach zero as n approaches infinity, since a_n does.

The series $\sum_{n=1}^{\infty} b_n$ and $\sum_{n=1}^{\infty} c_n$ either both converge or both diverge, since $\sum_{n=1}^{\infty} (b_n - c_n)$ converges [by Proposition 1.1(b)]. They cannot both converge since $\sum_{n=1}^{\infty} (b_n + c_n) = \sum_{n=1}^{\infty} |a_n|$ diverges. Therefore, both diverge. Since they are series of nonnegative terms, Lemma 2.1 says that their partial sums are unbounded.

Now we are ready to rearrange the series. Assume for convenience that $K > 0$. Start by adding together enough positive a_n's (in order) to get a sum just greater than K. (We can do this, because the partial sums of the positive terms are unbounded.) Now add just enough negative a_n's to get a sum just less than K. Next, add enough terms to make the sum greater than K, again, and so on. We can continue the procedure forever, since the partial sums of $\sum_{n=1}^{\infty} b_n$ and $\sum_{n=1}^{\infty} c_n$ are both unbounded. Furthermore, the sum converges to K, for the following reason: Suppose ε is any positive number and choose N so that $|a_n| < \varepsilon$ for $n > N$. Suppose further that we have progressed in our construction of the rearrangement far enough so that all of the terms a_n, $n \leq N$, have been included. Our procedure now is to add positive terms until the sum is greater than K; thus the partial sums will be less than $K + \varepsilon$ (since all remaining terms are less than ε in absolute value). Similarly, we add negative terms until the sum is less than K; thus the partial sums will be greater than $K - \varepsilon$. This clearly implies that the kth partial sums differ from K by less than ε for large k, and hence that the series converges to K.

The proofs for $K < 0$ and for a divergent series are similar and are left as Exercises 4 and 5.

As an example, consider

$$\sum_{n=1}^{\infty} (-1)^{n-1} \frac{1}{n} = 1 - \frac{1}{2} + \frac{1}{3} - \frac{1}{4} + \cdots$$

This series converges conditionally (to $\ln 2$). If we wish to rearrange the terms to make it converge to 1, the proof says that we begin with

$$1 + \tfrac{1}{3} - \tfrac{1}{2} + \tfrac{1}{5} - \tfrac{1}{4} + \tfrac{1}{7} + \tfrac{1}{9} - \tfrac{1}{6} + \cdots$$

Notice that $1 + \frac{1}{3} > 1$, $1 + \frac{1}{3} - \frac{1}{2} < 1$, $1 + \frac{1}{3} - \frac{1}{2} + \frac{1}{5} > 1$, and so on.

Note. If $\sum_{n=1}^{\infty} a_n$ converges absolutely, any rearrangement of the series gives the same sum. See Exercise 6.

How can one tell if a non-absolutely convergent series converges? The following theorem provides an answer in many cases.

Theorem 4.3 (Dirichlet) *Let $\{a_n\}$ and $\{b_n\}$ be two sequences such that*

(a) $\{a_n\}$ *is nonincreasing and* $\lim_{n\to\infty} a_n = 0$.
(b) *The sequence of partial sums* $\{\sum_{j=1}^{n} b_j\}$ *is bounded.*

Then $\sum_{n=1}^{\infty} a_n b_n$ *converges.*

Remark. A remarkable aspect of this theorem is that we do not assume that either $\sum_{n=1}^{\infty} a_n$ or $\sum_{n=1}^{\infty} b_n$ converges. We prove the theorem below.

The next result is a useful consequence of this theorem.

Corollary (Alternating Series Test) *If $\{a_n\}$ is a nonincreasing sequence of positive numbers which converge to 0, then $\sum_{n=1}^{\infty} (-1)^{n+1} a_n$ converges.*

Proof. Let $b_n = (-1)^{n+1}$. The partial sums $\sum_{j=1}^{n} b_j$ are $1, 0, 1, 0, \ldots$, and are certainly bounded. Thus the theorem applies.

Proof of Theorem 4.3. The proof uses a technique called *partial summation.* Let $s_n = \sum_{j=1}^{n} b_j$, and set $s_0 = 0$. Then $b_n = s_n - s_{n-1}$ for $n \geq 1$ (this explains the definition of s_0), and

$$\begin{aligned}
\sum_{n=1}^{N} a_n b_n &= \sum_{n=1}^{N} a_n(s_n - s_{n-1}) = \sum_{n=1}^{N} a_n s_n - \sum_{n=1}^{N} a_n s_{n-1} \\
&= \sum_{n=1}^{N} a_n s_n - \sum_{n=0}^{N-1} a_{n+1} s_n \\
&= \left(\sum_{n=1}^{N} a_n s_n - \sum_{n=1}^{N} a_{n+1} s_n \right) - a_1 s_0 + a_{N+1} s_N \\
&= \sum_{n=1}^{N} (a_n - a_{n+1}) s_n + a_{N+1} s_N - a_1 s_0
\end{aligned}$$

(Note that we have replaced the original sum by one in which the b's are summed and we take differences of the a's. This is analogous to the formula for integration by parts:

$$\int_a^b u(x)v'(x)\,dx = u(x)v(x)\Big|_a^b - \int_a^b v(x)u'(x)\,dx$$

in which v' is integrated and u is differentiated. Hence the name *partial summation.*)

In the case at hand, let $t_N = \sum_{n=1}^{N} a_n b_n$ and $u_N = \sum_{n=1}^{N} (a_n - a_{n+1}) s_n$. Then

$$t_N = u_N + a_{N+1} s_N - a_1 s_0 = u_N + a_{N+1} s_N$$

since $s_0 = 0$. We need to show that t_N converges. Since a_{N+1} approaches zero and $\{s_N\}$ is bounded, $a_{N+1} s_N$ approaches zero; thus we are done if we can show that $\{u_N\}$ converges or, equivalently, that the series

$$\sum_{n=1}^{\infty} (a_n - a_{n+1}) s_n$$

converges. We now prove that it converges absolutely.

As the sequence $\{s_n\}$ is bounded, we can find a number K such that $|s_n| \leq K$ for all n. Furthermore, $a_n \geq a_{n+1}$, by hypothesis; thus

$$|(a_n - a_{n+1}) s_n| \leq |a_n - a_{n+1}| K = (a_n - a_{n+1}) K$$

However, the series $\sum_{n=1}^{\infty} (a_n - a_{n+1})$ clearly converges (see Exercise 1 of Section 1). Therefore $\sum_{n=1}^{\infty} K(a_n - a_{n+1})$ converges, too, and so $\sum_{n=1}^{\infty} (a_n - a_{n+1}) s_n$ converges absolutely (by the comparison test, Theorem 2.2). This completes the proof.

Note. While the alternating series test follows easily from Theorem 4.3, it can also be proved directly (see Exercise 7). The proof given here is really rather complicated, since, as we have just seen, the proof of Theorem 4.3 is rather difficult.

Exercises

1. Which of the following series converge absolutely, which converge conditionally, and which diverge?

(a) $\sum_{k=1}^{\infty} (-1)^k \frac{1}{3k}$

(b) $\sum_{k=2}^{\infty} (-1)^k \frac{1}{k(\ln k)}$

(c) $\sum_{n=1}^{\infty} (-1)^n \frac{1}{n^{5/3}}$

(d) $\sum_{n=0}^{\infty} \sin \frac{n\pi}{2}$

(e) $\sum_{k=1}^{\infty} \frac{k^2}{(-2)^k}$

(f) $\sum_{n=1}^{\infty} \frac{\sin (n\pi/2)}{2}$

(g) $\sum_{k=1}^{\infty} \frac{(-1)^k (k+2)}{k^2 + 3k - 1}$

(h) $\sum_{n=1}^{\infty} \frac{(-n)^n}{n!}$

(i) $\sum_{n=2}^{\infty} \frac{(-1)^n (2n^2 + 7)}{n^4 + 3n^2 - 6}$

(j) $\sum_{n=1}^{\infty} \frac{(-3)^n}{n!}$

(k) $1 - \frac{1}{2} + \frac{1}{3} - \frac{1}{4} + \frac{1}{9} - \frac{1}{6} + \frac{1}{27} - \frac{1}{8} + \frac{1}{81} - \frac{1}{10} + \cdots$

(l) $\sum_{n=1}^{\infty} (-1)^n \sin \frac{1}{n}$

(m) $\sum_{n=1}^{\infty} \frac{\sin n}{2^n}$

(n) $\sum_{n=2}^{\infty} \frac{(-1)^n}{\ln n}$

(o) $\sum_{n=2}^{\infty} \frac{(-1)^n n}{\sqrt{(n+1)(n+2)}}$

2. Explain the fallacy in the following reasoning:

$$\begin{aligned}\ln 2 &= 1 - \tfrac{1}{2} + \tfrac{1}{3} - \tfrac{1}{4} + \tfrac{1}{5} - \tfrac{1}{6} + \cdots \qquad \text{(This is true—see Exercise 8.)}\\ &= (1 + \tfrac{1}{3} + \tfrac{1}{5} + \cdots) - (\tfrac{1}{2} + \tfrac{1}{4} + \tfrac{1}{6} + \cdots)\\ &= (1 + \tfrac{1}{2} + \tfrac{1}{3} + \tfrac{1}{4} + \cdots) - 2(\tfrac{1}{2} + \tfrac{1}{4} + \tfrac{1}{6} + \cdots)\\ &= (1 + \tfrac{1}{2} + \tfrac{1}{3} + \cdots) - (1 + \tfrac{1}{2} + \tfrac{1}{3} + \cdots)\\ &= 0 = \ln 1\end{aligned}$$

Thus $2 = 1$.

3. Show that if $\sum_{n=1}^{\infty} a_n$ converges absolutely, then $|\sum_{n=1}^{\infty} a_n| \le \sum_{n=1}^{\infty} |a_n|$.
4. Show that if $\sum_{n=1}^{\infty} a_n$ converges conditionally, then one can rearrange the terms of the series so that it converges to any negative number.

*5. Show that if $\sum_{n=1}^{\infty} a_n$ converges conditionally, then one can rearrange the terms of the series so that it diverges.

*6. Show that if $\sum_{n=1}^{\infty} a_n$ converges absolutely, then any rearrangement of the terms does not change the sum. (*Hint*: Given ε, choose N so that $\sum_{n=N+1}^{\infty} |a_n| < \varepsilon/2$; given a rearrangement, $\sum_{n=1}^{\infty} b_n$, pick M so that the terms $b_1, \ldots, b_m$ include all the terms $a_1, \ldots, a_N$.)

*7. Prove the alternating series test directly by showing that the even partial terms increase to a limit and that the odd partial sums decrease to the same limit.

*8. (a) Show that

$$\sum_{k=1}^{2n} \frac{(-1)^{k-1}}{k} = \sum_{k=n+1}^{2n} \frac{1}{k}$$

(*Hint*: Group all the fractions of the form $\pm 1/2q$, q a fixed odd number.)

(b) Conclude that

$$\sum_{k=1}^{\infty} \frac{(-1)^{k-1}}{k} = \ln 2$$

(See problem 6 of Section 3.)

*9. (a) Show that

$$\sum_{k=1}^{n} \sin k = \frac{\cos(x/2) - \cos[(2n+1)x/2]}{2\sin(x/2)}$$

when x is not a multiple of 2π. (Use mathematical induction.)

(b) Show that $\sum_{k=1}^{\infty} (\sin kx)/k$ converges for all x.

*10. Show that $\int_0^\infty [(\sin x)/x]\,dx$ converges. {See Exercise 15 of Section 10.4 for the definition of these improper integrals. Note that $\lim_{x\to 0} [(\sin x)/x] = 1$.} {*Hint*: Prove first that $\lim_{n\to\infty} \int_0^{\pi n} [(\sin x)/x]\,dx$ (n an integer) exists.}

*11. Let $\{a_n\}$ be a decreasing sequence of positive terms converging to 0, and let $s = \sum_{k=1}^\infty (-1)^{k-1} a_k$, $s_n = \sum_{k=1}^n (-1)^{k-1} a_k$. Show that $|s - s_n| < a_{n+1}$.

*12. Compute $\lim_{N\to\infty} \int_0^N \int_0^N e^{-xt} \sin x\,dx\,dt$ in two ways and thereby deduce the value of $\int_0^\infty [(\sin x)/x]\,dx$.

*13. Show that $\int_0^\infty \sin x^2\,dx$ and $\int_0^\infty \cos x^2\,dx$ both converge.

14. (Abel's theorem) Let $\{a_n\}$ and $\{b_n\}$ be sequences such that

(i) $\sum_{n=1}^{\infty} a_n$ converges.

(ii) $b_1 \geq b_2 \geq b_3 \geq \cdots$.

(iii) $b_n \geq 0$ for all n.

Prove that $\sum_{n=1}^\infty a_n b_n$ converges. [*Hint*: Let $t_n = \sum_{j=1}^n a_j b_j$. Fix m, and for $n > m$, let $s_n = \sum_{j=m+1}^n a_j$. Use partial summation to prove that $t_n - t_m = \sum_{j=m+1}^n (b_j - b_{j+1}) s_j + b_{n+1} s_n$. Show that if there is a constant c such that $|s_j| < c$ for $j > m$, then $|t_n - t_m| < c b_m$. Now show that there is a number N such that if $m > N$, then $|s_j| < \varepsilon$ for all $\varepsilon > 0$. Conclude that $\{t_n\}$ is a Cauchy sequence, and hence that $\{t_n\}$ converges.]

*15. Prove the analogue of Theorem 4.1 for improper integrals $\int_a^\infty f(x)\,dx$.

17

Infinite Series of Functions

We now turn to the study of infinite series of functions. The most obvious notion of convergence for a series of functions is pointwise convergence. However, this turns out to be less important than another sort of convergence, uniform convergence. We define this new notion, use it to prove a number of theorems, and apply these theorems to the study of power series. We conclude with a brief introduction to Fourier series. These are sums of sine and cosine functions; they are important in both pure and applied mathematics.

1. Uniform Convergence

We begin by formally defining the most obvious notion of convergence for a series of functions. (There is a similar definition for sequences.)

Suppose that U is a subset of $\mathbf{R}^n$, and let $\{f_j\}$ be a sequence of functions defined from U to $\mathbf{R}$. We say that $\sum_{j=1}^{\infty} f_j$ *converges pointwise* to f (or that $\sum_{j=1}^{\infty} f_j = f$) if for every point $\mathbf{v} \in U$, $\sum_{j=1}^{\infty} f_j(\mathbf{v}) = f(\mathbf{v})$.

Although this definition of convergence for functions probably seems the most natural, it has some disadvantages; the limit function f may be quite badly behaved even though the functions f_n are smooth. Here is one example (others will follow). Let

$$f_j(x) = x^j - x^{j-1} \qquad x \in [0, 1]$$

Each function f_j is, of course, continuous. Now let

$$s_n(x) = \sum_{j=1}^{n} f_j(x)$$

It is easy to check by induction that

$$s_n(x) = x - x^{n+1}$$

When $0 \leq x < 1$, x^{n+1} approaches 0 as n approaches ∞, and therefore $s_n(x)$ approaches x. When $x = 1$, however, $s_n(x) = 0$ for all n, and $s_n(x)$ approaches 0. Therefore

$$\sum_{j=1}^{\infty} f_j(x) = f(x)$$

where
$$f(x) = \begin{cases} x & \text{if } 0 \leq x < 1 \\ 0 & \text{if } x = 1 \end{cases}$$

That is, the pointwise sum of the continuous functions f_j is discontinuous–an awkward state of affairs.

To rectify matters, we need another definition. Let $\{g_n\}$ be a sequence of real- (or complex-) valued functions defined on the set U. We say that the sequence $\{g_n\}$ *converges uniformly* on U to g if for every $\varepsilon > 0$ we can choose an integer N such that for every $\mathbf{v} \in U$, $|g(\mathbf{v}) - g_n(\mathbf{v})| < \varepsilon$ if $n > N$.

Note the difference between uniform convergence and pointwise convergence: in uniform convergence, the same N works for all $\mathbf{v}$. The following example (very much like the one above) may make this difference clearer.

Let $g_n(x) = x^n$ on the interval [0, 1], and let

$$g(x) = \begin{cases} 0 & \text{if } 0 \leq x < 1 \\ 1 & \text{if } x = 1 \end{cases}$$

Then g_n approaches g pointwise; that is, $g_n(x)$ approaches $g(x)$ for all x. For $x = 0$ or 1, this is easy [since $g_n(x) = g(x)$ for all n]; if $0 < x < 1$, we need to show that for every $\varepsilon > 0$ there is an N such that $x^n (= |g_n(x) - g(x)|) < \varepsilon$ for $n > N$. We shall verify this for one value of ε, $\varepsilon = \frac{1}{4}$. Given x, we need to find n such that if $n > N$, then $x^n < \frac{1}{4}$. Taking logarithms, we need

$$n \ln x < \ln \tfrac{1}{4} = -\ln 4.$$

That is,

$$n > -\frac{\ln 4}{\ln x}$$

(since $\ln x < 0$, the inequality reverses). We simply pick N to be the biggest integer greater than $-(\ln 4)/(\ln x)$.

This choice of N depends on x in an essential way. As x gets close to 1, $\ln x$ becomes close to 0, and N needs to be large. In fact, no single value of N works for all x. Figure 1.1 may make this clearer. Since $f_n(x) = 1$ when $x = 1$, $f_n(x) \geq \frac{1}{4}$ when x is sufficiently close to 1. (How close "sufficiently close" is depends on n.) Therefore $f_n(x)$ cannot be $< \frac{1}{4}$ for all $x < 1$ simultaneously. It follows that $\{f_n\}$ does not converge uniformly.

Now consider the same functions restricted to $[0, \frac{1}{2}]$. Then $-\ln x \geq \ln 2$ for all $x \in (0, \frac{1}{2}]$, and therefore $N = (\ln 4)/(\ln 2) = 2$ works for all x. Similarly, if $\varepsilon > 0$, then $x^n < \varepsilon$ if

$$n \ln x < \ln \varepsilon$$

or if

$$n > \frac{-\ln \varepsilon}{-\ln x} \geq -\frac{\ln \varepsilon}{\ln 2}$$

since $-\ln x \geq \ln 2$. Thus $N = (-\ln \varepsilon)/(\ln 2)$ suffices in the definition of convergence for all $x \in [0, \frac{1}{2}]$, and the sequence $\{x^n\}$ converges uniformly to 0 on $[0, \frac{1}{2}]$.

There is a similar definition for series: we say that the series $\sum_{j=1}^{\infty} f_j$ *converges uniformly* to f on U (where f_n and f are functions on U) if the sequence of partial sums $\sum_{j=1}^{n} f_j = s_n$ converges uniformly to f.

We need methods of determining when series of functions converge uniformly. The simplest test for uniform convergence is the following:

Theorem 1.1 (The Weierstrass M-Test) *Let $\sum_{j=1}^{\infty} f_j$ be a series of functions defined on U. Suppose that there are numbers M_j such that*

(a) $|f_j(\mathbf{v})| \leq M_j$ *for all $x \in U$ and all integers $j \geq 1$.*

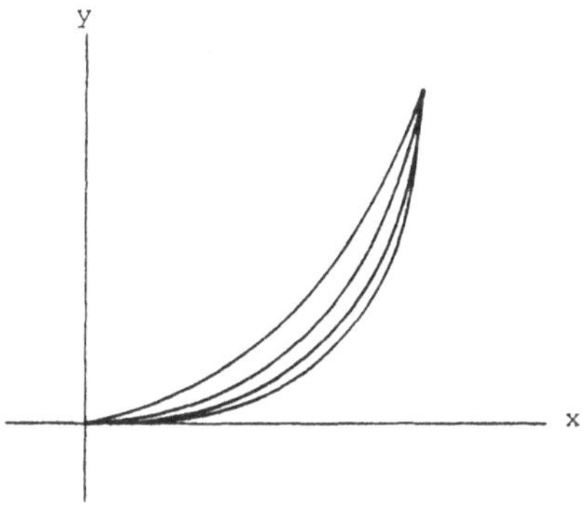

Figure 1.1 The graphs of $f_n(x) = x^n$.

(b) $\sum_{j=1}^{\infty} M_j$ *converges.*

Then $\sum_{j=1}^{\infty} f_j$ *converges uniformly on* U.

Proof. Let $s_n = \sum_{j=1}^{n} f_j$, and let $t_n = \sum_{j=1}^{n} M_j$, $t = \sum_{j=1}^{\infty} M_j$. We show first of all that $\sum_{j=1}^{n} f_j$ converges pointwise. This is fairly easy. Given $\mathbf{v} \in U$, we know that $|f_j(\mathbf{v})| \le M_j$. Therefore $\sum_{j=1}^{\infty} f_j(\mathbf{v})$ converges absolutely, by Theorem 2.2 of Chapter 16. Let $f(\mathbf{v}) = \sum_{j=1}^{\infty} f_j(\mathbf{v})$.

Given $\varepsilon > 0$, there exists an integer N such that if $n > N$, $|t - t_n| < \varepsilon$. We claim that if $n > N$, then $|f(\mathbf{v}) - s_n(\mathbf{v})| < \varepsilon$ for all $\mathbf{v} \in U$. The proof is again straightforward:

$$\begin{aligned} |f(\mathbf{v}) - s_n(\mathbf{v})| &= \left| \sum_{j=n+1}^{\infty} f_j(\mathbf{v}) \right| \\ &\le \sum_{j=n+1}^{\infty} |f_j(\mathbf{v})| \qquad \text{(by Exercise 3 of Section 16.4)} \\ &\le \sum_{j=n+1}^{\infty} M_j \qquad \text{(by Theorem 2.2 of Chapter 16)} \\ &= |t - t_n| < \varepsilon \end{aligned}$$

Since this result holds for all $\mathbf{v} \in U$, the series converges uniformly.

We next investigate some properties of uniform convergence.

Theorem 1.2 *Let* $\sum_{j=1}^{\infty} f_j = f$ *be a uniformly convergent series of functions on* U. *If the functions* f_n *are all continuous, then* f *is continuous as well.*

Proof. Let s_n be the nth partial sum of the series, and let $\mathbf{v}_0$ be a point in U. Given $\varepsilon > 0$, we can choose n such that

$$|s_n(\mathbf{v}) - s(\mathbf{v})| < \frac{\varepsilon}{3} \qquad \text{all } \mathbf{v} \in U$$

Since s_n is a continuous function, we can choose $\delta > 0$ such that if $\|\mathbf{v} - \mathbf{v}_0\| < \delta$, then $|s_n(\mathbf{v}) - s_n(\mathbf{v}_0)| < \varepsilon/3$. Then if $\|\mathbf{v} - \mathbf{v}_0\| < \delta$, we have

$$\begin{aligned} |s(\mathbf{v}) - s(\mathbf{v}_0)| &\le |s(\mathbf{v}) - s_n(\mathbf{v})| + |s_n(\mathbf{v}) - s_n(\mathbf{v}_0)| + |s_n(\mathbf{v}_0) - s(\mathbf{v}_0)| \\ &< \frac{\varepsilon}{3} + \frac{\varepsilon}{3} + \frac{\varepsilon}{3} = \varepsilon \end{aligned}$$

which proves that s is continuous at $\mathbf{v}_0$.

Remark. The previous examples show that the theorem does not hold if one assumes only that the functions converge pointwise.

Exercises

1. Which of the following series of functions converge uniformly on the given intervals?

   (a) $\sum_{n=1}^{\infty} nx^n(1-x)$ on $[0, 1]$

   (b) $\sum_{n=1}^{\infty} \frac{x^2}{x^2+n^2}$ on $(-\infty, \infty)$

   (c) $\sum_{n=1}^{\infty} \frac{\sin nx}{2^n}$ on $(-\infty, \infty)$

   (d) $\sum_{n=0}^{\infty} xe^{-nx}$ on $[0, \infty)$

   (e) $\sum_{n=0}^{\infty} 2^n x^n$ on $[0, \frac{1}{2})$

   (f) $\sum_{n=0}^{\infty} \frac{x^n}{n!}$ on $[-5, 5]$

   (g) $\sum_{n=0}^{\infty} \frac{x^n}{n!}$ on $(-\infty, \infty)$

   *(h) $\sum_{n=1}^{\infty} x^n g(x)$ on $[0, 1]$

   where $g\colon [0, 1] \to \mathbf{R}$ is a continuous function with $g(1) = 0$.
2. Show that $\zeta(s) = \sum_{n=1}^{\infty} n^{-s}$ converges uniformly on the interval $[a, \infty)$, for $a > 1$.
*3. Show that every continuous function on $[a, b]$ is the uniform limit of a sequence of step functions.
4. Prove that if $\sum_{n=1}^{\infty} f_n$ converges uniformly on two closed intervals I_1 and I_2, then it converges uniformly on their union.
*5. Modify the proof of Theorem 4.3 of Chapter 16 to prove the *Dirichlet–Hardy test*: if $\{a_n(\mathbf{v})\}$ and $\{b_n(\mathbf{v})\}$ are two sequences of functions such that $a_n \to 0$ uniformly, the sequence of partial sums $\{\sum_{j=1}^{\infty} b_j\}$ is uniformly bounded (there is a number K such that $|\sum_{j=1}^{n} b_j(\mathbf{v})| \le K$ for all n and $\mathbf{v}$), and $\{a_n(\mathbf{v})\}$ is nonincreasing for every $\mathbf{v}$, then $\sum_{n=1}^{\infty} a_n(\mathbf{v})b_n(\mathbf{v})$ converges uniformly.
*6. Show that if $\varepsilon > 0$, then $\sum_{k=1}^{\infty} (\sin kx)/k$ converges uniformly on the interval $[\varepsilon, 2\pi - \varepsilon]$. (See Exercise 9 of Section 16.4.)
*7. Let $f(x, y)$ be a function of two variables defined for all $(x, y) \in S \times [0, \infty)$, where S is some subset of the reals. We say that $\int_0^{\infty} f(x, y)\,dy$ *converges uniformly* to $g(x)$ on the set S if for every $\varepsilon > 0$ there is a number N such that if $t > N$, then

$$\left| \int_0^t f(x, y)\,dy - g(x) \right| < \varepsilon \qquad \text{all } x \in S$$

Show that if f is a continuous function of x and y and $\int_0^\infty f(x, y)\,dy$ converges uniformly to $g(x)$ on S, then g is continuous.

2. Differentiation and Integration of Series of Functions

Let U be an interval in $\mathbf{R}$, and let $\sum_{j=1}^\infty f_j = f$ be a series of functions from U to $\mathbf{R}$ which converge. In this section, we discuss the question of when we can differentiate and integrate the series termwise; that is, when the equations

$$\sum_{j=1}^{\infty} f_j' = f' \qquad \sum_{j=1}^{\infty} \int_a^b f_j = \int_a^b f$$

hold.

We need not restrict ourselves to functions on $\mathbf{R}$, of course; we could work with sets $U \subset \mathbf{R}^n$. However, the results for $\mathbf{R}$ can be used to prove similar theorems for $\mathbf{R}^n$ (see Exercise 3 for an example).

We begin with integration.

Theorem 2.1 *Suppose that $\sum_{j=1}^\infty f_j$ is a series of continuous functions converging uniformly to f on the finite interval $[a, b]$. Then*

$$\sum_{j=1}^{\infty} \int_a^x f_j(t)\,dt = \int_a^x f(t)\,dt \tag{2.1}$$

uniformly for $x \in [a, b]$.

Proof. Let $s_n(x)$ be the nth partial sum of the f_j's; we need to show that $\lim_{n\to\infty} \int_a^x s_n(t)\,dt = \int_a^x f(t)\,dt$ uniformly. Theorem 1.2 says that f is continuous and therefore integrable. Given $\varepsilon > 0$, we may choose N such that if $n > N$, then $|s_n(x) - f(x)| < \varepsilon/(b-a)$ for all $x \in [a, b]$. For such an n,

$$\int_a^x |s_n(t) - f(t)|\,dt < \int_a^x \frac{\varepsilon}{b-a}\,dt = \varepsilon \frac{x-a}{b-a} < \varepsilon$$

for all $x \in [a, b]$. This proves the theorem.

Remarks

1. There is no real need to make the functions f_j continuous; if they are simply integrable, then so is f (see Exercise 6), and the theorem holds.
2. There is a theorem like Theorem 2.1 for uniformly convergent sequences, with essentially the same proof (see Exercise 1). This same remark holds for the other theorems cited in this section; every theorem for series has an analogue for sequences, and vice versa. In the future, we shall use both results, regardless of which we have proved.
3. We *cannot* always integrate termwise if the series only converges pointwise;

Exercise 2 gives an example. We do not, however, need something as stringent as uniform convergence to allow us to integrate termwise. One stronger result is the *theorem of bounded convergence*: if $\{s_n\}$ is a sequence of functions such that s_n converges pointwise to s, and if
(a) The functions s and s_n are all integrable on $[a, b]$.
(b) There is a number K such that for all $x \in [a, b]$ and all n,

$$|s_n(x)| \leq K$$

then $\int_a^x s_n(t)\,dt$ converges to $\int_a^x s(t)\,dt$ uniformly on $[a, b]$. The proof is not easy.* In Exercise 2, of course, the functions $\{s_n\}$ are not bounded.

4. There is another reason Theorem 2.1 need not hold for pointwise convergence: even if every function $\{s_n\}$ is integrable and s_n converges pointwise to s, s need not be integrable. (Exercise 5 gives an example.) In this case, of course, the conclusion of Theorem 2.1 does not even make sense.

We next take up differentiation of series of functions. Here we can prove only a much weaker result.

Theorem 2.2 *Suppose that $\sum_{j=1}^{\infty} f_j$ is a series of continuously differentiable functions on $[a, b]$ converging pointwise to f, and that $\sum_{j=1}^{\infty} f_j'$ converges uniformly on $[a, b]$ to some function g. Then $f'(x) = g(x)$ for all $x \in [a, b]$.*

Proof. We proceed directly, using Theorem 2.1 and the fundamental theorem of calculus. Theorem 2.1 tells us that

$$\begin{aligned}\int_a^x g(t)\,dt &= \sum_{j=1}^{\infty} \int_a^x f_j'(t)\,dt \\ &= \sum_{j=1}^{\infty} [f_j(x) - f_j(a)] \\ &= \sum_{j=1}^{\infty} f_j(x) - \sum_{j=1}^{\infty} f_j(a) = f(x) - f(a)\end{aligned}$$

The fundamental theorem of calculus says that $f'(x) = g(x)$, as claimed.

Remarks

1. Though we did not assume it, it turns out that $\sum_{j=1}^{\infty} f_j(x)$ converges uniformly on $[a, b]$. See Exercise 8.
2. It is not always true that if $\sum_{j=1}^{\infty} f_j = f$ uniformly, then $\sum_{j=1}^{\infty} f_j' = f'$

*A proof can be found in Jeffreys and Jeffreys, *Methods of Mathematical Physics*, Cambridge University Press, 1956, p. 691.

uniformly. In fact, the series can even diverge at points, and f need not be differentiable. One example is given in Exercise 6.

The contrast between Theorems 2.1 and 2.2 is an illustration of a useful principle: integration tends to smooth functions out, while differentiation makes them bumpier. Another example of this principle is that if f is small, so is its integral, but its derivative may be big. [Consider, for example, $f(x) = 10^{-10} \sin 10^{100} x$.] There are various ways of making this principle precise; they amount to saying that differentiation is, in a certain sense, a discontinuous operation, while integration is continuous.

Exercises

1. State and prove the analogue of Theorem 2.1 for sequences.
2. Show that on the interval [0, 1], the sequence of functions $\{2nxe^{-nx^2}\}$ converges pointwise to 0, but $\int_0^1 2nxe^{-nx^2}\,dx$ does not converge to zero. (In view of Exercise 5, Section 16.1, one could easily make up an example involving series.)
3. Let S be a rectangle in $\mathbf{R}^n$ and let $\sum_{n=1}^{\infty} f_n = s$, the convergence being uniform on S. Show that $\sum_{n=1}^{\infty} \int_S f_n(x)\,dx = \int_S s(x)\,dx$.
4. Let $f(x) = \sum_{n=1}^{\infty} x/(1 + n^2x^2)$. For what values of x can one differentiate termwise to find $f'(x)$?

*5. Let

$$f_0(x) = \begin{cases} 1 & \text{if } x = 0 \\ 0 & \text{otherwise} \end{cases}$$

$$f_1(x) = \begin{cases} 1 & \text{if } x = 1 \\ 0 & \text{otherwise} \end{cases}$$

$$f_2(x) = \begin{cases} 1 & \text{if } x = \frac{1}{2} \\ 0 & \text{otherwise} \end{cases}$$

and if $2^{n-1} < j \leq 2^n$, so that $j = 2^{n-1} + k$ (with $1 < k \leq 2^{n-1}$),

$$f_j(x) = \begin{cases} 1 & \text{if } x = \dfrac{2k-1}{2^n} \\ 0 & \text{otherwise} \end{cases}$$

Show that $\sum_{j=1}^{\infty} f_j = s$, where

$$s(x) = \begin{cases} 1 & \text{if } x \text{ is a fraction of the form } \dfrac{a}{2^n},\ a = 0, 1, 2, \ldots, 2^n \\ 0 & \text{otherwise} \end{cases}$$

Show also that

$$\int_0^1 f_j(x)\,dx = 0 \qquad \text{for all } j$$

but that s is not integrable.

*6. Let $\sum_{j=1}^{\infty} f_j$ be a series of integrable functions on an interval $[a, b]$ which converges uniformly to s. Show that s is integrable and that

$$\sum_{j=1}^{\infty} \int_a^b f_j(x)\,dx = \int_a^b s(x)\,dx$$

{*Hint*: Show that if $|s_n(x) - s(x)| < \varepsilon$ for $x \in [a, b]$ and P is any partition of $[a, b]$, then $U(s, P) < U(s_n, P) + \varepsilon(b - a)$.}

7. Show that the series $\sum_{n=1}^{\infty} 2^{-n} \sin(10^n x)$ converges uniformly on $\mathbf{R}$, but that the series formed by differentiating this series termwise diverges at 0. (This new series actually diverges everywhere, and the series gives an example of a continuous function with no derivative at any point—though proving this last fact is not so easy.)
8. Prove that if $\sum_{n=1}^{\infty} f_n$ satisfies the hypotheses of Theorem 2.2, then it converges uniformly.

*9. State and prove the analogues of Theorems 2.1 and 2.2 for infinite integrals. (See Exercise 7 of Section 1 for the appropriate definition.)

3. Power Series

In this section, we shall deal primarily with functions defined on subsets of $\mathbf{R}$, leaving various generalizations as exercises.

A power series centered at x_0 is a series of the form

$$\sum_{n=0}^{\infty} c_n(x - x_0)^n = c_0 + c_1(x - x_0) + c_2(x - x_0)^2 + \cdots \tag{3.1}$$

The first term may appear to be undefined when $x = x_0$, since 0^0 has no meaning. By convention, however, $(x - x_0)^0$ is 1 even when $x = x_0$.

Our first task is to determine for which values of x the series (3.1) converges. The following lemma gives the key fact.

Lemma 3.1 *Suppose that* $|x_2 - x_0| > |x_1 - x_0|$. *If* (3.1) *converges for* $x = x_2$, *it converges absolutely for* $x = x_1$.

Proof. Since $\sum_{n=0}^{\infty} c_n(x_2 - x_0)^n$ converges, its terms tend to 0 (Proposition 1.3, Chapter 16). By Proposition 6.5 of Chapter 3, they are bounded; that is, we can find a number K such that $|c_n(x_2 - x_0)^n| \le K$ for all n. It follows that

$$|c_n(x_1 - x_0)^n| = |c_n(x_2 - x_0)^n| \left|\frac{x_2 - x_0}{x_1 - x_0}\right|^n \le K\left|\frac{x_2 - x_0}{x_1 - x_0}\right|^n$$

Since $\sum_{n=0}^{\infty} K|(x_2 - x_0)/(x_1 - x_0)|^n$ is a convergent geometric series, $\sum_{n=0}^{\infty} c_n(x_1 - x_0)^n$ converges absolutely by the comparison test (Theorem 2.2, Chapter 16).

Proposition 3.2 *There is a number $R \geq 0$ (R may be ∞) such that the series (3.1) converges absolutely when $|x - x_0| < R$ and diverges when $|x - x_0| > R$.*

The number R is called the *radius of convergence* of the series; if it is ∞, the series converges for all x.

Proof. Let S be the set of all numbers $s \geq 0$ such that there is a number x with $|x - x_0| = s$ for which the series converges. [Equivalently, $S = \{|x - x_0| : (3.1) \text{ converges at } x\}$.] S is not empty, since the series converges at $x = x_0$, and therefore $0 \in S$.

There are two cases. If S is unbounded, then for any number x we can find $x_1 \in S$ such that $|x_1 - x_0| > |x - x_0|$. By Lemma 3.1, the series converges absolutely at x. We can therefore set $R = \infty$.

If S is bounded, it has a least upper bound R. Then if $|x - x_0| > R$, the series diverges at x (otherwise $|x - x_0| \in S$ and R is not an upper bound). If $0 < |x - x_0| < R$, then we can find a number x_1 such that $|x - x_0| < |x_1 - x_0|$ and the series converges at x_1. By Lemma 3.1, the series converges absolutely at x. Thus R satisfies the conditions of the proposition.

Remarks

1. If $R = 0$, the series diverges for all $x \neq x_0$. An example of such a series is $\sum_{n=0}^{\infty} n!x^n$. [See Exercise 1(j).]
2. At the points where $|x - x_0| = R$, we can say little about the series. Here are some examples of different behavior for series, all of which have radius of convergence 1 (as the next proposition will show):

   (a) $\sum_{n=0}^{\infty} x^n$ diverges whenever $|x| = 1$.
   (b) $\sum_{n=1}^{\infty} x^n/n^2$ converges absolutely wherever $|x| = 1$.
   (c) $\sum_{n=1}^{\infty} x^n/n$ diverges when $x = 1$ and converges conditionally if $x = -1$.

The next proposition gives a simple way of computing the radius of convergence for many power series.

Proposition 3.3 *Suppose that $\sum_{n=0}^{\infty} c_n(x - x_0)^n$ is a power series such that $\lim_{n \to \infty} |c_{n+1}/c_n| = q$. Then the radius of convergence of the series is $1/q$.*

Proof. If $|x - x_0| < 1/q$, then

$$\lim_{n \to \infty} \left| \frac{c_{n+1}(x - x_0)^{n+1}}{c_n(x - x_0)^n} \right| = \lim_{n \to \infty} \left| \frac{c_{n+1}}{c_n} \right| |x - x_0|$$

$$= q|x - x_0| < 1$$

and the series converges by Theorem 2.2 of Chapter 16.

If $|x - x_0| > 1/q$, then

$$\lim_{n \to \infty} \left| \frac{c_{n+1}(x - x_0)^{n+1}}{c_n(x - x_0)^n} \right| > 1$$

and the series diverges by Theorem 2.2 of Chapter 16.

Note. If $q = 0$, then this proof shows that the radius of convergence of the power series is ∞.

There is a general formula for the radius of convergence of a power series (Exercise 4), but it is not very useful in practice.

In order to apply the theorems of the previous section, we need to know something about the uniform convergence of power series. The next two results answer our need.

Theorem 3.4 *Let $\sum_{n=0}^{\infty} c_n(x - x_0)^n$ be a power series with radius of convergence R. If $r < R$, then the series converges uniformly on the set $\{X : |x - x_0| \le r\}$.*

Remark. We cannot assert that the series converges uniformly on the set where the series converges; that may well be false if the series does not converge at the radius of convergence. It is true, however, that if the series converges at $x - x_0 = R$, then it converges uniformly for $0 \le x - x_0 \le R$. This result is known as *Abel's theorem*. (See Exercise 5.)

Proof. For convenience, we assume $x_0 = 0$. Choose a number r_1 with $r < r_1 < R$. The series $\sum_{n=0}^{\infty} |c_n r_1^n|$ converges, and if $|x| \le r_1$,

$$|c_n x^n| \le |c_n r_1^n|$$

Now Theorem 1.1 says that the series converges uniformly for $|x| \le r_1$.

Proposition 3.5 *Let $f(x) = \sum_{n=0}^{\infty} c_n(x - x_0)^n$ be a power series with radius of convergence R, and let $g(x) = \sum_{n=0}^{\infty} (n + 1)c_{n+1}(x - x_0)^n$. [The series for $g(x)$ is obtained by differentiating the series for $f(x)$ termwise.] Then the series for g also has radius of convergence R.*

Proof. Once again, we take $x_0 = 0$. Let the radius of convergence of the series for g be R_1. We need to prove that $R = R_1$. There are two parts.

(a) We first prove that $R \le R_1$. We need to show that if $|x| < R$, then the series for g converges absolutely at x. Choose a number r with $|x| < r < R$. The series $\sum_{n=0}^{\infty} c_n r^n$ converges, and, just as in Lemma 2.1, the numbers $c_n r^n$ are bounded by some number K. Thus

$$|(n+1)c_{n+1}x^n| = \frac{(n+1)}{r} c_{n+1} r^{n+1} \left|\frac{x}{r}\right|^n \leq \frac{n+1}{r} K \left|\frac{x}{r}\right|^n$$

The series $\sum_{n=0}^{\infty} [(n+1)/r]K|x/r|^n$ converges, by the ratio test (the ratio of successive terms is $[(n+2)/(n+1)]|x/r|$ which converges to $|x/r| < 1$). Therefore $\sum_{n=0}^{\infty} (n+1)c_{n+1}x^n$ converges absolutely, as claimed.

(b) Now we show that $R_1 \leq R$. We need to know that if $|x| < R_1$, then the series for f converges absolutely at x. We know that

$$\sum_{n=0}^{\infty} (n+1)c_{n+1}x^n$$

converges absolutely. Multiply by x; then we see that

$$\sum_{n=0}^{\infty} (n+1)c_{n+1}x^{n+1} = \sum_{n=1}^{\infty} nc_n x^n$$

converges absolutely, and therefore (by the comparison test) that

$$\sum_{n=1}^{\infty} c_n x^n$$

converges absolutely. Now add the term c_0 to the series to get the result.

Theorem 3.6 *The power series $\sum_{n=0}^{\infty} c_n(x-x_0)^n$ can be differentiated and integrated termwise in any interval $[a, b]$ where $|a - x_0|$ and $|b - x_0|$ are both less than R, the radius of convergence.*

Proof. Since all of the series in question converge uniformly on the interval, we simply apply Theorems 2.1 and 2.2.

Exercises

1. Find the radius of convergence of each of the following series and check the convergence at the endpoints of the corresponding interval.

(a) $\sum_{n=0}^{\infty} x^n$

(b) $\sum_{n=1}^{\infty} \frac{x^n}{n}$

(c) $\sum_{n=0}^{\infty} (-1)^n \frac{x^{2n}}{(2n)!}$

(d) $\sum_{n=0}^{\infty} (-1)^{n+1} \frac{x^{2n+1}}{(2n+1)!}$

(e) $\sum_{n=0}^{\infty} \frac{x^n}{2^n}$

(f) $\sum_{n=1}^{\infty} nx_n$

(g) $\sum_{n=1}^{\infty} n^2 x^n$

(h) $\sum_{n=1}^{\infty} c^{n^2} x^n \qquad 0 < c < \infty$

(i) $\sum_{n=0}^{\infty} \frac{\alpha(\alpha-1)\cdots(\alpha-n+1)}{n!} x^n$ (j) $\sum_{n=0}^{\infty} n!\,x^n$

$\alpha > 0$ and not an integer

2. Given that $\sum_{n=0}^{\infty} x^n = 1/(1-x)$ where it converges, find power series expressions for

(a) $\frac{1}{1+x}$ (b) $\ln(1+x)$

(c) $\frac{1}{1+x^2}$ (d) $\arctan \frac{1}{1+x^2}$

3. We have stated that a power series $\sum_{n=0}^{\infty} c_n(x-x_0)^n$ can be integrated or differentiated termwise on the interval $|x-x_0| < R$, where R is the radius of convergence. However, the power series generally does not converge uniformly on that open interval. Give an example where it does not converge uniformly on $|x-x_0| < R$.

*4. Show that the radius of convergence of $\sum_{n=0}^{\infty} c_n x^n$ is $\liminf |c_n|^{-1/n}$. (See Exercise 14 of Section 16.2 and Exercise 12 of Section 7.2.)

*5. Abel's theorem states that if the power series $\sum_{n=0}^{\infty} c_n(x-x_0)^n$ converges when $x - x_0 = R$, the radius of convergence, then it converges uniformly on the interval $[x_0, x_0+R]$. (A similar result holds for $x - x_0 = -R$.) Here is a sketch of the proof. (See Exercise 14 of Section 16.4 for a similar proof.)
(a) Show that we may assume $x_0 = 0$ and $R = 1$.
(b) Given $\varepsilon > 0$, choose N such that if $n > N$, then $|\sum_{j=n+1}^{\infty} c_j| < \varepsilon/4$. Now apply partial summation (see the proof of Theorem 4.3, Chapter 16) to $\sum_{j=n+1}^{m} c_j x^j$. Show that if $t_k = \sum_{j=n+1}^{k} c_j$, then $|t_k| < \varepsilon/2$ and

$$\left|\sum_{j=n+1}^{m} c_j x^j\right| < \frac{\varepsilon}{2}(1-x^m) + t_m x^{-m} < \frac{\varepsilon}{2}$$

Conclude from this that the series converges uniformly on the interval [0, 1].

*6. An analogue of power series, using integrals, is the Laplace transform. Let $f\colon [0, \infty) \to R$ be continuous. We define the *Laplace transform* of f, $\mathscr{L}f$, by

$$\mathscr{L}f(s) = \int_0^{\infty} f(t)e^{-st}\,dt$$

when the integral converges. (See Section 16.1, Exercise 12.)
(a) Suppose that $\mathscr{L}f(s_0)$ exists. Show that $\mathscr{L}f(s)$ exists for $s > s_0$. [*Hint*: Write $\mathscr{L}f(s) = \int_0^{\infty} f(t)e^{-s_0t}e^{-(s-s_0)t}\,dt$ and integrate by parts.]
(b) In the above situation, let $s_1 > s_0$. Show that the integral for $\mathscr{L}(s)$ converges *uniformly* on $[s_1, \infty)$.
(c) Let $g(t) = tf(t)$ and assume $|f(t)e^{-s_0t}|$ is bounded for $t \geq 0$. Show that on $[s_1, \infty)$, $s_1 > s_0$,

$$\frac{d}{ds}(\mathscr{L}f)(s) = (\mathscr{L}g)(s)$$

(d) Show that if f is of class C^1 and $|f(t)e^{-s_0 t}|$ is bounded for $t \geq 0$, then

$$(\mathscr{L}f')(s) = s(\mathscr{L}f)(s) - f(0)$$

4. Taylor's Series

Let $f(x) = \sum_{n=0}^{\infty} c_n(x - x_0)^n$, where the power series has a nonzero radius of convergence. What can we say about the relationship between f and the c_n?

To begin with, set $x = x_0$; we see that

$$f(x_0) = c_0$$

Next, using Theorem 3.6,

$$g'(x) = \sum_{n=1}^{\infty} nc_n(x - x_0)^{n-1}$$

and so

$$f'(x_0) = c_1$$

By repeatedly differentiating, we find that

$$f^{(n)}(x_0) = n!\, c_n$$

That is, if f is given by a power series about x_0, then

$$f(x) = \sum_{n=0}^{\infty} \frac{f^{(n)}(x_0)}{n!}(x - x_0)^n$$

In particular, f is infinitely differentiable at x_0.

Now we try to reverse the process: given a function f which is infinitely differentiable at x_0, we ask if

$$f(x) = \sum_{k=0}^{\infty} \frac{f^{(k)}(x_0)}{k!}(x - x_0)^k$$

One answer is provided by using Taylor's theorem (Theorem 2.1, Chapter 6):

$$f(x) = \sum_{k=0}^{n} \frac{f^{(k)}(x_0)}{k!}(x - x_0)^k + R_n(x) \tag{4.1}$$

where

$$R_n(x) = \frac{f^{(n+1)}(x_1)}{(n+1)!}(x - x_0)^{n+1}$$

for some number x_1 between x_0 and x. If the sequence $\{|R_n(x)|\}$ converges to zero, then

$$\sum_{k=0}^{\infty} \frac{f^{(k)}(x_0)}{k!}(x - x_0)^k = f(x)$$

It is not always true that the sequence $R_n(x)$ converges to zero. (See Exercise 5.) For most familiar functions, however, it is true, and

$$\sum_{k=0}^{\infty} \frac{f^{(k)}(x_0)}{k!}(x - x_0)^k = f(x)$$

on some interval about x_0. Any function for which this result holds is called *analytic near* x_0, and the formula

$$\sum_{k=0}^{\infty} \frac{f^{(k)}(x_0)}{k!}(x - x_0)^k = f(x)$$

is called the *Taylor's series* expansion of f at x_0. (When $x_0 = 0$, the corresponding series is sometimes called *Maclaurin's series*, for no good reason.)

Here are two examples.

1. Let $f(x) = e^x$ and $x_0 = 0$. Then $f^{(n)}(x) = e^x$, $f^{(n)}(0) = 1$, and if $x \geq 0$,

$$|R_n(x)| = \frac{e^{x_1}}{(n+1)!}x^n \leq \frac{e^x x^n}{(n+1)!}$$

which does converge to zero. [One proof of convergence is this: $\sum_{n=0}^{\infty} e^x x^n/(n+1)!$ converges, by the ratio test, and therefore its terms converge to zero.] Therefore

$$e^x = \sum_{n=0}^{\infty} \frac{x^n}{n!}$$

for $x \geq 0$. When $x < 0$, a similar analysis shows that

$$|R_n(x)| \leq \frac{|x^{n+1}|}{(n+1)!}$$

which also converges to 0. Thus for all x,

$$e^x = \sum_{n=0}^{\infty} \frac{x^n}{n!}$$

2. Let $f(x) = (1 + x)^\alpha$, $x_0 = 0$ (α need not be an integer). Some calculation (Exercise 3) shows that the Taylor's series of f is

$$\sum_{n=0}^{\infty} \frac{\alpha(\alpha - 1) \cdots (\alpha - n + 1)}{n!} x^n \tag{4.2}$$

which has radius of convergence 1 (unless α is a nonnegative integer, in which case the terms for $n = \alpha, \alpha + 1, \ldots$ are all 0). Estimating the remainder is not so easy, but a trick lets us show that the series converges to $(1 + x)^\alpha$ when $|x| < 1$. Let $g(x)$ be the series (4.2). Then

$$g'(x) = \sum_{n=1}^{\infty} \frac{\alpha(\alpha-1)\cdots(\alpha-n+1)}{(n-1)!}x^{n-1} = \sum_{n=0}^{\infty} \frac{\alpha(\alpha-1)\cdots(\alpha-n)}{n!}x^n$$

and

$$xg'(x) = \sum_{n=1}^{\infty} \frac{\alpha(\alpha-1)\cdots(\alpha-n+1)}{(n-1)!}x^{n} = \sum_{n=0}^{\infty} \frac{\alpha(\alpha-1)\cdots(\alpha-n+1)n}{n!}x^n$$

(The $n = 0$ term of the last series is 0, so that we can add it on.) Thus, for $|x| < 1$,

$$(1+x)g'(x) = \sum_{n=0}^{\infty} \frac{\alpha(\alpha-1)\cdots(\alpha-n+1)}{n!}(\alpha-n+n)x^n = \alpha g(x)$$

Let $h(x) = g(x)/(1+x)^{\alpha}$. Then

$$h'(x) = \frac{(1+x)^{\alpha}g'(x) - \alpha(1+x)^{\alpha-1}g(x)}{(1+x)^{2\alpha}} = \frac{(1+x)g'(x) - \alpha g(x)}{(1+x)^{\alpha+1}} = 0$$

so h is constant. Since $h(0) = 1$, we know that $h(x) = 1$ for all x (with $|x| < 1$), and

$$g(x) = (1+x)^{\alpha}$$

as claimed.

Devices like these can be used to verify that the Taylor series for most common functions converge to the functions (at least on some interval). There are some other tricks which can be used to compute the Taylor series of some functions; here is an example. (Others will be found in the exercises.)

3. From the formula for the geometric series, we know that, for $|t| < 1$,

$$\frac{1}{1+t^2} = \sum_{n=0}^{\infty} (-1)^n t^{2n} \tag{4.3}$$

Integrate both sides of (4.3) from 0 to x (where $|x| < 1$); from Theorem 3.6, we get

$$\begin{aligned}\arctan x &= \int_0^x \frac{1}{1+t^2}\,dt \\ &= \sum_{n=0}^{\infty} \int_0^x (-1)^n t^2\,dt \\ &= \sum_{n=0}^{\infty} (-1)^n \frac{x^{2n+1}}{2n+1}\end{aligned}$$

or $$\arctan x = x - \frac{x^3}{3} + \frac{x^5}{5} - \frac{x^7}{7} + \cdots$$

Exercises

1. Determine the Taylor series for the following functions about $x = 0$.

   (a) $\sin x$ (b) $\cos x$
   (c) $\sinh x$ (d) $\cosh x$

   Here sinh and cosh are the hyperbolic sine and hyperbolic cosine defined by

   $$\sinh x = \frac{e^x - e^{-x}}{2} \qquad \cosh x = \frac{e^x + e^{-x}}{2}$$

2. Determine the Taylor series for the function $f(x)$ about $x = x_0$, where

   (a) $f(x) = \sin x$, $x_0 = \pi/3$ (b) $f(x) = \cos x$, $x_0 = \pi/3$
   (c) $f(x) = 3x^3 + 7x^2 + 4x - 1$, $x_0 = -1$ (d) $f(x) = e^x$, $x_0 = 1$

3. Verify that (4.2) is the Taylor series for $(1 + x)^\alpha$ about $x = 0$.
4. Determine the Taylor series for $f(x) = 1/(1 + x^2)$ about $x = 0$. Compare the answer with (4.3).
5. Show that $1/(1 + x) = \sum_{n=0}^{\infty} (-1)^n x^n$ ($|x| < 1$). Use this fact to find a series expansion for $\ln (1 + x)$.
6. Compute the first four nonzero terms of the Taylor series for
   (a) $f(x) = e^{x^2}$, about 0.
   (b) $f(x) = \sin x^2$, about 0.
   (c) $f(x) = \sin^2 x$, about 0.
   (d) $f(x) = \tan x$, about 0.
7. Show that if k is a positive integer or 0, then (for $|x| < 1$)

   (a) $$\sum_{n=0}^{\infty} n(n + 1) \cdots (n + k - 1)x^n = \frac{k!x}{(1 - x)^{k+1}}$$

   (b) $$\sum_{n=0}^{\infty} n(n - 1) \cdots (n - k + 1)x^n = \frac{k!x^k}{(1 - x)^{k+1}}$$

   (*Hint for (a)*: Use induction and Theorem 3.6.)

*8. Let $f(x) = \sum_{n=0}^{\infty} a_n x^n$ have radius of convergence $>r$. Show that there is a constant K such that if $|x| < r$ and $k \geq 0$ is an integer, then

   $$|f^{(k)}(x)| \leq \frac{Kk!}{(1 - |x|/r)^{k+1}} \left(\frac{1}{r}\right)^k$$

   [*Hint*: We know that $\sum_{n=0}^{\infty} a_n r^n$ converges; thus there is a number K with $|a_n r^n| < K$ for all n. Compute the series for $f^{(k)}(x)$ and use Exercise 7.]

*9. Let $f(x) = \sum_{n=0}^{\infty} a_n x^n$ and $g(x) = \sum_{n=0}^{\infty} b_n x^n$, both with radius of convergence greater than zero.

(a) Show that the Taylor series for $f(x)\,g(x)$ is

$$\sum_{n=0}^{\infty} c_n x^n$$

where
$$c_n = \sum_{j=0}^{n} a_j b_{n-j}$$

(b) Show that it has a radius of convergence greater than zero.
(c) Show that for x sufficiently close to 0, $\sum_{n=0}^{\infty} c_n x^n = f(x)\,g(x)$.
(It is actually true that this last formula holds whenever the series for f and g both converge absolutely, but proving this is somewhat harder.) [*Hint*: Use Leibniz' rule:

$$\frac{d^k}{dx^k}(fg) = \sum_{j=0}^{k} \frac{k!}{j!\,(k-j)!}\frac{d^j f}{dx^j}\frac{d^{k-j} g}{dx^{k-j}}$$

For (b) and (c), use Exercise 8 to estimate the derivatives of fg. In (c), use this estimate to show that the remainder term approaches 0 for small enough x.]

10. Use the result of Exercise 9(a) to compute the following Taylor series:
(a) $\dfrac{1}{(1-x)^2}$
(b) $\sin x \cos x$
(c) $e^x \sin x$ (first four nonzero terms)
(d) $e^x/(1-x)$ (first four nonzero terms)
(e) $(1+x)\ln(1+x)$ (first four nonzero terms)

11. Let $f(x) = \sum_{n=0}^{\infty} a_n x^n$ and $g(x) = \sum_{n=0}^{\infty} b_n x^n$ with $b_0 \neq 0$. It is a fact (which we shall not prove here) that the Taylor series about 0 of $f(x)/g(x)$ can be computed by doing an "infinite long division": one divides $b_0 + b_1 x + b_2 x^2 + \cdots$ into $a_0 + a_1 x + a_2 x^2 + \cdots$, obtaining

$$\frac{a_0}{b_0} + \left(\frac{a_1}{b_0} - \frac{b_1 a_0}{b_0^2}\right)x + \cdots$$

Use this procedure to find the first four nonzero terms of the Taylor series for
(a) $\tan x$
(b) $\sec x$
about $x = 0$.

12. (a) Let $\alpha = \arctan \frac{1}{2}$ and $\beta = \arctan \frac{1}{3}$. Show that

$$\tan(\alpha + \beta) = 1$$

and hence that $\pi/4 = \alpha + \beta = \arctan \frac{1}{2} + \arctan \frac{1}{3}$.
(b) Use (4.4) and the result of (a) to compute π to three decimal places.
(c) Show, by a procedure like that in (a), that

$$\frac{\pi}{4} = 4 \arctan \frac{1}{5} - \arctan \frac{1}{239}$$

[This formula and (4.4) combine to give a formula which has often been used to compute π.]

13. Let $f(x) = \sum_{n=0}^{\infty} a_n x^n$ and $g(x) = \sum_{n=0}^{\infty} b_n x^n$ with $b_0 = 0$. One can get the Taylor series for $f \circ g$ about 0 by substituting the series for g into the series for f:

$$f(g(x)) = \sum_{n=0}^{\infty} a_n \left(\sum_{k=0}^{\infty} b_k x^k \right)^n$$

This series has a positive radius of convergence (if the series for f and g do) and converges to $f \circ g$ near 0. We shall not prove this here.

Use the above result to compute the Taylor series about 0 of
(a) e^{x^2}
(b) $\cos x^2$
(c) $\cos(\sin x)$ (the first three nonzero terms will suffice).

*14. Let

$$f(x) = \begin{cases} e^{-1/x^2} & x \neq 0 \\ 0 & x = 0 \end{cases}$$

Show that f is infinitely differentiable for all x and that $f^{(n)}(x) = 0$ for all x. (At 0, use difference quotients to compute the derivative, since f is not given by a formula there.) Thus the Taylor series for f at 0 is 0, and cannot converge to f near 0.

5. Fourier Series

In this section, we deal with a different sort of infinite series, Fourier series. The subject of Fourier series is still an important area of research, and we cannot hope to do justice to it here. We shall instead give some examples, describe some of the interesting results in the subject, and prove one of them.

A *Fourier series* is one of the form

$$a_0 + \sum_{j=1}^{\infty} (a_j \cos jx + b_j \sin jx) \tag{5.1}$$

If this series converges to a function f, we might expect to be able to compute the coefficients $a_0, a_j, b_j, 1 \leq j < \infty$, in terms of f. Our first result substantiates this expectation.

Proposition 5.1 *Suppose that the series* (5.1) *converges uniformly to a function f on the interval $[-\pi, \pi]$. Then, for any $k > 0$,*

$$a_0 = \frac{1}{2\pi} \int_{-\pi}^{\pi} f(x)\, dx$$

$$a_k = \frac{1}{\pi} \int_{-\pi}^{\pi} f(x) \cos kx \, dx$$

$$b_k = \frac{1}{\pi} \int_{-\pi}^{\pi} f(x) \sin kx \, dx$$

Proof. We begin with a lemma.

Lemma 5.2 *For any pair i, j of integers, we have*

$$\int_{-\pi}^{\pi} \sin ix \sin jx \, dx = \begin{cases} 0 & \text{if } i \neq j \\ \pi & \text{if } i = j \end{cases}$$

$$\int_{-\pi}^{\pi} \cos ix \cos jx \, dx = \begin{cases} 0 & \text{if } i \neq j \\ \pi & \text{if } i = j \end{cases}$$

and

$$\int_{-\pi}^{\pi} \sin ix \cos jx \, dx = 0$$

The proof of this lemma is elementary; we leave it to the reader. (See Exercise 8.)

We can now prove Proposition 5.1. To begin with,

$$\int_{-\pi}^{\pi} f(x) \, dx = \int_{-\pi}^{\pi} \left(a_0 + \sum_{j=1}^{\infty} a_j \cos jx + b_j \sin jx \right) dx$$

$$= \int_{-\pi}^{\pi} a_0 + \sum_{j=1}^{\infty} a_j \int_{-\pi}^{\pi} \cos jx \, dx + \sum_{j=1}^{\infty} b_j \int_{-\pi}^{\pi} \sin kx \, dx$$

by Theorem 1.3. However,

$$\int_{-\pi}^{\pi} \cos kx \, dx = \int_{-\pi}^{\pi} \sin kx \, dx = 0$$

for $j \neq 0$. Thus, we have

$$\int_{-\pi}^{\pi} f(x) \, dx = 2\pi a_0$$

which is the first identity of Proposition 5.1. Similarly,

$$\int_{-\pi}^{\pi} f(x) \cos kx \, dx = a_0 \int_{-\pi}^{\pi} \cos kx \, dx + \sum_{j=1}^{\infty} a_j \int_{-\pi}^{\pi} \cos jx \cos kx \, dx$$

$$+ \sum_{j=1}^{\infty} b_j \int_{-\pi}^{\pi} \sin jx \cos kx \, dx = \pi a_k$$

by Theorem 1.3 and Lemma 5.2. The remaining identities,

$$\int_{-\pi}^{\pi} f(x) \sin kx\, dx = \pi b_k$$

are proved in the same way.

Suppose that the series (5.1) converges for $x \in [-\pi, \pi)$. Then, in fact, the series converges for all x. To see this, we note that if x is any real number, we can find a unique integer n such that $y = x - 2n\pi \in [-\pi, \pi)$. It follows that

$$\cos jx = \cos j(y + 2n\pi) = \cos jy$$

and

$$\sin jx = \sin j(y + 2n\pi) = \sin jy$$

Thus the series (5.1) evaluated at x is the same as the series (5.1) evaluated at y. However, this series evaluated at y converges since $y \in [-\pi, \pi)$.

If $f(x)$ is the function determined by (5.1), then

$$f(x + 2\pi) = f(x)$$

for all x. Such a function is said to be *periodic with period* 2π.

If $g: [-\pi, \pi) \to \mathbf{R}$ is any function, we can extend g to a periodic function f by defining $f(x) = g(y)$, where $y = x - 2n\pi \in [-\pi, \pi)$, n an integer. For example, if g: $[-\pi, \pi) \to \mathbf{R}$ is given by $g(x) = x$, then the corresponding periodic function has the graph shown in Figure 5.1.

We assume, for the remainder of this chapter, that all functions to be discussed are periodic with period 2π.

Suppose f is a periodic function which is integrable on the interval $[-\pi, \pi]$. We define the *Fourier series of the function f* to be the Fourier series whose coefficients are given by the equations in Proposition 5.1. The numbers a_k and b_k are called the *Fourier coefficients* of f.

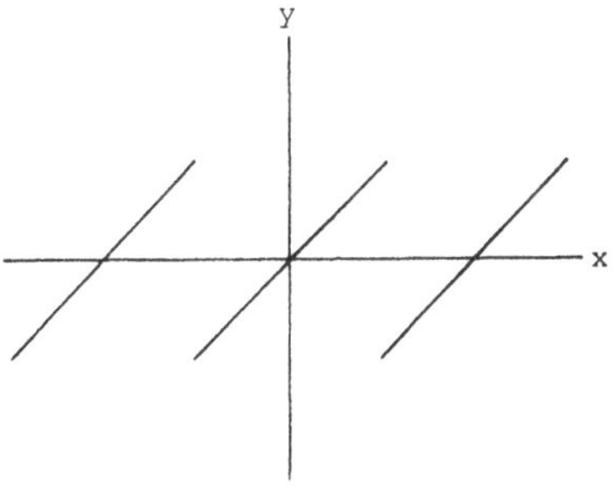

Figure 5.1 The periodic extension of $f(x) = x$.

Here are two examples.

1. Let f be the periodic function with $f(x) = x$ for $x \in [-\pi, \pi)$. (See Figure 5.1.) The Fourier coefficients for f are given by Proposition 5.1:

$$a_0 = \frac{1}{2\pi} \int_{-\pi}^{\pi} x\,dx$$

$$a_k = \frac{1}{\pi} \int_{-\pi}^{\pi} x \cos kx\,dx$$

$$b_k = \frac{1}{\pi} \int_{-\pi}^{\pi} x \sin kx\,dx$$

Now

$$\int_{-\pi}^{\pi} x\,dx = \frac{x^2}{2}\bigg|_{-\pi}^{\pi} = 0$$

so that $a_0 = 0$. To evaluate a_k, $k > 0$, we integrate by parts; let

$$u = x \qquad dv = \cos kx\,dx$$

so that

$$du = dx \qquad v = \frac{1}{k} \sin kx$$

Thus

$$\begin{aligned} \int_{-\pi}^{\pi} x \cos kx\,dx &= uv\bigg|_{-\pi}^{\pi} - \int_{-\pi}^{\pi} v\,du \\ &= \frac{1}{k} x \sin kx\bigg|_{-\pi}^{\pi} - \frac{1}{k}\int_{-\pi}^{\pi} \sin kx\,dx \\ &= 0 + \frac{1}{k^2} \cos kx\bigg|_{-\pi}^{\pi} = 0 \end{aligned}$$

and $a_k = 0$. Similarly, we have

$$b_k = (-1)^{k+1} \frac{2}{k}$$

Therefore, the Fourier series for the function $f(x) = x$ is

$$2 \sum_{k=1}^{\infty} (-1)^{k+1} \frac{\sin kx}{k} = 2\left(\frac{\sin x}{1} - \frac{\sin 2x}{2} + \frac{\sin 3x}{3} - \cdots\right)$$

2. Let f be the periodic function with

$$f(x) = \begin{cases} -1 & -\pi \le x < 0 \\ 1 & 0 \le x < \pi \end{cases}$$

(See Figure 5.2.) Again, $a_k = 0$ for all $k \ge 0$ and

$$b_k = \begin{cases} 0 & \text{if } k \text{ is even} \\ \dfrac{4}{\pi k} & \text{if } k \text{ is odd} \end{cases}$$

Thus the Fourier series for this function is

$$\frac{4}{\pi}\sum_{j=0}^{\infty} \frac{\sin(2j+1)x}{2j+1} = \frac{4}{\pi}\left(\frac{\sin x}{1} + \frac{\sin 3x}{3} + \frac{\sin 5x}{5} + \cdots\right)$$

Now that we have defined the Fourier series of a periodic function, it is natural to ask when the Fourier series of a function converges to the function. It turns out, for instance, that the Fourier series of a continuous function need not converge pointwise to the function. (The examples are somewhat complicated.) The problem of convergence of Fourier series is a difficult one and is still a subject of research in mathematics. We shall concern ourselves with a comparatively simple result.

In order to state our theorem, we need to define a piecewise differentiable function. Roughly speaking, such a function is made up of chunks of continuously differentiable functions. More precisely, a periodic function with period 2π is said to be *piecewise differentiable* if we can find a partition

$$-\pi = x_0 < x_1 < \cdots < x_n = \pi$$

of the interval $[-\pi, \pi]$ and functions $g_1, \ldots, g_n$ such that, for each $k = 1, \ldots, n$,

(i) g_k is continuously differentiable on an open interval containing $[x_{k-1}, x_k]$.
(ii) $f(x) = g_k(x)$ for all $x \in (a_{k-1}, a_k)$.

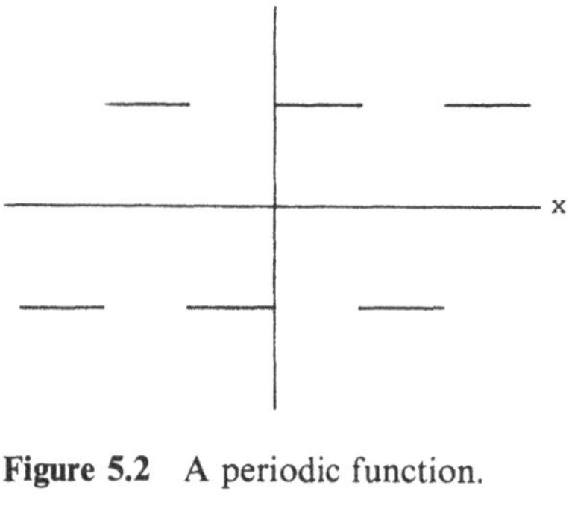

Figure 5.2 A periodic function.

Examples

1. Let

$$f(x) = \begin{cases} -1 & \text{if } -\pi < x < 0 \\ 1 & \text{if } 0 < x < \pi \\ 0 & \text{if } x = 0 \text{ or } \pi \end{cases}$$

   and extend f to be periodic with period 2π. Then we may partition $[-\pi, \pi]$ as $-\pi < 0 < \pi$, and let $g_1(x) = -1, g_2(x) = 1$. Clearly, f agrees with g_1 on $(-\pi, 0)$, and with g_2 on $(0, \pi)$. Notice that there is no restriction on f at the endpoints, $-\pi$, 0, and π.
2. Let $f(x) = |x|$ for $-\pi \leq x \leq \pi$. Again, we may partition $[-\pi, \pi]$ as $-\pi < 0 < \pi$; we let $g_1(x) = -x$ and $g_2(x) = x$.

As these examples show, a piecewise differentiable function consists of pieces which are continuously differentiable, but which need not meet differentiably or even continuously.

Let f be a piecewise differentiable function,

$$-\pi = x_0 < x_1 < \cdots < x_n = \pi$$

the partition of $[-\pi, \pi]$, and $g_1, \ldots, g_n$ the corresponding functions. At the point x_k, the function f need not have a limit. However, it does have a "one-sided limit." That is, if $x > x_k$ and x approaches x_k, then $f(x)$ approaches $g_{k+1}(x_k)$ (since g_{k+1} is defined on an open interval containing $[x_k, x_{k+1}]$). To make this precise, we say that

$$\lim_{x \to a^+} f(x) = c_0$$

means that for every $\varepsilon > 0$, there is a $\delta > 0$ such that if $0 < x - a < \delta$, then $|f(x) - c_0| < \varepsilon$. Similarly,

$$\lim_{x \to a^-} f(x) = c_1$$

means that for every $\varepsilon > 0$, there is a $\delta > 0$ such that if $0 < a - x < \delta$, then $|f(x) - c_1| < \varepsilon$.

If f is the function defined in Example 1,

$$f(x) = \begin{cases} -1 & -\pi < x < 0 \\ 0 & x = 0 \text{ or } \pi \\ 1 & 0 < x < \pi \end{cases}$$

then

$$\lim_{x \to 0^+} f(x) = 1$$

$$\lim_{x \to 0^-} f(x) = -1$$

We write $f(a^+)$ for $\lim_{x\to a^+} f(x)$ and $f(a^-)$ for $\lim_{x\to a^-} f(x)$. It is not difficult to show that $\lim_{x\to a} f(x)$ exists if and only if $f(a^+)$ and $f(a^-)$ both exist and are equal. (See Exercise 9.)

We can now state our theorem on the convergence of Fourier series.

Theorem 5.3 *Let $f(x)$ be a piecewise differentiable function. Then*

$$\lim_{n\to\infty}\left[a_0 + \sum_{j=1}^{n} (a_j \cos jx + b_j \sin jx)\right] = \frac{1}{2}[f(x^+) + f(x^-)]$$

where a_j and b_j are defined in Proposition 5.1.

Note that $f(x^+) = f(x^-) = f(x)$ whenever f is continuous at x. Thus $f(x) = \frac{1}{2}[f(x^+) + f(x^-)]$ whenever f is continuous at x.

The proof of Theorem 5.3 is given in the next section.

We close this section with a remark about the Fourier series of a more general function.

Suppose f is a continuous periodic function with period 2π. As we indicated above, the Fourier series of such a function need not converge to f. However, we can obtain some information about f from its Fourier series.

Theorem 5.4 (Parseval's Theorem) *Let f be a continuous periodic function with period 2π. Then*

$$\int_{-\pi}^{\pi} [f(x)]^2\,dx = 2\pi a_0^2 + \pi \sum_{j=1}^{\infty} (a_j^2 + b_j^2)$$

where a_0, a_j, b_j, $1 \le j < \infty$, are defined in Proposition 5.1.

We shall not give a proof of Parseval's theorem; we shall, however, sketch a proof in Exercises 3 to 9 of Section 6.

Exercises

1. Let $f(x)$ be an *even* function [that is, $f(x) = f(-x)$]. Prove that the Fourier coefficients b_k of f must vanish and (if $k > 0$)

$$a_k = \frac{2}{\pi}\int_0^{\pi} f(x)\cos kx\,dx$$

2. Let $f(x)$ be an *odd* function [that is, $f(x) = -f(-x)$]. Prove that the Fourier coefficients a_k of f must vanish and

$$b_k = \frac{2}{\pi}\int_0^{\pi} f(x)\sin kx\,dx$$

3. Find the Fourier series for the function f, where

$$f(x) = |x| \qquad \text{for } x \in [-\pi, \pi)$$

4. Find the Fourier series for the function f, where

$$f(x) = x^2 \qquad \text{for } x \in [-\pi, \pi)$$

5. Find the Fourier series for the function f, where

$$f(x) = x \cos x \qquad \text{for } x \in [-\pi, \pi)$$

6. Find the Fourier series for the function f, where

$$f(x) = e^{ax} \qquad \text{for } x \in [-\pi, \pi)$$

7. Find the Fourier series for the function f, where

$$f(x) = (x^2 - \pi^2)^2 \qquad \text{for } x \in [-\pi, \pi)$$

8. Prove Lemma 5.2. (*Hint*: Use the addition formulas for sines and cosines to re-express $\sin ix \sin jx$, $\cos ix \cos jx$, and $\sin jx \cos jx$.)
9. Prove that $\lim_{x \to a} f(x)$ exists if and only if $f(a^+)$ and $f(a^-)$ both exist and are equal.
10. Use the Fourier series for the function $f(x) = x$ to give a proof that

$$\frac{\pi}{4} = 1 - \frac{1}{3} + \frac{1}{5} - \cdots$$

11. Prove that

$$\frac{\pi^2}{8} = 1 + \frac{1}{3^2} + \frac{1}{5^2} + \cdots$$

(*Hint*: Use Exercise 3.)

*12. Prove that

$$\sum_{n=1}^{\infty} \frac{1}{n^2} = \frac{\pi^2}{6} \qquad \sum_{n=1}^{\infty} \frac{1}{n^4} = \frac{\pi^4}{90}$$

*13. Prove that

$$1 - \frac{1}{2^2} + \frac{1}{3^2} - \frac{1}{4^2} + \cdots = \frac{\pi^2}{12}$$

6. The Proof of Theorem 5.3

We now prove Theorem 5.3. For simplicity, we assume f differentiable; the proof for f piecewise differentiable is similar (and is left as Exercise 10). We begin by obtaining a more workable expression for the partial sums.

Let a_0, a_j, b_j, $1 \leq j < \infty$, be the Fourier coefficients of the piecewise differentiable function f, and let

$$S_n(x) = a_0 + \sum_{j=1}^{n} (a_j \cos jx + b_j \sin jx)$$

be the nth partial sum. We must show that

$$\lim_{n\to\infty} S_n(x) = f(x)$$

According to Proposition 5.1 and the addition formula for cosines, we have

$$\begin{aligned} a_j \cos jx + b_j \sin jx &= \int_{-\pi}^{\pi} f(t)(\cos jt \cos jx + \sin jt \sin jx)\,dt \\ &= \int_{-\pi}^{\pi} f(t) \cos j(t-x)\,dt \end{aligned}$$

Therefore $$S_n(x) = \frac{1}{\pi}\int_{-\pi}^{\pi} f(t)\left[\frac{1}{2} + \sum_{j=1}^{n} \cos j(t-x)\right] dt$$

Now, if we change variables, setting $u = t - x$, we have

$$\begin{aligned} S_n(x) &= \int_{-\pi-x}^{\pi-x} f(u+x)\left(\frac{1}{2} + \sum_{j=1}^{n} \cos ju\right) du \\ &= \int_{-\pi}^{\pi} f(u+x)\left(\frac{1}{2} + \sum_{j=1}^{n} \cos ju\right) du \end{aligned}$$

since the integrand is periodic.

We define $D_n\colon [-\pi, \pi] \to \mathbf{R}$ by

$$D_n(u) = \begin{cases} \dfrac{\sin[(2n+1)u/2]}{2\pi \sin(u/2)} & u \neq 0 \\[2ex] \dfrac{2n+1}{2\pi} & u = 0 \end{cases} \tag{6.1}$$

This function is called the *nth Dirichlet kernel.* Note that

$$\lim_{n\to 0} \frac{\sin[(2n+1)u/2]}{2\pi \sin(u/2)} = \frac{2n+1}{2\pi}$$

(Use L'Hôpital's rule; for example, see Exercise 4, Section 6.2.) Thus D_n is continuous.

Lemma 6.1 *For all $u \in [-\pi, \pi]$, we have*

$$D_n(u) = \frac{1}{\pi}\left(\frac{1}{2} + \sum_{j=1}^{n} \cos ju\right)$$

The proof is a straightforward induction; we give it at the end of the section. (See also Exercise 2.)

Lemma 6.2 *The integral of the Dirichlet kernel is given by*

$$\int_{-\pi}^{\pi} D_n(u)\,du = 1$$

Proof. This follows from Lemma 6.1 and the fact that

$$\int_{-\pi}^{\pi} \cos ju\,du = 0$$

for $j \neq 0$.

Using Lemmas 6.1 and 6.2, we see from equation (6.1) that

$$
\begin{aligned}
S_n(x) - f(x) &= \int_{-\pi}^{\pi} f(x+u)D_n(u)\,du - f(x)\int_{-\pi}^{\pi} D_n(u)\,du \\
&= \int_{-\pi}^{\pi} [f(x+u) - f(x)]D_n(u)\,du \qquad (6.2)\\
&= \frac{1}{\pi}\int_{-\pi}^{\pi} \frac{f(x+u) - f(x)}{2\sin(u/2)} \sin[(2n+1)u/2]\,du
\end{aligned}
$$

Define $F: [-\pi, \pi] \to \mathbf{R}$ by

$$F(u) = \begin{cases} \dfrac{f(x+u) - f(x)}{2\sin(u/2)} & u \neq 0 \\ f'(x) & u = 0 \end{cases}$$

where x is fixed.

Lemma 6.3 *The function* $F: [-\pi, \pi] \to \mathbf{R}$ *defined above is continuous.*

Proof. If $u \neq 0$, then F is clearly continuous since $\sin(u/2) \neq 0$. To show that F is continuous at $u = 0$, we prove that

$$\lim_{u \to 0} F(u) = F(0)$$

First of all,

$$
\begin{aligned}
\lim_{u \to 0} F(u) &= \lim_{u \to 0} \frac{f(x+u) - f(x)}{2\sin(u/2)} \\
&= \lim_{u \to 0} \frac{f(x+u) - f(x)}{u} \cdot \frac{u/2}{\sin(u/2)}
\end{aligned}
$$

Now,

$$\lim_{u\to 0}\frac{\sin u}{u} = \lim_{u\to 0}\frac{\sin u - \sin 0}{u - 0}$$
$$= \frac{d}{du}\sin u\bigg|_{u=0}$$
$$= 1$$

Thus
$$\lim_{u\to 0}\frac{\sin (u/2)}{u/2} = 1$$

and so
$$\lim_{u\to 0}\frac{u/2}{\sin u/2} = 1$$

Therefore
$$\lim_{u\to 0} F(u) = \lim_{u\to 0}\frac{f(x+u)-f(x)}{u}\cdot\lim_{u\to 0}\frac{u/2}{\sin(u/2)}$$
$$= f'(x)$$

which completes the proof of Lemma 6.3.

Using equation (6.2) we have

$$S_n(x) - f(x) = \int_{-\pi}^{\pi} F(u)\sin[(2n+1)u/2]\,du$$

Theorem 5.3 is now a consequence of the following lemma.

Lemma 6.4 (Riemann-Lebesgue Lemma) *Let $f\colon [-\pi, \pi] \to \mathbf{R}$ be continuous. Define λ_s (for s real) by*

$$\lambda_s = \int_{-\pi}^{\pi} f(t)\sin st\,dt$$

Then
$$\lim_{s\to\infty}\lambda_s = 0$$

Proof. We prove the lemma first in the special case where f is of class C^1. For this, we use integration by parts. Let

$$u = f(t) \qquad \text{and} \qquad dv = \sin st\,dt \qquad s \neq 0$$

Then

$$du = f'(t)\,dt \qquad \text{and} \qquad v = -\frac{1}{s}\cos st$$

Thus

$$\int_{-\pi}^{\pi} f(t)\sin st\,dt = \frac{1}{s}\left[f(-\pi)\cos(-s\pi) - f(\pi)\cos s\pi + \int_{-\pi}^{\pi} f'(t)\cos st\,dt\right]$$

$$\leq \frac{1}{s}\left[\,|f(-\pi)| + |f(\pi)| + \int_{-\pi}^{\pi} |f'(t)|\,dt\right] \tag{6.3}$$

since $\cos|st| \leq 1$. But

$$|f(-\pi)| + |f(\pi)| + \int_{-\pi}^{\pi} |f'(t)|\,dt$$

is clearly bounded, so (6.3) tends to zero as s increases. This proves the lemma in the special case.

The general case is more difficult. Suppose that for every ε we can find a function g of class C^1 such that for all $x \in [-\pi, \pi]$,

$$|f(x) - g(x)| < \frac{\varepsilon}{4\pi}$$

Then, letting $\mu_s = \int_{-\pi}^{\pi} g(t)\sin st\,dt$, we have

$$|\lambda_s| = \int_{-\pi}^{\pi} f(t)\sin st\,dt \leq \left|\int_{-\pi}^{\pi} [f(t) - g(t)]\sin st\,dt\right| + \left|\int_{-\pi}^{\pi} g(t)\sin st\,dt\right|$$

$$\leq \int_{-\pi}^{\pi} |f(t) - g(t)|\,|\sin st|\,dt + |\mu_s| \tag{6.4}$$

$$< \int_{-\pi}^{\pi} \frac{\varepsilon}{4\pi}\,dt + |\mu_s| = \frac{\varepsilon}{2} + |\mu_s|$$

From the special case above, $\mu_s \to 0$ as $s \to \infty$. Thus we can pick N such that

$$|\mu_s| < \frac{\varepsilon}{2} \qquad \text{if } s > N$$

But then (6.4) says that

$$|\lambda_s| < \varepsilon \qquad \text{for } s > N$$

and hence λ_s approaches 0.

To complete the proof of Lemma 6.4 (and Theorem 5.3), we need to find g. By uniform continuity, we may find δ such that

$$|x - y| < \delta \Rightarrow |f(x) - f(y)| < \frac{\varepsilon}{4\pi}$$

Now define

$$g(x) = \frac{1}{\delta}\int_{x}^{x+\delta} f(t)\,dt$$

[When $x > \pi - \delta$, this definition does not make sense; to get around this problem, we define $g(x) = f(\pi)$ for $x > \pi$.] Then g is continuously differentiable, by the fundamental theorem of calculus, and

$$\begin{aligned}|g(x) - f(x)| &= \left|\frac{1}{\delta}\int_x^{x+\delta} [f(t) - f(x)]\,dt\right| \qquad \left[\text{since } f(x) = \frac{1}{\delta}\int_x^{x+\delta} f(x)\,dt\right]\\ &\le \frac{1}{\delta}\int_x^{x+\delta} |f(t) - f(x)|\,dt\\ &< \frac{1}{\delta}\int_x^{x+\delta} \frac{\varepsilon}{4\pi}\,dt\\ &= \frac{\varepsilon}{4\pi}\end{aligned}$$

and the proof is finished.

We conclude with a proof of Lemma 6.1. The proof is by induction. First of all,

$$\begin{aligned}\frac{1}{2} + \cos u &= \frac{\sin(u/2)}{\sin(u/2)}\left(\frac{1}{2} + \cos u\right)\\ &= \frac{\sin(u/2) + 2\sin(u/2)\cos u}{2\sin(u/2)}\end{aligned}$$

Now
$$\sin\left(v + \frac{u}{2}\right) = \sin\frac{u}{2}\cos v + \cos\frac{u}{2}\sin v$$

and
$$\sin\left(v - \frac{u}{2}\right) = -\sin\frac{u}{2}\cos v + \cos\frac{u}{2}\sin v$$

so that

$$2\sin\frac{u}{2}\cos v = \sin\left(v + \frac{u}{2}\right) - \sin\left(v - \frac{u}{2}\right) \tag{6.5}$$

Therefore (with $v = u$),

$$\begin{aligned}\frac{1}{2} + \cos u &= \frac{\sin(u/2) + \sin(u + u/2) - \sin(u/2)}{2\sin(u/2)}\\ &= \frac{\sin(1 + \frac{1}{2})u}{2\sin(u/2)}\end{aligned}$$

which shows that (b) holds when $n = 1$. If the lemma holds when $n = k$ so that

$$D_k(u) = \frac{1}{\pi}\frac{\sin(k+\frac{1}{2})u}{2\sin\frac{1}{2}u}$$

then

$$\begin{aligned} D_{k+1}(u) &= D_k(u) + \frac{1}{\pi}\cos(k+1)u \\ &= \frac{1}{\pi}\left[\frac{\sin(k+\frac{1}{2})u}{2\sin\frac{1}{2}u} + \cos(k+1)u\right] \end{aligned}$$

Let $v = (k+1)\,u$ in (6.5):

$$\begin{aligned} D_{k+1}(u) &= \frac{1}{\pi}\left[\frac{\sin(k+\frac{1}{2})u + 2\cos(k+1)u\sin\frac{1}{2}u}{2\sin\frac{1}{2}u}\right] \\ &= \frac{1}{\pi}\left\{\frac{\sin(k+\frac{1}{2})u + [\sin(k+1+\frac{1}{2})u + \sin(k+1-\frac{1}{2})u]}{2\sin\frac{1}{2}u}\right\} \\ &= \frac{1}{\pi}\frac{\sin(k+1+\frac{1}{2})u}{2\sin\frac{1}{2}u} \end{aligned}$$

as was to be checked.

Exercises

1. Prove that if f is continuous, then

$$\lim_{s\to\infty}\int_{-\pi}^{\pi} f(t)\cos st\,dt = 0$$

(This is also part of the Riemann-Lebesgue lemma.)

2. Use the fact that

$$\cos kx \sin\frac{x}{2} = \frac{1}{2}\left[\sin\left(\frac{2k+1}{2}x\right) - \sin\left(\frac{2k-1}{2}x\right)\right]$$

to give a proof of Lemma 6.1.

3. Let the periodic function f (period 2π) be continuously differentiable, with Fourier coefficients $a_0, a_j, b_j\,(1 \le j < \infty)$; let the Fourier coefficients of f' be $c_0, c_j, d_j\ (1 \le j < \infty)$. Show that

$$\begin{aligned} c_0 &= 0 \\ c_j &= nb_j \\ d_j &= -nc_j \end{aligned}$$

4. Show that if f (periodic with period 2π) is of class C^2, then the Fourier series of f converges absolutely and uniformly to f. (*Hint*: Use the result of Section 1.)

*5. (a) Suppose that f is continuous and periodic (period 2π). Show that there are periodic, continuously differentiable functions f_n such that $f = \lim_{n\to\infty} f_n$, where the limit is uniform on $[-\pi, \pi]$. [*Hint*: Consider $g_\varepsilon(x) = (1/\varepsilon)\int_x^{x+\varepsilon} f(t)\,dt$, for appropriate ε.]

(b) Show that $f = \lim_{n\to\infty} g_n$, where the limit is uniform on $[-\pi, \pi]$ and the functions g_n are periodic and of class C^2.

6. Show that if the Fourier series of f converges absolutely and uniformly to f, then

$$\int_{-\pi}^{\pi} [f(x)]^2\,dx = 2\pi a_0^2 + \pi \sum_{j=1}^{\infty} (a_j^2 + b_j^2)$$

(This is Parseval's theorem for these functions.)

7. Let V be the vector space of continuous periodic functions with period 2π, and define

$$\langle f, g\rangle = \int_{-\pi}^{\pi} f(x)\,g(x)\,dx$$

(a) Show that $\langle\,,\,\rangle$ satisfies the properties (a), (b), and (c) of Proposition 1.1 of Chapter 5. We think of $\langle\,,\,\rangle$ as an inner product on V.

(b) Show that the functions $\varphi_0(x) = 1/\sqrt{2\pi}$, $\varphi_j(x) = (1/\sqrt{\pi})\cos jx$, $\psi_j(x) = (1/\sqrt{\pi})\sin jx$ form an *orthonormal set*, relative to $\langle\,,\,\rangle$. That is, $\langle \varphi_i, \varphi_j\rangle = \langle \psi_i, \psi_j\rangle = \delta_{ij}$ and $\langle \varphi_i, \psi_j\rangle = 0$, where $\delta_{ij} = 1$ if $i = j$ and 0 otherwise.

(c) Let the Fourier coefficients of f be a_0, a_j, b_j. Show that the functions $g_0(x) = a_0$, $g_j(x) = a_j\cos jx$, $h_j(x) = b_j\sin x$ $(1 \le j \le n)$, and $f_n(x) = f(x) - g_0(x) - \sum_{j=1}^{n}[g_j(x) + h_j(x)]$ are mutually orthogonal.

(d) Show that

$$\int_{-\pi}^{\pi} [f(x)]^2\,dx = \int_{-\pi}^{\pi} [g_0(x)]^2\,dx + \int_{-\pi}^{\pi} [f_n(x)]^2\,dx$$
$$+ \sum_{j=1}^{n}\left\{\int_{-\pi}^{\pi} [g_j(x)]^2\,dx + \int_{-\pi}^{\pi} [h_j(x)]^2\,dx\right\}$$
$$\ge 2\pi a_0^2 + \pi\sum_{j=1}^{n}(a_j^2 + b_j^2)$$

(This result is known as *Bessel's inequality*.) Exercise 7 of Section 4.2 may provide a hint.

*8. Let f be a continuous periodic function with period 2π. Let a_0, a_j, b_j be the Fourier coefficients of f, and define

$$G_n(x_0, x_1, \ldots, x_n, y_1, \ldots, y_n) = \left\| f - x_0 - \sum_{j=1}^{n}(x_j\cos jx + y_j\sin jx)\right\|^2$$

(Here, $\|\varphi\|^2 = \langle\varphi, \varphi\rangle$.) Show that G_n has a minimum at

$$(a_0, a_1, \ldots, a_n, b_1, \ldots, b_n)$$

and that

$$G_n(a_0, a_1, \ldots, a_n, b_1, \ldots, b_n) = \|f\|^2 - \left[2\pi a_0^2 + \pi \sum_{j=1}^{n} (a_j^2 + b_j^2)\right]$$

*9. (a) Let f be a continuous function with period 2π. Show that for every $\varepsilon > 0$ there exists an integer n and numbers $x_0, x_1, \ldots, x_n, y_1, \ldots, y_n$ such that

$$|G_n(x_0, x_1, \ldots, x_n, y_1, \ldots, y_n)| < \varepsilon$$

where G_n is defined as in Exercise 8. [*Hint*: Let δ be a small number, and choose f_0 to be of class C^2, with $|f(x) - f_0(x)| < \delta$ for all x, as in Exercise 3. Then $\|f - f_0\|^2 < \delta^2\pi$. Now use the triangle inequality and Exercise 4.]

(b) Show that for the n of part (a),

$$\int_{-\pi}^{\pi} [f(x)]^2 - 2\pi a_0^2 - \pi \sum_{j=1}^{n} (a_j^2 + b_j^2) < \varepsilon$$

(c) Prove Parseval's theorem (Theorem 5.4).

*10. Prove Theorem 5.3 for piecewise differentiable functions. {The proof is like the one given, but one estimates $s_n(x) - \frac{1}{2}[f(x^+) + f(x^-)]$. One also needs a Riemann-Lebesgue lemma for intervals other than $[-\pi, \pi]$.}

*11. Suppose that f is periodic and continuous. Show that on $[-\pi, \pi]$, f is the uniform limit of polynomials. (This result is essentially what is known as *Weierstrass' approximation theorem.*) *Hint*: Given ε, use Exercise 5 to find a periodic function of class C^2 uniformly within $\varepsilon/3$ of f. Use Exercise 4 to approximate f uniformly by a finite linear combination of sines and cosines. Then use Taylor's theorem.)

Appendix A

Determinants*

In this appendix, we define the determinant of an $n \times n$ matrix and derive some of its properties. To begin with, it will be convenient to think of the determinant somewhat differently.

Definition The *determinant* is a function Det from n-tuples of vectors in $\mathbf{R}^n$ to $\mathbf{R}$,

$$\text{Det}: \mathbf{R}^n \times \mathbf{R}^n \times \cdots \times \mathbf{R}^n \to \mathbf{R}$$

with the following properties:

(A.1) If all of the vectors but one are fixed, Det is linear in the remaining variable.

(A.2) If any two of the vectors $\mathbf{v}_1, \ldots, \mathbf{v}_n$ are equal, then $\text{Det}(\mathbf{v}_1, \ldots, \mathbf{v}_n) = 0$.

(A.3) If $\mathbf{e}_1, \ldots, \mathbf{e}_n$ is the standard basis for $\mathbf{R}^n$, then $\text{Det}(\mathbf{e}_1, \mathbf{e}_2, \ldots, \mathbf{e}_n) = 1$.

As we shall see below, these properties determine the function Det uniquely.

For most of this discussion, we do not need to use property (A.3). We shall deal, therefore, with a function Δ which has properties (A.1) and (A.2) but not necessarily (A.3). Whatever properties we prove for Δ, apply, of course, to Det as well.

*For a detailed treatment of determinants, see L. Corwin and R. Szczarba, *Calculus in Vector Spaces*, Marcel Dekker, New York, 1979.

Proposition A.1 *Let* $\Delta\colon \mathbf{R}^n \times \cdots \times \mathbf{R}^n \to \mathbf{R}$ *be a function satisfying* (*A*.1) *and* (*A*.2). *Then interchanging any two of* $\mathbf{v}_1, \ldots, \mathbf{v}_n$ *reverses the sign of* Δ.

[For instance, if $n = 4$, then $\Delta(\mathbf{v}_1, \mathbf{v}_4, \mathbf{v}_3, \mathbf{v}_2) = -\Delta(\mathbf{v}_1, \mathbf{v}_2, \mathbf{v}_3, \mathbf{v}_4)$.]

Proof. For convenience, we interchange $\mathbf{v}_1$ and $\mathbf{v}_2$; the same proof works for any two vectors.

We know by property (A.2) that

$$\Delta(\mathbf{v}_1 + \mathbf{v}_2, \mathbf{v}_1 + \mathbf{v}_2, \mathbf{v}_3, \ldots, \mathbf{v}_n) = 0$$

However, using property (A.1) twice, we have

$$\begin{aligned}\Delta(\mathbf{v}_1 + \mathbf{v}_2, \mathbf{v}_1 + \mathbf{v}_2, \mathbf{v}_3, \ldots, \mathbf{v}_n) &= \Delta(\mathbf{v}_1 + \mathbf{v}_2, \mathbf{v}_1, \mathbf{v}_3, \ldots, \mathbf{v}_n) \\ &\quad + \Delta(\mathbf{v}_1 + \mathbf{v}_2, \mathbf{v}_2, \mathbf{v}_3, \ldots, \mathbf{v}_n) \\ &= \Delta(\mathbf{v}_1, \mathbf{v}_1, \mathbf{v}_3, \ldots, \mathbf{v}_n) + \Delta(\mathbf{v}_2, \mathbf{v}_1, \mathbf{v}_3, \ldots, \mathbf{v}_n) \\ &\quad + \Delta(\mathbf{v}_1, \mathbf{v}_2, \mathbf{v}_3, \ldots, \mathbf{v}_n) + \Delta(\mathbf{v}_2, \mathbf{v}_2, \mathbf{v}_3, \ldots, \mathbf{v}_n) \\ &= 0 + \Delta(\mathbf{v}_2, \mathbf{v}_1, \mathbf{v}_3, \ldots, \mathbf{v}_n) + \Delta(\mathbf{v}_1, \mathbf{v}_2, \mathbf{v}_3, \ldots, \mathbf{v}_n) + 0\end{aligned}$$

by property (A.2). Thus

$$\Delta(\mathbf{v}_2, \mathbf{v}_1, \mathbf{v}_3, \ldots, \mathbf{v}_n) = -\Delta(\mathbf{v}_1, \mathbf{v}_2, \mathbf{v}_3, \ldots, \mathbf{v}_n)$$

as claimed.

Theorem A.2 *Properties* (*A*.1), (*A*.2), *and* (*A*.3) *uniquely determine the function* Det. *In fact, if* Δ *satisfies properties* (*A*.1), (*A*.2), *and*

$$\text{(A.3}') \qquad \Delta(\mathbf{e}_1, \ldots, \mathbf{e}_n) = c$$

then $\Delta = c$ Det.

Proof. First of all, suppose that Δ satisfies (A.1), (A.2), and (A.3′). Then we can determine $\Delta(\mathbf{v}_1, \ldots, \mathbf{v}_n)$ whenever $\mathbf{v}_1, \ldots, \mathbf{v}_n$ are all in the standard basis. For if two of the vectors $\mathbf{v}_1, \ldots, \mathbf{v}_n$ are identical, then $\Delta(\mathbf{v}_1, \ldots, \mathbf{v}_n) = 0$. If $\mathbf{v}_1, \ldots, \mathbf{v}_n$ are all different, then they are the vectors $\mathbf{e}_1, \ldots, \mathbf{e}_n$ in some other order. We can evaluate $\Delta(\mathbf{v}_1, \ldots, \mathbf{v}_n)$ by interchanging pairs of vectors until they are in the order $\mathbf{e}_1, \ldots, \mathbf{e}_n$, using property (A.1) to evaluate Δ.

Next, we can evaluate $\Delta(\mathbf{v}_1, \ldots, \mathbf{v}_n)$ when $\mathbf{v}_2, \ldots, \mathbf{v}_n$ are in the standard basis and $\mathbf{v}_1$ is any vector in $\mathbf{R}^n$. For if $\mathbf{v}_1 = \sum_{i=1}^n x_i\mathbf{e}_i$, then property (A.1) says that

$$\Delta(\mathbf{v}_1, \mathbf{v}_2, \ldots, \mathbf{v}_n) = \sum_{i=1}^{n} x_i\Delta(\mathbf{e}_i, \mathbf{v}_2, \mathbf{v}_3, \ldots, \mathbf{v}_n)$$

We can evaluate each of the terms in the sum, since all the vectors are in the standard basis.

We continue the same way: we can evaluate $\Delta(\mathbf{v}_1, \ldots, \mathbf{v}_n)$ when $\mathbf{v}_3, \ldots \mathbf{v}_n$ are in the standard basis and $\mathbf{v}_1$, $\mathbf{v}_2$ are any vectors in $\mathbf{R}^n$; if $\mathbf{v}_2 = \sum_{i=1}^{n} y_i \mathbf{e}_i$, then

$$\Delta(\mathbf{v}_1, \mathbf{v}_2, \ldots, \mathbf{v}_n) = \sum_{i=1}^{n} y_i \Delta(\mathbf{v}_1, \mathbf{e}_i, \mathbf{v}_3, \ldots, \mathbf{v}_n)$$

and we already know how to evaluate each term in the sum. By induction (the induction step is clear, but complicated to write out), we see that $\Delta(\mathbf{v}_1, \ldots, \mathbf{v}_n)$ is uniquely determined for all vectors $\mathbf{v}_1, \ldots, \mathbf{v}_n$ in $\mathbf{R}^n$.

In particular, Det is uniquely determined. Moreover, c Det satisfies properties (A.1), (A.2), and (A.3′), and therefore (since Δ is uniquely determined) $\Delta = c$ Det.

This proof also provides a method of evaluating Det. It turns out to be a very inefficient method.

We should emphasize one fact about Theorem A.2. It does *not* show that any function satisfying properties (A.1), (A.2), and (A.3) exists; it just shows that at most one such function does. It is conceivable that properties (A.1), (A.2), and (A.3) are somehow inconsistent. Our next result states that this is not the case.

Theorem A.3 *The function* Det *really does exist.*

Before proving this theorem, we derive some of the properties of the determinant. We first introduce a slight notational change. From now on, we shall also think of Det as a function of $n \times n$ matrices. If A is an $n \times n$ matrix, we regard it as a row of n column vectors: if

$$A = \begin{bmatrix} a_{11} & \cdots & a_{1n} \\ \vdots & & \vdots \\ a_{n1} & \cdots & a_{nn} \end{bmatrix} \qquad A = (A_1, \ldots, A_n)$$

where $A_1 = \begin{bmatrix} a_{11} \\ \vdots \\ a_{n1} \end{bmatrix} \quad A_2 = \begin{bmatrix} a_{12} \\ \vdots \\ a_{n2} \end{bmatrix} \quad \cdots \quad A_n = \begin{bmatrix} a_{1n} \\ \vdots \\ a_{nn} \end{bmatrix}$

We then identify each of these column vectors with a vector in $\mathbf{R}^n$ and define Det A to be Det$(A_1, \ldots, A_n)$. Property (A.3) now says that Det $I = 1$, where I is the identity matrix.

The next result asserts that the determinant of a product is the product of the determinants.

Proposition A.4 *If A and B are $n \times n$ matrices, then*

$$\operatorname{Det} AB = (\operatorname{Det} A)(\operatorname{Det} B)$$

Proof. Define a function Δ on $n \times n$ matrices by

$$\Delta(B) = \operatorname{Det} AB$$

We show that Δ satisfies (A.1), (A.2), and (A.3′) with $c = \operatorname{Det} A$. Proposition A.4 will then follow from Theorem A.2.

Suppose $B = (B_1, \ldots, B_n)$ is a row of column vectors, or equivalently, $n \times 1$ matrices. Then

$$AB = (AB_1, \ldots, AB_n)$$

so that

$$\Delta(B) = \operatorname{Det}(AB_1, \ldots, AB_n)$$

The fact that Δ satisfies property (A.2) follows immediately from the fact that Det does; if two of the B_i are the same, so are two of the AB_i.

To see that Δ satisfies (A.1), we proceed directly. For instance,

$$\begin{aligned}\Delta(B_1 + C_1, B_2, \ldots, B_n) &= \operatorname{Det}(A(B_1 + C_1), AB_2, \ldots, AB_n)\\ &= \operatorname{Det}(AB_1 + AC_1, AB_2, \ldots, AB_n)\\ &= \operatorname{Det}(AB_1, AB_2, \ldots, AB_n)\\ &\quad + \operatorname{Det}(AC_1, AB_2, \ldots, AB_n)\\ &= \Delta(B_1, B_2, \ldots, B_n) + \Delta(C_1, B_2, \ldots, B_n)\end{aligned}$$

Theorem A.2 now states that $\Delta(B) = c \operatorname{Det} B$, where $c = \Delta(I)$. However

$$\Delta(I) = \operatorname{Det} AI = \operatorname{Det} A$$

Thus

$$\operatorname{Det} AB = \Delta(B) = c \operatorname{Det} B = (\operatorname{Det} A)(\operatorname{Det} B)$$

as claimed.

An immediate and useful corollary is Proposition A.5.

Proposition 1.5 *If A is invertible, then* $\operatorname{Det} A \neq 0$ *and* $\operatorname{Det} A^{-1} = (\operatorname{Det} A)^{-1}$.

Proof. If A is invertible, let B be its inverse. Then

$$1 = \operatorname{Det} I = \operatorname{Det} AB = (\operatorname{Det} A)(\operatorname{Det} B)$$

Thus, $\operatorname{Det} A \neq 0$ and

$$\operatorname{Det} A^{-1} = \operatorname{Det} B = (\operatorname{Det} A)^{-1}$$

This proposition has a converse.

Proposition A.6 *If A is not invertible,* Det $A = 0$.

We omit the proof of this result. However, we give a formula for A^{-1} when Det $A \neq 0$ in Exercise 11.

Another useful fact is the following.

Proposition A.7 *Let A be an $n \times n$ matrix. Adding a multiple of one column of A to a different column does not affect the value of* Det A.

For instance,

$$\text{Det}(A_1, A_2, \ldots, A_n) = \text{Det}(A_1 + cA_2, A_2, \ldots, A_n)$$

Proof. $\text{Det}(A_1 + cA_2, A_2, \ldots, A_n)$

$$= \text{Det}(A_1, A_2, \ldots, A_n) + \text{Det}(cA_2, A_2, \ldots, A_n)$$

$$= \text{Det}(A_1, \ldots, A_n) + c\,\text{Det}(A_2, A_2, \ldots, A_n)$$

and the second term is 0, by property (A.2). The general case follows in the same way.

We now prove Theorem A.3 by defining the determinant of an $n \times n$ matrix. The definition is by induction on n, the size of the matrix. If $n = 1$, we define

$$\text{Det}\,(a_{11}) = a_{11}$$

Now, assume that we have defined the determinant of an $n \times n$ matrix so that (A.1), (A.2), and (A.3) hold and let $A = (a_{ij})$ be an $(n + 1) \times (n + 1)$ matrix. Let A_{ij} (where $1 \leq i, j \leq n + 1$) be the $n \times n$ matrix obtained by deleting the row and column containing a_{ij}. Now fix a value of $i\,(1 \leq i \leq n)$, and define

$$\text{Det}\,A = \sum_{j=1}^{n+1} (-1)^{i+j} a_{ij}\,\text{Det}\,A_{ij}$$

Notice that the definition makes sense, since A_{ij} is an $n \times n$ matrix and the determinant is assumed to be defined for such matrices.

For example,

$$\text{Det}\begin{bmatrix} a_{11} & a_{12} \\ a_{21} & a_{22} \end{bmatrix} = a_{11}a_{22} - a_{12}a_{21}$$

and

$$\operatorname{Det}\begin{bmatrix} a_{11} & a_{12} & a_{13} \\ a_{21} & a_{22} & a_{23} \\ a_{31} & a_{32} & a_{33} \end{bmatrix} = a_{11}\operatorname{Det}\begin{bmatrix} a_{22} & a_{23} \\ a_{32} & a_{33} \end{bmatrix}$$

$$- a_{12}\operatorname{Det}\begin{bmatrix} a_{21} & a_{23} \\ a_{31} & a_{33} \end{bmatrix} + a_{13}\operatorname{Det}\begin{bmatrix} a_{21} & a_{22} \\ a_{31} & a_{32} \end{bmatrix}$$

$$= a_{11}(a_{22}a_{33} - a_{23}a_{32})$$

$$- a_{12}(a_{21}a_{33} - a_{23}a_{31}) + a_{13}(a_{21}a_{32} - a_{22}a_{31})$$

The easiest property to check is (A.3). Suppose that $A = I_{n+1}$ is the identity matrix. If $i \neq j$, the elements a_{ii} and a_{jj}, both of which are equal to 1, are deleted in forming A_{ij}. Since I_{n+1} has exactly $(n + 1)$ nonzero entries, A_{ij} has $(n - 1)$ nonzero entries in this case. Since A_{ij} has n columns, one of the columns consists entirely of zeros. Thus, if $i \neq j$, $\operatorname{Det} A_{ij} = 0$ by property (A.1). If $i = j$, then $A_{ii} = I_n$, the $n \times n$ identity matrix (try an example to see why). Therefore

$$\begin{aligned} \operatorname{Det} I_{n+1} &= \sum_{j=1}^{n+1} (-1)^{i+j} a_{ij} \operatorname{Det} A_{ij} \\ &= (-1)^{i+i} a_{ii} \operatorname{Det} A_{ii} \\ &= \operatorname{Det} I_n = 1 \end{aligned}$$

as required.

Property (A.1) is not too difficult, either. To simplify notation, we will prove only that Det is linear in the first column. Let B be the matrix

$$B = \begin{bmatrix} b_{11} & a_{12} & \cdots & a_{1,n+1} \\ b_{21} & a_{22} & \cdots & a_{2,n+1} \\ \vdots & \vdots & & \vdots \\ b_{n+1,1} & a_{n+1,2} & \cdots & a_{n+1,n+1} \end{bmatrix}$$

and let

$$C = \begin{bmatrix} a_{11} + b_{11} & a_{12} & \cdots & a_{1,n+1} \\ a_{21} + b_{21} & a_{22} & \cdots & a_{2,n+1} \\ \vdots & \vdots & & \vdots \\ a_{n+1,1} + b_{n+1,1} & a_{n+1,2} & \cdots & a_{n+1,n+1} \end{bmatrix}$$

We need to show that

$$\operatorname{Det} C = \operatorname{Det} A + \operatorname{Det} B$$

Let B_{ij} be the matrix obtained by deleting the ith row and jth column from B, and define C_{ij} similarly. Then our definition says that

(A.4) $$\operatorname{Det} C = (-1)^{i+1}(a_{i1} + b_{i1}) \operatorname{Det} C_{i1} + \sum_{j=2}^{n+1} (-1)^{i+j} a_{ij} \operatorname{Det}(C_{ij})$$

Since the first column is deleted to obtain A_{i1}, B_{i1}, and C_{i1}, it is clear that they are identical:

$$A_{i1} = B_{i1} = C_{i1}$$

If $j > 1$, the matrices A_{ij}, B_{ij}, and C_{ij} agree except for the first column; the first column of C_{ij} is the sum of the first columns of A_{ij} and B_{ij}. Because (A.1) holds for $n \times n$ determinants,

$$\operatorname{Det} C_{ij} = \operatorname{Det} A_{ij} + \operatorname{Det} B_{ij} \qquad \text{if } j > 1$$

Substituting these results in (A.4), we find that

$$\begin{aligned}
\operatorname{Det} C &= \left[(-1)^{i+1} a_{i1} \operatorname{Det} A_{i1} + \sum_{j=2}^{n+1} (-1)^{i+j} a_{ij} \operatorname{Det} A_{ij} \right] \\
&\quad + \left[(-1)^{i+1} b_{i1} \operatorname{Det} B_{i1} + \sum_{j=2}^{n+1} (-1)^{i+j} a_{ij} \operatorname{Det} B_{ij} \right] \\
&= \operatorname{Det} A + \operatorname{Det} B
\end{aligned}$$

as can be seen by using the determinant formula for Det A and Det B.

To finish the proof that (A.1) holds, we need to show that multiplying any column by a constant multiplies the determinant by the same constant. This proof is just like the one we have just done; we leave the details as an exercise.

We are left with (A.2). Again, we will do a special case: if the first two columns agree, Det $A = 0$. Thus we are assuming that

$$A = \begin{bmatrix} a_{11} & a_{11} & a_{13} & \cdots & a_{1,n+1} \\ a_{21} & a_{21} & a_{23} & \cdots & a_{2,n+1} \\ \vdots & \vdots & \vdots & & \vdots \\ a_{n+1,1} & a_{n+1,1} & a_{n+1,3} & \cdots & a_{n+1,n+1} \end{bmatrix}$$

Now it is easy to see that $A_{i1} = A_{i2}$ and

(A.5) If $j > 2$, the first two columns of A_{ij} are the same. Hence

$$\begin{aligned}
\operatorname{Det} A &= \sum_{j=1}^{n+1} (-1)^{i+j} a_{ij} \operatorname{Det} A_{ij} \\
&= (-1)^{i+1} a_{i1} \operatorname{Det} A_{i1} + (-1)^{i+2} a_{i2} \operatorname{Det} A_{i1} \\
&\quad + \sum_{j=3}^{n+1} (-1)^{i+j} a_{ij} \operatorname{Det} A_{ij}
\end{aligned}$$

The first two terms have opposite signs, and therefore sum to 0. The others are all 0 by the inductive hypothesis and statement (A.5). Therefore Det $A = 0$ and the proof of Theorem A.3 is finished.

This result does not depend on the value of i—that is, on the row we used to expand the determinant. Consequently, the following corollary holds.

Corollary Det *is linear in each row: that is, if we fix every row of* A *but one,* Det A *is linear in that remaining row.*

Proof. To see that Det A is linear in the ith row, simply use the formula:

$$\operatorname{Det} A = \sum_{j=1}^{n+1} (-1)^{i+j} a_{ij} \operatorname{Det} A_{ij}$$

The matrices A_{ij} have no entries from the ith row of A, and are therefore fixed. It is now clear that Det A is a linear function of the numbers a_{ij}, which proves the corollary.

Exercises

1. Compute the determinant of A when

(a) $A = \begin{bmatrix} 1 & 7 \\ 3 & 4 \end{bmatrix}$ (b) $A = \begin{bmatrix} 2 & 2 \\ -7 & 4 \end{bmatrix}$

(c) $A = \begin{bmatrix} 1 & 0 & 2 \\ -1 & 3 & 0 \\ 2 & 1 & 1 \end{bmatrix}$ (d) $A = \begin{bmatrix} -1 & 2 & 2 \\ 0 & 3 & -1 \\ 4 & 2 & -4 \end{bmatrix}$

(e) $A = \begin{bmatrix} 2 & 4 & -2 & 0 \\ -2 & 0 & 0 & 9 \\ -1 & -2 & 1 & 1 \\ 7 & 0 & 1 & -5 \end{bmatrix}$ (f) $A = \begin{bmatrix} 3 & 6 & 0 & 1 \\ 1 & 4 & -1 & 0 \\ 1 & 1 & -2 & 6 \\ 2 & 1 & 0 & -2 \end{bmatrix}$

2. Prove that Det $\lambda A = \lambda^n$ Det A for any $n \times n$ matrix A and any number λ.
3. Show by example that Det$(A + B)$ need not equal Det A + Det B.
4. Determine the inverse of the matrix

$$A = \begin{bmatrix} 2 & 1 \\ 3 & 2 \end{bmatrix}$$

5. Determine the inverse of the matrix

$$A = \begin{bmatrix} a & b \\ c & d \end{bmatrix}$$

when it exists.

6. Let $f\colon \mathbf{R} \to M_n(\mathbf{R})$ be a differentiable function from the reals to the $n \times n$ real matrices, and let $g(x) = \operatorname{Det} f(x)$. We can write $f(x)$ as $(A_1(x), \ldots, A_n(x))$,

where $A_i(x)$ is the ith column of $f(x)$. Show that

$$g'(x) = \sum_{i=1}^{n} \text{Det}(A_1(x), \ldots, A_{i-1}(x), A_i'(x), A_{i+1}(x), \ldots, A_n(x))$$

(*Hint*: Check the formula for $n = 1$ and 2 by brute force. Then use induction, plus the description of Det given in the proof of Theorem A.3.)

7. The matrix $A = (a_{ij})$ is said to be *upper triangular* if $a_{ij} = 0$ whenever $i > j$. Show that $\text{Det}\, A = a_{11}a_{22} \cdots a_{nn}$ for any upper triangular matrix.

*8. Let $T: \mathbf{R}^n \to \mathbf{R}^n$ be a linear transformation. The *adjoint* of T, T^*, is the linear transformation from $\mathbf{R}^n \to \mathbf{R}^n$ such that for any vectors $\mathbf{v}$ and $\mathbf{w}$ in $\mathbf{R}^n$,

$$\langle T\mathbf{v}, \mathbf{w} \rangle = \langle \mathbf{v}, T^*\mathbf{w} \rangle$$

(a) Show that at most one linear transformation T^* has this property.
(b) Let T be given by a matrix $A = (a_{ij})$. Show that T^* is given by the matrix A^* formed from A by interchanging rows and columns:

$$A^* = \begin{bmatrix} a_{11} & \cdots & a_{n1} \\ \vdots & & \vdots \\ a_{1n} & \cdots & a_{nn} \end{bmatrix} = (a_{ji})$$

(A^* is sometimes called the *transpose* of A, and sometimes written A^t or ${}^tA$.)
(c) Show that $\text{Det}\, A^* = \text{Det}\, A$.

*9. The linear transformation $T: \mathbf{R}^n \to \mathbf{R}^n$ is called *orthogonal* if $\langle T\mathbf{v}, T\mathbf{w} \rangle = \langle \mathbf{v}, \mathbf{w} \rangle$ for all vectors $\mathbf{v}, \mathbf{w} \in \mathbf{R}^n$.
(a) Show that T is orthogonal if and only if $TT^* = I$.
(b) Show that if T is orthogonal and A is the matrix of T, then $\text{Det}\, A = \pm 1$.
(c) Show that $T: \mathbf{R}^n \to \mathbf{R}^n$ is orthogonal if and only if $\| T\mathbf{v} \| = \|\mathbf{v}\|$ for all vectors $\mathbf{v} \in \mathbf{R}^n$. (*Hint*: Use the polarization identity—see Exercise 9 of Section 5.1.)

10. Finish the proof of Theorem A.3 by checking that the other part of property (A.1) holds.

11. Let $A = (a_{ij})$ be an $n \times n$ matrix with $\text{Det}\, A \neq 0$. Define the $n \times n$ matrix $B = (b_{ij})$ by

$$b_{ij} = (-1)^{i+j} (\text{Det}\, A)^{-1}\, \text{Det}\, A_{ji}$$

(a) Prove that $B = A^{-1}$ when $n = 2$ (that is, prove $AB = BA = I$).
(b) Prove that $B = A^{-1}$ when $n = 3$. (In fact, $B = A^{-1}$ for all values of n. Thus, if $\text{Det}\, A \neq 0$, then A is invertible.)

Appendix B

The Proof of the General Inverse Function Theorem

We now prove Theorem 2.3 of Chapter 9. The proofs of statements (a) and (c) are analogous to the proofs of the corresponding statements of Theorem 2.2, Chapter 9. However, the proof of statement (b) is quite different; this is due to the fact that the intermediate value theorem has no analogue in higher dimensions.

We begin with another extension of the mean value theorem.

Proposition B.1 *Let $B = B_r(\mathbf{v}_0)$ be the open ball of radius r about $\mathbf{v}_0$ and $f: B \to \mathbf{R}^m$ a function with continuous partial derivatives on B and component functions $f_1, \ldots, f_m$: $B \to \mathbf{R}$. Then, for any $\mathbf{v}_1$, $\mathbf{v}_2 \in B$, there are points $\mathbf{c}_1, \ldots, \mathbf{c}_m$ on the line segment from $\mathbf{v}_1$ to $\mathbf{v}_2$ such that*

$$f(\mathbf{v}_1) - f(\mathbf{v}_2) = L(\mathbf{v}_1 - \mathbf{v}_2)$$

where $L: \mathbf{R}^n \to \mathbf{R}^m$ is the linear transformation with matrix (a_{ij}) given by

$$a_{ij} = \frac{\partial f_i}{\partial x_j}(\mathbf{c}_i)$$

This result is proved by applying Proposition 4.1 of Chapter 6 to each of the component functions f_i: $U \to \mathbf{R}$, $1 \leq i \leq m$. We leave the details to the reader.

We can now prove statement (a) of Theorem 2.3 of Chapter 9. [Compare this proof with the proof of statement (a) of Theorem 2.2, Chapter 9.]

Proposition B.2 *Let U_1 be an open subset in $\mathbf{R}^n$ and $f: U_1 \to \mathbf{R}^n$ a function with* *a continuous derivative. Suppose* $\mathbf{v}_0$ *is a point in* U_1 *such that* $D_{\mathbf{v}_0}f$ *is an isomorphism. Then* f *is injective on some open set* U *with* $\mathbf{v}_0 \in U \subset U_1$.

Proof. Let $W = \mathbf{R}^n \times \cdots \times \mathbf{R}^n$ be the n-fold Cartesian product of $\mathbf{R}^n$ with itself with inner product

$$\langle(\mathbf{v}_1, \ldots, \mathbf{v}_n), (\mathbf{w}_1, \ldots, \mathbf{w}_n)\rangle = \sum_{j=1}^{n} \langle \mathbf{v}_j, \mathbf{w}_j \rangle$$

Let $\tilde{U}_1 = U_1 \times \cdots \times U_1 \subset W$ and define $G: \tilde{U}_1 \to \mathbf{R}$ by

$$G(\mathbf{v}_1, \ldots, \mathbf{v}_n) = \operatorname{Det} b_{ij}$$

where
$$b_{ij} = \frac{\partial f_i}{\partial x_j}(\mathbf{v}_i)$$

Now, G is continuous since $G(\mathbf{v}_1, \ldots, \mathbf{v}_n)$ is the sum of terms of the form

$$\frac{\partial f_1}{\partial x_{i_1}}(\mathbf{v}_1)\, \frac{\partial f_2}{\partial x_{i_2}}(\mathbf{v}_2) \cdots \frac{\partial f_n}{\partial x_{i_n}}(\mathbf{v}_n)$$

and each of these is a continuous function. Furthermore, $G(\mathbf{v}_0, \ldots, \mathbf{v}_0) = \operatorname{Det}(D_{\mathbf{v}_0}f)$ which is nonzero. Thus we can find a $\delta > 0$ such that

$$|G(\mathbf{v}_1, \ldots, \mathbf{v}_n) - G(\mathbf{v}_0, \ldots, \mathbf{v}_0)| < |G(\mathbf{v}_0, \ldots, \mathbf{v}_0)|$$

whenever $\|\mathbf{v}_i - \mathbf{v}_0\| < \delta$, for all $i = 1, \ldots, n$. As a result, $G(\mathbf{v}_1, \ldots, \mathbf{v}_n) \neq 0$ whenever $\mathbf{v}_1, \ldots, \mathbf{v}_n \in B_\delta(\mathbf{v}_0)$.

Suppose now that $\mathbf{v}_1, \mathbf{v}_2 \in B_\delta(\mathbf{v}_0)$ with $f(\mathbf{v}_1) = f(\mathbf{v}_2)$. According to Proposition B.1, we can find points $\mathbf{c}_1, \ldots, \mathbf{c}_n$ on the line segment from $\mathbf{v}_1$ to $\mathbf{v}_2$ such that

$$0 = f(\mathbf{v}_1) - f(\mathbf{v}_2) = L(\mathbf{v}_1 - \mathbf{v}_2)$$

where L is the linear transformation with matrix (b_{ij}),

$$b_{ij} = \frac{\partial f_i}{\partial x_j}(\mathbf{c}_i)$$

However, since $\mathbf{v}_1, \mathbf{v}_2 \in B_\delta(\mathbf{v}_0)$, it follows that $\mathbf{c}_1, \ldots, \mathbf{c}_n \in B_\delta(\mathbf{v}_0)$, so that $G(\mathbf{c}_1, \ldots, \mathbf{c}_n) = \operatorname{Det} L \neq 0$. Thus L is an isomorphism and $L(\mathbf{v}_1 - \mathbf{v}_2) = \mathbf{0}$ can only happen if $\mathbf{v}_1 - \mathbf{v}_2 = \mathbf{0}$. Therefore $\mathbf{v}_1 = \mathbf{v}_2$ whenever $f(\mathbf{v}_1) = f(\mathbf{v}_2)$ on $B_\delta(\mathbf{v}_0)$ and Proposition B.2 is proved with $U = B_\delta(\mathbf{v}_0)$.

It follows that the restriction of f to U has an inverse. We show next that this inverse is defined on an open set, thus proving (b).

Proposition B.3 *Let* U *be an open set in* $\mathbf{R}^n$ *and* $f: U \to \mathbf{R}^n$ *a function. Suppose* *that the derivative of f is continuous and that $D_{\mathbf{v}}f: \mathbf{R}^n \to \mathbf{R}^n$ is an isomorphism for all $\mathbf{v} \in U$. Then the image V of U under f is an open set.*

Proof. Let $\mathbf{v}_0$ be an arbitrary point in U and set $\mathbf{w}_0 = f(\mathbf{v}_0)$. We must show that there is some open ball about $\mathbf{w}_0$ contained entirely in V.

Since U is open, $\mathbf{v}_0 \in U$, and $D_{\mathbf{v}_0}f$ is an isomorphism, we can find an $r > 0$ such that the closed ball $\bar{B} = \bar{B}_r(\mathbf{v}_0)$ is contained in U and f is injective on $\bar{B}$. Let C be the boundary of $\bar{B}$,

$$C = \{\mathbf{v} \in U : \|\mathbf{v} - \mathbf{v}_0\| = r\}$$

Then, both $\bar{B}$ and C are compact (since they are both closed and bounded; see Theorem 4.1, Chapter 7), so that both $f(\bar{B})$ and $f(C)$ are also compact (by Proposition 3.2, Chapter 7). Let $d > 0$ be the number defined by

$$d = \text{glb}\,\{\|\mathbf{w} - \mathbf{w}_0\| : \mathbf{w} \in f(C)\}$$

We now need the following:

Lemma B.4 *The number d defined above is positive.*

Proof. Note, first of all, that $\mathbf{w}_0 \notin f(C)$. For, if $f(\mathbf{v}_1) = \mathbf{w}_0$ for $\mathbf{v}_1 \in C$, then we would have $\mathbf{v}_0 = \mathbf{v}_1$ since f is injective on $\bar{B}$ and $f(\mathbf{v}_0) = \mathbf{w}_0$. However, $\mathbf{v}_0 \notin C$; so this is impossible.

Now define a function $h: C \to \mathbf{R}$ by

$$h(\mathbf{v}) = \|f(\mathbf{v}) - \mathbf{w}_0\|$$

As h is continuous and C is compact, h has a minimum (Theorem 3.3, Chapter 7). This minimum is nonnegative (since h is), and is not 0 [otherwise $\mathbf{w} \in f(C)$]. But it is easy to see that this minimum value is d, and the lemma is proved.

Proposition B.3 is now a consequence of the next result.

Lemma B.5 *Let $s = \frac{1}{3}d$. Then $B_s(\mathbf{w}_0) \subset f(U)$.*

Proof. Let $\mathbf{w}_1$ be any point in $B_s(\mathbf{w}_0)$ and define $\varphi: \bar{B} \to \mathbf{R}$ by

$$\varphi(\mathbf{v}) = \|f(\mathbf{v}) - \mathbf{w}_1\|^2$$

Since $\bar{B}$ is compact and φ is continuous, φ attains its maximum and minimum on $\bar{B}$ by Proposition 3.3 of Chapter 7. Let $\mathbf{v}_1 \in \bar{B}$ be the point at which φ attains its minimum. We will show that $f(\mathbf{v}_1) = \mathbf{w}_1$, so that $\mathbf{w}_1 \in f(\bar{B}) \subset f(U)$ and the lemma follows.

To prove that $f(\mathbf{v}_1) = \mathbf{w}_1$, we first note that $\mathbf{v}_1 \notin C$. For if $\mathbf{v}_1 \in C$, we have

$$\begin{aligned}\|f(\mathbf{v}_1) - \mathbf{w}_1\| &= \|f(\mathbf{v}_1) - \mathbf{w}_0 + \mathbf{w}_0 - \mathbf{w}_1\| \\ &\geq \|f(\mathbf{v}_1) - \mathbf{w}_0\| - \|\mathbf{w}_0 - \mathbf{w}_1\| \\ &\geq d - \tfrac{1}{3}d = \tfrac{2}{3}d\end{aligned}$$

Thus $\varphi(\mathbf{v}_1) \geq \frac{4}{9}d^2$. However $\varphi(\mathbf{v}_0) \leq \frac{1}{9}d^2$ [since $\mathbf{w}_1 \in B_s(\mathbf{w}_0)$], so that $\varphi(\mathbf{v}_1)$ cannot be a minimum value for φ.

Therefore $\mathbf{v}_1$ is in the interior of $\bar{B}$, and it follows from Theorem 1.1 in Chapter 8 that $D_{\mathbf{v}_1}\varphi$ is the zero transformation. To complete the proof, we need to compute $D_{\mathbf{v}_1}\varphi$.

Lemma B.6 *If φ and f are as above, then*

$$(D_{\mathbf{v}}\varphi)(\mathbf{w}) = 2\langle (D_{\mathbf{v}}f)\mathbf{w}, f(\mathbf{v}) - \mathbf{w}_1\rangle$$

Proof. Let $\mathbf{w}_1 = (a_1, \ldots, a_n)$ and $f(\mathbf{v}) = (f_1(\mathbf{v}), \ldots, f_n(\mathbf{v}))$. Then

$$\varphi(\mathbf{v}) = \sum_{j=1}^{n} [f_j(\mathbf{v}) - a_j]^2$$

Using the chain rule, we see that

$$\frac{\partial \varphi}{\partial x_i}(\mathbf{v}) = 2\sum_{j=1}^{n} \frac{\partial f_j}{\partial x_i}(\mathbf{v})\,[f_j(\mathbf{v}) - a_j]$$

Thus, if $\mathbf{w} = (b_1, \ldots, b_n)$,

$$\begin{aligned}(D_{\mathbf{v}}\varphi)(\mathbf{w}) &= 2\sum_{i=1}^{n}\sum_{j=1}^{n} b_i \frac{\partial f_j}{\partial x_i}(\mathbf{v})\,[f_j(\mathbf{v}) - a_j] \\ &= 2\langle (D_{\mathbf{v}}f)\mathbf{w}, f(\mathbf{v}) - \mathbf{w}_1\rangle\end{aligned}$$

We may rewrite this formula as

$$\text{(B.1)} \qquad (D_{\mathbf{v}}\varphi)(\mathbf{w}) = 2\langle \mathbf{w}, (D_{\mathbf{v}}f)^*(f(\mathbf{v}) - \mathbf{w}_1)\rangle$$

Since $D_{\mathbf{v}_1}\varphi$ is the zero transformation, we must have $(D_{\mathbf{v}_1}\varphi)(\mathbf{w}) = 0$ for all $\mathbf{w} \in \mathbf{R}^n$. It follows from Lemma 3.7 [or equation (B.1)] that

$$\langle (D_{\mathbf{v}_1}f)(f(\mathbf{v}_1) - \mathbf{w}_1, \mathbf{w}\rangle = 0$$

for all $\mathbf{w} \in \mathbf{R}^n$. It follows that

$$(D_{\mathbf{v}_1}f)(f(\mathbf{v}_1) - \mathbf{w}_1) = \mathbf{0}$$

However, $D_{\mathbf{v}_1}f$ is an isomorphism for all $\mathbf{v} \in U_1$ and therefore $(D_{\mathbf{v}_1}f)^*$ is also. Thus $f(\mathbf{v}_1) - \mathbf{w}_1$ must be the zero vector. This completes the proof of Lemma B.5 and of Proposition B.3.

It remains to prove (c). We begin with a lemma.

Lemma B.7 *Let U be an open subset of $\mathbf{R}^n$ and $f: U \to \mathbf{R}^n$ an injective function with a continuous derivative. Suppose further that $D_{\mathbf{v}} f$ is an isomorphism for all $\mathbf{v} \in U$, and let V be the image of U under f (which is open by Proposition B.3). Let $\mathbf{w}_0$ be a point in V. Then there are numbers m and δ_1, both positive, such that*

$$\| f^{-1}(\mathbf{w}) - f^{-1}(\mathbf{w}_0) \| \leq \frac{1}{m} \|\mathbf{w} - \mathbf{w}_0\|$$

for all $\mathbf{w}$ such that $\|\mathbf{w} - \mathbf{w}_0\| < \delta_1$.

Proof. Let $\mathbf{w}_0 = f(\mathbf{v}_0)$, $\mathbf{v}_0 \in U$. As in the proof of Proposition B.2, let $\tilde{U} = U \times U \times \cdots \times U$ (n factors), a subset of $\mathbf{R}^n \times \cdots \times \mathbf{R}^n$ (n factors). Define $F: \tilde{U} \to M_n(\mathbf{R})$ by

$$F(\mathbf{v}_1, \ldots, \mathbf{v}_n) = (b_{ij})$$

where

$$b_{ij} = \frac{\partial f}{\partial x_j}(\mathbf{v}_i)$$

and define $G: \tilde{U} \to \mathbf{R}$ by

$$G(\mathbf{v}_1, \ldots, \mathbf{v}_n) = \operatorname{Det} F(\mathbf{v}_1, \ldots, \mathbf{v}_n)$$

Then F and G are continuous, and $G(\mathbf{v}_0, \ldots, \mathbf{v}_0) \neq 0$. Therefore we can find a number $\delta_0 > 0$ such that $B_{\delta_0}(\mathbf{v}_0) \subseteq U$ and such that

$$G(\mathbf{v}_1, \ldots, \mathbf{v}_n) \neq 0$$

whenever $\|\mathbf{v}_i - \mathbf{v}_0\| < \delta_0$ for all $i = 1, \ldots, n$.

Let $U_1 = B_{\delta_0/2}(\mathbf{v}_0)$, and let $\bar{U}_1$ be the closure of U_1. Then $V_1 = f(U_1)$ is an open set, by Proposition B.3, and $\bar{U}_1$ is compact. Let $\tilde{S} = \bar{U}_1 \times \cdots \times \bar{U}_1$, and let $C = \{\mathbf{v} \in \mathbf{R}^n : \|\mathbf{v}\| = 1\}$. Then S and C are compact, and $\tilde{S} \times C$ is compact. (See Exercise 3 of Section 7.1.) Define the function $H: \tilde{S} \times C \to \mathbf{R}$ by

$$H(\mathbf{v}_1, \ldots, \mathbf{v}_n, \mathbf{v}) = \| F(\mathbf{v}_1, \ldots, \mathbf{v}_n)(\mathbf{v}) \|$$

for $\mathbf{v}_1 \in \bar{U}_1, \ldots, \mathbf{v}_n \in \bar{U}_1$, and $\mathbf{v} \in C$. Then H is continuous (because F is), and by Theorem 3.3 of Chapter 7, H attains its minimum value m on $\tilde{S} \times C$. Moreover, $m \neq 0$, since otherwise there would be a point $(\mathbf{v}_1, \ldots, \mathbf{v}_n, \mathbf{v})$ in $\tilde{S} \times C$ with

$$0 = \| F(\mathbf{v}_1, \ldots, \mathbf{v}_n)(\mathbf{v}) \|$$

and hence

$$\mathbf{0} = F(\mathbf{v}_1, \ldots, \mathbf{v}_n)(\mathbf{v})$$

Since $F(\mathbf{v}_1, \ldots, \mathbf{v}_n)$ is invertible for every point $(\mathbf{v}_1, \ldots, \mathbf{v}_n)$ in $\tilde{S}$ (because its determinant is nonzero) and $\mathbf{v} \neq \mathbf{0}$, this is impossible.

Let δ_1 be a number such that $f(U_1)$ contains $B_{\delta_1}(\mathbf{w}_0)$. [By Proposition B.3, $f(U_1)$ is open, and so δ_1 exists.] If $\|\mathbf{w} - \mathbf{w}_0\| < \delta_1$, then we can find a point $\mathbf{v} \in U_1$ such that $f(\mathbf{v}) = \mathbf{w}$. From Proposition B.1, there are points $\mathbf{c}_1, \ldots, \mathbf{c}_n$, all in U_1, such that

$$\mathbf{w} - \mathbf{w}_0 = f(\mathbf{v}) - f(\mathbf{v}_0) = F(\mathbf{c}_1, \ldots, \mathbf{c}_n)(\mathbf{v} - \mathbf{v}_0)$$

If $\mathbf{v} = \mathbf{v}_0$, then the lemma is certainly true. If not, then let

$$\mathbf{x} = \frac{\mathbf{v} - \mathbf{v}_0}{\|\mathbf{v} - \mathbf{v}_0\|}$$

so that $\mathbf{x} \in C$. Then

$$f(\mathbf{v}) - f(\mathbf{v}_0) = \|\mathbf{v} - \mathbf{v}_0\| F(\mathbf{c}_1, \ldots, \mathbf{c}_n)(\mathbf{x})$$

and so

$$\begin{aligned} \|\mathbf{w} - \mathbf{w}_0\| &= \| f(\mathbf{v}) - f(\mathbf{v}_0)\| = \|\mathbf{v} - \mathbf{v}_0\| \, \| F(\mathbf{c}_1, \ldots, \mathbf{c}_n)(\mathbf{x})\| \\ &\geqq \|\mathbf{v} - \mathbf{v}_0\| m \end{aligned}$$

This proves the lemma.

We could now prove that f^{-1} is continuous. (See Exercise 5.) However, we can also prove directly that f^{-1} is differentiable, and we do so.

Proposition B.8 *Let U be an open subset of $\mathbf{R}^n$ and $f: U \to \mathbf{R}^n$ an injective function with a continuous derivative. Suppose further that $D_{\mathbf{v}} f$ is an isomorphism for all $\mathbf{v} \in U$. Let V be the image of U under f, let $\mathbf{v}_0$ be a point in U, and let $\mathbf{w}_0 = f(\mathbf{v}_0)$. Then f^{-1}, the inverse to f on V, is differentiable, and*

$$D_{\mathbf{w}_0}(f^{-1}) = (D_{\mathbf{v}_0} f)^{-1}$$

Proof. Let $L = D_{\mathbf{v}_0} f$. We need to show that

$$\text{(B.2)} \qquad \lim_{\mathbf{w} \to \mathbf{w}_0} \frac{\|f^{-1}(\mathbf{w}) - f^{-1}(\mathbf{w}_0) - L^{-1}(\mathbf{w} - \mathbf{w}_0)\|}{\|\mathbf{w} - \mathbf{w}_0\|} = 0$$

Let ε be any positive number with $\varepsilon < 1/2$. If $\mathbf{v} \in U$, $\mathbf{v} = f^{-1}(\mathbf{w})$, then

$$\begin{aligned} \|f^{-1}(\mathbf{w}) - f^{-1}(\mathbf{w}_0) - L^{-1}(\mathbf{w} - \mathbf{w}_0)\| &= \| L^{-1}(f(\mathbf{v}) - f(\mathbf{v}_0)) - L(\mathbf{v} - \mathbf{v}_0)\| \\ &\leqq M \| f(\mathbf{v}) - f(\mathbf{v}_0) - L(\mathbf{v} - \mathbf{v}_0)\| \end{aligned}$$

where M is chosen so that

$$\| L^{-1}\mathbf{w}\| \leq M \|\mathbf{w}\|$$

for all $\mathbf{w} \in \mathbf{R}^n$. (See Proposition 5.1 of Chapter 4.)

Choose m and δ_1 as in Lemma B.7, so that if $\|\mathbf{w} - \mathbf{w}_0\| < \delta_1$, then

$$\|f^{-1}(\mathbf{w}) - f^{-1}(\mathbf{w}_0)\| \leq \frac{1}{m+1} \|\mathbf{w} - \mathbf{w}_0\|$$

Since $L = D_{\mathbf{v}_0} f$, we can find $\delta_0 > 0$ such that

$$\text{(B.3)} \qquad \| f(\mathbf{v}) - f(\mathbf{v}_0) - L(\mathbf{v} - \mathbf{v}_0)\| < \frac{\varepsilon M}{m} \|\mathbf{v} - \mathbf{v}_0\|$$

whenever $\|\mathbf{v} - \mathbf{v}_0\| < \delta_0$. Furthermore,

$$\|\mathbf{v} - \mathbf{v}_0\| = \| f^{-1}(\mathbf{w}) - f^{-1}(\mathbf{w}_0)\|$$

Now the proof is a matter of putting things together. Choose $\delta > 0$ such that $\delta < \delta_1$ and $\delta < m\delta_0$. Then if $0 < \|\mathbf{w} - \mathbf{w}_0\| < \delta$, we have

$$\begin{aligned}\|\mathbf{v} - \mathbf{v}_0\| &= \| f^{-1}(\mathbf{w}) - f^{-1}(\mathbf{w}_0)\| \\ &< \frac{1}{m} \|\mathbf{w} - \mathbf{w}_0\| \leqq \frac{1}{m}(m\delta_0) = \delta_0\end{aligned}$$

Therefore (B.3) holds, and

$$\begin{aligned}\| f^{-1}(\mathbf{w}) - f^{-1}(\mathbf{w}_0) - L^{-1}(\mathbf{w} - \mathbf{w}_0)\| &\leqq \| f(\mathbf{v}) - f(\mathbf{v}_0) - L(\mathbf{v} - \mathbf{v}_0)\| \\ &< M\frac{\varepsilon m}{M} \|\mathbf{v} - \mathbf{v}_0\| \\ &= \varepsilon m \| f^{-1}(\mathbf{w}) - f^{-1}(\mathbf{w}_0)\| < \varepsilon \|\mathbf{w} - \mathbf{w}_0\|\end{aligned}$$

Since ε is arbitrary, this implies (B.2). Thus Proposition B.8 is proved, and the proof of Theorem 2.3 of Chapter 9 is complete.

Exercises

1. Prove Proposition B.1.
2. Let $W = \mathbf{R}^n \times \cdots \times \mathbf{R}^n$ be the Cartesian product of n copies of $\mathbf{R}^n$ with inner product

$$\langle(\mathbf{w}_1, \ldots, \mathbf{w}_n), (\mathbf{v}_1, \ldots, \mathbf{v}_n)\rangle = \sum_{i=1}^{n} \langle \mathbf{v}_i, \mathbf{w}_i\rangle$$

   Let U be an open subset of $\mathbf{R}^n$ and $\tilde{U} = U \times \cdots \times U \subset W$. Prove that $\tilde{U}$ is open.
3. Prove the analogue of Exercise 2 for closed sets.
4. Prove that the function G defined in the proof of Proposition B.2 is continuous. Prove also that there is a $\delta > 0$ such that $G(\mathbf{v}_1, \ldots, \mathbf{v}_n) \neq 0$ whenever $\mathbf{v}_1, \ldots, \mathbf{v}_n \in B_\delta(\mathbf{v}_0)$.

5. Use Lemma B.7 to prove directly that f^{-1} is continuous (without necessarily proving differentiablity).
6. Let C and D be disjoint subsets of $\mathbf{R}^n$, C compact and D closed. Let $d \in \mathbf{R}$ be defined by

$$d = \text{glb}\{\|\mathbf{v} - \mathbf{w}\| : \mathbf{v} \in C, \mathbf{w} \in D\}$$

Prove that $d > 0$. (See the proof of Lemma B.5.)
7. Give an example to show Exercise 6 is not true if we only assume C closed.

Solutions to Selected Exercises

CHAPTER 1

Section 1

1. (a) $\{1, 2, 3, 4, 5, 6, 7, 10, 20, 21\}$
   (c) $\{2, 4, 6, 10, 20\}$
   (e) $\{1, 3, 5, 7\}$
   (f) $\{n : n$ is an even positive integer and $n \neq 4, 6$, or $20\}$
8. (a) and (c) are functions.

Section 2

4. $(\exists \varepsilon > 0)(\forall$ integers $N)(\exists n > N)\,|a_n - a| \geq \varepsilon$.
5. $(\exists \varepsilon > 0)(\forall \delta > 0)(\exists x, y \in \mathbf{R})\,|x - y| < \delta$ and $|f(x) - f(y)| \geq \varepsilon$.

Section 4

4. (a) $(-1, \frac{13}{3})$
   (c) $(-\frac{2}{3}, 2)$
5. (a) $(-\infty, 3) \cup (4, \infty)$
   (c) $(-\infty, -1] \cup [\frac{11}{3}, \infty)$

Section 5

1. (b) and (e) are equivalence relations [assuming in (e) that $a \sim b$ only if *both* parents are the same].

CHAPTER 2

Section 2

1. (a) (3, 12)
   (c) (0, 6)
   (e) (0, 0)
   (g) (−6, 21)

(i) (0, 0)
(k) (10, 11)
(m) (3, 2)
(o) (2, 22)

2. (a) $1(1, 1) + 1(1, 2)$
(c) $(\frac{11}{2} + a)(1, 1) - 2a(2, 1) + (a - \frac{3}{2})(3, 1)$, a any real number
3. (a) $(1, 1) = 2(2, 1) - (3, 1)$
4. (b), (e), (g), (i), and (j) are linearly dependent.

Section 3

1. (a) (5, 17, 9)
(c) (5, 3, 1)
(e) (0, 0, 0)
(g) (0, 0, 0)
2. (b), (d), (e), (g), and (h) are linearly dependent.

Section 4

1. (a) $\sqrt{13}$
(c) $\sqrt{5}$
(e) $\sqrt{51}$
(g) $\sqrt{15}$
3. (a) $\sqrt{2}$
(c) 3
(e) $\sqrt{15}$

Section 5

1. (a), (c), and (g) are linear transformations.
3. (b), (c), (d), and (f) are linear transformations.

Section 6

1. (a) $[1 \quad -7]$
(c) $[4 \quad -6]$
(e) $[-4 \quad 1]$
2. (a) $\begin{bmatrix} 4 & -1 \\ 2 & 3 \end{bmatrix}$
(c) $\begin{bmatrix} 0 & 0 \\ 1 & 0 \end{bmatrix}$
(e) $\begin{bmatrix} 0 & 1 \\ 1 & 0 \end{bmatrix}$
(g) $\begin{bmatrix} 1 & 2 \\ 1 & 2 \end{bmatrix}$
3. (a) $\begin{bmatrix} 1 & 2 & 0 \\ 2 & -1 & 1 \end{bmatrix}$
(c) $\begin{bmatrix} 1 \\ 4 \\ -1 \\ -4 \end{bmatrix}$
(e) $\begin{bmatrix} 1 & 0 & -4 \\ -1 & 0 & 6 \end{bmatrix}$
4. (a) $A(x) = (2x, 3x)$

(c) $C(x_1, x_2) = (2x_1, -x_1 - 2x_2)$
(e) $E(x_1, x_2) = (2x_1, 6x_1 + 4x_2, 4x_1 + 6x_2)$
(g) $G(x_1, x_2, x_3) = (x_2 + x_3, 2x_1 + x_2 + 7x_3, -4x_1 + 3x_2 + 3x_3)$
(i) $J(x_1, x_2, x_3, x_4) = (x_1 + x_4, x_1 + x_2, x_2 + x_3, x_3 + x_4)$

5. (a) $(2, -7)$
(c) $(4, 0)$
6. (a) $(11, 14)$
(c) $(53, 66)$
7. (a) $(1, 2, 4, 3)$
(c) $(7, 6, 0, 1)$

Section 7

1. (a) $(Q + S)(x_1, x_2) = (x_1 + 2x_2, 2x_2)$
(c) $(2S + T)(x_1, x_2) = (x_1 + 7x_2, x_1 + 9x_2)$
(e) $(5T + 4Q)(x_1, x_2) = (9x_1 + x_2, 19x_1 + 25x_2)$
(g) $(2S - 4Q + 3T)(x_1, x_2) = (13x_2 - x_1, 3x_1 + 9x_2)$
2. (a) $(S + T)(x_1, x_2, x_3, x_4) = (2x_1 + x_2 - x_3, x_1 + x_2 - x_3 + x_4, x_1 + x_4, 2x_2 + 2x_3)$
(b) $7S(x_1, x_2, x_3, x_4) = (7x_2, 7x_1, 7x_4, 7x_3)$

3. (a) $\begin{bmatrix} 2 & -3 \\ 0 & 3 \end{bmatrix}$

(c) $\begin{bmatrix} 11 & 4 \\ 7 & 15 \end{bmatrix}$

(e) $\begin{bmatrix} 2 & 4 \\ 3 & 2 \end{bmatrix}$

4. (a) $\begin{bmatrix} 1 & 1 & 3 \\ 4 & -1 & 1 \\ 7 & 3 & 2 \\ 3 & 5 & -2 \end{bmatrix}$ (c) $\begin{bmatrix} 2 & 3 & 7 \\ 10 & -4 & 4 \\ 20 & 6 & 5 \\ 9 & 9 & -8 \end{bmatrix}$

Section 8

1. (a) $(S \circ T)(x_1, x_2) = (2x_2 - x_1, x_1)$
(c) $(S \circ T)(y_1, y_2, y_3) = (3y_1 + 2y_2 + 4y_3, y_1 + 2y_2, 2y_1 - 2y_2)$

2. (a) $\begin{bmatrix} 2 & 3 \\ 6 & 8 \end{bmatrix}$

(c) $\begin{bmatrix} 16 & 56 \\ -4 & -14 \end{bmatrix}$

(e) $\begin{bmatrix} b & a \\ d & c \end{bmatrix}$

7. $A^{-1} = \begin{bmatrix} 1 & -1 \\ -1 & 2 \end{bmatrix}$

10. (a) $\begin{bmatrix} -4 & 3 \\ 3 & -2 \end{bmatrix}$

(b) Not invertible.

Section 9

1. (a) $\begin{bmatrix} 26 & 39 \\ 17 & 40 \end{bmatrix}$

(c) $\begin{bmatrix} 4 & 17 \\ 4 & 2 \\ 4 & 11 \end{bmatrix}$

(e) $\begin{bmatrix} 6 & 4 & 2 & 8 \\ -18 & 12 & 15 & -6 \\ 21 & 7 & 42 & -56 \end{bmatrix}$

(g) [20]

(i) $\begin{bmatrix} -30 & 30 & -6 \\ -10 & -5 & -21 \\ -16 & 43 & 31 \end{bmatrix}$

2. (a) $A^2 = \begin{bmatrix} 7 & 15 \\ 10 & 22 \end{bmatrix}$

(c), (g), and (i) make no sense.

(e) $BA = \begin{bmatrix} 9 & 19 \\ -9 & -15 \end{bmatrix}$

4. (a) $(9, 5, -3)$ (c) $(-3, -4, 27)$
5. (a) $(6, 0, 0, 8, 1)$ (c) $(1, -7, 18, 37, 45)$

12. $\begin{bmatrix} 1 & 0 & -a \\ 0 & 1 & 0 \\ 0 & 0 & 1 \end{bmatrix}$

CHAPTER 3

Section 5

1. (a), (b), (c), (d), (f), and (g) are open.
2. (a), (b), (e), (f), and (j) are closed.
3. (b), (c), and (f) are open; (a), (d), and (g) are closed; (e) and (h) are neither.
13. The sets 1(a), (b), (c), (d), (f), (g), 2(d), (g), (i), are open; so $\text{Int}(S) = S$ for these sets.
    1(e) $\varnothing$
    1(j) $\{(x, y) : y - x < 6\}$
    2(a) $\{(x, y) : 2x + 3y - 7 > 0\}$
    2(e) $\varnothing$
    2(j) $\{(x, y) : y > |x|\}$

15. The sets 1(e), (h), (i), (j), 2(a), (b), (e), (f), (j) are closed so $\mathrm{Cl}(S) = S$ for these sets.
   1(a) $\{(x, y) : x \geq y\}$
   1(f) $\{(x, y) : \|(x, y)\| \leq 3\}$
   2(c) $\{(x, y) : x^2 + y^2 \leq 1\}$
   2(h) $\{(x, y) : x \geq 3 \text{ or } y \leq 2\}$
20. 1(a) $\{(x, y) : x = y\}$
   1(c) $\{(x, y) : x = y\}$
   1(e) $\{(x, y) : y = 3\}$
   1(g) $\{(x, y) : |x| = 2\}$
   1(i) $\{(x, y) : x^2 + y^2 = 9\}$
   2(a) $\{(x, y) : 2x + 3y - 7 = 0\}$
   2(c) $\{(x,y) : x^2 + y^2 = 0 \text{ or } 1\}$
   2(e) $\{(x, y) : |x| = |y|\}$
   2(g) $\{(x, y) : 2x - y = 4\}$
   2(i) $\{(x, y) : \|(x, y)\| = 7\}$

Section 6
1. (a) Diverges.
   (c) Converges to $\frac{1}{2}$.
   (e) Converges to 1.
   (g) Diverges.

CHAPTER 4

Section 3
1. (a) $\dfrac{\partial f}{\partial x_1} = 2x_1x_2 \qquad \dfrac{\partial f}{\partial x_2} = x_1^2 + 4x_2$

   (c) $\dfrac{\partial f}{\partial x_1} = \dfrac{1}{x_1^2 + x_2^2} - \dfrac{2x_1(x_1 + x_2)}{(x_1^2 + x_2^2)^2} \qquad \dfrac{\partial f}{\partial x_2} = \dfrac{1}{x_1^2 + x_2^2} - \dfrac{2x_2(x_1 + x_2)}{(x_1^2 + x_2^2)^2}$

   (e) $\dfrac{\partial f}{\partial x_1} = \left(\dfrac{1}{2\sqrt{x_1}} + x_2\sqrt{x_1}\right) e^{x_2 + x_1x_2} \qquad \dfrac{\partial f}{\partial x_2} = (1 + x_1)\sqrt{x_1}\, e^{x_2 + x_1x_2} \qquad \dfrac{\partial f}{\partial x_3} = 0$

2. (a) $\begin{bmatrix} -\sin x_1 \\ \cos x_1 \end{bmatrix}$

   (c) $\begin{bmatrix} x_2 & x_1 & 0 \\ 2x_1 & x_3 & x_2 \\ 0 & x_3^2 & 2x_2x_3 \end{bmatrix}$

   (e) $\begin{bmatrix} 2x_1 & x_3 & x_2 \\ \dfrac{1}{1 + (x_1 + x_3)^2} & 0 & \dfrac{1}{1 + (x_1 + x_3)^2} \\ -x_2e^{x_1} & -e^{x_1} & 1 \end{bmatrix}$

Section 4
1. (a) $[2xy^2(y^2 + y - 1) \qquad x^2y(4y^2 + 3y - 2)]$

(c) $[4x^3y^2 - 2xy^4 \quad 2x^4y - 4x^2y^3]$

(e) $\begin{bmatrix} (1+2xy)yz^3e^{2xy} & (1+2xy)xz^3e^{2xy} & 3xyz^2e^{xy} \\ (1+xy)yz^2e^{xy} & (1+xy)xz^2e^{xy} & 2xyze^{xy} \end{bmatrix}$

(g) $2xyz[yz \; zx \; xy] + ze^{xy}[yz + xy^2z \;\; xz + x^2yz \;\; 2xy] - 2e^{2xy}[yz^2 \;\; xz^2 \;\; x]$

(i) $\begin{bmatrix} 3x^2y - y^3 & x^3 - 3xy^2 \\ ye^{x^2-y^2}(1+2x^2) & xe^{x^2-y^2}(1-2y^2) \end{bmatrix}$

4. $\dfrac{\partial F}{\partial r} = \dfrac{\partial f}{\partial x}\cos\theta + \dfrac{\partial f}{\partial y}\sin\theta \qquad \dfrac{\partial F}{\partial \theta} = -r\sin\theta\dfrac{\partial f}{\partial x} + r\cos\theta\dfrac{\partial f}{\partial y}$

5. $\dfrac{\partial F}{\partial \rho} = \dfrac{\partial f}{\partial x}\sin\varphi\cos\theta + \dfrac{\partial f}{\partial y}\sin\varphi\sin\theta + \dfrac{\partial f}{\partial z}\cos\varphi$

$\dfrac{\partial F}{\partial \theta} = -\dfrac{\partial f}{\partial x}\rho\sin\varphi\sin\theta + \dfrac{\partial f}{\partial y}\rho\sin\varphi\cos\theta$

$\dfrac{\partial F}{\partial \varphi} = \dfrac{\partial f}{\partial x}\rho\cos\varphi\cos\theta + \dfrac{\partial f}{\partial y}\rho\cos\varphi\sin\theta - \dfrac{\partial f}{\partial z}\rho\sin\varphi$

6. Let $f(x) = (f_1(x), f_2(x), f_3(x), f_4(x))$;

$$(g\circ f)' = \frac{\partial g}{\partial x_1}f_1' + \frac{\partial g}{\partial x_2}f_2' + \frac{\partial g}{\partial x_3}f_3' + \frac{\partial g}{\partial x_4}f_4'$$

CHAPTER 5

Section 1

1. (a) 1
   (c) -5
   (e) 6
2. (a) 3
   (c) -4
   (e) 41
3. (a) $\cos^{-1}\left(\dfrac{1}{\sqrt{55}}\right)$ (c) $\cos^{-1}\left(\dfrac{-5}{\sqrt{35}}\right)$
4. (a) $\cos^{-1}\left(\dfrac{1}{2\sqrt{3}}\right)$ (c) $\cos^{-1}\left(\dfrac{3}{2\sqrt{21}}\right)$

Section 3

1. (a) $y + z = 2$
   (c) $2x_1 - x_3 - x_4 = 13$
   (e) $2x_1 - 3x_2 + 2x_3 + x_4 + 6x_5 = 2$
3. $x_1 + x_2 - x_3 - x_4 = 0$

Section 4

There are many answers to each part of Exercises 1, 2, and 3; we give one each.

1. (a) $x = z, y = 0$ (c) $x = z, y = 2x$
2. (a) $x_1 = x_3, x_2 = x_4 = 0$ (c) $x_2 = 2x_1, x_1 + x_3 = 0, x_4 = 0$
3. (a) $x_1 = x_3, x_2 = x_4$ (c) $x_1 + x_2 = x_4, x_3 = 2x_1 - x_2$
4. (a), (c), and (d) are linearly independent.
5. We give the equation parametrically; other forms are possible.
   (a) $\mathbf{v} = t(1, 1, 1)$
   (c) $\mathbf{v} = (1, 0, 1. 0) + t(1, 2, 3, 4)$
   (e) $\mathbf{v} = (1, 0, -1, 2, 0) + t(2, 0, 0, -1, 4)$

Section 5

1. (a) 1 (c) $\dfrac{\sqrt{3}-4}{2}$
2. (a) 0 (c) $\dfrac{-2}{3}$
3. (a) 0 (c) $\left(1 - \dfrac{\sqrt{3}}{2}\right) \sin 2$
4. (a) $\dfrac{\sqrt{2}}{2}$ (c) $2 \cos 2$

Section 6

1. (a) $(4, 1), \left(\dfrac{4}{\sqrt{17}}, \dfrac{1}{\sqrt{17}}\right)$

   (c) $(2, -4, 1), \left(\dfrac{2}{\sqrt{21}}, \dfrac{-4}{\sqrt{21}}, \dfrac{1}{\sqrt{21}}\right)$

   (e) $(6, 1, 0, 4), \dfrac{1}{\sqrt{53}}(6, 1, 0, 4)$

   (g) $(2x_1, 2x_2, \ldots, 2x_n) = 2\mathbf{v}; \dfrac{\mathbf{v}}{\|\mathbf{v}\|}$(if $\mathbf{v} \neq 0$).

Section 7

1. (a) $x_3 = 2x_1 + 5x_2 - 7$
   (c) $x_4 = 2x_1 - 2x_2 - 1$
   (e) $x_5 = -x_4$
   (g) $(x_2, x_3) = (2x_1 - 3, x_1 + 5)$
   (i) $(x_2, x_3, x_4) = \left(\sqrt{2} + \dfrac{\pi\sqrt{2}}{4} - x_1\sqrt{2}, x_1\sqrt{2} - \dfrac{\pi\sqrt{2}}{4} + \sqrt{2}, x_1\right)$

(k) $(x_3, x_4, x_5) = \left(\frac{\sqrt{2}}{2}x_1 - \frac{\pi\sqrt{2}}{6}x_2 + \frac{\pi^2\sqrt{2}}{24}, \frac{\pi}{8}x_1 + \frac{\sqrt{3}}{2}x_2 - \frac{\pi^2}{24},\right.$
$\left.\frac{\pi}{4}x_1 + \frac{\pi}{3}x_2 - \frac{\pi^2}{12}\right)$

CHAPTER 6

Section 1

1. (a) $2y, 2x, 0$
   (c) $(4x^3 + 6x)ye^{x^2}, (2x^2 + 1)e^{x^2}, 0$
   (e) $-\frac{2}{9}y^2(xy + y^2)^{-5/3}; \frac{1}{3}(xy + y^2)^{-2/3}\left[1 - \frac{2y(x + 2y)}{3(xy + y^2)}\right];$
   $\frac{2}{9}(xy + y^2)^{-8/3}[\frac{5}{3}(x + 2y)^2y - 2(x + 3y)(xy + y^2)]$
   (g) $\frac{2xy}{\sqrt{1 - x^4y^2}}\left(3 + \frac{2x^4y^2}{1 - x^4y^2}\right) \quad \frac{x^2}{\sqrt{1 - x^4y^2}}\left(3 + \frac{2x^4y^2}{1 - x^4y^2}\right)$
   $\frac{x^6y}{(1 - x^4y^2)^{3/2}}\left[7 + \frac{6x^4y^2}{(1 - x^4y^2)}\right]$
2. (a) $\begin{bmatrix} 0 & 0 \\ 0 & 0 \end{bmatrix}$ (c) $\begin{bmatrix} 0 & 1 \\ 1 & 0 \end{bmatrix}$
   (e) Undefined (g) $\begin{bmatrix} 0 & 0 \\ 0 & 0 \end{bmatrix}$
3. (a) $\begin{bmatrix} 0 & 1 & 0 \\ 1 & 0 & 0 \\ 0 & 0 & 0 \end{bmatrix}$
   (c) $\begin{bmatrix} 0 & 0 & 0 \\ 0 & 0 & 0 \\ 0 & 0 & 0 \end{bmatrix}$
   (e) $\begin{bmatrix} 0 & 0 & 1 & 0 \\ 0 & 0 & 0 & 1 \\ 1 & 0 & 0 & 0 \\ 0 & 1 & 0 & 0 \end{bmatrix}$
5. $\frac{\partial^2}{\partial x^2}(f \circ g) = f''(t)\left(\frac{\partial g}{\partial x}\right)^2 + f'(t)\frac{\partial^2 g}{\partial x^2}$
   $\frac{\partial^2}{\partial y \partial x}(f \circ g) = \frac{\partial^2}{\partial x \partial y}(f \circ g) = f''(t)\frac{\partial g}{\partial x}\frac{\partial g}{\partial y} + f'(t)\frac{\partial^2 g}{\partial x \partial y}$

$$\frac{\partial^2}{\partial y^2}(f\circ g) = f''(t)\left(\frac{\partial g}{\partial y}\right)^2 + f'(t)\frac{\partial^2 g}{\partial y^2}$$

All derivatives of g are evaluated at (x, y); $t = g(x, y)$.

6. If $g(t) = (g_1(t), \ldots, g_n(t))$, then

$$\frac{d^2}{dt^2}(f\circ g) = \sum_{j=1}^{n}\left[\frac{\partial f}{\partial x_j}g_j''(t) + \sum_{i=1}^{n}\frac{\partial^2 f}{\partial x_i \partial x_j}g_i'(t)g_j'(t)\right]$$

[All derivatives of f are evaluated at $g(t)$.]

7. $$\frac{\partial^2 F}{\partial r^2} = \cos^2\theta\frac{\partial^2 f}{\partial x_1^2} + 2\sin\theta\cos\theta\frac{\partial^2 f}{\partial x_1\partial x_2} + \sin^2\theta\frac{\partial^2 f}{\partial x_2^2}$$

$$\frac{\partial^2 F}{\partial \theta^2} = r^2\sin^2\theta\frac{\partial^2 f}{\partial x_1^2} + r^2\cos^2\theta\frac{\partial^2 f}{\partial x_2^2} - 2r^2\sin\theta\cos\theta\frac{\partial^2 f}{\partial x_1\partial x_2} - r\cos\theta\frac{\partial f}{\partial x_1} - r\sin\theta\frac{\partial f}{\partial x_2}$$

$$\frac{\partial^2 F}{\partial\theta\,\partial r} = \frac{\partial^2 F}{\partial r\,\partial\theta} = -r\sin\theta\cos\theta\frac{\partial^2 f}{\partial x_1^2} + r\sin\theta\cos\theta\frac{\partial^2 f}{\partial x_2^2} + r(\cos^2\theta - \sin^2\theta)\frac{\partial^2 f}{\partial x_1\partial x_2} - \sin\theta\frac{\partial f}{\partial x_1} + \cos\theta\frac{\partial f}{\partial x_2}$$

10. $$\Delta f = \frac{\partial^2 F}{\partial\rho^2} + \frac{1}{\rho^2\sin^2\varphi}\frac{\partial^2 F}{\partial\theta^2} + \frac{1}{\rho^2}\frac{\partial^2 F}{\partial\varphi^2} + \frac{1}{\rho}\frac{\partial F}{\partial\rho} + \frac{1}{\rho\sin\varphi}\frac{\partial F}{\partial\theta}$$

Section 2

1. (a) $\sum_{j=1}^{n}\frac{(-1)^{j-1}}{j}x^j = x - \frac{x^2}{2} + \cdots \pm \frac{x^n}{n}$

(c) For $n = 2m - 1$ or $2m$,

$$\sum_{j=1}^{m}\frac{(-1)^{j-1}}{2j-1}x^{2j-1} = x - \frac{x^3}{3} + \frac{x^5}{5} + \cdots \pm \frac{x^{2m-1}}{2m-1}$$

(e) $e^2\left[1 + (x-2) + \frac{(x-2)^2}{2} + \frac{(x-2)^3}{6} + \frac{(x-2)^4}{24} + \frac{(x-2)^5}{120}\right]$

(g) $1 + rx + \frac{r(r-1)}{2}x^2 + \frac{r(r-1)(r-2)}{6}x^3$

2. (a) .199 (c) .732 (e) 1.414

Section 3

1. (a) $3 + x_1x_2 + x_2^3$

(c) $-27 + [-3(x_1 - 1) + 28(x_2 + 3)] + [(x_1 - 1)(x_2 + 3) - 9(x_2 + 3)^2] + (x_2 + 3)^3$

(e) 1

(g) $2 + \left[\left(1 + \frac{\pi}{2}\right)(x_1 - 2) + 8\left(x_2 - \frac{\pi}{8}\right)\right] + \left[\left(\frac{\pi}{4} + \frac{\pi^2}{16}\right)(x_1 - 2)^2 + (8 + 2\pi)(x_1 - 2)\left(x_2 - \frac{\pi}{8}\right) + 16\left(x_2 - \frac{\pi}{8}\right)^2\right]$

Section 4

1. (a) $x_1x_2x_3$

(c) $-2 + [-2(x_1 - 1) - 1(x_2 - 2) + 2(x_3 + 1)] + [(x_2 - 2)(x_3 + 1) + 2(x_1 - 1)(x_3 + 1) - (x_1 - 1)(x_2 - 2)]$

(e) $2x_2 + (x_1 - 2)x_2 + 2(x_3 - 1)x_2 + (x_1 - 2)(x_3 - 1)x_2 - \frac{4}{3}x_2^3$

(g) x_1

CHAPTER 7

Section 2

1. (a) lub $= 2$; no glb

(c) glb $= \frac{\pi}{6}$; no lub

(e) lub $= \ln 3$; no glb

Section 4

1. A, D, E, G, H, I, and K are compact.

Section 6

1. B, C, E, F, J, and K are connected.

CHAPTER 8

Section 1

1. (a) Interior $= (a, b)$; boundary $= \{a, b\}$

(c) Interior $= \varnothing$; boundary $= \{(x, y) \in \mathbf{R}^2 : xy = 1\}$

(e) Interior $= \varnothing$; boundary $= \{(x, y) \in \mathbf{R}^2 : x > 0$ and $y = \sin(1/x)$, or $x = 0$, $-1 \le y \le 1\}$

(g) Interior $= \{(x, y, z) \in \mathbf{R}^3 : x + y + z < 0\}$; boundary $= \{(x, y, z) \in \mathbf{R}^3 : x + y + z = 0\}$

4. (a) 1, 2

(c) $\pm\pi/2, \pm 3\pi/2, \pm 5\pi/2, \ldots$

(e) (0, 0)

(g) (0, 1)

(i) $\{(x, y) \in \mathbf{R}^2 : x + y = n\pi$ for some integer $n\}$

5. (a) Local minimum

(b) and (c) Neither

(d) Local maximum

6. $\frac{\sqrt{3}}{2}$

7. $\sqrt{17}$

9. (b) The minimum solves the equations

$$a \sum_{j=1}^{n} x_j^2 + b \sum_{j=1}^{n} x_j y_j + c \sum_{j=1}^{n} x_j = \sum_{j=1}^{n} x_j z_j$$

$$a \sum_{j=1}^{n} x_j y_j + b \sum_{j=1}^{n} y_j^2 + c \sum_{j=1}^{n} y_j = \sum_{j=1}^{n} y_j z_j$$

$$a \sum_{j=1}^{n} x_j + b \sum_{j=1}^{n} y_j + nc = \sum_{j=1}^{n} z_j$$

12. $\sqrt{21}$

Section 2

1. (a) $x^2 + 4xy + y^2$
   (c) $3x^2 + 4xy + 4y^2$
   (e) $3x^2 + 4xy + 7y^2$
2. (c) and (e) are positive definite; (d) is negative definite; (a), (b), and (f) are indefinite.
3. (a) $x^2 + y^2 + z^2$
   (c) $-4x^2 - 3y^2 + 2xz + 4yz - 5z^2$
   (e) $4x_1^2 + 4x_1x_2 + 5x_2^2 + 6x_2x_3 + 5x_3^2 + 2x_2x_4 + 3x_4^2$
4. (a), (b), (e), and (f) are positive definite; (c) is negative definite; (d) is indefinite.
5. (a) $\begin{bmatrix} 1 & 3 \\ 3 & -3 \end{bmatrix}$

   (c) $\begin{bmatrix} 1 & \frac{1}{2} \\ \frac{1}{2} & 1 \end{bmatrix}$

   (e) $\begin{bmatrix} 6 & -\frac{5}{2} \\ -\frac{5}{2} & 6 \end{bmatrix}$

6. (c) and (e) are positive definite; (d) is negative definite; (a), (b), and (f) are indefinite.
7. (a) $\begin{bmatrix} 1 & 0 & 0 \\ 0 & -2 & 0 \\ 0 & 0 & 4 \end{bmatrix}$

   (c) $\begin{bmatrix} 0 & 1 & 2 \\ 1 & 0 & 3 \\ 2 & 3 & 0 \end{bmatrix}$

   (e) $\begin{bmatrix} 1 & -3 & 0 & 0 \\ -3 & 1 & \frac{1}{2} & 0 \\ 0 & \frac{1}{2} & -1 & 0 \\ 0 & 0 & 0 & 1 \end{bmatrix}$

8. (b) is positive definite; (d) is negative definite; (a), (c), and (e) are indefinite.

Section 3

1. (a) Local minimum at (2, 0).
   (c) Local maximum at (0, 0).
   (e) Local minimum at (2, 1).
   (g) Local minimum at (2, 0); local maximum at $(-2, 8)$; saddles at $(1, -1)$ and (3, 3).
2. For $a > 0$, a saddle at (0, 0) and a minimum at (a, a).
   For $a < 0$, a saddle at (0, 0) and a maximum at (a, a).
   For $a = 0$, (0, 0) is a nonextreme critical point.
6. 5 in × 5 in × 5 in
7. $\frac{4}{9} \times \frac{2}{3} \times \frac{4}{3}$
8. $\sqrt[3]{\frac{2000}{1001}}$ in $\times \sqrt[3]{\frac{2000}{1001}}$ in $\times \sqrt[3]{\frac{1000(1001)^2}{4}}$ in, or roughly $1\frac{1}{4}$ in × $1\frac{1}{4}$ in × $52\frac{1}{2}$ ft.
9. $x = 4, y = 8, z = 12$
10. $x = 18, y = 9, z = 6$
11. Bob picks $x = 4$, Larry picks $y = -3$, and Larry pays Bob $5.
12. The coefficients solve

$$a \sum_{j=1}^{n} x_j^2 + b \sum_{j=1}^{n} x_j + cn = \sum_{j=1}^{n} y_j$$

$$a \sum_{j=1}^{n} x_j^3 + b \sum_{j=1}^{n} x_j^2 + c \sum_{j=1}^{n} x_j = \sum_{j=1}^{n} x_j y_j$$

$$a \sum_{j=1}^{n} x_j^4 + b \sum_{j=1}^{n} x_j^3 + c \sum_{j=1}^{n} x_j^2 = \sum_{j=1}^{n} x_j^2 y_j$$

Section 4

1. 36, at $(-3, -3, 9)$
2. Maxima at $\pm(\sqrt{2}, \sqrt{2})$; minima at $\pm(\sqrt{2}, -\sqrt{2})$
3. $(-\frac{1}{3}, -\frac{2}{3}, \frac{2}{3})$
9. $\frac{2\sqrt{3}}{3} \times \sqrt{3} \times \frac{5\sqrt{3}}{3}$

Section 5

11. -175
12. $4 + 2\sqrt{2}$
13. 1
14. $a\sqrt{3} \times b\sqrt{3} \times c\sqrt{3}$
15. (a) Radius $= 10/\sqrt[3]{2\pi}$ cm; height $= 20/\sqrt[3]{2\pi}$ cm.
    (b) Radius = height $= 10/\sqrt[3]{\pi}$ cm.
16. (a) $10\sqrt[3]{4} \times 10\sqrt[3]{4} \times 10\sqrt[3]{4}$ cm.
    (c) $20\sqrt[3]{3} \times 10\sqrt[3]{3} \times \frac{20}{3}\sqrt[3]{3}$ cm.

17. $10\sqrt[3]{\frac{3}{2}} \times 10\sqrt[3]{\frac{3}{2}} \times 10\sqrt[3]{12}$ cm.
18. 20 of French, 50 of German, and 100 of California wine
19. $w = 2r\sqrt{3}/3$, $t = 2r\sqrt{6}/3$
22. (a) $1/p + 1/q$

Section 6
1. Maximum $= \frac{1}{3}$, at $(-\frac{2}{3}, -\frac{2}{3}, \frac{1}{3})$; minimum $= -1$, at $(0, 0, -1)$.
3. $-(1 + \sqrt{2})$
5. $(\frac{4}{3}, \frac{4}{3}, \frac{4}{3})$
7. $\dfrac{\sqrt{5}}{4}$
9. $x_1 = y_1$, $x_2 = y_2$

CHAPTER 9

Section
1. (a) Inverse is $g(y) = \sqrt{y - 6}$, $y \in (10, 42)$.
   (c) No inverse.
   (e) Inverse is $g(y) = \sqrt[3]{y + 1}$, $y \in (-344{,}999)$.
   (g) Inverse is $g(y) = \arccos y$, $y \in (-1, 1)$.
   (i) No inverse.
2. Matrix of inverses:
   (a) $[\frac{1}{3}]$

   (c) $\begin{bmatrix} 1 & -2 \\ 0 & 1 \end{bmatrix}$

   (e) $\begin{bmatrix} 0 & 1 \\ 1 & -1 \end{bmatrix}$

   (g) $\begin{bmatrix} a & -b \\ -c & d \end{bmatrix}$

   (i) $\begin{bmatrix} 0 & 1 & 0 \\ 0 & 0 & 1 \\ 1 & 0 & 0 \end{bmatrix}$

Section 2
3. (a) $(f^{-1})'(y) = \dfrac{-1}{\sqrt{1 - y^2}}$

   (c) $(f^{-1})'(y) = \dfrac{1}{y\sqrt{y^2 - 1}}$

   (e) $(f^{-1})'(y) = \dfrac{-1}{\sqrt{1 + 4y}}$

(g) $(f^{-1})'(y) = \dfrac{1}{\sqrt{y^2+1}}$

4. (a) $(-\infty, \frac{7}{2})$

(c) $\left(\dfrac{\pi}{2}, \dfrac{3\pi}{2}\right)$

(e) $\left(-\dfrac{\pi}{2}, \dfrac{\pi}{2}\right)$

(g) $(-\infty, 0)$

5. (a) $-\frac{1}{5}$

(c) sec 3

(e) 1

(g) $\dfrac{-1}{2e}$

Section 4

1. (a) $-\frac{4}{13}$

(c) $\frac{1}{3}$

(e) $\frac{3}{8}$

2. (a) $-\frac{13}{4}$

(c) 3

(e) $\frac{8}{3}$

3. (a) At $\pm(1, 0)$.

4. At (0, 0).

5. (b) All points except (0, 0) and $(\pm a\sqrt{2}, 0)$; $x(a^2 - x^2 - y^2)/y(a^2 + x^2 + y^2)$.

Section 5

1. (a) $\dfrac{\partial z}{\partial x} = 3, \dfrac{\partial z}{\partial y} = -\dfrac{3}{2}$

(c) $\dfrac{\partial z}{\partial x} = \dfrac{\partial z}{\partial y} = 0$

(e) No function.

2. (a) $\dfrac{\partial z}{\partial x} = +6 \quad \dfrac{\partial z}{\partial y} = -\dfrac{5}{2} \quad \dfrac{\partial w}{\partial x} = -2 \quad \dfrac{\partial^2 z}{\partial x \partial y} = -18$

(c) No function.

3. Any two can be regarded as functions of the third.

CHAPTER 10

Section 2

2. (a) $\frac{2}{3}(5\sqrt{5} - 2\sqrt{2})$

(c) $e - 1$

(e) $\frac{\pi}{8}$

4. (a) 1

(c) $\frac{\pi}{4} - \frac{1}{2}\ln 2$

(e) $\frac{1}{4}e^{5\pi/6}(\sqrt{3} + 1) + \frac{1}{2}$

Section 3

5. (a) $\frac{3}{2}$ (c) $\frac{2}{3}$

9. $\frac{1}{3}$

Section 4

5. (a) 2 (c) 1

CHAPTER 11

Section 1

1. (a) 2

(c) 12

(e) $\frac{e^2 - e^{-2}}{2}(e^2 - 1) = (e^2 - 1)\sinh 2$

(g) $2\ln 2 - 1$

2. (a) 288

(c) $\sqrt{3} - 1$

(e) $\frac{3\sqrt{3}}{4}$

3. (a) 2 (c) $\frac{16}{3}$

(e) $2\ln 2 - 1$ (g) 288

(i) $\frac{\sqrt{3} - 1}{4}$ (k) 0

Section 2

1. (a) $\frac{1016}{105}$

(c) $\frac{568}{945}$

(e) $\frac{2}{15}$

(g) $\frac{1}{24}$

(i) $\frac{32}{945}$

2. (a) $\frac{1}{24}$

(c) 0

(e) $\frac{2971}{3780}\sqrt{2}$

3. (a) 0

(c) -2

Section 3

1. (a) $-\frac{57}{4}$ (c) $\frac{1}{2}\sin 4 - \frac{1}{4}\sin 8$

   (e) $\frac{1}{2}$ (g) $\frac{3}{2}\ln(1+\sqrt{2})$

2. (a) $\frac{1}{8}$ (c) $\frac{\pi}{2}$

   (e) 243π (g) $\frac{128a^5}{15}$

5. $\frac{185}{24}$
7. 144
9. $16 - 8\sqrt{2}$
10. (a) $(1, 1, \frac{1}{2})$
11. $(0, 0, \frac{8}{15})$

Section 5

1. (a) $\frac{x\sin x + \cos x - 1}{x^2}$

   (c) $\frac{-1}{x^2+5x+6}$

5. $e^{x^3+y^3}$
7. $\frac{\partial z}{\partial x} = \frac{\partial z}{\partial y} = -1$

Section 5

1. (a) $\frac{15\pi}{32}$ (c) $\frac{15\pi}{256}$

2. (a) 0 (c) $\frac{31}{60}(4-\sqrt{2})$

3. $\frac{3}{8}$

5. (a) $12\pi\ln 2$ (c) $\frac{3}{2}\pi\ln 2$

6. (a) 0 (c) $\frac{9}{2}\left(\frac{\sqrt{3}-1}{2}\right)\left(\sqrt{3}-1-\frac{\pi}{12}\right)$

8. (a) $2\pi\left(5\sqrt{5}-\frac{1}{5}\right)$ (c) $\pi(2-\sqrt{2})\left(5\sqrt{5}-\frac{1}{5}\right)$

9. (a) 0 (c) 0

CHAPTER 12

Section 2

1. (a) $\frac{1211\sqrt{2}}{3}$

   (c) $\frac{1}{20}(37^{5/2} - 17^{5/2})$

   (e) $\frac{-219}{10}$

2. (a) $-\frac{117}{2}\sqrt{6}$

   (c) $\frac{2e^{10}}{15} + \frac{e^8}{12} + \frac{2e^7}{7} + \frac{e^5}{5} - \frac{59}{84}$

   (e) $-1/5$

3. (a) $4\sqrt{2}$ (c) $4 + 2\sqrt{2}$

4. (a) $\frac{49}{4}$ units

5. (a) $\frac{\sqrt{2}}{6}$ (c) 0

Section 3

1. (a) -7

   (c) $\frac{19}{3}$

   (e) $\frac{96}{5}$

2. (a) $\frac{1}{2}$ (c) $-\pi$

3. (a) $\frac{11}{15}$ (c) 2

4. (a) 0 (c) $\frac{825}{8}$

5. (a) $-\frac{549}{20}$ (c) $\frac{16}{3}$

6. 0, 0, 0

Section 4

1. (a) $x^2y^3, -4$

(c) $x^2 + xy + z,\ -2$

(e) $\sin xyz, -\frac{\sqrt{3}}{2}$

(g) $\frac{1}{3}(x + 2y)^3,\ 0$

Note: The potentials are not unique, since one can add an arbitrary constant.

Section 5

1. (a) 6
   (c) -6
   (e) -3
2. (a) Nonconservative
   (c) Nonconservative
   (e) Conservative, e^{xy}

The potentials are not unique; one can add an arbitrary constant.

CHAPTER 13

Section 1

1. (a) $(-1, 1, -1)$
   (c) $(-1, 7, 5)$
   (e) $(-2, -26, 22)$
3. (a) -43
   (c) -72
   (e) $(36,\ -2,\ -24)$

Section 2

3. $F(\mathbf{v}) = \frac{\sqrt{4 - 3\|\mathbf{v}\|^2}}{4 + \|\mathbf{v}\|^2}(x, y, 2) - (0, 0, 2);\ \mathbf{v} = (x, y)$

5. $z = 2x + 4y - 5$

Section 3

2. $2\pi\left(\sqrt{3} - \frac{2}{3}\sqrt{2}\right)$

4. (c) $16\pi\sqrt{5}$
6. $4\pi r^2$
9. $4\pi^2 ar$

Section 4

(a) $2 + \frac{\pi}{2}$

(c) $\frac{293}{768}\sqrt{5} - \frac{1}{24} - \frac{1}{512}\ln(2 + \sqrt{5})$

2. (a) $1 + \frac{\pi}{2}$

(c) $-\frac{1}{3}$

Section 5

3. (a) $(xz - y, -xy, -x)$
   (c) $(1, \sin x - 1, \cos x)$

CHAPTER 14

Section 1

1. (a) $2x_2^3\,dx_1 + x_1x_2\,dx_2 - x_2\,dx_3$
   (c) $-x_1\,dx_1\,dx_2 + dx_1\,dx_3 - x_2x_3\,dx_2\,dx_3$
   (e) $2x_2x_3^3\,dx_1\,dx_2\,dx_3$
2. (a) $x_2^2\,dx_1 + 3x_3^2\,dx_2 + (x_2 \sin x_1 + 3x_2x_4)\,dx_3 + 2x_2\,dx_4$
   (c) $(x_4 - x_1)\sin x_1\,dx_1\,dx_3 + (x_2^2 + 2x_4 - 2x_1)\,dx_1\,dx_4 + x_2 \sin x_1\,dx_3\,dx_4$
   (e) $(x_4 - x_1)x_3^2 \sin x_1\,dx_1\,dx_2\,dx_3 + x_3^2(x_2^2 + 2x_4 - 2x_1)\,dx_1\,dx_2\,dx_4 +$ $x_2x_4(x_2^2 + 2x_4 - 2x_1)\,dx_1\,dx_3\,dx_4 - x_2x_3^2 \sin x_1\,dx_2\,dx_3\,dx_4$
   (i) $(1 + x_2)x_4(x_1x_2^2x_3^2 + 2x_1x_3^2x_4 - 2x_1^2x_3^2 + x_2^4 + 2x_2^2x_4 -$ $2x_1x_2^2)\,dx_1\,dx_2\,dx_3\,dx_4$

Section 3

1. (a) $y\,dx + x\,dy + 2z\,dz$
   (c) $(y - x)\,dx\,dy + yz^2\,dx\,dz + (xz^2 + 2y)\,dy\,dz$
   (e) $(z^2 + 2xy)\,dx\,dy\,dz$
2. (a) $x_3\,dx_1 - 2x_2x_4\,dx_2 + x_1\,dx_3 - x_2^2\,dx_4$
   (c) $-x_1 \cos x_1x_2\,dx_1\,dx_2 - 2x_1\,dx_1\,dx_4 - 2x_2\,dx_2\,dx_4 + x_3 \sin x_4\,dx_3\,dx_4$
   (e) $-x_1\,dx_1\,dx_3\,dx_4$

Section 5

1. (a) $4x^2y^2(x^2 + y^2)$
   (c) $(8x^3y^2 - 2x^2y - 2y^3)\,dx + (8x^2y^3 - 2x^3 - 2xy^2)\,dy$
   (e) $4(x^2 - y^2) \sin 2xy(x^2 + y^2)\,dx\,dy$
2. (a) xy^3z^2
   (c) $z^2y^3\,dx + (xy^2z^2 - xyz)\,dy - xy^2\,dz$
   (e) $zy \sin(xy^2z)\,dx\,dy + y^2 \sin(xy^2z)\,dx\,dz + xy \sin(xy^2z)\,dy\,dz$
3. (a) $x_1x_3(2x_1 + x_2)(x_2 - x_3)$
   (c) $2x_1x_3\,dx_1 + (x_1x_3 - 2x_1 - x_2)\,dx_2 + (2x_1 + x_2)\,dx_3$
   (e) $[-2(2x_1 + x_2)^3 - x_3(2x_1 + x_2)(x_2 - x_3)^2 + x_3 \sin(2x_1 + x_2)]dx_1dx_2 +$ $[2x_1(2x_1 + x_2)(x_2 - x_3)^2 + 2(2x_1 + x_2)^3 - x_3 \sin(2x_1 + x_2)]\,dx_1\,dx_3 +$ $[x_1(2x_1 + x_2)(x_2 - x_3)^2 + (2x_1 + x_2)^3 - x_1 \sin(2x_1 + x_2)]\,dx_2\,dx_3$
   (g) $(x_3 - 2x_1)(2x_1^2x_3 + x_1x_2x_3 - x_2^2 - x_3^2 + 2x_2x_3)dx_1dx_2dx_3$
4. (a) $x_1^2x_2^4x_3^2$
   (c) $2x_1x_2x_3^2\,dx_1 - 3x_1^2x_2^2x_3\,dx_2 + x_1^2x_2x_3\,dx_3$
   (e) $(2x_1^3x_2^6x_3^3 - 6x_1^7x_2^2x_3^4)\,dx_1\,dx_2 + 2x_1^3x_2^7x_3^2\,dx_2\,dx_3 +$ $[3x_1^8x_2^2x_3^3 - x_1^4x_2^6x_3^2 - 3x_2^3 \sin(x_1^2x_3)]dx_2dx_3$
   (g) $(x_2^6 - x_1^2x_2x_3^2)6x_1x_2^3x_3\,dx_1\,dx_2\,dx_3$

CHAPTER 15

Section 1

1. (a) 2

(c) $\frac{1}{2}$
(e) $\frac{5}{4}$
(g) $\frac{1}{2}$
(i) $-\frac{13}{63}$

2. (a) $\frac{2}{3}$
(c) $-\frac{23}{36}$

3. (a) $\frac{2}{5}$
(c) $\frac{17}{120}$

Section 2

1. (a) $\frac{5}{18}$ (c) $-\frac{5}{6}$
2. (a) $\frac{3}{2}$ (c) $-\frac{10}{3}$
3. (a) $-\frac{5}{6}$ (c) $\frac{14}{3}$
4. (a) $-\frac{353}{720}$ (c) $\frac{173}{72}$

Section 4

1. 0

CHAPTER 16

Section 2

1. (a), (c), (d), (e), (f), (g), and (i) converge; the others diverge.

Section 3

1. (c), (d), (e), (g), (j), (l), (m), (n), and (o) converge; the others diverge.
2. Convergence for $r > 1$; divergence for $r \leq 1$.
9. Convergence if and only if $b > a + 1$.

Section 4

1. (a), (b), (g), (l), and (n) converge conditionally; (d), (f), (h), (k), and (o) diverge.

CHAPTER 17

Section 1

1. (c), (f), and (h) converge uniformly.

Section 2

4. For $x \neq 0$.

Section 3

1. (a) $R = 1$; diverges at ± 1.
(c) $R = \infty$.
(e) $R = 2$; diverges at ± 2.
(g) $R = 1$; diverges at ± 1.
(i) $R = 1$; converges at ± 1.

2. (a) $\sum_{n=0}^{\infty} (-1)^n x^n$

(c) $\sum_{n=0}^{\infty} (-1)^n x^{2n}$

Section 4

1. (a) $\sum_{n=0}^{\infty} \frac{(-1)^n x^{2n+1}}{(2n+1)!}$

   (c) $\sum_{n=0}^{\infty} \frac{x^{2n+1}}{(2n+1)!}$

2. (a) $\frac{\sqrt{3}}{2} + \frac{1}{2}\left(x - \frac{\pi}{3}\right) - \frac{\sqrt{3}}{4}\left(x - \frac{\pi}{3}\right)^2 - \frac{1}{12}\left(x - \frac{\pi}{3}\right)^3 + \frac{\sqrt{3}}{48}\left(x - \frac{\pi}{3}\right)^4 + \cdots$

   (The coefficient of $(x - \pi/3)^n$ is $\sqrt{3}/2n!$, $1/2n!$, $-\sqrt{3}/2n!$, or $-1/2n!$, depending on whether the remainder when n is divided by 4 is 0, 1, 2 or 3.)

   (c) $-1 - (x+1) - 2(x+1)^2 + 3(x+1)^3$

4. $\sum_{n=0}^{\infty} (-1)^n x^{2n}$

6. (a) $1 + x^2 + \frac{x^4}{2} + \frac{x^6}{6} + \cdots$

   (c) $x^2 - \frac{x^4}{3} + \frac{2x^6}{45} - \frac{x^8}{315} + \cdots$

10. (a) $\sum_{n=0}^{\infty} (n+1)x^n$

    (c) $x - x^2 + \frac{x^3}{3} - \frac{x^5}{30} + \cdots$

    (e) $x + \frac{x^2}{2} - \frac{x^3}{6} + \frac{x^4}{12} - \cdots$

11. (a) $x + \frac{x^3}{3} + \frac{2x^5}{15} + \frac{17x^7}{315} + \cdots$

13. (a) $\sum_{n=0}^{\infty} \frac{x^{2n}}{n!}$ (c) $1 - \frac{x^2}{2} + \frac{5x^4}{24} + \cdots$

Section 5

3. $\frac{\pi}{2} - \frac{4}{\pi} \sum_{n=0}^{\infty} \frac{1}{(2n+1)^2} \cos(2n+1)x$

5. $-\frac{1}{2} \sin x + 2 \sum_{n=2}^{\infty} \frac{(-1)^n n}{n^2 - 1} \sin nx$

7. $\frac{8}{15}\pi^4 + 48 \sum_{n=1}^{\infty} \frac{(-1)^{n-1}}{n^4} \cos nx$

Appendix A

1. (a) -17 (c) -11 (e) -8

4. $\begin{bmatrix} 2 & -1 \\ -3 & 2 \end{bmatrix}$

Index

C

D

K

L

M

N

O

P

Q

R

S

ISBN 0-8247-6962-7

EAN

9 780824 769628

90000>

Printed in the United States
by Baker & Taylor Publisher Services